APPROXIMATE SOLUTION OF OPERATOR EQUATIONS

M. A. Krasnosel'skii, G. M. Vainikko, P. P. Zabreiko,
Ya. B. Rutitskii, V. Ya. Stetsenko

APPROXIMATE SOLUTION
OF OPERATOR EQUATIONS

translated by

D. LOUVISH

Israel Program for Scientific Translations

IЍ🙰ЍI

WOLTERS-NOORDHOFF PUBLISHING GRONINGEN

ISBN: 90 01 50403 5
Library of Congress Catalog Card Number: 78–184992

Original title "Priblizhennoe reshenie operatornykh uravnenii" published in 1969 in Moscow.

Text set on Monophoto Times Roman 10/12 by Keter Publishing House, Ltd., Jerusalem
Printed in the Netherlands

Contents

Chapter 2

Linear equations

Chapter 3

Equations with smooth operators

Chapter 4

Projection methods

Chapter 5

Small solutions of operator equations

Preface

One of the most important chapters in modern functional analysis is the theory of approximate methods for solution of various mathematical problems. Besides providing considerably simplified approaches to numerical methods, the ideas of functional analysis have also given rise to essentially new computation schemes in problems of linear algebra, differential and integral equations, nonlinear analysis, and so on.

The general theory of approximate methods includes many known fundamental results. We refer to the classical work of Kantorovich; the investigations of projection methods by Bogolyubov, Krylov, Keldysh and Petrov, much furthered by Mikhlin and Pol'skii; Tikhonov's methods for approximate solution of ill-posed problems; the general theory of difference schemes; and so on.

During the past decade, the Voronezh seminar on functional analysis has systematically discussed various questions related to numerical methods; several advanced courses have been held at Voronezh University on the application of functional analysis to numerical mathematics. Some of this research is summarized in the present monograph. The authors' aim has not been to give an exhaustive account, even of the principal known results.

The book consists of five chapters.

In the first chapter we study iterative processes: conditions for convergence, estimates of convergence rate, effect of round-off errors, etc. Much attention is paid to the convergence of iterative processes under conditions incompatible with the contracting mapping principle (the theory of concave operators, the role of uniformly convex norms, and so on). The second chapter studies linear problems: methods for approximate solution of linear equations, estimates for the spectral radius of a linear operator, approximate determination of eigenvalues, etc. The theory of semiordered spaces plays an important role. The third chapter considers equations with smooth nonlinear operators, employing ideas close to those of Kantorovich. Considerable attention is paid to the situations arising in approximate methods which utilize various simplified formulas. Topological methods are proposed for *a posteriori* error estimates. Much of Chapters 1 to 3 borrows from the above-mentioned advanced courses, which were given alternately by Krasnosel'skii and Rutitskii; a few sections in the first and second chapters were written by Krasnosel'skii and Stetsenko.

Chapter 4 is devoted to a systematic theory of projection methods (method of least squares, the methods of Galerkin, Galerkin–Petrov, et al.) as applied to the approximate solution of linear and nonlinear equations, and approximate determination of eigenvalues. Most of this chapter was written by G. M. Vainikko.

The fifth and last chapter considers approximate methods in a difficult field of nonlinear analysis—the theory of branching of small solutions. The authors have seen fit to present a short account of the basic theory of formal power series. This chapter was written by Zabreiko and Krasnosel'skii.

The book includes a large number of exercises, ranging from the simple to the very difficult.

G. A. Bezsmertnykh, N. N. Gudovich, A. Yu. Levin, E. A. Lifshits, V. B. Melamed and A. I. Perov offered valuable remarks and advice in discussing various parts of the book. Several important remarks were made by L.V. Kantorovich and G.P. Akilov after reading the manuscript. The authors are deeply indebted to all those mentioned.

The authors

Successive approximations

§ 1. Existence of the fixed point of a contraction operator

1.1. *Contraction operators.* Let A be an operator defined on a set $\mathfrak{M}$ in a Banach space E, satisfying a Lipschitz condition

$$\|Ax - Ay\| \leqq q \|x - y\| \quad (x, y \in \mathfrak{M}). \tag{1.1}$$

If $q < 1$, A is called a *contraction operator* (or *contracting operator*).

Theorem 1.1. (*Contracting mapping principle*). *Let $\mathfrak{M}$ be a closed set and assume that the contraction operator A maps $\mathfrak{M}$ into itself: $A\mathfrak{M} \subset \mathfrak{M}$. Then A has a unique fixed point x^* in $\mathfrak{M}$; in other words, the equation*

$$x = Ax \tag{1.2}$$

has a unique solution x^* in $\mathfrak{M}$.

Proof. Consider the following nonnegative functional on $\mathfrak{M}$:

$$\Phi(x) = \|x - Ax\|. \tag{1.3}$$

Let x_n be a minimizing sequence for $\Phi(x)$:

$$\lim_{n \to \infty} \Phi(x_n) = \inf_{x \in \mathfrak{M}} \Phi(x) = \alpha.$$

Obviously,

$$\alpha \leqq \Phi(Ax_n) = \|Ax_n - A^2x_n\| \leqq q\|x_n - Ax_n\| = q\Phi(x_n),$$

and hence $\alpha \leqq q\alpha$. It follows that $\alpha = 0$.

Since

$$\|x_n - x_m\| \leqq \|x_n - Ax_n\| + \|Ax_n - Ax_m\| + \|Ax_m - x_m\| \leqq$$

$$\leqq \Phi(x_n) + q\|x_n - x_m\| + \Phi(x_m),$$

it follows that

$$\|x_n - x_m\| \leqq \frac{\Phi(x_n) + \Phi(x_m)}{1 - q} \to 0 \qquad (n, m \to \infty),$$

i.e., x_n is a Cauchy sequence. Its limit x^* belongs to $\mathfrak{M}$ ($\mathfrak{M}$ is closed!). The functional $\Phi(x)$ is continuous, since by (1.1)

$$|\Phi(x) - \Phi(y)| = |\,\|x - Ax\| - \|y - Ay\|\,| \leqq$$
$$\leqq \|x - y\| + \|Ax - Ay\| \leqq (1 + q)\|x - y\|.$$

Therefore

$$\Phi(x^*) = \lim_{n \to \infty} \Phi(x_n) = 0,$$

and this means that x^* is a solution of equation (1.2).

Suppose that y^* is another solution of (1.2) in $\mathfrak{M}$. Then

$$\|x^* - y^*\| = \|Ax^* - Ay^*\| \leqq q\|x^* - y^*\|$$

and so $\|x^* - y^*\| = 0$, i.e., $x^* = y^*$. ∎

The most commonly studied contraction operators are defined on the entire space E or on some ball T. In the latter case, it is convenient to replace Theorem 1.1 by the following special case:

Theorem 1.2. *Let* A *be a contraction operator on a closed ball* $T = \{x : \|x - x_0\| \leqq r\}$ *in a Banach space* E, *and let*

$$\|Ax_0 - x_0\| \leqq (1 - q)r. \tag{1.4}$$

Then A *has a unique fixed point in* T.

To prove this theorem it suffices to verify the inclusion $AT \subset T$, which is obvious: if $\|x - x_0\| \leqq r$, then (1.1) and (1.4) imply the inequality

$$\|Ax - x_0\| \leqq \|Ax - Ax_0\| + \|Ax_0 - x_0\| \leqq q\|x - x_0\| + (1 - q)r \leqq r.$$

Exercise 1.1. Let the function $K(t, s; u)$ be jointly continuous in $t, s \in [a, b]$, $-\infty < u < \infty$, and assume that

$$|K(t, s; u) - K(t, s; v)| \leqq M(\rho)|u - v| \qquad (|u|, |v| \leqq \rho).$$

Show that the operator

$$Ax(t) = \int_a^b K[t, s; x(s)]\, ds$$

is defined on the space C of continuous functions on $[a, b]$ and satisfies a Lipschitz condition (1.1) on every ball of C. Under what conditions is it a contraction operator on the ball $\|x\| \leqq \rho_0$?

Exercise 1.2. Under the assumptions of Theorem 1.1, show that the sequence $Ax_n = x_{n+1}$ $(n = 0,1,2,\ldots)$ minimizes the functional (1.3) for arbitrary $x_0 \in \mathfrak{M}$.

1.2. *Use of an equivalent norm.* Suppose that, apart from the basic norm $\| \cdot \|$, the space E has another norm, which we shall denote by $\| \cdot \|_*$. Recall that the norms $\| \cdot \|$ and $\| \cdot \|_*$ are said to be *equivalent* if there exist positive constants m and M such that

$$m \|x\| \leqq \|x\|_* \leqq M \|x\| \qquad (x \in E).$$

When the basic norm is replaced by an equivalent one, convergent sequences of elements in the space remain convergent, closed (open) sets remain closed (open), etc.

If the operator A satisfies a Lipschitz condition with respect to one norm, this is also true for any equivalent norm. Now it often happens that A, though not a contraction, satisfies a Lipschitz condition with respect to the original norm; however, by suitable construction of an equivalent norm, A becomes a contraction. The principal difficulty in investigating concrete equations usually involves constructing this special norm.

As an example, consider a system of ordinary differential equations

$$\frac{d\xi_i}{dt} = f_i(t, \xi_1, \ldots, \xi_n) \qquad (i = 1, \ldots, n),$$

or, in vector notation,

$$\frac{dx}{dt} = f(t, x), \qquad\qquad (1.5)$$

where $x(t) = \{\xi_1(t), \ldots, \xi_n(t)\}$ is a vector-function with values in n-space R^n. The problem of solving the system (1.5) with the initial condition

$$x(0) = 0 \qquad\qquad (1.6)$$

is known to be equivalent to that of solving the vector integral equation

$$x(t) = \int_0^t f[s, x(s)]\, ds. \qquad\qquad (1.7)$$

Assume that $f(t, x)$ is jointly continuous in t, x and satisfies a Lipschitz condition in x:

$$|f(t, x) - f(t, y)| \leq L |x - y| \qquad (x, y \in R^n ; \ 0 \leq t \leq T). \quad (1.8)$$

Here and above $|x|$ denotes the length of the vector x in R^n.

Consider the operator A defined by the right member of equation (1.7):

$$Ax(t) = \int_0^t f[s, x(s)] \, ds. \quad (1.9)$$

It is easily seen that, for any $\tau \ (0 < \tau \leq T)$, this operator is defined in the space C_τ of continuous vector-functions on $[0, \tau]$ with the norm

$$\|x\|_\tau = \max_{0 \leq t \leq \tau} |x(t)|. \quad (1.10)$$

It follows from (1.8) that

$$\|Ax - Ay\|_\tau \leq L\tau \|x - y\|_\tau \qquad (x, y \in C_\tau).$$

Therefore, if

$$L\tau < 1,$$

the operator A satisfies the assumptions of Theorem 1.1. Thus the problem (1.5)–(1.6) has a unique solution $x_*(t)$ on the interval $0 \leq t < \min\{T, 1/L\}$.

In fact, the problem (1.5)–(1.6) has a solution on the entire interval $[0, T]$. To prove this, we may again use Theorem 1.1, introducing a special norm in the space C_T. Set

$$\|x\|_* = \max_{0 \leq t \leq T} e^{-L_1 t} |x(t)| \qquad (x \in C_T), \quad (1.11)$$

where $L_1 > 0$. Clearly,

$$e^{-L_1 T} \|x\|_T \leq \|x\|_* \leq \|x\|_T \qquad (x \in C_T),$$

i.e., the norms $\|x\|_*$ and $\|x\|_T$ are equivalent. Now by (1.8)

$$\|Ax - Ay\|_* = \max_{0 \leq t \leq T} \left| e^{-L_1 t} \int_0^t \{f[s, x(s)] - f[s, y(s)]\} \, ds \right| \leq$$

$$\leq L \max_{0 \leq t \leq T} \int_0^t e^{L_1 (s-t)} e^{-L_1 s} |x(s) - y(s)| \, ds \leq$$

$$\leqq L \, \|x - y\|_* \, \max_{0 \leqq t \leqq T} \int_0^t e^{L_1 (s-t)} \, ds = \frac{L}{L_1} (1 - e^{-L_1 T}) \, \|x - y\|_* \, .$$

Setting $L_1 = L$, we see that the operator (1.9) is a contraction in the norm (1.11), the constant being

$$q = 1 - e^{-L_1 T} < 1 \, .$$

Thus the problem (1.5)–(1.6) has a unique solution $x_*(t)$, defined for all $t \in [0, T]$.

Another example is the equation

$$x(t) = \int_0^t K(t, s) f[s, x(s)] \, ds + \phi(t) \, , \tag{1.12}$$

where $x(t)$ is an unknown function, $\phi(t)$ a given continuous function on $[0, T]$, $K(t, s)$ and $f(s, x)$ are jointly continuous for $0 \leqq t, s \leqq T, -\infty < x < \infty$ and moreover

$$|f(t, x) - f(t, y)| \leqq L \, |x - y| \qquad (-\infty < x, y < \infty). \tag{1.13}$$

The right member of (1.12) obviously defines an operator A on the space C of continuous functions on $[0, T]$. It satisfies the inequality

$$\|Ax - Ay\|_* \leqq \frac{KL}{L_1} (1 - e^{-L_1 T}) \, \|x - y\|_* \qquad (x, y \in C),$$

where $\| \cdot \|_*$ is the norm (1.11) and

$$K = \max_{0 \leqq t, s \leqq T} |K(t, s)| \, .$$

If $L_1 \geqq KL$, then A is a contraction in the norm (1.11). It follows from Theorem 1.1 that equation (1.12) has a unique solution on $[0, T]$.

Exercise 1.3. Show that in the arguments of subsection 1.2 the norm (1.11) may be replaced by the norm

$$\|x\|_{**} = \max_{0 \leqq t \leqq T} e^{-L_2 t^2} |x(t)|$$

(for what values of L_2?).

Exercise 1.4. Prove that the linear Volterra integral equation

$$x(t) = \int_0^t K(t, s) x(s) \, ds + \phi(t)$$

with continuous kernel $K(t, s)$ (and continuous $\phi(t)$) has a unique summable solution.

1.3. *Relative uniqueness of the solution.* As a rule, concrete problems lead to equations not associated with any well-defined function space. The same equations may be considered with an operator defined in different subsets $\mathfrak{M}$ of different spaces E. Suppose that the uniqueness of the solution of an equation has somehow been proved (say, using Theorem 1.1) in a subset $\mathfrak{M}$ of a space E. Naturally, this does not imply that there are no other solutions in the space E, or that there are no solutions outside E.

As an example, consider the equation

$$x(t) = \int_0^1 x^2(s)\, ds. \tag{1.14}$$

The operator A defined by the right member of this equation satisfies the assumptions of Theorem 1.1 in any ball $\|x\| \leqq \rho\, (\rho < \frac{1}{2})$ in the space C of continuous functions on $[0, 1]$. Thus the unique solution of equation (1.14) in the ball $\|x\| < \frac{1}{2}$ is the trivial solution. However, there is another continuous solution, $x(t) \equiv 1$.

Instead of equation (1.14), we could have considered the scalar equation $x = x^2$, which satisfies the assumptions of Theorem 1.1 on any interval $[-\rho, \rho]\,(\rho < \frac{1}{2})$, but it has two solutions, $x = 0$ and $x = 1$. The scalar equation $x = -x^3$ has a unique solution (the trivial solution) in the space of real numbers, and three solutions in the space of complex numbers.

Now consider the homogeneous Volterra integral equation

$$x(t) = \int_0^t K(t, s)\, x(s)\, ds \tag{1.15}$$

with kernel

$$K(t, s) = \begin{cases} se^{1/t^2 - 1}, & \text{if} \quad 0 \leqq s \leqq te^{1 - 1/t^2}, \\ t, & \text{if} \quad te^{1 - 1/t^2} \leqq s \leqq 1. \end{cases}$$

The kernel $K(t, s)$ is continuous on the square $0 \leqq t, s \leqq 1$ (verify!). Equation (1.15) therefore has no nontrivial summable solutions (see Exercise 1.4). However, the equation has a nonsummable solution:

$$x(t) = \begin{cases} 0, & \text{if} \quad t = 0, \\ 1/t, & \text{if} \quad 0 < t \leqq 1. \end{cases}$$

This example is due to Urysohn.

1.4. *Spectral radius of a linear operator.* If A is a linear operator, it is a contraction if and only if its norm is smaller than 1. This raises the problem of constructing equivalent norms with respect to which a given linear operator A has as small a norm as possible.

It is proved in the theory of linear operators (see, e.g., Kantorovich and Akilov [1], p. 153) that the limit

$$\rho_0 = \lim_{n \to \infty} \sqrt[n]{\|A^n\|} \tag{1.16}$$

exists and is finite. The number ρ_0 is called the *spectral radius* of the bounded linear operator A. Clearly,

$$\rho_0 \leqq \|A\| .$$

The spectral radius ρ_0 is characterized by the fact that the inequality

$$|\lambda| > \rho_0$$

implies that the operator $(A - \lambda I)^{-1}$ exists and is bounded. In particular, if A is a compact linear operator, then $\rho_0 = |\lambda_0|$, where λ_0 is an eigenvalue of A with maximum absolute value. Here one must consider both real and complex eigenvalues of the operator A; we recall that a number $\lambda = \sigma + i\tau$ is an *eigenvalue* of an operator A defined in a real Banach space E if there exist $x, y \in E$ such that

$$Ax = \sigma x - \tau y, \qquad Ay = \tau x + \sigma y \qquad (\|x\| + \|y\| > 0).$$

Exercise 1.5. Show that the spectral radius of a linear operator is invariant with respect to equivalent norms.

Exercise 1.6. Let

$$Ax(t) = \int_a^t K(t, s) \, x(s) \, ds$$

be a Volterra integral operator with continuous kernel $K(t, s)$ ($a \leqq t, s \leqq b$). Show that, as an operator on C, A has spectral radius zero.

Hint. First prove the inequality

$$\|A^n\| \leqq \frac{K^n (b - a)^n}{n!},$$

where $K = \max\limits_{a \leqq t, s \leqq b} |K(t, s)|$.

Exercise 1.7. Let $\beta_n (n = 1, 2, \ldots)$ be a given monotone decreasing numerical sequence converging to zero. Construct a linear operator A such that $\sqrt[n]{\|A^n\|} = \beta_n$ ($n = 1, 2, \ldots$).

We shall now construct an equivalent norm in the space E such that the norm of the linear operator A is arbitrarily close to its spectral radius. Let $\varepsilon > 0$ be given, and determine n such that

$$\sqrt[n]{\|A^n\|} \leqq \rho_0 + \varepsilon .$$

Now set

$$\|x\|_* = (\rho_0 + \varepsilon)^{n-1} \|x\| + (\rho_0 + \varepsilon)^{n-2} \|Ax\| + \ldots + \|A^{n-1}x\| . \quad (1.17)$$

Clearly,

$$(\rho_0 + \varepsilon)^{n-1} \|x\| \leq \|x\|_* \leq$$

$$\leq [(\rho_0 + \varepsilon)^{n-1} + (\rho_0 + \varepsilon)^{n-2} \|A\| + \ldots + \|A^{n-1}\|] \|x\|,$$

i.e., the norms $\|\cdot\|$ and $\|\cdot\|_*$ are equivalent. A simple calculation shows that

$$\|A\|_* = \sup_{\|x\|_* \leq 1} \|Ax\|_* \leq \rho_0 + \varepsilon.$$

Since $\rho_0 \leq \|A\|_*$ in any norm, we have

$$\rho_0 \leq \|A\|_* \leq \rho_0 + \varepsilon. \blacksquare$$

It follows in particular that for any $\varepsilon > 0$ one can construct an equivalent norm in C such that the norm of the linear Volterra integral operator (whose spectral radius is zero; see Exercise 1.6) is smaller than ε.

The norm of a selfadjoint linear operator A defined on a Hilbert space is equal to its spectral radius. In the general case, there are linear operators for which there is no equivalent norm such that $\|A\|_* = \rho_0$. An example is the nonzero Volterra integral operator.

Exercise 1.8. Let A be a compact linear operator on a space E. Assume that the invariant subspace of any eigenvalue of A whose absolute value is equal to the spectral radius consists solely of eigenvectors. Construct an equivalent norm in E for which $\|A\|_* = \rho_0$.

Exercise 1.9. Let A be a compact linear operator defined on a space E. Assume that at least one eigenvalue λ, equal in absolute value to the spectral radius ρ_0, has not only eigenvectors but also generalized eigenvectors.* Show that the inequality $\|A\|_* > \rho_0$ holds for any equivalent norm in E.

1.5. *Operators which commute with contraction operators.* The following simple observation is often useful: Let B be an operator which maps a closed set $\mathfrak{M}$ in a Banach space E into itself, and A an operator which satisfies the conditions of the contracting mapping principle on $\mathfrak{M}$; if B commutes with A ($AB = BA$), then the fixed point of A is also a fixed point of B.

Indeed, if $Ax^* = x^*$, then

$$ABx^* = BAx^* = Bx^*,$$

that is, Bx^* is a fixed point of A, and therefore $Bx^* = x^*$.

---

* g_0 is a *generalized eigenvector* of a linear operator A, corresponding to the eigenvalue λ_0, if $Ag_0 \neq \lambda_0 g_0$, but $(A - \lambda_0 I)^n g_0 = 0$ for some $n > 1$.

For example, assume that B maps a closed set $\mathfrak{M}$ into itself, and that some power $A = B^n$ is a contraction. Then $AB = B^{n+1} = BA$, and therefore B has a fixed point in $\mathfrak{M}$. In this case B has a unique fixed point in $\mathfrak{M}$, since all fixed points of B are also fixed points of $A = B^n$.

The only assumption employed in these arguments is that the equation $Ax = x$ is uniquely solvable in $\mathfrak{M}$. Hence

Theorem 1.3. *Let the equation* $Ax = x$ *have a unique solution in* $\mathfrak{M}$. *Assume that* $B\mathfrak{M} \subset \mathfrak{M}$ *and* B *commutes with* A *on* $\mathfrak{M}$.

Then the equation $x = Bx$ *has at least one solution in* $\mathfrak{M}$.

Exercise 1.10. Consider the integral operator

$$Ax(t) = \int_0^t K(t, s) x(s) \, ds + \phi(t)$$

with continuous kernel $K(t, s)$ (and continuous $\phi(t)$); show that certain powers of A are contractions in the space C of continuous functions on $[0, 1]$.

In a certain sense, the use of powers A^n of a linear operator A to prove existence theorems, which is based on application of the contracting mapping principle, is the same as construction of an equivalent norm. Assume that in some equivalent norm $\| \cdot \|_*$ the operator A satisfies the contraction condition. This means that $\|Ax - Ay\|_* \leqq q\|x - y\|_*$, where $q < 1$, and

$$m \|x\| \leqq \|x\|_* \leqq M \|x\| \qquad (x \in E).$$

Then, for all n such that

$$\frac{Mq^n}{m} \leqq q_0 < 1,$$

we have the inequality

$$\|A^n x - A^n y\| \leqq \frac{1}{m} \|A^n x - A^n y\|_* \leqq \frac{q^n}{m} \|x - y\|_* \leqq \frac{Mq^n}{m} \|x - y\| \leqq q_0 \|x - y\|,$$

that is, the n-th power of A is a contraction operator in the old norm.

Conversely, assume that some power of A is a contraction operator on $\mathfrak{M}$:

$$\|A^n x - A^n y\| \leqq q_0 \|x - y\| \qquad (q_0 < 1 \, ; \ x, y \in \mathfrak{M}).$$

Define a new norm in E:

$$\|x\|_* = q_0^{(n-1)/n} \|x\| + q_0^{(n-2)/n} \|Ax\| + \ldots + \|A^{n-1}x\|.$$

It is easily verified that

$$\|Ax - Ay\|_* \leqq q_0^{1/n} \|x - y\|_* \qquad (x, y \in \mathfrak{M}).$$

1.6. *Case of a compact set.* Generally speaking, the contracting mapping principle is no longer valid if A is not a contraction operator on $\mathfrak{M}$ but

$$\|Ax - Ay\| < \|x - y\| \qquad (x, y \in \mathfrak{M} \; ; \quad x \neq y). \qquad (1.18)$$

Thus, for example, the mapping

$$Ax = x + \frac{1}{x}$$

maps the half-line $[1, \infty)$ into itself and satisfies (1.18), but has no fixed point in this set.

However, the principle does remain valid if the set $\mathfrak{M}$ is compact. To prove this, we return to the functional (1.3)

$$\Phi(x) = \|x - Ax\|.$$

This functional is continuous (since A is). It therefore assumes a minimum value at some point x^* of the compact set $\mathfrak{M}$. Now this minimum is zero, since otherwise (1.18) would imply that

$$\Phi(Ax^*) = \|Ax^* - A^2x^*\| < \|x^* - Ax^*\| = \min_{x \in \mathfrak{M}} \Phi(x).$$

Thus $\Phi(x^*) = 0$, and so x^* is a fixed point of A. Its uniqueness follows directly from (1.1). These arguments imply the following theorem.

Theorem 1.4. *Assume that the operator A maps a closed set $\mathfrak{M}$ into a compact subset of itself and satisfies there condition* (1.18).

Then A has a unique fixed point in $\mathfrak{M}$.

Note that an operator A satisfying the assumptions of this theorem need not be a contraction (neither on $\mathfrak{M}$ nor on $A\mathfrak{M}$). An example is the mapping

$$Ax = x - \frac{x^2}{2}$$

of the interval $[0, 1]$ into itself.

Exercise 1.11. Replace condition (1.18) by the weaker condition

$$\|Ax - Ay\| \leq \|x - y\| \qquad (x, y \in \mathfrak{M}).$$

Let $\mathfrak{M}$ be either an interval $[a, b]$, a disk $x^2 + y^2 \leq R^2$, or an annulus $r^2 \leq x^2 + y^2 \leq R^2$. Find conditions in these three cases under which the operator A always has a fixed point.

Exercise 1.12. Let c_0 be the space of null sequences $x = \{\xi_1, \ldots, \xi_n, \ldots\}$ $(\xi_n \to 0)$ with norm $\|x\| = \max_n |\xi_n|$. Define an operator A in this space by

$$Ax = \left\{ \tfrac{1}{2}(1 + \|x\|), \tfrac{3}{4}\xi_1, \tfrac{7}{8}\xi_2, \ldots, \left(1 - \frac{1}{2^{n+1}}\right)\xi_n, \ldots \right\}.$$

Show that A maps the ball $\|x\| \leq 1$ into itself, satisfies condition (1.18) there, but has no fixed points.

Exercise 1.13. Do the same for the operator A defined in c_0 by $Ax = \{\eta_1, \ldots, \eta_n, \ldots\}$, where $\eta_n = \dfrac{n-1}{n}\xi_n + \dfrac{1}{n}\sin n$ $(n = 1, 2, \ldots)$.

Exercise 1.14. Do the same for the operator A defined in $C[0, 1]$ by $Ax(t) = (1 - t)x(t) + t\sin(1/t)$.

1.7. Estimating the Lipschitz constant.
By the mean-value theorem

$$f(x) - f(y) = f'(c)(x - y)$$

any differentiable scalar function $f(x)$ satisfies a Lipschitz condition (1.1) on the interval $[a, b]$, with constant

$$q = \sup_{a \leq x \leq b} |f'(x)|.$$

An analogous statement holds for differential operators, though they do not satisfy the mean-value theorem.

Consider a nonlinear operator A defined on a set $\mathfrak{M} \subset E$. Recall that A is said to be *Fréchet-differentiable* at a point $x_0 \in \mathfrak{M}$ if there exists a bounded linear operator B such that

$$\lim_{\|h\| \to 0} \frac{\|A(x_0 + h) - Ax_0 - Bh\|}{\|h\|} = 0. \tag{1.19}$$

The operator B is called the *derivative of the operator A at the point x_0*, denoted by $A'(x_0)$.

For example, let $K(t, s; u)$ and $K'_u(t, s; u)$ be jointly continuous in $a \leq t, s \leq b$; $|u| \leq R$. Then the integral operator

$$Ax(t) = \int_a^b K[t, s; x(s)]\, ds \tag{1.20}$$

is defined on the ball $\|x\| \leq R$ in the space C of functions continuous on $[a, b]$, and is differentiable there as an operator with values in C. The derivative of the operator (1.20) is easily seen to be defined by

$$A'(x_0) h(t) = \int_a^b K'_u[t, s; x_0(s)] h(s) \, ds .\tag{1.21}$$

Exercise 1.15. Suppose that for every fixed $h \in E$ the quotient $[A(x_0 + th) - Ax_0]/t$ converges weakly to Bh as $t \to 0$, where B is a bounded linear operator; A is then said to be Gâteaux-differentiable at the point x_0, and the operator B is called the *weak Gâteaux derivative* of A. It is usually denoted by the same symbol $A'(x_0)$ as the Fréchet derivative.

Assume that the Gâteaux derivative $A'(x)$ exists in a certain neighborhood of the point x_0, and moreover $\|A'(x) - A'(x_0)\| \to 0$ as $\|x - x_0\| \to 0$. Show that A is Fréchet-differentiable at x_0.

Assume that the set $\mathfrak{M}$ on which the operator A is defined is convex. Let

$$q_0 = \sup_{x \in \mathfrak{M}} \|A'(x)\| .\tag{1.22}$$

Then, for any $x, y \in \mathfrak{M}$ and any linear functional l

$$l(Ax - Ay) = l[A'(y + \theta(x - y))(x - y)] \leqq \|l\| \cdot q_0 \cdot \|x - y\| .$$

Thus,

$$\|Ax - Ay\| \leqq q_0 \|x - y\| (x, y \in \mathfrak{M}) .\tag{1.23}$$

Hence a general method for estimating the Lipschitz constant q in (1.1)— any estimate for the norm of the derivative provides the required estimate.

As an example, we again consider the operator (1.20). On the ball $\|x\| \leqq R$ it satisfies a Lipschitz condition with constant q, provided

$$|K'_u(t, s; u)| \leqq \frac{q}{b - a}   (a \leqq t, s \leqq b; \, |u| \leqq R) .\tag{1.24}$$

Exercise 1.16. Show that if A is differentiable on a convex set then the smallest Lipschitz constant q in (1.1) coincides with the number (1.22).

Exercise 1.17. Show by an example that if q_0 is defined by (1.22) the estimate (1.23) is false if the set $\mathfrak{M}$ is not convex.

Exercise 1.18. Let the operator A be Gâteaux-differentiable (see Exercise 1.15) on the interval $[x, y]$. Discuss the validity of the formula

$$\|Ax - Ay\| = \|A'[(1 - \theta) x + \theta y] (x - y)\| ,$$

where $\theta = \theta(x, y)$ is some number in the interval $(0, 1)$.

1.8. *Equations with uniform contraction operators.* Let T be a ball $\|x - x_0\| \leqq \beta$ in a Banach space E, and S a ball $\|z - z_0\| \leqq \alpha$ in a Banach space E_1. An operator $F(x; z)$, depending on a parameter $z \in S$ and defined on the space E, is called a *uniform contraction operator* if, for all $z \in S$,

$$\|F(x; z) - F(y; z)\| \leqq q \|x - y\| \qquad (x, y \in T), \qquad (1.25)$$

where $q < 1$ is independent of z.

In this subsection we shall assume that for every $z \in S$ the uniform contraction operator $F(x; z)$ maps the ball T into itself. Theorem 1.1 then implies that the equation

$$x = F(x; z) \qquad (1.26)$$

has a unique solution $x_* = x_*(z) \, (z \in S)$ in T. Let us examine the properties of the solution $x_*(z)$.

Theorem 1.5. Let the operator $F(x; z)$ be continuous in z for every fixed x. Then the solution $x_*(z)$ is continuous.

Proof. By (1.25),

$$\|x_*(z_0) - x_*(z)\| = \|F[x_*(z_0); z_0] - F[x_*(z); z]\| \leqq$$

$$\leqq \|F[x_*(z_0); z_0] - F[x_*(z_0); z]\| + \|F[x_*(z_0); z] - F[x_*(z); z]\| \leqq$$

$$\leqq \|F[x_*(z_0); z_0] - F[x_*(z_0); z]\| + q \|x_*(z_0) - x_*(z)\|,$$

and hence

$$\|x_*(z_0) - x_*(z)\| \leqq \frac{1}{1 - q} \|F[x_*(z_0); z_0] - F[x_*(z); z]\|.$$

The continuity of the operator $F(x; z)$ therefore implies that of the function $x_*(z)$ at the point x_0. ∎

The properties of the solution $x_*(z)$ of equation (1.26) may be investigated as follows.

Assume that the functions $x(z) \, (z \in S)$ with values in T which possess a certain property (!) form a space R which is complete in some metric. The operator $F(x; z)$ generates a natural superposition operator on this space:

$$\tilde{F}x(z) = F[x(z); z]. \qquad (1.27)$$

If this operator maps R into itself, one can attempt to prove the existence of a fixed point. If there is a fixed point, it coincides with $x_*(z)$ (since the solution is unique!). Thus $x_*(z)$ belongs to R, and so possesses property (!). Let us illustrate this procedure.

An operator $F(x; z)$ $(x \in T; z \in S)$ is said to be *weakly continuous* if, whenever $x_n \xrightarrow{w} x_0$ and $z_n \xrightarrow{w} z_0$,

$$F(x_n; z_n) \xrightarrow{w} F(x_0; z_0).$$

Theorem 1.6. *If the operator* $F(x; z)$ *is weakly continuous, so is the solution* $x_*(z)$ *of equation* (1.26).

Proof. Let R denote the set of bounded weakly continuous functions $x(z)$ with values in T. It is easily seen that R is a closed set in the complete space of all bounded functions on S with norm

$$\|x\| = \sup_{z \in S} \|x(z)\|.$$

It follows from the assumptions of our theorem and from (1.25) that the operator (1.27) leaves R invariant and is a contraction on R. By Theorem 1.1, this operator has a fixed point in R, which coincides with the solution $x_*(z)$ of equation (1.26). As an element of R, the solution $x_*(z)$ is weakly continuous. ■

Exercise 1.19. Assume that the operator $F(x; z)$ has the following property: if $z_n \xrightarrow{w} z$ and $\|x_n - x\| \to 0$, then $\|F(x_n; z_n) - F(x; z)\| \to 0$. Show that the solution $x_*(z)$ of equation (1.26) is strongly continuous, i.e., $z_n \xrightarrow{w} z$ implies $\|x_*(z_n) - x_*(z)\| \to 0$.

Exercise 1.20. Assume that $\|z_n - z\| \to 0, x_n \xrightarrow{w} x$ imply $F(x_n; z_n) \to F(x; z)$. Show that the solution $x_*(z)$ of equation (1.26) is a weakly continuous function, i.e., if $\|z_n - z\| \to 0$ then $x_*(z_n) \xrightarrow{w} x_*(z)$.

Exercise 1.21. Prove the classical theorems on the continuity and differentiability of the solutions of ordinary differential equations with respect to their initial values, by the method used to prove Theorem 1.6.

1.9. *Local implicit function theorem.* Let E, E_1 and E_2 be Banach spaces, $f(x; z)$ an operator defined for $\|z - z_0\| \leqq a$ $(z, z_0 \in E_1)$ on the ball $\|x - x_0\| \leqq b$ in the space E, with values in the space E_2. Let

$$f(x_0; z_0) = 0, \tag{1.28}$$

so that x_0 is a solution of the equation

$$f(x; z) = 0 \tag{1.29}$$

for $z = z_0$. If equation (1.29) has solutions $x(z)$ near x_0 for all z near z_0, the equation is said to define an implicit function $x(z)$.

The question of the existence and properties of an implicit function defined by an equation of type (1.29) arises in various problems of analysis. Proofs of the existence of implicit functions may be based on the contracting mapping principle. One such theorem will be presented here.

Theorem 1.7. Assume that the operator $f(x; z)$ has the following properties:

a) $f(x; z)$ is jointly continuous in x, z for $\|z - z_0\| \leqq a$, $\|x - x_0\| \leqq b$ and $f(x_0; z_0) = 0$.

b) In the above domain, the operator $f(x; z)$ has a Fréchet derivative $f'_x(x; z)$ which is continuous in the operator norm at $(x_0; z_0)$.

c) The linear operator $f'_x(x_0; z_0)$ has a continuous inverse.

Then there exist numbers α, $\beta > 0$ such that, for any z in the ball $\|z - z_0\| \leqq \alpha$, equation (1.29) has a unique solution $x_*(z)$ in the ball $\|x - x_0\| \leqq \beta$. The function $x_*(z)$ is continuous.

Proof. The equation

$$x = x - [f'_x(x_0; z_0)]^{-1} f(x, z)$$

is equivalent to equation (1.29). The existence and continuity of the implicit function $x_*(z)$ will follow if we show that there exist numbers α, β such that the operator

$$F(x; z) = x - [f'_x(x_0; z_0)]^{-1} f(x, z),$$

which is jointly continuous, is a uniform contraction for $\|x - x_0\| \leqq \beta$, $\|z - z_0\| \leqq \alpha$.

It follows from condition b) that the operator $f(x; z)$ is Fréchet-differentiable with respect to x, and that the derivative

$$F'_x(x; z) = I - [f'_x(x_0; z_0)]^{-1} f'_x(x; z)$$

is jointly continuous (in the operator norm). Since $F'_x(x_0; z_0) = 0$, one can find $\beta > 0$ such that, if $\|x - x_0\| \leqq \beta$, $\|z - z_0\| \leqq \beta$, then

$$\|F'_x(x; z)\| \leqq q < 1. \tag{1.30}$$

Since $F(x_0; z_0) = x_0$, there exists a positive number $\alpha \leqq \beta$ such that if $\|z - z_0\| \leqq \alpha$

$$\|F(x_0; z) - x_0\| \leqq (1 - q)\beta. \ \blacksquare$$

This inequality implies that if $\|z - z_0\| \leqq \alpha$ then the operator $F(x; z)$ maps the ball $\|x - x_0\| \leqq \beta$ into itself. It follows from (1.30) that for these values of z

$$\|F(x; z) - F(y; z)\| \leqq q\|x - y\| \qquad (\|x - x_0\|, \ \|y - y_0\| \leqq \beta).$$

Exercise 1.22. Let A be a nonlinear operator on a Banach space E which satisfies a Lipschitz condition (1.1) on the ball $T = \{x: \|x\| \leqq \rho\}$. Let $A\theta = \theta$, where θ is the zero of E. The resolvent $R(\alpha)$ of A is defined by the identity

$$R(\alpha)f = \alpha AR(\alpha)f + f.$$

1) Show that when $|\alpha| < 1/q$ the resolvent $R(\alpha)$ is defined on the ball $\|f\| \leqq (1 - |\alpha|q)\rho$.
2) Show that

$$\frac{1}{1 + |\alpha|q}\|f - g\| \leqq \|R(\alpha)f - R(\alpha)g\| \leqq \frac{1}{1 - |\alpha|q}\|f - g\|.$$

3) Show that

$$\|R(\alpha)f - R(\beta)f\| \leqq \frac{q \cdot \|f\| \cdot |\alpha - \beta|}{(1 - |\alpha|q)(1 - |\beta|q)}.$$

§ 2. Convergence of successive approximations

2.1. *Successive approximations.* We now continue our study of the equation

$$x = Ax \tag{2.1}$$

where the operator A is defined in a Banach space E. Consider the sequence

$$x_n = Ax_{n-1} \qquad (n = 1, 2, \ldots), \tag{2.2}$$

where x_0 is some initial element. Of course, this construction is possible only when x_0 belongs to the domains of all the operators $A^n (n = 1, 2, \ldots)$.

If the sequence x_n converges to some element x^* and the operator A is continuous at x^*, it follows from (2.2) that x^* is a solution of equation (2.1). Thus, a necessary condition for convergence of the sequence x_n is that equation (2.1) be solvable. However, we are consdering the sequence x_n from a different viewpoint—it may be regarded as a source of increasingly better approximations to the solution of equation (2.1). Our aim is therefore to determine for what equations (2.1) the successive approximations (2.2) converge.

2.2. *Equations with a contraction operator.* The successive approximations (2.2) will converge, for example, if the conditions of Theorem 1.1 hold.

Theorem 2.1. *Let the operator A map a closed set* $\mathfrak{M} \subset E$ *into itself, and assume that it is a contraction operator:*

$$\|Ax - Ay\| \leq q \|x - y\| \qquad (x, y \in \mathfrak{M}; q < 1). \tag{2.3}$$

Then for any initial element $x_0 \in \mathfrak{M}$ *the successive approximations* (2.2) *converge to the unique (by Theorem 1.1) solution* x^* *of equation* (2.1).

Proof. It follows from (2.2), (2.3) and the equality $x^* = Ax^*$ that

$$\|x_n - x^*\| = \|Ax_{n-1} - Ax^*\| \leq q \|x_{n-1} - x^*\|,$$

and so

$$\|x_n - x^*\| \leq q^n \|x_0 - x^*\| \qquad (n = 1, 2, \ldots). \tag{2.4}$$

Thus

$$\lim_{n \to \infty} \|x_n - x^*\| = 0. \ \blacksquare$$

Inequality (2.4) describes the convergence rate. Note that it is convenient only when an *a priori* estimate of $x_0 - x^*$ is available. An estimate of this kind may be derived from the inequality

$$\|x_0 - x^*\| \leq \|x_0 - Ax_0\| + \|Ax_0 - Ax^*\| \leq \|x_0 - Ax_0\| + q \|x_0 - x^*\|$$

which implies that

$$\|x_0 - x^*\| \leq \frac{1}{1 - q} \|x_0 - Ax_0\|. \tag{2.5}$$

Inequalities (2.4) and (2.5) imply that

$$\|x_n - x^*\| \leq \frac{q^n}{1 - q} \|x_0 - Ax_0\| \qquad (n = 1, 2, \ldots). \tag{2.6}$$

Inequalities (2.4) and (2.6) may be used to determine the number of successive approximations needed to solve equation (2.1) with given accuracy. For example, the inequality $\|x_n - x^*\| < \delta$ will certainly hold if

$$n > \frac{1}{\ln q} \ln \frac{\delta (1 - q)}{\|x_0 - Ax_0\|}. \tag{2.7}$$

Note that (2.3) is not a necessary condition for the convergence of the successive approximations (2.2).

Exercise 2.1. Let A be an operator mapping a closed set $\mathfrak{M}$ in a complete space E into itself, and let some power A^k be a contraction operator. Prove that the successive approximations (2.2) converge to the unique fixed point of A in $\mathfrak{M}$.

Exercise 2.2. Let A be an operator mapping a compact set $\mathfrak{M} \subset E$ into itself, such that

$$\|Ax - Ay\| < \|x - y\| \qquad (x, y \in \mathfrak{M} ; \ x \neq y).$$

Show that the successive approximations (2.2) converge to a solution of equation (2.1).

Exercise 2.3. Let $f(x)$ be a continuous function on $[0, 1]$ such that $0 \leqq f(x) \leqq 1$ for $x \in [0, 1]$. Define successive approximations x_n by the formula

$$x_n = x_{n-1} + \frac{1}{n} [f(x_{n-1}) - x_{n-1}]. \tag{2.8}$$

Show that the sequence (2.8) converges for any $x_0 \in [0, 1]$ to a solution of the equation $x = f(x)$.

Formula (2.8) is also applicable when $f(x)$ is a mapping of the ball $\|x\| \leqq r$ in a Banach space E into itself. Show that the approximations (2.8) may not converge, even when E is only two-dimensional.

2.3. *Linear equations.* In general, the estimates (2.4) and (2.6) cannot be improved, as evidenced by equation (2.1) with the operator $Ax \equiv qx$. However, additional assumptions may guarantee more rapid convergence.

First consider the linear equation

$$x = Bx + f. \tag{2.9}$$

If $\|B\| < 1$, Theorem 2.1 implies that the successive approximations

$$x_n = Bx_{n-1} + f \qquad (n = 1, 2, \ldots) \tag{2.10}$$

converge to a solution x^* of equation (2.9). We now prove a more precise result.

Theorem 2.2. If the spectral radius $\rho(B)$ of the operator B satisfies the inequality $\rho(B) < 1$, then the successive approximations (2.10) converge to a solution x^* of equation (2.9), and for any ε, $0 < \varepsilon < 1 - \rho(B)$,

$$\|x_n - x^*\| \leqq c(\varepsilon)\,[\rho(B) + \varepsilon]^n\,\|x_0 - Bx_0 - f\|. \qquad (2.11)$$

Proof. Following subsection 1.4, we introduce a norm $\|\cdot\|_*$ in E such that

$$m(\varepsilon)\,\|x\| \leqq \|x\|_* \leqq M(\varepsilon)\,\|x\| \qquad (x \in E) \qquad (2.12)$$

and

$$\|Bx\|_* \leqq [\rho(B) + \varepsilon]\,\|x\|_* \qquad (x \in E). \qquad (2.13)$$

It follows from (2.13) that equation (2.9) may be regarded as equation (2.1) with a contraction operator. The approximations (2.10) therefore converge to x^*. Formulas (2.13) and (2.6) imply the estimate

$$\|x_n - x^*\|_* \leqq \frac{[\rho(B) + \varepsilon]^n}{1 - \rho(B) - \varepsilon}\,\|x_0 - Bx_0 - f\|_*.$$

This estimate and (2.12) now imply (2.11). ∎

Exercise 2.4. Assume that the spectral radius $\rho(B)$ of the linear operator B is smaller than 1. Show that the solution x^* of equation (2.9) may be represented by a series

$$x^* = f + Bf + B^2f + \ldots$$

which converges in the norm (*Neumann series*).

Exercise 2.5. It is clear from the proof of Theorem 2.2 that (2.11) may be replaced by a sharper estimate

$$\|x_n - x^*\| \leqq c\rho^n(B)\,\|x_0 - Bx_0 - f\|, \qquad (2.14)$$

if the spectral radius $\rho(B)$ coincides with the norm of the operator B with respect to some equivalent norm. It is not always possible to construct a norm with this property (Exercise 1.9). Show that the estimate (2.14) is valid if B is a compact operator and $\rho(B) > 0$. Is this statement true when B is not compact?

If

$$\|x_n - x^*\| \leqq cq^n \qquad (n = 1, 2, \ldots),$$

the approximations x_n are said to *converge to x^* at the rate of a geometric progression with quotient q*. Theorem 2.2 states that the successive approximations (2.10) converge to a solution at the rate of a geometric progression whose quotient is arbitrarily close to $\rho(B)$. In particular,

if $\rho(B) = 0$ (B is then said to be *quasinilpotent* or *Volterra*), the approximations (2.10) converge at the rate of a progression with arbitrarily small
quotient.

2.4. *Factorial convergence.* Consider the integral equation

$$x(t) = \int_a^t K(t, s)\, x(s)\, ds + f(t) \tag{2.15}$$

with continuous kernel $K(t, s)$ ($a \leqq t, s \leqq b$). Since the spectral radius of
the operator

$$Bx(t) = \int_a^t K(t, s)\, x(s)\, ds$$

is zero, it follows that equation (2.15) has a unique continuous solution
$x^*(t)$ for any continuous function $f(t)$. This solution is the limit of successive approximations

$$x_n(t) = \int_a^t K(t, s)\, x_{n-1}(s)\, ds + f(t). \tag{2.16}$$

Let $\delta_n(t) = |x_n(t) - x^*(t)|$. Then

$$\delta_n(t) \leqq \int_a^t \delta_{n-1}(s)\, ds \qquad (K = \max_{a \leqq t, s \leqq b} |K(t, s)|)$$

and so

$$\delta_n(t) \leqq \frac{K^n (t - a)^n}{n!} \max_{a \leqq t \leqq b} \delta_0(t).$$

Consequently,

$$\|x_n - x^*\| = \max_{a \leqq t \leqq b} \delta_n(t) \leqq c \frac{K^n (b - a)^n}{n!} \qquad (n = 1, 2, \ldots). \tag{2.17}$$

One result of this estimate is that the successive approximations
(2.16) converge to the solution $x^*(t)$ more rapidly than any geometric
progression. This has been proved before, by other arguments. However,
inequality (2.17) provides more information about the rate of convergence. When inequality (2.17) is true, the convergence is said to be
factorial.

Note that factorial convergence has been demonstrated here for a special type of equation.

Exercise 2.6. Construct an example of a linear Volterra operator B for which the convergence of the successive approximations (2.10) is not factorial.

Exercise 2.7. Let $c_n (n = 1, 2, \ldots)$ be a numerical sequence such that

$$\lim_{n \to \infty} \sqrt[n]{c_n} = 0.$$

Construct a linear Volterra operator for which the successive approximations (2.10) satisfy the condition

$$\lim_{n \to \infty} \frac{\|x_n - x^*\|}{c_n} = \infty.$$

2.5. *Nonlinear equations.* We now consider the nonlinear equation

$$x = Ax. \tag{2.18}$$

Theorem 2.3. *Let the operator A be Fréchet-differentiable at a point x^* which is a solution of equation (2.18). Let ρ_0 denote the spectral radius of the linear operator $A'(x^*)$, and assume that $\rho_0 < 1$. Then the successive approximations*

$$x_n = Ax_{n-1} \qquad (n = 1, 2, \ldots) \tag{2.19}$$

converge to x^*, provided the initial approximation x_0 is sufficiently close to x^*, and then

$$\|x_n - x^*\| \leqq c(x_0; \varepsilon)(\rho_0 + \varepsilon)^n, \tag{2.20}$$

where ε is an arbitrary positive number.

Proof. Introduce an equivalent norm $\| \cdot \|_*$ in E,

$$m(\varepsilon) \|x\| \leqq \|x\|_* \leqq M(\varepsilon) \|x\| \qquad (x \in E),$$

such that

$$\|A'(x^*) h\|_* \leqq \left(\rho_0 + \frac{\varepsilon}{2} \right) \|h\|_* \qquad (h \in E).$$

It follows from the definition of the Fréchet derivative that there exists $\delta > 0$ such that, whenever $\|x - x^*\|_* < \delta$,

$$\|Ax - Ax^* - A'(x^*)(x - x^*)\|_* \leqq \frac{\varepsilon}{2} \|x - x^*\|_*.$$

Therefore, if $\|x - x^*\|_* < \delta$, then

$$\|Ax - x^*\|_* \leqq \|Ax - Ax^* - A'(x^*)(x - x^*)\|_* +$$
$$+ \|A'(x^*)(x - x^*)\|_* \leqq (\rho_0 + \varepsilon)\|x - x^*\|_* . \qquad (2.21)$$

If $\rho_0 + \varepsilon < 1$ this inequality means, in particular, that the operator A maps the ball $\|x - x^*\|_* < \delta$ into itself. Assume that some approximation x_{n_0} is in this ball. Inequality (2.21) then implies that, for all $n > n_0$,

$$\|x_n - x^*\|_* \leqq (\rho_0 + \varepsilon)^{n - n_0 - 1} \|x_{n_0} - x^*\|_* .$$

Returning to the old norm, we get (2.20). ∎

The estimate (2.20) cannot be improved without additional assumptions.

It follows from (2.20) that the successive approximations (2.19) converge to x^* more rapidly than a geometric progression with arbitrarily small quotient, provided the derivative $A'(x^*)$ vanishes. This remark may be strengthened if not only

$$\|Ax - Ax^*\| = o(\|x - x^*\|),$$

but also

$$\|Ax - Ax^*\| \leqq L\|x - x^*\|^2 . \qquad (2.22)$$

For example, this condition holds if the operator A is Fréchet-differentiable in a neighborhood of the point x^* and the derivative $A'(x)$ satisfies a Lipschitz condition

$$\|A'(x) - A'(y)\| \leqq L\|x - y\| . \qquad (2.23)$$

In fact, let l denote a normalized linear functional on E such that

$$l(Ax - Ax^*) = \|Ax - Ax^*\| .$$

Then, by the usual mean-value theorem for scalar functions,

$$\|Ax - Ax^*\| = l\{A'[(1 - \theta)x^* + \theta x](x - x^*)\} \leqq$$
$$\leqq \|A'[(1 - \theta)x^* + \theta x]\| \cdot \|x - x^*\| \qquad (\theta \in (0, 1)) .$$

Since $A'(x^*) = 0$,

$$\|A'[(1 - \theta)x^* + \theta x]\| = \|A'[(1 - \theta)x^* + \theta x] - A'(x^*)\| \leqq L\|x - x\|^* .$$

Thus inequality (2.22) is valid.* Let U be a neighborhood $\|x - x^*\| \leqq r$, $Lr < 1$, of x^* in which (2.22) holds. Assume that $x_{n_0} \in U$. Then the successive approximations (2.19) also belong to U for $n > n_0$. Inequality (2.22) then implies a chain of inequalities

$$\|x_n - x^*\| \leqq L \|x_{n-1} - x^*\|^2 \leqq L \cdot L^2 \|x_{n-2} - x^*\|^4 \leqq \dots$$

that is,

$$\dots \leqq L^{1 + 2 + \dots + 2^{n - n_0 - 1}} \|x_n - x^*\|^{2^{n - n_0}},$$

$$\|x_n - x^*\| \leqq \frac{1}{L} (L \|x_{n_0} - x^*\|)^{2^{n - n_0}}. \tag{2.24}$$

Under these conditions, the successive approximations (2.19) converge to x^*, provided the initial approximation is sufficiently close to x^*. Thus, the number

$$\delta = L \|x_{n_0} - x^*\|$$

may be assumed arbitrarily small. Consequently, the rate of convergence of the successive approximations (2.19) is described by an inequality

$$\|x_n - x^*\| \leqq c(\delta, x_0) \delta^{2^{n - n_0(\delta)}}, \tag{2.25}$$

where δ is an arbitrarily small positive number.

The rate of convergence described by (2.25) is much more rapid than convergence of a geometric progression with arbitrarily small quotient or factorial convergence.

Exercise 2.8. Does inequality (2.25) imply the estimate

$$\|x_n - x^*\| \leqq c_1(\delta_1, x_0) \delta_1^{2^n} ?$$

Exercise 2.9. How is (2.25) affected if (2.22) is replaced by the less restrictive assumption

$$\|Ax - Ax^*\| \leqq L \|x - x^*\|^{1 + \alpha},$$

where $\alpha \in (0, 1)$?

2.6. Effect of the initial approximation on estimate of convergence rate.

We have described various estimates of the error $x_n - x^*$. For certain

---

* Inequality (2.23) implies a stronger inequality than (2.22):

$$\|Ax - Ax^*\| \leqq (L/2) \|x - x^*\|^2.$$

initial approximations these estimates are too liberal. For example, for the linear equation

$$x = Ax,$$

where A is a symmetric matrix whose spectrum consists of two numbers $q_1, q_2 \in (0, 1)$, the successive approximations $x_n = Ax_{n-1}$ converge to the trivial solution at a rate q_1^n if the initial approximation has the property $Ax_0 = q_1 x_0$, and at a rate q_2^n if $Ax_0 = q_2 x_0$.

Thus, an "apt" choice of the initial approximation may sometimes yield successive approximations which converge more rapidly than guaranteed by the general estimates. Were the convergence more rapid in an absolute majority of cases (a phrase which must, of course, be given a precise probability-theoretic formulation), the rate of convergence would have to be characterized, in a great number of problems of approximate solution, by estimates other than (2.11). Unfortunately, this is usually not the case. A first example is the equation

$$x = Bx, \tag{2.26}$$

where B is a selfadjoint linear operator on a real Hilbert space H. Assume that

$$Bx = b_1 P_1 x + B_2 P_2 x, \tag{2.27}$$

where P_1 is the orthogonal projection on some subspace H_1, $P_2 = I - P_1$ is the orthogonal projection on the orthogonal complement H_2 of H_1, $b_1 = \|B\|$, and B_2 is a selfadjoint operator defined in H_2 such that $\|B_2\| = b_2 < b_1$.

Successive approximations $x_n = Bx_{n-1}$ to the trivial solution of the equation $x = Bx$ are defined by

$$x_n = b_1^n P_1 x_0 + B_2^n P_2 x_0 \qquad (n = 1, 2, \ldots).$$

Let $\|P_1 x_0\| > 0$. Then

$$b_1^n \|P_1 x_0\| \leqq \|x_n\| \leqq b_1^n \|x_0\|.$$

Thus, for practically all initial approximations (excluding the case $P_1 x_0 = 0$), the rate of convergence is precisely that of a geometric progression with quotient b_1.

A similar result holds for the nonlinear equation $x = Ax$ if

$$Ax = Bx + Dx, \tag{2.28}$$

where B is defined by (2.27) and D satisfies the condition

$$\|Dx\| \leqq \gamma(\rho)\,\|x\| \qquad (\|x\| \leqq \rho), \tag{2.29}$$

where $\gamma(\rho)$ is a continuous nonnegative function, monotone decreasing for $\rho \to +0$, such that $\gamma(0) = 0$. By (2.27) and (2.29), the successive approximations

$$x_n = Bx_{n-1} + Dx_{n-1} \tag{2.30}$$

satisfy the inequality

$$\|x_n\| \leqq [b_1 + \gamma(\|x_{n-1}\|)]\,\|x_{n-1}\|\,. \tag{2.31}$$

Therefore, for sufficiently small x_0,

$$\|x_n\| \leqq (b_1 + \gamma_n)\,\|x_{n-1}\| \qquad (n = 1, 2, \ldots), \tag{2.32}$$

where $\gamma_n \to 0$. This inequality means that the successive approximations converge more rapidly than any geometric progression with quotient greater than b_1. We claim that, for "almost all" sufficiently small x_0, the approximations (2.30) converge more slowly than any geometric progression with fixed quotient $b_1 - \varepsilon \ (\varepsilon > 0)$.

Let $\alpha(\rho)$ be a continuous function, monotone decreasing for $\rho \to +0$, such that $\alpha(0) = 0$ and

$$\lim_{\rho \to 0} \frac{\gamma(\rho)}{\alpha(\rho)} = 0 \tag{2.33}$$

(for example, set $\alpha(\rho) = \sqrt{\gamma(\rho)}$). We can then choose $\rho_0 > 0$ such that, whenever $\rho \leqq \rho_0$,

$$\frac{b_1 - \left[\gamma(\rho) + \dfrac{\gamma(\rho)}{\alpha(\rho)}\right]}{b_2 + \gamma(\rho) + \alpha(\rho)\,\gamma(\rho)} > 1\,. \tag{2.34}$$

By (2.29), we may also assume that

$$\|Ax\| \leqq \frac{b_1 + 1}{2}\,\|x\| \qquad (\|x\| \leqq \rho_0)\,. \tag{2.35}$$

Let $\mathfrak{N}(\rho_0)$ denote the set of all points $x \in H$ such that

$$\|P_1 x\| \geqq \alpha(\|x\|)\,\|P_2 x\|, \qquad \|x\| \leqq \rho_0\,. \tag{2.36}$$

It is easy to see that for small ρ_0 the set $\mathfrak{N}(\rho_0)$ accounts for "most" of the ball $\|x\| \leqq \rho_0$. The operator A maps $\mathfrak{N}(\rho_0)$ into itself. In fact, for $x \in \mathfrak{N}(\rho_0)$,

$$\|P_1 Ax\| \geqq \|P_1 Bx\| - \|P_1 Dx\| \geqq b_1 \|P_1 x\| - \gamma(\|x\|) \|x\| \geqq$$

$$\geqq b_1 \|P_1 x\| - \gamma(\|x\|)\left[1 + \frac{1}{\alpha(\|x\|)}\right] \|P_1 x\| \qquad (2.37)$$

and

$$\|P_2 Ax\| \leqq \|P_2 Bx\| + \|P_2 Dx\| \leqq b_2 \|P_2 x\| + \gamma(\|x\|) \|x\| \leqq$$

$$\leqq \frac{b_2}{\alpha(\|x\|)} \|P_1 x\| + \gamma(\|x\|)\left[1 + \frac{1}{\alpha(\|x\|)}\right] \|P_1 x\|. \qquad (2.38)$$

Thus (2.34) implies that

$$\|P_1 Ax\| \geqq \alpha(\|x\|) \|P_2 Ax\|,$$

which in turn, together with (2.35), implies that

$$\|P_1 Ax\| \geqq \alpha(\|Ax\|) \|P_2 Ax\|.$$

Hence the point Ax satisfies the first inequality of (2.36). The second follows directly from (2.35).

Thus, if $x_0 \in \mathfrak{N}(\rho_0)$, all approximations (2.30) belong to $\mathfrak{N}(\rho_0)$. It follows from (2.37) that, for any nonzero $x_0 \in \mathfrak{N}(\rho_0)$, the projections $P_1 x_n$ are different from zero for all approximations (2.30), and in fact

$$\|P_1 x_n\| \geqq (b_1 - \gamma_n) \|P_1 x_{n-1}\| \qquad (n = 1, 2, \ldots), \qquad (2.39)$$

where $\gamma_n \to 0$. In particular, for any $\varepsilon > 0$ there exists n_0 such that for $n > n_0$,

$$\|P_1 x_n\| \geqq (b_1 - \varepsilon) \|P_1 x_{n-1}\|,$$

whence

$$\|x_n\| \geqq \|P_1 x_n\| \geqq (b_1 - \varepsilon)^n \frac{\|P_1 x_{n_0}\|}{(b_1 - \varepsilon)^{n_0}} \qquad (n = 1, 2, \ldots). \qquad (2.40)$$

We leave it to the reader to prove the following inequality, which will be used in the sequel: if $x_0 \in \mathfrak{N}(\rho_0)$, then

$$\lim_{n \to \infty} \frac{\|P_2 x_n\|}{\|P_1 x_n\|} = 0. \qquad (2.41)$$

Arguments similar to those above apply to the case of equations in a Banach space.

2.7. *Acceleration of convergence.* We continue our investigation of the equation $x = Ax$ with an operator (2.28) in a Hilbert space H. We shall impose upon D $(D\theta = \theta)$ a condition stricter than (2.29):

$$\|Dx - Dy\| \leq \gamma(\rho)\|x - y\| \qquad (\|x\|, \|y\| \leq \rho). \qquad (2.42)$$

Equality (2.41) has a simple geometric interpretation: the sequence x_n is "tangent" to the subspace H_1. It turns out that the angle between the vectors $x_{n+1} - x_n$ and the subspace H_1 also tends to zero.

To simplify matters, we shall assume that the subspace H_1 is one-dimensional; let e be a unit vector in H_1. Then the angle ϕ between the vector x and the subspace H_1 is defined by

$$\cos \phi = \frac{(x, e)}{\|x\|}. \qquad (2.43)$$

Let $x \in \mathfrak{N}(\rho_0)$ (see subsection 2.6), $(x, e) > 0$. Then

$$(Ax, e) = b_1(x, e) + (Dx, e) \geq b_1\|P_1x\| - \gamma(\|x\|)\|x\| > 0.$$

It follows that, if $x_0 \in \mathfrak{N}(\rho_0)$ and $(x_0, e) > 0$, all approximations (2.30) satisfy the inequality $(x_n, e_0) > 0$.

Clearly,

$$(x_n - x_{n+1}, e) = (1 - b_1)(P_1x_n, e) - (Dx_n, e) \geq$$

$$\geq (1 - b_1)\|P_1x_n\| - \gamma(\|x_n\|)\|x_n\|$$

and

$$\|x_n - x_{n+1}\| \leq \|P_1x_n - P_1x_{n+1}\| + \|P_2x_n\| + \|P_2x_{n+1}\| \leq$$

$$\leq (1 - b_1)\|P_1x_n\| + \|P_2x_n\| + b_2\|P_2x_n\| + \gamma(\|x_n\|)\|x_n\|.$$

Hence

$$1 \geq \frac{(x_n - x_{n+1}, e)}{\|x_n - x_{n+1}\|} \geq \frac{1 - b_1 - \gamma(\|x_n\|)\dfrac{\|x_n\|}{\|P_1x_n\|}}{1 - b_1 + (1 + b_2)\dfrac{\|P_2x_n\|}{\|P_1x_n\|} + \gamma(\|x_n\|)\dfrac{\|x_n\|}{\|P_1x_n\|}}.$$

The right member of this inequality tends to 1, and it follows that

$$\lim_{n \to \infty} \frac{(x_n - x_{n+1}, e)}{\|x_n - x_{n+1}\|} = 1. \tag{2.44}$$

This relation means that the angle between $x_n - x_{n+1}$ and the subspace H_1 tends to zero.

We now claim that

$$\lim_{n \to \infty} \frac{(x_{n+1} - x_n, x_n - x_{n-1})}{\|x_n - x_{n-1}\|^2} = b_1. \tag{2.45}$$

Obviously (this is our first application of (2.42)),

$$\|x_{n+1} - x_n - b_1(x_n - x_{n-1})\| \leqq$$

$$\leqq \|B(x_n - x_{n-1}) - b_1(x_n - x_{n-1})\| + \|Dx_n - Dx_{n-1}\| \leqq$$

$$\leqq \|P_2B(x_n - x_{n-1}) - b_1P_2(x_n - x_{n-1})\| + \gamma(\|x_{n-1}\|) \|x_n - x_{n-1}\| \leqq$$

$$\leqq (b_1 + b_2) \|P_2(x_n - x_{n-1})\| + \gamma(\|x_{n-1}\|) \|x_n - x_{n-1}\|,$$

whence

$$\left| \frac{(x_{n+1} - x_n, x_n - x_{n-1})}{\|x_n - x_{n-1}\|^2} - b_1 \right| \leqq \frac{\|x_{n+1} - x_n - b_1(x_n - x_{n-1})\|}{\|x_n - x_{n-1}\|} \leqq$$

$$\leqq (b_1 + b_2) \frac{\|P_2(x_n - x_{n-1})\|}{\|x_n - x_{n-1}\|} + \gamma(\|x_{n-1}\|).$$

By (2.44), the right member of this inequality tends to zero as $n \to \infty$, which proves (2.45).

By (2.44) and (2.45), we have the following approximate equalities for large n:

$$x_{n+1} - x_n \approx b_1(x_n - x_{n-1}), \quad x_{n+2} - x_{n-1} \approx b_1(x_{n+1} - x_n), \dots$$

But then the exact solution x^* of the equation $x = Ax$ may be expressed approximately as

$$x^* = x_n + (x_{n+1} - x_n) + (x_{n+2} - x_{n+1}) + \dots \approx$$

$$\approx x_n + (x_{n+1} - x_n) + b_1(x_{n+1} - x_n) + b_1^2(x_{n+1} - x_n) + \dots,$$

whence

$$x^* \approx x_n + \frac{x_{n+1} - x_n}{1 - b_1} \tag{2.46}$$

or

$$x^* \approx x_n + \frac{x_{n+1} - x_n}{1 - b_1^2} + \frac{b_1(x_{n+1} - x_n)}{1 - b_1^2}.$$

It follows from the last equality that

$$x^* \approx x_n + \frac{x_{n+2} - x_n}{1 - b_1^2} \tag{2.47}$$

since $b_1(x_{n+1} - x_n) \approx x_{n+2} - x_{n+1}$.

By (2.45), the number b_1 in (2.46) may be replaced by its approximation

$$b_1 \approx \frac{(x_{n+1} - x_n, x_n - x_{n-1})}{\|x_n - x_{n-1}\|^2}. \tag{2.48}$$

This gives the following approximate formula for x^*:

$$x^* \approx x_n - \frac{(x_n - x_{n-1}, x_n - x_{n-1})}{(x_{n+1} - 2x_n + x_{n-1}, x_n - x_{n-1})}(x_{n+1} - x_n). \tag{2.49}$$

This formula gives a better approximation to the solution than x_{n+1}, but only under the assumptions adopted above. In particular, the number (2.48) may be negative, the derivative B of the operator A may not be selfadjoint, and so on. Formula (2.49) is suitable for the most general cases. Moreover, it even gives good approximations to x^* in certain cases for which ordinary successive approximations diverge; for example, when the operator A is "weakly nonlinear," and the entire spectrum of the derivative $A'(x^*)$ with the exception of a single point lies inside the unit disk.

The above reasoning carries over to equations in Banach spaces, though the lack of an inner product complicates matters somewhat. For Banach spaces, therefore, it is sometimes more convenient to compute an approximation of $|b_1|$ rather than b_1. Obviously,

$$|b_1| \approx \frac{\|x_{n+2} - x_{n+1}\|}{\|x_{n+1} - x_n\|}.$$

Formula (2.47) then implies the approximate formula

$$x^* \approx x_n + \frac{x_{n+2} - x_n}{1 - \left(\dfrac{\|x_{n+2} - x_{n+1}\|}{\|x_{n+1} - x_n\|}\right)^2}. \qquad (2.50)$$

To determine b_1, apply some linear functional $l(x)$ to the approximate equality $b_1(x_{n+1} - x_n) \approx x_{n+2} - x_{n+1}$. Then

$$b_1 \approx \frac{l(x_{n+2} - x_{n+1})}{l(x_{n+1} - x_n)}$$

and (2.46) gives the approximate formula

$$x^* \approx x_n - \frac{l(x_{n+1} - x_n)}{l(x_{n+2} - 2x_{n+1} + x_n)}(x_{n+1} - x_n). \qquad (2.51)$$

The "quality" of the formula depends essentially on the choice of the functional $l(x)$.

Formulas of type (2.49), (2.50) and (2.51) are known as *convergence-accelerating formulas*.* When used only once, they are given the natural name *formulas for improving the last approximation*. However, they may also be employed as follows: Find several ordinary approximations (2.2), then improve the last approximation with formula (2.49) or (2.50); then use this improved approximation as a new initial approximation, and so on.

2.8. *Distribution of errors.*** Computations employing the iteration procedure (2.2) are usually continued as long as the correction $x_{n+1} - x_n$ exceeds a certain given value. The value of the error $x_n - x^*$ may nevertheless be different, and it is natural to ask whether the probability distribution of the error norms may be determined.

Will the most probable errors not be considerably smaller than the maximum errors? As it turns out, this is not the case.

In this subsection we consider a system of algebraic linear equations

$$x = Ax + b, \qquad (2.52)$$

---

* Formulas of this kind for linear equations were first proposed by Lyusternik [2].
** Krasnosel'skii and Krein [2].

where A is a positive definite matrix all of whose eigenvalues λ_i ($i = 1, \ldots,$ n) are smaller than unity, b is given and x is the unknown (m-dimensional) vector. We shall show that the maximum errors are also the most probable, if the successive approximations are defined by the recurrence relation

$$x_{n+1} = Ax_n + b. \tag{2.53}$$

We denote

$$\delta_n = Ax_n + b - x_n, \quad \varepsilon_n = x^* - x_n.$$

Clearly,

$$\varepsilon_n = (I - A)^{-1}\delta_n \quad (n = 1, 2, \ldots). \tag{2.54}$$

Subtracting (2.53) from (2.52), we get

$$\varepsilon_{n+1} = A\varepsilon_n \quad (n = 1, 2, \ldots), \tag{2.55}$$

which implies the equality

$$\delta_n = x_{n+1} - x_n = (I - A)\varepsilon_n = (I - A)A^n\varepsilon_0. \tag{2.56}$$

Let us stipulate that the last iteration is the p-th if the correction-vector $\delta_p = x_{p+1} - x_p$ is in a certain preassigned neighborhood G of zero (ball, cube, etc.). By (2.56), this means that the error ε_p is in the set $G_0 = (I - A)^{-1}G$. Now if the matrix A has eigenvalues λ near 1, the eigenvalues $(1 - \lambda)^{-1}$ of the matrix $(I - A)^{-1}$ will be very large, and the domain G_0 will "stretch" along the directions of the corresponding eigenvectors of A. The error ε_p may thus be large, irrespective of how small the diameter of G is. As we shall show below, the probability of such large errors is, under natural assumptions, greater than that of small errors.

Denote $G_n = A^{-n}G_0$ ($n = 1, 2, \ldots$). If the p-th iteration is the last one, it follows from (2.56) that the initial error ε_0 is in the domain G_p, but $\varepsilon_0 \notin G_{p-1}$, since otherwise the procedure would have ended at an earlier step.

We limit the discussion to the simple case in which the domain G is a ball of radius α (qualitatively speaking, our conclusions will of course be valid in the general case), i.e., we assume that the procedure ends when the length of the vector $\delta_p = x_{p+1} - x_p$ (the square root of the sum of squares of the corrections) is smaller than α. The domain G_0 is then an ellipsoid containing the ball G, with semi-axes of length $\alpha/(1 - \lambda_i)$ ($i =$

$1, 2, \ldots, m$) directed along the corresponding eigenvectors of the matrix A.

Let G_{-1} denote the ellipsoid AG_0. The lengths of its semi-axes are $\alpha\lambda_i/(1 - \lambda_i)$ $(i = 1, 2, \ldots, m)$. As indicated above, the procedure will end if $\varepsilon_0 \in G_p$ and $\varepsilon_0 \notin G_{p-1}$, i.e., when the vector ε_0 is in the layer $G_p \setminus G_{p-1}$. The final error will then be in the layer $A^p(G_p \setminus G_{p-1}) = G_0 \setminus G_{-1}$. Thus the final error ε_p will belong to G_{-1} only if ε_0 belongs to G_0.

Now the probability of the final error being in a volume element Δ_0 of the layer $G_0 \setminus G_{-1}$ is the sum of the probabilities of the initial error being in the elements $\Delta_n = A^{-n}\Delta_0$ $(n = 0, 1, 2, \ldots)$.

Let us assume (to simplify matters) that the initial error ε_0 is uniformly distributed in a ball T of sufficiently large radius R. Then the probability $P(\Delta_n)$ of the initial error being in the element $\Delta_n \subset T$ is proportional to the volume of the element. Obviously,

$$\text{Volume } \Delta_n = \frac{\text{Volume } \Delta_0}{\lambda_1^n \ldots \lambda_m^n}.$$

Thus the probability $P(\Delta_0)$ that the final error will be in the element Δ_0 is given by

$$P(\Delta_0) = \frac{\text{Volume } \Delta_0}{\text{Volume } T} \sum_{k=0}^{n_0} \lambda_1^{-k}\lambda_2^{-k} \ldots \lambda_m^{-k},$$

where n_0 is the smallest natural number such that $\Delta_{n_0+1} \not\subset T$. It follows from this formula that the probability density of the final error at a point of the layer $G_0 \setminus G_{-1}$ will be greater, the later this point leaves the ball T under successive applications of the operation A^{-1}.

The layer $G_0 \setminus G_{-1}$ may be split into segments $F_1, F_2, \ldots$ (the first few of which may be empty), each consisting of the points leaving the ball T after exactly $k + 1$ iterations of A^{-1}. The set F_k is then the intersection of the layers $G_0 \setminus G_{-1}$ and $A^{k-1}T \setminus A^kT$. The probability density of the final error is constant on each set F_k, and increases with increasing k.

Theorem 2.4. Assume that the iteration procedure (2.53) *ends when the length of the correction vector becomes smaller than* α, *and that the initial error is uniformly distributed over a ball* T *of radius* R. *Then the probability* P *that the length of the final error* ε *will satisfy the inequality*

$$\eta\frac{\alpha\lambda_m}{1 - \lambda_m} \leqq \|\varepsilon\| \leqq \frac{\alpha}{1 - \lambda_m}, \tag{2.57}$$

where λ_m is the greatest eigenvalue of the matrix A, tends to unity as $R \to \infty$ for any $\eta < 1$.

Proof. Let e_i $(i = 1, \ldots, m)$ be eigenvectors of the matrix A corresponding to the eigenvalues λ_i $(i = 1, \ldots, m)$, indexed in order of magnitude.

Let us estimate the probability that the final error

$$\varepsilon = \sum_{i=1}^{m} \xi_i e_i$$

will not satisfy (2.57), i.e.,

$$\|\varepsilon\|^2 = \sum_{i=1}^{m} \xi_i^2 < \left(\eta \frac{\alpha \lambda_m}{1 - \lambda_m} \right)^2. \tag{2.58}$$

Assume first that this error is in the layer $G_0 \setminus G_{-1}$, i.e.,

$$\sum_{i=1}^{m} \xi_i^2 \frac{(1 - \lambda_i)^2}{\lambda_i^2} \geq \alpha^2. \tag{2.59}$$

Let

$$\eta < \eta_1 < 1, \tag{2.60}$$

and let $\lambda_1, \ldots, \lambda_s$ be the eigenvalues of A such that

$$\frac{1 - \lambda_m}{\lambda_m} \leq \eta_1 \frac{1 - \lambda_i}{\lambda_i} \qquad (i = 1, \ldots, s). \tag{2.61}$$

The eigenvalues $\lambda_{s+1}, \ldots, \lambda_m$ will then satisfy the inequality

$$\eta_1 \frac{1 - \lambda_i}{\lambda_i} < \frac{1 - \lambda_m}{\lambda_m} \qquad (i = s + 1, \ldots, m). \tag{2.62}$$

By virtue of (2.59) and (2.62),

$$\sum_{i=1}^{s} \xi_i^2 \frac{(1 - \lambda_i)^2}{\lambda_i^2} \geq \alpha^2 - \sum_{i=s+1}^{m} \xi_i^2 \frac{(1 - \lambda_i)^2}{\lambda_i^2} \geq$$

$$\geq \alpha^2 - \left(\frac{1 - \lambda_m}{\eta_1 \lambda_m} \right)^2 \sum_{i=s+1}^{m} \xi_i^2,$$

whence, by (2.58),

$$\sum_{i=1}^{s} \xi_i^2 \frac{(1 - \lambda_i)^2}{\lambda_i^2} \geq \alpha^2 \left[1 - \left(\frac{\eta}{\eta_1} \right)^2 \right], \tag{2.63}$$

and since

$$\frac{1 - \lambda_i}{\lambda_i} \leqq \frac{1 - \lambda_1}{\lambda_1} \qquad (i = 1, \ldots, m),$$

it follows that

$$\sum_{i=1}^{s} \xi_i^2 \geqq \frac{\lambda_1^2 \alpha^2}{(1 - \lambda_1)^2} \left[1 - \left(\frac{\eta}{\eta_1} \right)^2 \right]. \tag{2.64}$$

It follows from (2.64) that the vector $A^{-k}\varepsilon$ is not in the ball T when

$$k > \frac{\ln \dfrac{\lambda_1 \alpha}{(1 - \lambda_1)R} \sqrt{1 - \left(\dfrac{\eta}{\eta_1} \right)^2}}{\ln \lambda_s}. \tag{2.65}$$

In fact, if k satisfies (2.65) then, by (2.64),

$$\|A^{-k}\varepsilon\|^2 = \sum_{i=1}^{m} \frac{\xi_i^2}{\lambda_i^{2k}} \geqq \frac{1}{\lambda_s^{2k}} \sum_{i=1}^{s} \xi_i^2 \geqq$$

$$\geqq \frac{1}{\lambda_s^{2k}} \cdot \frac{\lambda_1^2 \alpha^2}{(1 - \lambda_1)^2} \left[1 - \left(\frac{\eta}{\eta_1} \right)^2 \right] > R^2.$$

Thus, the final error can satisfy the conditions (2.58) and (2.59) only if the initial error is in the ellipsoid $G_k = A^{-k}G_0$, where $k = k(R)$ is a fixed natural number satisfying (2.65). Inequality (2.65) then implies that

$$\lim_{R \to \infty} k(R) = \infty. \tag{2.66}$$

It is easily seen that $k(R)$ may be chosen to satisfy the additional inequality

$$\lambda_s^k R \geqq \frac{\alpha \lambda_1 \lambda_s}{1 - \lambda_1} \sqrt{1 - \left(\frac{\eta}{\eta_1} \right)^2}. \tag{2.67}$$

If the final error is in G_{-1}, then the initial error is in G_0 and, consequently, also in G_k. Therefore, the probability P of the initial error being in the ellipsoid G_k is greater than the probability that the final error satisfies (2.57).

On the other hand, the initial error must be in the ball T. It follows that the ratio of the volume of the intersection $G_k \cap T$ to the volume of T

is greater than the probability that the final error does not satisfy (2.57).

Let E_1 be the linear subspace spanned by the eigenvectors $e_1, \ldots, e_s$ and E_2 that spanned by $e_{s+1}, \ldots, e_m$. The volume of $G_k \cap T$ is clearly smaller than the product of the s-dimensional volume of the ball $T_1 = T \cap E_1$ and the volume of the $(n-s)$-dimensional ellipsoid $G_k^* = G_k \cap E_2$. Since

$$\text{Volume } T_1 = c_s R^s,$$

$$\text{Volume } G_k^* = \frac{c_{n-s}\alpha^{n-s}}{(1-\lambda_{s+1})\ldots(1-\lambda_m)} \cdot \frac{1}{\lambda_{s+1}^k \ldots \lambda_m^k},$$

it follows that

$$\text{Volume } G_k \cap T < \frac{MR^s}{\lambda_{s+1}^k \ldots \lambda_m^k} \leqq \frac{MR^s}{\lambda_{s+1}^{(m-s)k}} = \frac{MR^m}{(\lambda_s^k R)^{m-s}}\left(\frac{\lambda_s}{\lambda_{s+1}}\right)^{(m-s)k}$$

where

$$M = \frac{c_s c_{n-s}\alpha^{n-s}}{(1-\lambda_{s+1})\ldots(1-\lambda_m)}, \qquad j!! \cdot c_j = \begin{cases} (2\pi)^{j/2} & \text{for even } j, \\ 2^{(j+1)/2}\pi^{(j-1)/2} & \text{for odd } j, \end{cases}$$

and, by (2.67),

$$\text{Volume } G_k \cap T < MR^m \left\{ \frac{1-\lambda_1}{\alpha\lambda_1\lambda_s\sqrt{1-\left(\dfrac{\eta}{\eta_1}\right)^2}} \right\}^{m-s} \left(\frac{\lambda_s}{\lambda_{s+1}}\right)^{(m-s)k}. \qquad (2.68)$$

On the other hand,

$$\text{Volume } T = c_m R^m;$$

therefore, by (2.68),

$$P_1 < \frac{M}{c_m}\left\{ \frac{1-\lambda_1}{\alpha\lambda_1\lambda_s\sqrt{1-\left(\dfrac{\eta}{\eta_1}\right)^2}} \right\}^{m-s} \left(\frac{\lambda_s}{\lambda_{s+1}}\right)^{(m-s)k} = M_1\left(\frac{\lambda_s}{\lambda_{s+1}}\right)^{(m-s)k(R)} \qquad (2.69)$$

Letting $R \to \infty$ in (2.69), we see that the probability of condition (2.57) being false tends to zero as $R \to \infty$. ∎

Theorem 2.4 reflects a property of the successive errors $x_n - x^*$ established in subsection 2.6: for large n, these errors have approximately the same direction as an eigenvector corresponding to the greatest eigenvalue.

In his paper "On the Distribution of Errors in Iterational Solution of a System of Linear Algebraic Equations" (*Soobshcheniya AN Gruzinskoi SSR*, **50**, No. 2, 1968), D.G. Peradze studies the distribution of errors in detail for the case in which the matrix A in (2.52) is not symmetric.

2.9. Effect of round-off errors. We now return to the equation

$$x = Ax \tag{2.70}$$

in an arbitrary Banach space. E. We shall assume that the operator A maps a closed set $\mathfrak{M} \subseteq E$ into itself and is a contraction on $\mathfrak{M}$:

$$\|Ax - Ay\| \leqq q \, \|x - y\| \qquad (x, y \in \mathfrak{M} \, ; \, q < 1) \,. \tag{2.71}$$

In actual computation of successive approximations

$$x_n = Ax_{n-1} \qquad (n = 1, 2, \ldots),$$

exact determination of x_n from the previous approximation x_{n-1} is as a rule impossible, since computation of the values of the operator A involves various approximate formulas (numerical integration, etc.), and numerical computations inevitably involve round-off errors. The only possible general assertion is that the total error in application of the operator A does not exceed some number δ in the norm of the space E.

Thus, in actual computation of successive approximations

$$x_n = Ax_{n-1} + h_n \qquad (n = 1, 2, \ldots), \tag{2.72}$$

where $h_n \in E$ is an unknown random element, though we have an estimate

$$\|h_n\| \leqq \delta \qquad (n = 1, 2, \ldots) \,. \tag{2.73}$$

The successive approximations (2.72) may no longer converge to the solution x^* of equation (2.70). Nevertheless, they generally satisfy the inequality

$$\overline{\lim_{n \to \infty}} \, \|x_n - x^*\| \leqq \frac{\delta}{1 - q} \,. \tag{2.74}$$

This formula may be used to find an approximate solution, provided the required accuracy is at most $\delta/(1 - q)$.

Let us assume that all the successive approximations (2.72) belong to the set $\mathfrak{M}$ on which A is a contraction operator. Formulas (2.71) and

(2.72) then imply the estimate

$$\|x_{n+1} - x^*\| \leqq q \|x_n - x^*\| + \delta.$$

Hence

$$\|x_{n+1} - x^*\| \leqq q^2 \|x_{n-1} - x^*\| + q\delta + \delta \leqq \ldots$$

$$\ldots \leqq q^{n+1} \|x_0 - x^*\| + q^n\delta + q^{n-1}\delta + \ldots + \delta,$$

and so

$$\|x_{n+1} - x^*\| < q^{n+1} \|x_0 - x^*\| + \frac{\delta}{1 - q}, \tag{2.75}$$

which implies (2.74).

§ 3. Equations with monotone operators

3.1. *Statement of the problem.* In the preceding sections, the convergence of the successive approximations

$$x_n = Ax_{n-1} \tag{3.1}$$

to a solution x^* of the equation

$$x = Ax \tag{3.2}$$

followed from the fact that A was a contraction operator. However, the approximations (3.1) may also converge in other cases. In particular, there are certain classes of operators A, defined on a set $\mathfrak{M}$ in a Banach space E, with the following property: for $x_0 \in \mathfrak{M}$ and sufficiently large n, the successive approximations (3.1) belong to another set $\mathfrak{M}_0 \subset \mathfrak{M}$ on which A is a contraction operator.

We consider a simple example. Let $f(x)$ be a function which is continuous, concave, nonnegative, and monotone increasing on an interval $[a, b]$, $a < f(x) < b$. Let x^* be the solution of the equation $x = f(x)$ (Fig. 3.1). Since $f(x)$ is concave, the successive approximations $x_n = f(x_{n-1})$ converge to x^* for any $x_0 \in [a, b]$. And though $f(x)$ may not satisfy a Lipschitz condition in $[a, b]$ with constant smaller than unity, it *will* satisfy a condition of this type in a certain neighborhood of the point x^*.

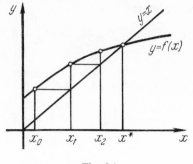

Fig. 3.1.

Now assume that $f(x)$ is nonnegative, continuous and monotone increasing for $0 \leq a \leq x \leq b$, while the quotient $f(x)/x$ is monotone decreasing. The geometric meaning of this condition is simple: the ray $y = kx$ through the origin cuts the curve $y = f(x)$ in at most one point $x(k)$, and for $x > x(k)$ the curve is under the ray (Fig. 3.2). This class of functions (which we call *quasiconcave*) includes, in particular, all concave functions. It is not difficult to see that if $f(x)$ is quasiconcave the successive approximations $x_n = f(x_{n-1})$ converge to the solution x^* of the equation $x = f(x)$, provided it exists (see Fig. 3.2). If $f'(x^*) = 1$, the function $f(x)$ is not a contraction operator in any neighborhood of x^*, and the successive approximations x_n converge more slowly than any convergent geometric progression.

In the sequel we shall show how the concepts of concavity and quasi-concavity extend to operators.

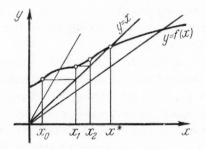

Fig. 3.2.

Assume that an operator A leaves a set $\mathfrak{M}$ invariant and has a unique fixed point x^* on $\mathfrak{M}$. Denoting $B = A^k$, we assume further that the successive approximations $x_n = Bx_{n-1}$ converge to x^* for any initial approximation $x_0 \in \mathfrak{M}$. Then, as is easily seen, the approximations $x_n = Ax_{n-1}$ also converge to x^*. We may thus formulate conditions for the con vergence of successive approximations which require the operator A^k to belong to the class of concave or quasiconcave operators for some k.

3.2. *Cones in Banach spaces.** Let x and y be points in a Banach space E; then the set of elements $tx + (1 - t)y$ $(0 \leqq t \leqq 1)$ is called the *segment* joining x and y, denoted by $[x, y]$. A set is said to be *convex* if, together with any two points, it contains the entire segment joining them. The *ray* through a point $x \in E$ $(x \neq 0)$ is defined to be the set of points tx $(t \geqq 0)$.

A closed convex set $K \subset E$ is called a *cone* if, together with any point it contains the ray through x, and moreover x, $-x \in K$ implies that $x = \theta$. The basic examples of cones are the sets of nonnegative functions in various function spaces $(C, L_p$, etc.), and the set of vectors with non-negative components.

Any cone K in a space E defines a *semi-order* relation: we write $x \leqslant y$ if $y - x \in K$. The relation $\leqslant$ satisfies the usual properties of the relation $\leqq$: inequalities may be multiplied by positive numbers; if $x \leqslant y$ and $y \leqslant z$ then $x \leqslant z$; if $x \leqslant y$ and $y \leqslant x$ then $x = y$; and inequalities are preserved by limit processes.

The set of elements x such that $v_0 \leqslant x \leqslant w_0$ is called a *conic segment*, denoted by $\langle v_0, w_0 \rangle$.

A sequence x_n is said to be *monotone increasing* if $x_1 \leqslant x_2 \leqslant \ldots \leqslant x_n \leqslant \ldots$. The sequence x_n is said to be *bounded* (*above*) if there is a fixed z such that $x_n \leqslant z$ $(n = 1, 2, \ldots)$. The sequence x_n is said to be *bounded in norm* if $\|x_n\| \leqq M$ $(n = 1, 2, \ldots)$. A cone K is *regular* if any bounded monotone sequence converges in norm, *completely regular* if any mono-tone sequence which is bounded in norm converges in norm. It turns

---

* In this subsection we define elementary concepts related to spaces semi-ordered by a cone (see Kantorovich, Vulikh and Pinsker [1], Krein and Rutman [1], Vulikh [1], Krasnosel'skii [8, 11]). All concepts and propositions used in this book which belong to the theory of cones and the theory of positive operators are discussed in detail in Krasnosel'skii's monograph [11].

out that *a completely regular cone is regular*. In a finite-dimensional space, every cone is completely regular. The cone of nonnegative functions in L_p ($1 \leqq p < \infty$) is completely regular, but the cone of nonnegative functions in the space C is not even regular. An example of a completely regular cone in C is any cone K_α ($0 < \alpha < 1$) consisting of all functions $x(t)$ such that

$$\alpha \max_\tau | x(\tau) | \leqq x(t)$$

(prove this statement!).

An operator A which leaves a cone K invariant ($AK \subset K$) is said to be *positive*. The operator A is said to be *monotone* if $x \leqslant y$ implies that $Ax \leqslant Ay$. In subsection 3.3 we shall indicate a class of positive operators A for which the equation $x = Ax$ may be solved by successive approximations.

Let E be n-space, K the cone of vectors with nonnegative components. Then the matrix $A = (a_{ij})$ defines a positive operator if and only if all its elements are nonnegative.*

Exercise 3.1. Let A be a nonsingular matrix with nonnegative elements, the elements of whose inverse are also nonnegative. Show that each row (column) of A contains exactly one nonzero element.

As another example, we consider the linear integral operator

$$Ax(t) = \int_\Omega G(t, s) x(s) \, ds \tag{3.3}$$

with nonnegative kernel $G(t, s)$. This operator is clearly positive on the cone of nonnegative functions in the space in which it is defined.

Exercise 3.2. Show that a linear operator is positive if and only if it is monotone.

An example of a nonlinear positive operator is the Hammerstein integral operator

$$Ax(t) = \int_\Omega G(t, s) f[s, x(s)] \, ds \,, \tag{3.4}$$

where Ω is a bounded closed set in a finite-dimensional space. Assume that the kernel $G(t, s)$ is nonnegative on $\Omega \times \Omega$ and $f(t, x) \geqq 0$ for

---

* Matrices with nonnegative elements are studied in detail in Gantmakher [1].

$x \geq 0$. Then the operator A is positive on the cone K of nonnegative functions of its domain of definition.* A is then monotone if $f(t, x)$ is not a decreasing function of x.

Exercise 3.3. Assume that the kernel $G(t, s)$ of the operator (3.4) is continuous and

$$0 < m \leq G(t, s) \leq M \qquad (t, s \in \Omega). \tag{3.5}$$

Let $f(t, x)$ be nonnegative and continuous in all its variables $t \in \Omega, x \geq 0$.

Show that if $M\alpha \leq m$ the operator (3.4) is positive on every cone K_α in the space C of continuous functions on Ω. Show that this operator is monotone on K_α if $f(t, x)$ is not a decreasing function of x.

Exercise 3.4. Consider the system of ordinary differential equations

$$\frac{dx}{dt} = Ax, \tag{3.6}$$

where $A = (a_{ij})$ is a constant matrix. The solution of (3.6) satisfying the initial condition

$$x(0) = x_0 \tag{3.7}$$

is determined by the formula

$$x(t) = e^{At}x_0. \tag{3.8}$$

Show that the linear operators $e^{At}(t > 0)$ are positive on the cone of vectors with nonnegative coordinates, if the off-diagonal elements of A are nonnegative: $a_{ij} \geq 0$ for $i \neq j$.

Exercise 3.5. Consider the nonlinear system of differential equations

$$\frac{dx_i}{dt} = f_i(t, x_1, \ldots, x_n) \qquad (i = 1, \ldots, n) \tag{3.9}$$

where f_i are continuous functions. Assume that the solutions of this system are uniquely determined by every initial condition (3.7) and defined for all $t > 0$. The solutions may be expressed as

$$x(t) = U(t)x_0. \tag{3.10}$$

The operator $U(t)$ is said to be a *translation* along the trajectories of the differential equation.

Show that the operators $U(t)$ $(t > 0)$ are positive on the cone of vectors with nonnegative components, if, for each $i = 1, 2, \ldots, n$,

$$f_i(t, x_1, \ldots, x_{i-1}, 0, x_{i+1}, \ldots, x_n) \geq 0 \qquad (x_j \geq 0; t \geq 0). \tag{3.11}$$

---

* Conditions under which various nonlinear integral operators are defined in various function spaces and have desirable properties (continuity, compactness, etc.) have been studied by many authors. The simplest theorems are due to Kantorovich and Akilov [1]. For more general results, see the monographs of Krasnosel'skii, Zabreiko, Pustyl'nik and Sobolevskii [1], Krasnosel'skii and Rutitskii [1], Zaanen [1], and Zabreiko [2]. For a study of positive integral operators, see Krasnosel'skii [8, 11].

Exercise 3.6. Show that the operators $U(t)\,(t > 0)$ are monotone on the cone of vectors with nonnegative components, if the right members of the system (3.9) have the property of off-diagonal monotonicity, i.e., for each $i = 1, 2, \ldots, n$ and each fixed $z \geqq 0$, the inequalities

$$0 \leqq x_1 \leqq y_1, \ldots, 0 \leqq x_{i-1} \leqq y_{i-1}, 0 \leqq x_{i+1} \leqq y_{i+1}, \ldots, 0 \leqq x_n \leqq y_n$$

imply the inequalities

$$f_i(t, x_1, \ldots, x_{i-1}, z, x_{i+1}, \ldots, x_n) \leqq f_i(t, y_1, \ldots, y_{i-1}, z, y_{i+1}, \ldots, y_n).$$

3.3. *Solvability of equations with monotone operators.* We begin with a simple example.

Given a positive monotone operator A, assume that there are elements v_0 and w_0, $v_0 \preccurlyeq w_0$, such that

$$Av_0 \succcurlyeq v_0, \quad Aw_0 \preccurlyeq w_0. \tag{3.12}$$

Then the conic segment $\langle v_0, w_0 \rangle$ is invariant under the operator A. Form sequences

$$v_n = Av_{n-1}, \qquad w_n = Aw_{n-1} \qquad (n = 1, 2, \ldots). \tag{3.13}$$

By (3.12), the first of these sequences is monotone increasing and bounded above, while the second is monotone decreasing and bounded below. Both sequences are therefore convergent if the cone K is regular.

If the operator A is continuous, we can let $n \to \infty$ in (3.13), with the obvious notation

$$v^* = Av^*, \qquad w^* = Aw^*, \tag{3.14}$$

where the limits $v^* = \lim_{n \to \infty} v_n$ and $w^* = \lim_{n \to \infty} w_n$ may be different.

The above arguments show that, to prove that an equation (3.2) with a continuous monotone operator A is solvable and to construct convergent successive approximations, it suffices to establish the existence of elements v_0 and w_0 satisfying (3.12) and to construct one of them.

We have thus proved

Theorem 3.1. *If the cone K is regular and the operator A is continuous and monotone on the conic segment $\langle v_0, w_0 \rangle$, mapping it into itself, then A has at least one fixed point on $\langle v_0, w_0 \rangle$, and the sequences (3.13) converge to fixed points of the operator A.*

The regularity of the cone K was used in Theorem 3.1 only in proving that the sequences (3.13) converge. This restriction may therefore be relaxed if the convergence of the sequences (3.13) can be established by other arguments. For example, it is easy to see that bounded monotone sequences are convergent if they are compact. Therefore:

Theorem 3.2. If the operator A is continuous and monotone on the conic segment $\langle v_0, w_0 \rangle$, mapping it into a compact subset of itself, then A has at least one fixed point on $\langle v_0, w_0 \rangle$. The sequences (3.13) then converge to fixed points of the operator A.

Under the assumptions of this theorem, the compactness of the set $A \langle v_0, w_0 \rangle$ may be replaced by a less restrictive condition, viz., the compactness of one of the sets $A^k \langle v_0, w_0 \rangle$.

Another convergence test for the sequences (3.13) is the existence of an operator F, defined on $\langle v_0, w_0 \rangle$, such that

$$F(x + y) \succcurlyeq Fx + \alpha(\|y\|)z_0 \qquad (x, x + y \in \langle v_0, w_0 \rangle; \ y \in K), \qquad (3.15)$$

where $\alpha(r)$ is a nondecreasing positive function for $r > 0$, and z_0 is a nonzero element of the cone K.

In fact, assume that the first sequence (3.13) is not convergent. Then there exists a subsequence

$$v_0 \preccurlyeq v_{n_1} \preccurlyeq v_{n_2} \preccurlyeq \ldots \preccurlyeq v_{n_k} \preccurlyeq \ldots \preccurlyeq w_0$$

and a number $\varepsilon > 0$ such that

$$\|v_{n_{k+1}} - v_{n_k}\| > \varepsilon \qquad (k = 1, 2, \ldots).$$

Therefore

$$Fv_{n_{k+1}} \succcurlyeq Fv_{n_k} + \alpha(\varepsilon)z_0 \succcurlyeq \ldots \succcurlyeq Fv_{n_1} + K\alpha(\varepsilon)z_0 \qquad (k = 1, 2, \ldots),$$

and so

$$\frac{1}{k\alpha(\varepsilon)}(Fw_0 - Fv_0) \succcurlyeq z_0 \qquad (k = 1, 2, \ldots).$$

Letting $k \to \infty$, we get $z_0 \preccurlyeq \theta$—a contradiction.

The second sequence of (3.13) is treated similarly.

The reader should note that the conic segments $\langle v_0, w_0 \rangle$ may be unbounded (in norm). If they are bounded in norm for any v_0, w_0, the cone K is said to be *normal*.* The cones of nonnegative functions in the

---

* Krein calls a cone K *normal* if there exists $\delta > 0$ such that $\|x + y\| \geqq \delta$ for $x, y \in K$ and $\|x\| = \|y\| = 1$. Our definition is equivalent.

spaces C and L_p are normal. The simplest test for normality is given by the following proposition (Bakhtin [2]): If $\theta \leqslant x \leqslant y$ implies that $\|x\| \leq M \|y\|$ (where M is a constant), then the cone K is normal. The converse is also valid.

A frequent choice for the element v_0 in the conic segment $\langle v_0, w_0 \rangle$ is the zero of the space. To construct the element w_0 (or to prove its existence), one can use *majorants* of the operator A, i.e., operators B such that

$$Ax \leqslant Bx \qquad (x \in K). \tag{3.16}$$

If an element w_0 may be contructed for a majorant B such that $Bw_0 \leqslant w_0$, then *a fortiori* $Aw_0 \leqslant w_0$.

A convenient form for the majorant is

$$Bx = Dx + f, \tag{3.17}$$

where D is a positive linear operator and f an element of K. Assume that the spectral radius of the operator D is smaller than 1. Then the solution z^* of the equation

$$z = Dz + f$$

may, as we know, be represented by a series

$$z^* = f + Df + \ldots + D^n f + \ldots \tag{3.18}$$

which is convergent in norm. Obviously, $z^* \in K$. The inequality (3.16) then implies that $Az^* \leqslant z^*$.

One more remark is in order. To apply Theorems 3.1 or 3.2 it is sufficient that the operator (3.17) be a majorant of A on the conic segment $\langle \theta, z^* \rangle$.

As an example, consider the integral operator

$$x(t) = \int_\Omega G(t, s) f[s, x(s)] \, ds \tag{3.19}$$

where the kernel $G(t, s)$ is continuous and nonnegative and the function $f(t, x)$ is continuous and nondecreasing in x for $x \geq 0$. Assume that the function $f(t, x)$ satisfies the estimate

$$0 \leq f(t, x) \leq ax + b \qquad (x \geq 0), \tag{3.20}$$

where

$$a\rho_0 < 1, \tag{3.21}$$

ρ_0 denoting the spectral radius of the linear integral operator with kernel $G(t, s)$. Define an operator B by

$$Bx(t) = a \int_\Omega G(t, s) \, x(s) \, ds + Mb \, \text{mes} \, \Omega,$$

where $M \geqq G(t, s)$ for all $t, s \in \Omega$. B is clearly a majorant of the operator A defined by the right member of (3.19) on the cone K of nonnegative functions of the space C. The above reasoning implies that the conic segment of functions such that

$$0 \leqq x(t) \leqq z^*(t) \qquad (t \in \Omega),$$

where z^* is a fixed point of the operator B, is invariant under A.

It is obvious that A is monotone and compact. Theorem 3.2 therefore implies that equation (3.19) has at least one nonnegative solution, and that the successive approximations

$$x_n(t) = \int_\Omega G(t, s) f[s, x_{n-1}(s)] \, ds \qquad (n = 1, 2, \ldots; \ x_0(t) \equiv 0)$$

converge uniformly to one of these solutions.

Exercise 3.7. Consider the boundary value problem

$$Lx = f(t, x), \tag{3.22}$$

where L is an elliptic differential operator in a bounded domain, with boundary conditions for which the Green's function is nonnegative. Let $f(t, x)$ be continuous and nondecreasing in $x(x \geqq 0)$. Assume that

$$f(t, 0) \geqq 0, \quad \lim_{x \to \infty} \frac{f(t, x)}{x} = 0. \tag{3.23}$$

Show that the boundary value problem (3.22) has at least one nonnegative solution, and one solution is the limit of the successive approximations $x_n(t)$ defined by the linear equations

$$Lx_n(t) = f[t, x_{n-1}(t)] \qquad (n = 1, 2, \ldots; \ x_0(t) \equiv 0).$$

3.4. *Nontrivial positive solutions.* Assuming now that equation (3.2) with positive operator A has the trivial solution $(A\theta = \theta)$, we wish to determine whether the equation has nontrivial solutions in the cone K and whether the successive approximations (3.1) converge to these solutions. To be able to apply the arguments set forth in subsection 3.3, we must find a nonzero element v_0 such that $v_0 \preccurlyeq Av_0$. In trying to find elements v_0 it is natural to expect the values of Av_0 to depend mainly on the elements $A'(\theta)v_0$, provided $A'(\theta)$ exists. Various results may be proved in this way. We shall also use the concept of minorants.

An operator F is said to be a *minorant* of the operator A on a set $\mathfrak{M}$ if

$$Fx \leqslant Ax \qquad (x \in \mathfrak{M}). \qquad (3.24)$$

If $v_0 \leqslant Fv_0$, then *a fortiori* $v_0 \leqslant Av_0$. We may thus try to determine the element v_0 for a minorant F.

As an example, we again consider equation (3.19), assuming conditions (3.20) and (3.21) to hold and, in addition, $f(t, 0) \equiv 0$. The equation then has the trivial solution.

Let

$$f(t, x) \geqq \alpha x \qquad (0 \leqq x \leqq \delta \,;\, \alpha, \delta > 0). \qquad (3.25)$$

It follows from (3.25) that the linear operator

$$Fx(t) = \alpha \int_{\Omega} G(t, s)\, x(s)\, ds$$

is a minorant of the operator A defined by the right member of equation (3.19) on the intersection of the cone K of nonnegative functions of C and the ball $\|x\| \leqq \delta$. By the general theory of integral operators with positive kernel, there exists a nonzero nonnegative function $h(t)$ such that

$$Fh(t) = \rho(F)\, h(t).$$

Assume that $\rho(F) > 1$. Then, for all elements $v_0 = \lambda h$, the minorant F satisfies the condition $Fv \geqslant v_0$. Therefore, to prove that equation (3.19) has a nontrivial nonnegative solution and that the successive approximations

$$x_n(t) = \int_{\Omega} G(t, s) f[s, x_{n-1}(s)]\, ds \qquad (n = 1, 2, \ldots) \qquad (3.26)$$

converge to one of these solutions when $x_0(t) = \lambda h(t)$ (where λ is a small positive number), it suffices to find a function $w_0(t)$ which satisfies two conditions: $w_0(t) \geqq x_0(t)$ and $Aw_0(t) \leqq w_0(t)$. One possible choice is a solution of the equation

$$z(t) = a \int_{\Omega} G(t, s)\, z(s)\, ds + Mb \operatorname{mes} \Omega + x_0(t).$$

Exercise 3.8. Let A be a continuous and monotone operator on the cone K. Assume that

$$Ax \leqslant Bx + f \quad (x \in K),$$

$$Ax \geqslant B_1 x \quad (x \in K, \|x\| \leq \delta),$$

where $f \in K$, the operators B and B_1 are linear and their spectral radii satisfy the inequality

$$\rho(B) < 1 < \rho(B_1).$$

Assume that the operator B_1 has an eigenvector h in K corresponding to the eigenvalue $\rho(B_1)$.

Finally, assume that either the cone K is normal and the operator A compact, or K is regular and A continuous.

Prove that the equation $x = Ax$ has at least one nontrivial solution in K, which is the limit of successive approximations $x_n = Ax_{n-1}$ for suitable initial approximations. How may the initial approximations be constructed using the operators B and B_1?

3.5. *Equations with concave operators.* Let u_0 be a fixed element of the cone K. A positive monotone operator A is said to be u_0-*concave* if it satisfies the following two conditions:

1. For any nonzero $x \in K$,

$$\alpha u_0 \leqslant Ax \leqslant \beta u_0, \qquad (3.27)$$

where $\alpha = \alpha(x) > 0$, $\beta = \beta(x) > 0$.

2. With each x such that $\alpha u_0 \leqslant x \leqslant \beta u_0$ $(\alpha, \beta > 0)$ and each $t \in (0, 1)$ one can associate a positive number $\eta = \eta(x, t)$ such that

$$A(tx) \geqslant (1 + \eta)tAx. \qquad (3.28)$$

An example of a u_0-concave operator $(u_0(t) \equiv 1)$ on the cone K of nonnegative functions in the space C is

$$Ax(t) = \int_\Omega G(t, s)f[x(s)] \, ds, \qquad (3.29)$$

provided the function $f(x)$ is quasiconcave (see subsection 3.1) and the kernel $G(t, s)$ continuous and positive.

Equations with u_0-concave operators have several interesting properties. In particular, they have at most one nontrivial solution in K. The following theorem on the convergence of successive approximations is stated without proof.

Theorem 3.3.* *Let the operator A be continuous and u_0-concave, and the cone K regular. Assume that equation* (3.2) *has a nontrivial solution* x^* *in K. Then the successive approximations* (3.1) *converge to* x^* *for any nonzero initial approximation* $x_0 \in K$.

If A is compact,** it is sufficient to assume that K is normal. Consider the equation

$$x(t) = \int_\Omega G(t, s) f[x(s)] \, ds, \qquad (3.30)$$

where the function $f(x)$ is continuous, concave and nonnegative on $[0, \infty)$ and the kernel $G(t, s)$ is continuous and positive. Assume that the conditions stated in subsection 3.4 for the existence of a nontrivial non-negative solution $x^*(t)$ of equation (3.30) are satisfied. Since the operator (3.29) is u_0-concave, this solution is unique.

It follows from Theorem 3.3 that $x^*(t)$ is the limit of a uniformly convergent sequence

$$x_n(t) = \int_\Omega G(t, s) f[x_{n-1}(s)] \, ds,$$

where the initial approximation $x_0(t)$ is an arbitrary nonnegative function, not identically zero.

Theorem 3.3 preassumes that equation (3.2) has a nontrivial solution in K. Theorems 3.1 and 3.2 are convenient tools for proving the existence of such a solution. To do so one must construct a conic segment $\langle v_0, w_0 \rangle$ which is invariant under a monotone operator A. The following simple remark is useful in this respect.

Let the operator A be u_0-concave, and

$$Av \geqslant v, \qquad Aw \leqslant w, \qquad (3.31)$$

where v, w are nonzero elements of K. The inequalities

$$Av \leqslant \beta(v) u_0, \qquad \alpha(w) u_0 \leqslant Aw$$

---

* See Krasnosel'skii [11]. For more general results, see Bakhtin and Krasnosel'skii [3]. Theorem 3.3 stems from a construction of Urysohn [1], who also considered a special type of initial approximations for integral equations.

** A nonlinear operator A is said to be *compact* if it is continuous and maps any bounded set onto a compact set.

then imply

$$w \succcurlyeq \frac{\alpha(w)}{\beta(v)} v \,.$$

Thus, there exists a maximum positive t such that $w \succcurlyeq tv$; denote it by t_0. Assume that $t_0 < 1$. By the properties of u_0-concave operators, $A(t_0 v) \succcurlyeq t_0 Av$. Therefore,

$$w \succcurlyeq Aw \succcurlyeq A(t_0 v) \succcurlyeq t_0(1 + \eta)v \,,$$

which contradicts the fact that t_0 is maximal. Thus $t_0 \geqq 1$, and we have proved that (3.31) implies $v \preccurlyeq w$.

Thus, to find a conic segment which is invariant under a u_0-concave operator, it suffices to find an element v such that $Av \succcurlyeq v$ and an element w such that $Aw \preccurlyeq w$.*

Exercise 3.9. Suppose that condition (3.28) in the definition of a u_0-concave operator A has been replaced by a less restrictive condition $A(tx) \succcurlyeq tAx (0 \leqq t \leqq 1)$. Let $v \preccurlyeq Av$, $w \succcurlyeq Aw$ ($v, w \in K; v \neq \theta$). Find the conic segment which is invariant under A.

3.6. *Use of powers of operators.* The arguments of subsection 1.5 employing powers of operators carry over to operators which have an invariant cone.

Assume that some power A^p of a continuous operator A is u_0-concave. It then follows from Theorem 3.3 that the sequences $A^{pn}(A^i x_0)$ ($i = 0$, $1, \ldots, p - 1$) converge to the fixed point x^* of A^p as $n \to \infty$. Hence the sequence $x_n = A^n x_0$ ($n = 1, 2, \ldots$) also converges to x^*. Letting $n \to \infty$ in the equality $x_n = Ax_{n-1}$ ($n = 1, 2, \ldots$), we get $x^* = Ax^*$, showing that x^* is a solution of (3.2) which is the limit of successive approximations.

There are various classes of operators which have u_0-concave powers.** Note that a power A^p may be monotone even though the operator A itself is not. For example, if $x \preccurlyeq y$ implies that $Ax \succcurlyeq Ay$, the operator A^p is monotone for all even p.

As an example, consider the integral equation[†]

$$x(t) = \int_\Omega G(t, s) x^\nu(s) \, ds \,, \tag{3.32}$$

---

* For examples of the procedure, see Krasnosel'skii [11], and also Pokornyi [1, 3].
** See Bakhtin [4], Krasnosel'skii [11].
[†] Equations of this type have been studied by Marchenko [1], by other methods.

where the kernel $G(t, s)$ is continuous and positive. Let $v \in (-1, 0)$. We claim that *equation (3.32) has a unique positive solution which is the limit of the successive approximations*

$$x_{n+1}(t) = \int_\Omega G(t, s) x_n^v(s) \, ds \qquad (3.33)$$

for any positive initial approximation $x_0(t)$.

To prove this assertion, consider the operator

$$Ax(t) = \int_\Omega G(t, s) x^v(s) \, ds. \qquad (3.34)$$

Let

$$0 < m \le G(t, s) \le M \qquad (t, s \in \Omega).$$

The operator A maps every positive continuous function onto a function of $K_{m/M}$ (see subsection 3.2). In particular, A maps all nonzero elements of the cone $K_{m/(2M)}$ onto nonzero elements of $K_{m/M}$.

Now every nonzero function $z(t)$ in $K_{m/M}$ is an interior point of the cone $K_{m/(2M)}$, and so

$$\alpha u_0 \preccurlyeq z(t) \preccurlyeq \beta u_0,$$

where α, β are positive numbers, $u_0(t) \equiv 1$, and the semi-order is defined by the cone $K_{m/(2M)}$. In particular, for any nonzero function $x(t) \in K_{m/(2M)}$,

$$\alpha(x) u_0 \preccurlyeq A^2 x \preccurlyeq \beta(x) u_0, \qquad (3.35)$$

where $\alpha(x), \beta(x) > 0$.

Let x be a nonzero element of $K_{m/(2M)}$ and $t \in (0, 1)$. Since

$$A(tx) = t^v Ax,$$

we have

$$A^2(tx) = t^{v^2} A^2 x = t(1 + \eta) A^2 x, \qquad (3.36)$$

where $\eta > 0$. The operator A^2 is monotone, since if $x(t) \preccurlyeq y(t)$, then $Ax(t) \succcurlyeq Ay(t)$.

It follows that the operator A^2 is u_0-concave.

The cone $K_{m/(2M)}$ is regular. We claim that, for sufficiently small positive ε, the conic segments $\langle \varepsilon u_0, (1/\varepsilon) u_0 \rangle$ are invariant under A^2.

This will prove our assertion, since the function $Ax_0(t)$ belongs to one of these conic segments for any nonzero initial approximation $x_0(t)$.

Introduce the notation

$$\psi(t) = A^2 u_0 = \int_\Omega \int_\Omega G(t, s)\, G(s, s_1)\, ds_1\, ds.$$

It is clear that $\psi(t) \in K_{m/M}$, and so

$$\alpha_0 u_0 \preccurlyeq \psi(t) \preccurlyeq \beta_0 u_0,$$

where $\alpha_0, \beta_0 > 0$ and the semi-order is defined by the cone $K_{m/(2M)}$. Let

$$\varepsilon^{1 - v^2} \leqq \min\left\{\alpha_0,\, \frac{1}{\beta_0}\right\}.$$

With this ε, we have the inequalities

$$A^2(\varepsilon u_0) = \varepsilon^{v^2}\psi(t) \succcurlyeq \alpha_0 \varepsilon^{v^2} u_0 \succcurlyeq \varepsilon u_0$$

and

$$A^2\left(\frac{1}{\varepsilon}\, u_0\right) = \varepsilon^{-v^2}\psi(t) \preccurlyeq \beta_0 \varepsilon^{-v^2} u_0 \preccurlyeq \frac{1}{\varepsilon}\, u_0.$$

Consequently, the operator A^2 maps the conic segment $\langle \varepsilon u_0, (1/\varepsilon)\, u_0 \rangle$ into itself. Theorems 3.1 and 3.3 complete the proof. ∎

3.7. *The contracting mapping principle in metric spaces.** In the preceding sections we considered contraction operators on Banach spaces (or subsets of Banach spaces). All the theorems we have proved (except those involving linear operators) carry over without difficulty to metric spaces. We shall prove a more general statement.

Let R be a metric space with metric $\rho(x, y)$. An operator A is called a *generalized contraction* if

$$\rho(Ax, Ay) \leqq q(\alpha, \beta)\, \rho(x, y) \qquad (\alpha \leqq \rho(x, y) \leqq \beta), \tag{3.37}$$

where

$$q(\alpha, \beta) < 1 \tag{3.38}$$

for $\beta \geqq \alpha > 0$.

---

* Theorem 3.4 is due to Krasnosel'skii.

For example, A is a generalized contraction if

$$\rho(Ax, Ay) \leqq \rho(x, y) - \gamma[\rho(x, y)], \tag{3.39}$$

where $\gamma(u)$ is continuous and positive for $u > 0$.

Exercise 3.10. Construct an operator A which satisfies (3.37) but not (3.39).

Theorem 3.4. *If A is a generalized contraction operator on a complete metric space R, then the equation*

$$x = Ax \tag{3.40}$$

has a unique solution x^* in R. This solution is the limit of successive approximations

$$x_n = Ax_{n-1} \qquad (n = 1, 2, \ldots) \tag{3.41}$$

for any initial approximation $x_0 \in R$.

Proof. Consider the numerical sequence

$$\alpha_n = \rho(x_n, x_{n-1}) \qquad (n = 1, 2, \ldots).$$

It follows from (3.37) that it is a nonincreasing sequence; let α^* be its limit. If $\alpha^* > 0$, then, for sufficiently large N and for all $m = 1, 2, \ldots$, the inequality

$$\alpha_{N+m} \leqq [q(\alpha^*, \alpha^* + 1)]^m (\alpha^* + 1)$$

follows from (3.37), and this contradiction implies that $\alpha_n \to 0$.

Let $\varepsilon > 0$ be given, and choose N such that

$$\alpha_N \leqq \frac{\varepsilon}{2} \left[1 - q\left(\frac{\varepsilon}{2}, \varepsilon\right) \right].$$

We claim that the operator A maps the ball $\rho(x, x_N) \leqq \varepsilon$ into itself, which will imply that (3.41) is a Cauchy sequence. Indeed, if $\rho(x, x_N) \leqq \varepsilon/2$, then

$$\rho(Ax, x_N) \leqq \rho(Ax, Ax_N) + \alpha_N \leqq \rho(x, x_N) + \alpha_N < \varepsilon.$$

But if $\varepsilon/2 \leqq \rho(x, x_N) \leqq \varepsilon$, then, again,

$$\rho(Ax, x_N) \leqq \rho(Ax, Ax_N) + \alpha_N \leqq q\left(\frac{\varepsilon}{2}, \varepsilon\right) \varepsilon + \alpha_N < \varepsilon.$$

The limit x^* of the sequence (3.41) is a fixed point of A. Its uniqueness is obvious. ∎

Exercise 3.11. Show that a statement on non-accumulating errors analogous to that of subsection 2.7 holds for equations with generalized contraction operators.

The conditions given by Theorem 3.4 for the existence of a unique solution and the convergence of successive approximations to the solution are more general than the ordinary contracting mapping principle. It is easy to see, in particular, that Theorem 1.4 is a special case of Theorem 3.4.

Below we shall use Theorem 3.4 to construct a theory of equations with concave operators.

Note that the convergence of the approximations (3.41) is uniform in initial approximations over any ball

$$\rho(x, x^*) \leqq r \,.$$

The following theorem, stated without proof, generalizes the contracting mapping principle.

Two metrics $\rho(x, y)$ and $\rho_1(x, y)$ in a space R are said to be *equivalent* if any sequence which is Cauchy in one of them is Cauchy in the other.

Theorem 3.5. *Let A be a continuous operator which maps a complete metric space R into itself, the space having finite diameter with respect to a metric $\rho_0(x, y)$. Assume that A has a unique fixed point in R, and that the successive approximations* (3.41) *converge uniformly (in initial approximations $x_0 \in R$) to this fixed point. Then one can define an equivalent metric $\rho(x, y)$ in R for which A is a contraction operator:*

$$\rho(Ax, Ay) \leqq q \, \rho(x, y) \qquad (0 < q < 1) \,. \tag{3.42}$$

In particular, generalized contractions are contraction operators in certain equivalent metrics.

The statement of Theorem 3.5 is in effect a general principle. Suppose that, in studying some equation, we suspect that the solution may be obtained by successive approximations. We may then try to determine an equivalent metric for which the contracting mapping principle is applicable. Of course, Theorem 3.5 provides no "recipe" for the required equivalent metric, though this by no means depreciates from its role!

The statement of Theorem 3.5 was first proved, in a weakened form, by Levin and Lifshits [1], who proved that, under the assumptions of Theorem 3.5, the operator A is a generalized contraction in some equivalent metric. The full statement is due to V.M. Gershtein and B.N.

Sadovskii, who essentially employ the construction of the above-mentioned paper of Levin and Lifshits.

In a certain sense, Theorem 3.5 is analogous to a well-known statement on the existence of Lyapunov functions in the stability theory of differential equations (see, e.g., Krasovskii [1]).

Let A be a continuous operator that maps a bounded complete metric space R into itself and has a unique fixed point x^* in R. Assume that the successive approximations $x_n = A^n x_0$ converge to x^* for any initial approximation $x_0 \in R$. A question which arises naturally in the context of the constructions just studied is whether these approximations converge to x^* uniformly in $x_0 \in R$. The answer is in the negative, even when R is compact. Example: let R be the unit circle, whose points are defined by the polar angle $\phi (0 \leq \phi \leq 2\pi)$. Let A be the operator which maps each point ϕ_0 onto the value at $t = 1$ of the solution of the differential equation

$$\frac{d\phi}{dt} = (2\pi - \phi)\phi,$$

satisfiying the initial condition $\phi(0) = \phi_0$.

3.8. *On a special metric.** We now return to the study of nonlinear positive operators defined in a Banach space E with a distinguished cone K.

Let x_0 be a fixed nonzero element of K. We define the component $K(x_0)$ of the cone K generated by x_0 as the set of all elements $y \in K$ such that

$$x_0 \leqslant \alpha y, \qquad y \leqslant \alpha x_0 \tag{3.43}$$

for some $\alpha > 0$. It is clear that $y \in K(x)$ implies $x \in K(y)$. The cone K is therefore the union of disjoint components, each generated by any of its elements.

Exercise 3.12. What is the minimum possible number of components of a solid (i.e., containing interior points) cone in n-space? Construct a cone in n-space a) with countably many components, b) with a continuum of components.

---

* This metric was apparently first introduced by G. Birkhoff.

Let $\alpha(x, y)$ denote the smallest α satisfying inequality (3.43). It is obvious that $\alpha(x, y) \geqq 1$. Set

$$\rho(x, y) = \ln \alpha(x, y) \qquad (x, y \in K(x_0)). \tag{3.44}$$

Clearly, $\rho(x, y) \geqq 0$, and $\rho(x, y) = 0$ only if $x = y$. The symmetry of inequalities (3.43) implies that $\rho(y, x) = \rho(x, y)$. Finally, from

$$x \leqslant \alpha(x, y)y, \qquad y \leqslant \alpha(x, y)x,$$

$$y \leqslant \alpha(y, z)z, \qquad z \leqslant (y, z)y,$$

we get the inequalities

$$x \leqslant \alpha(x, y)\alpha(y, z)z, \qquad z \leqslant \alpha(y, z)\alpha(x, y)x$$

and hence

$$\alpha(x, z) \leqq \alpha(x, y)\alpha(y, z).$$

Therefore

$$\rho(x, z) \leqq \rho(x, y) + \rho(y, z) \qquad (x, y, z \in K(x_0)).$$

Thus $\rho(x, y)$ satisfies all the axioms for a metric, and it follows that every component may be regarded as a metric space.

As an example, consider the space C of functions continuous on $[a, b]$, K the cone of nonnegative functions. Suppose the component $K(x_0)$ is determined by some strictly positive function $x_0(t)$. Then $K(x_0)$ consists of all strictly positive functions. The metric (3.44) is then defined by

$$\rho(x, y) = \max_{a \leqq t \leqq b} \big| \ln x(t) - \ln y(t) \big|.$$

We now return to an abstract Banach space E. Again, let x_0 be a fixed nonzero element of K. Let E_{x_0} denote the set of all elements $y \in E$ which satisfy the inequality

$$-\alpha x_0 \leqslant y \leqslant \alpha x_0 \tag{3.45}$$

for some $\alpha = \alpha(y) \geqq 0$. We call the minimum α for which (3.45) holds the x_0-*norm* of y, denoted by $\|y\|_{x_0}$. It is easy to see that E_{x_0} is a linear space normed by $\|y\|_{x_0}$. One important property of the x_0-norm is its monotonicity: $-x \leqslant y \leqslant x$ $(x, y \in E_{x_0})$ implies that $\|y\|_{x_0} \leqq \|x\|_{x_0}$.

The component $K(x_0)$ is obviously contained in E_{x_0}.

Lemma 3.1. *If z_n is a Cauchy sequence in $K(x_0)$ with respect to the norm* (3.44), *then it is also a Cauchy sequence with respect to the x_0-norm.*

Proof. Since z_n is a Cauchy sequence, it is bounded; there exists $\mu > 0$ such that

$$\rho(z_n, x_0) \leqq \mu \qquad (n = 1, 2, \ldots).$$

This means that

$$e^{-\mu} x_0 \leqslant z_n \leqslant e^{\mu} x_0 \qquad (n = 1, 2, \ldots),$$

i.e.,

$$\|z_n\|_{x_0} \leqq e^{\mu} \qquad (n = 1, 2, \ldots). \tag{3.46}$$

For any two elements $x, y \in K(x_0)$, the inequalities

$$x \leqslant e^{\rho(x, y)} y, \qquad y \leqslant e^{\rho(x, y)} x$$

imply that

$$\left[e^{-\rho(x, y)} - 1 \right] y \leqslant x - y \leqslant \left[e^{\rho(x, y)} - 1 \right] y,$$

$$- \left[e^{\rho(x, y)} - 1 \right] x \leqslant x - y \leqslant - \left[e^{-\rho(x, y)} - 1 \right] x.$$

Therefore

$$\|x - y\|_{x_0} \leqq \left[e^{\rho(x, y)} - 1 \right] \min \left\{ \|x\|_{x_0}, \|y\|_{x_0} \right\}. \tag{3.47}$$

It follows from (3.46) and (3.47) that

$$\|z_n - z_m\|_{x_0} \leqq e^{\mu} \left[e^{\rho(z_n, z_m)} - 1 \right] \qquad (n, m = 1, 2, \ldots).$$

Since $\lim_{\rho \to 0} (e^{\rho} - 1) = 0$, this inequality implies the assertion of the lemma. ∎

Exercise 3.13. Let $x, y \in K(x_0)$. Prove the inequality

$$\rho(x, y) \leqq \ln \left[1 + \|x - y\|_{x_0} \max \left\{ \|x_0\|_x, \|x_0\|_y \right\} \right]. \tag{3.48}$$

Exercise 3.14. Construct a sequence in $K(x_0)$ which is Cauchy in the x_0-norm but not Cauchy in the metric (3.44).

The space E_{x_0} may not be complete, that is, it may contain Cauchy sequences which do not converge to a limit. However, it the cone K is normal, E_{x_0} is complete.*

* See Krasnosel'skii [11], p. 22.

Exercise 3.15. Let E be the space of continuously differentiable functions on $[0, 1]$, normed by

$$\|x\| = \max_{0 \leq t \leq 1} \{| x(t)| + | x'(t)|\} .$$

Show that if K is the cone of nonnegative functions and $x_0(t) \equiv 1$, then E_{x_0} is not complete.

Theorem 3.6. *If the cone K is normal, then each of its components $K(x_0)$ is complete in the metric* (3.44).

Proof. Let z_n $(n = 1, 2, \ldots,)$ be a Cauchy sequence in $K(x_0)$. By Lemma 3.1 this sequence is Cauchy in the x_0-norm, and since K is normal the space E_{x_0} is complete. Consequently, there exists an element $z^* \in E_{x_0}$ such that

$$\lim_{n \to \infty} \|z_n - z^*\|_{x_0} = 0 . \tag{3.49}$$

By definition, the normality of the cone means that every conic segment is bounded in norm. In particular, the inequalities

$$-x_0 \leqslant x \leqslant x_0$$

imply that $\|x\| \leq a$, where a is some constant. But then it follows from

$$- \|x\|_{x_0} x_0 \leqslant x \leqslant \|x\|_{x_0} x_0$$

that

$$\|x\| \leq a \|x\|_{x_0} \qquad (x \in E_{x_0}) . \tag{3.50}$$

Hence, by (3.49),

$$\lim_{n \to \infty} \|z_n - z^*\| = 0 . \tag{3.51}$$

Let $\varepsilon > 0$ be given, and choose $N(\varepsilon)$ such that, for $n, n + m > N(\varepsilon)$, we have $\rho(z_n, z_{n+m}) < \varepsilon$, i.e.,

$$z_n \leqslant e^\varepsilon z_{n+m} , \qquad z_{n+m} \leqslant e^\varepsilon z_n .$$

By (3.51), we may let $m \to \infty$ in these inequalities, and in the limit we get

$$z_n \leqslant e^\varepsilon z^* , \qquad z^* \leqslant e^\varepsilon z_n .$$

It follows that $z^* \in K(x_0)$ and $\rho(z_n, z^*) < \varepsilon$ for $n > N(\varepsilon)$. This means that the sequence z_n converges to z^* in the metric (3.44). ∎

3.9. *Equations with power nonlinearities.* The metric (3.44) may be used in studying equations with concave operators.

A simple example is the equation

$$x = Ax \tag{3.52}$$

where A is a monotone operator satisfying the condition*

$$A(\lambda x) \succcurlyeq \lambda^\nu Ax \qquad (x \in K \; ; \; 0 \le \lambda \le 1), \tag{3.53}$$

where $\nu \in (0, 1)$. Assume that the cone K is normal and one of its components $K(x_0)$ is invariant under A. Let $x, y \in K(x_0)$. Applying the operator A to the inequalities

$$x \succcurlyeq e^{-\rho(x,y)} y, \qquad y \succcurlyeq e^{-\rho(x,y)} x$$

we see from the monotonicity of A and from (3.53) that

$$Ax \succcurlyeq e^{-\nu\rho(x,y)} Ay, \qquad Ay \succcurlyeq e^{-\nu\rho(x,y)} Ax \, .$$

Therefore

$$\rho(Ax, Ay) \le \nu\rho(x, y) \qquad (x, y \in K(x_0)) \, .$$

The assumptions of the contracting mapping principle thus hold, and so equation (3.52) has a unique solution in $K(x_0)$; this solution may be obtained by successive approximations, which converge both in the metric (3.44) and in the basic norm of the space E (by Lemma (3.1).

We have already considered particular classes of operators satisfying (3.53) (subsection 3.6). Here we shall present more general examples.

A kernel $K(t, s)$ $(t, s \in \Omega$, where Ω is a bounded closed set in a euclidean space) is called *uniform* if it is nonnegative, and if, for any set $\Omega_1 \subset \Omega$ of nonzero measure,

$$\int_\Omega K(t, s)\, ds \le \beta(\Omega_1) \int_{\Omega_1} K(t, s)\, ds \qquad (t \in \Omega) \, . \tag{3.54}$$

Any kernel satisfying the inequalities

$$0 < m \le K(t, s) \le M < \infty \qquad (t, s \in \Omega) \tag{3.55}$$

is clearly uniform, since

$$\int_\Omega K(t, s)\, ds \le M \operatorname{mes} \Omega \le \frac{M}{m} \frac{\operatorname{mes} \Omega}{\operatorname{mes} \Omega_1} \int_{\Omega_1} K(t, s)\, ds \, .$$

---

* Equations of this kind are studied in Stetsenko and Imomnazarov [1].

Other examples of uniform kernels are cited in Krasnosel'skii [11] (the Green's function of the first boundary-value problem for second-order elliptic operators is a uniform kernel; see Krasnosel'skii and Sobolevskii [1]).

Let $K(t, s)$ be a uniform kernel defining a continuous integral operator on the space C of continuous functions on Ω (for example, let $K(t, s)$ be jointly continuous). Then the nonnegative function

$$u_0(t) = \int_\Omega K(t, s)\, ds \qquad (3.56)$$

is continuous. Assume that $u_0(t)$ is not identically zero.

Now consider the nonlinear integral operator

$$Ax(t) = \int_\Omega K(t, s) f[s, x(s)]\, ds, \qquad (3.57)$$

where the function $f(s, x)$ is jointly continuous, nonnegative for $x \geqq 0$, and strictly positive for $x > 0$ and almost all s. The kernel $K(t, s)$ is assumed to be uniform and to define a continuous linear operator on C.

Let K be the cone of nonnegative functions in C, and consider its component $K(u_0)$, where $u_0(t)$ is the function (3.56). Let $x(t) \in K$, $x(t) \not\equiv 0$. Then the function $f[s, x(s)]$ is positive on a set of positive measure. Therefore, there exist $\delta > 0$ and a set $\Omega_1 \subset \Omega$ such that mes $\Omega_1 > 0$ and

$$f[s, x(s)] \geqq \delta \qquad (s \in \Omega_1).$$

Then it follows from (3.54) that

$$Ax(t) \geqq \delta \int_{\Omega_1} K(t, s)\, ds \geqq \frac{\delta}{\beta(\Omega_1)} u_0(t) \qquad (t \in \Omega).$$

On the other hand, we have the obvious inequality

$$Ax(t) \leqq \max_{s \in \Omega} f[s, x(s)]\, u_0(t) \qquad (t \in \Omega).$$

Consequently, $Ax(t) \in K(u_0)$. This proves that A maps all nonzero functions of K onto elements of the component $K(u_0)$. In particular, $AK(u_0) \subset K(u_0)$.

A nonnegative function $f(s, x)$ $(s \in \Omega; x \geqq 0)$ is said to be v-*quasiconcave*
$(0 < v < 1)$ if it is nondecreasing in x and the function

$$h(s, x) = \frac{f(s, x)}{x^v} \qquad (s \in \Omega; x \geqq 0) \tag{3.58}$$

is nonincreasing in x. If $f(s, x)$ is v-quasiconcave, the operator (3.57) is
clearly monotone on K. It satisfies condition (3.53). In fact, let $x(t) \in K$,
$0 \leqq \lambda \leqq 1$. Let Ω_0 denote the support of the function $\lambda x(s)$. Then

$$A[\lambda x(t)] = \int_{\Omega_0} K(t, s) \frac{f[s, \lambda x(s)]}{[\lambda x(s)]^v} [\lambda x(s)]^v \, ds + \int_{\Omega \backslash \Omega_0} K(t, a) f(s, 0) \, ds \geqq$$

$$\geqq \lambda^v \int_{\Omega_0} K(t, s) f[s, x(s)] \, ds + \int_{\Omega \backslash \Omega_0} K(t, s) f(s, 0) \, ds \geqq \lambda^v A x(t).$$

The above arguments show that the equation

$$x(t) = \int_\Omega K(t, s) f[s, x(s)] \, ds \tag{3.59}$$

with uniform kernel and v-quasiconcave nonlinearity $f(s, x)$ has a unique
nontrivial nonnegative solution, which is the uniform limit of the approximations

$$x_n(t) = \int_\Omega K(t, s) f[s, x_{n-1}(s)] \, ds \tag{3.60}$$

for any nonzero nonnegative initial approximation $x_0(t)$.

Exercise 3.16. Let the kernel $K(t, s)$ of the operator (3.57) be continuous and positive.
Let $f(s, x)$ $(s \in \Omega; 0 < x < \infty)$ be jointly continuous, positive, and nonincreasing in x.
Finally, let v be a number in $(0, 1)$ such that $x^v f(s, x)$ is an increasing function of x. Show
that the operator A^2 maps the component $K(u_0)$ $(u_0(t) \equiv 1)$ of the cone K of nonnegative
functions in C into itself, is monotone and satisfies condition (3.53).

Exercise 3.17. Assume that $K(t, s)$ and $f(s, x)$ satisfy the conditions of Exercise 3.16.
Show that equation (3.59) has a unique positive solution, which is the uniform limit of the
successive approximations (3.60) for any positive initial approximation.

Exercise 3.18. Let $u_0(t)$ be an almost everywhere nonzero bounded nonnegative function. Assume that for almost all $t \in \Omega$

$$a u_0(t) \leqq \int_\Omega K(t, s) f[s, u_0(s)] \, ds \leqq b u_0(t)$$

where $a, b > 0$. Finally, assume the continuous kernel $K(t, s)$ nonnegative and the function

$f(s, x)$ $(s \in \Omega, x > 0)$ jointly continuous, nonnegative and nonincreasing for increasing x, while the function $x^\nu f(s, x)$ is increasing for some $\nu \in (0, 1)$. Show that equation (3.59) has a unique solution in the component $K(u_0)$ of the cone of nonnegative functions, and that this solution is the uniform limit of successive approximations (3.60).

3.10. *Equations with uniformly concave operators.* Let A be a positive monotone operator on a Banach space E with cone K, and u_0 a fixed nonzero element of K. The operator A is said to be *uniformly u_0-concave* if for every conic segment $\langle \mu u_0, \nu u_0 \rangle$ $(\mu, \nu > 0)$ and every interval $[a, b] \subset (0, 1)$ there exists $\eta = \eta (\mu, \nu, a, b) > 0$ such that

$$A(tx) \geqslant (1 + \eta) t A x \qquad (3.61)$$

for all $x \in \langle \mu u_0, \nu u_0 \rangle, t \in [a, b]$. We may assume without loss of generality that $\eta(\mu, \nu, a, b)$ is a nonincreasing function of b.

Denote the metric (3.44) by $\rho(x, y)$.

Lemma 3.2. If the operator A is uniformly u_0-concave on $\langle \mu u_0, \nu u_0 \rangle$, then

$$\rho(Ax, Ay) \leqq \rho(x, y) - \Delta[\rho(x, y)] \qquad (x, y \in \langle \mu u_0, \nu u_0 \rangle), \qquad (3.62)$$

where

$$\Delta(u) = \ln \left[1 + \eta \left(\mu, \nu, \frac{\mu}{\nu}, e^{-u} \right) \right]. \qquad (3.63)$$

Proof. It follows from

$$\mu u_0 \leqslant x \leqslant \nu u_0, \qquad \mu u_0 \leqslant y \leqslant \nu u_0$$

that

$$x \leqslant \frac{\nu}{\mu} y, \qquad y \leqslant \frac{\nu}{\mu} x.$$

Therefore

$$\mu/\nu \leqq e^{-\rho(x, y)} \qquad (x, y \in \langle \mu u_0, \nu u_0 \rangle).$$

Apply the operator A to the inequalities

$$e^{-\rho(x, y)} x \leqslant y, \qquad e^{-\rho(x, y)} y \leqslant x.$$

The monotonicity of A and inequality (3.61) imply that

$$Ay \geqslant \{1 + \eta [\mu, v, \mu/v, e^{-\rho(x,y)}]\} e^{-\rho(x,y)} Ax,$$

$$Ax \geqslant \{1 + \eta [\mu, v, \mu/v, e^{-\rho(x,y)}]\} e^{-\rho(x,y)} Ay,$$

whence

$$\rho(Ax, Ay) \leqq \rho(x, y) - \ln \{1 + \eta [\mu, v, v/\mu, e^{-\rho(x,y)}]\}. \blacksquare$$

Lemma 3.2 and Theorem 3.4 imply

Theorem 3.7.* *If A is a uniformly u_0-concave operator which maps a conic segment $\langle \mu u_0, v u_0 \rangle$ ($\mu, v > 0$) into itself, then the equation $x = Ax$ has a unique solution on $\langle \mu u_0, v u_0 \rangle$, and the successive approximations $x_n = Ax_{n-1}$ converge to this solution in norm for any initial approximation $x_0 \in \langle \mu u_0, v u_0 \rangle$.*

§ 4. Equations with nonexpansive operators

4.1. Nonexpansive operators. An operator A is said to be *nonexpansive* on a set $\mathfrak{M}$ if

$$\|Ax - Ay\| \leqq \|x - y\| \qquad (x, y \in \mathfrak{M}). \tag{4.1}$$

If A is a nonexpansive operator, the equation

$$x = Ax \tag{4.2}$$

may have no solutions, several solutions, or infinitely many solutions. If equation (4.1) is solvable, the problem is to approximate a solution. Below we shall show that in many cases this may be done by successive approximations of the form

$$x_{n+1} = \tfrac{1}{2}(Ax_n + x_n) \qquad (n = 0, 1, 2, \ldots). \tag{4.3}$$

These successive approximations may converge to different solutions of equation (4.1), depending on the choice of the initial approximation.

---

* This theorem was proved by an essentially different method in Bakhtin [7].

Essentially, the last subsections of § 3 follow the plan of Krasnosel'skii and Stetsenko [2], which discusses certain classes of uniformly concave integral operators.

Formula (4.3) defines ordinary successive approximations

$$x_{n+1} = Fx_n \qquad (n = 0, 1, 2, \ldots) \tag{4.4}$$

for the operator F defined by

$$Fx = \tfrac{1}{2}(Ax + x). \tag{4.5}$$

It is clear that F is also nonexpansive.

A Banach space E is said to be *strictly convex* if $\|x + y\| < \|x\| + \|y\|$ for $x \neq \alpha y, \alpha > 0$. It is easily verified that a Hilbert space is strictly convex. Indeed, if $x \neq \alpha y$, then

$$\|x + y\|^2 = \|x\|^2 + 2(x, y) + \|y\|^2 < (\|x\| + \|y\|)^2.$$

The spaces L_p $(1 < p < \infty)$ are strictly convex (this follows from the so-called Clarkson inequalities; see, e.g., Sobolev [1]).

Exercise 4.1. Prove that the space C of continuous functions on $[a, b]$ and the space L_1 of summable functions are not strictly convex.

Exercise 4.2. Let E be a separable Banach space with norm $\|x\|$. Prove that for any $\varepsilon > 0$ one can introduce a norm $\|x\|_*$ such that $\|x\| \leqq \|x\|_* \leqq (1 + \varepsilon) \|x\|$ $(x \in E)$, with respect to which E is strictly convex.

Theorem 4.1. *Let A be a nonexpansive operator which maps a closed convex set T in a strictly convex Banach space E into a compact subset of T. Then, for any $x_0 \in T$, the successive approximations (4.3) converge to a solution of equation (4.2).*

Proof. The existence of solutions of equation (4.2) follows from Schauder's principle.

Let T_0 denote the closure of the convex hull of the union of AT and the point x_0. A well-known theorem of Mazur (see, e.g., Dunford and Schwartz [1]) implies that T_0 is compact. The approximations (4.3) clearly belong to T_0.

Let x_{n_k} be a subsequence of (4.3) that converges to a point z, and assume that z is not a solution of equation (4.2). Let x^* be some solution. It follows from the strict convexity of E and from (4.1) that

$$\|x^* - Fz\| = \tfrac{1}{2} \|(x^* - z) + (Ax^* - Az)\| < \|x^* - z\|.$$

Let $\|x^* - z\| - \|x^* - Fz\| = 2\delta > 0$ and $\|z - x_{n_{k_0}}\| < \delta$. Then for all $m > 0$

$$\left\| x^* - x_{n_{k_0}+m} \right\| \leqq \left\| x^* - x_{n_{k_0}+1} \right\| \leqq \left\| x^* - Fz \right\| + \left\| Fz - Fx_{n_{k_0}} \right\| \leqq$$

$$\leqq \left\| x^* - Fz \right\| + \left\| z - x_{n_{k_0}} \right\| \leqq \left\| x^* - z \right\| - \delta.$$

These inequalities show that the subsequence x_{n_k} cannot converge to z, and this contradiction proves that z is a solution of equation (4.2). It now follows from (4.1) that the entire sequence x_n converges to this solution. ∎

Exercise 4.3. Assume that the space E is *uniformly convex* (for every $\varepsilon > 0$ there exists $\delta > 0$ such that, if $\|x\| = \|y\| = 1$ and $\|x - y\| > \varepsilon$, then $\|x + y\| \leqq 2 - \delta$). Assume that the set $\mathfrak{M}$ of solutions of equation (4.2) is not empty. Finally, let A be a nonexpansive operator with the following property: for any $\eta > 0$, if $\rho(x, \mathfrak{M}) \equiv \inf_{y \in \mathfrak{M}} \|x - y\| \geqq \eta$, then $\|x - Ax\| \geqq \xi$, where $\xi = \xi(\eta) > 0$. Prove that the statement of Theorem 4.1 is valid.

Exercise 4.4. Prove that if E is strictly convex and the operator A is nonexpansive, then the set $\mathfrak{M}$ of solutions of equation (4.2) is convex.

Exercise 4.5. Prove that the statements of Theorem 4.1 and Exercise 4.3 remain true if the successive approximations are defined not by (4.3) but by

$$x_{n+1} = \theta x_n + (1 - \theta) A x_n \qquad (n = 0, 1, 2, \ldots), \tag{4.6}$$

where θ is a fixed number in $(0, 1)$.

Exercise 4.6. Construct an operator A satisfying the assumption of Theorem 4.1, for which the successive approximations $x_{n+1} = Ax_n$ converge only when the initial approximation x_0 is a solution of equation (4.2).

Theorem 4.1 was proved for uniformly convex spaces by Krasnosel'skii [5] and for strictly convex spaces by Edelstein [1]. There is an extensive literature on nonexpansive operators (see, e.g., Browder [1], Browder and Petryshyn [1, 2], Kirk [1], and especially the interesting survey of Opial [2]).

4.2. *Example.* Consider the equation

$$x(t) = \int_\Omega R\left[t, s, x(s) \right] ds. \tag{4.7}$$

We shall assume that $R(t, s, x)$ is continuous in x and

$$\left| R(t, s, x) \right| \leqq K(t, s)(a + b|x|) \qquad (t, s \in \Omega; \ -\infty < x < \infty), \tag{4.8}$$

where the kernel $K(t, s)$ defines a compact linear operator K in L_2. Then (see, e.g., Krasnosel'skii, Zabreiko, Pustyl'nik and Sobolevskii [1]) the operator A defined by the right member of equation (4.7) is defined in L_2 and compact. Assume that

$$b\rho(K) < 1, \tag{4.9}$$

where $\rho(K)$ is the spectral radius of K. By subsection 1.4, for any $\varepsilon > 0$ we can define an equivalent norm $\|x\|_\varepsilon$ in L_2 such that

$$\|Kx\|_\varepsilon \leqq [\rho(K) + \varepsilon] \|x\|_\varepsilon \qquad (x \in L_2).$$

Set

$$\varepsilon = \frac{1 - b\rho(K)}{2b}.$$

It then follows from (4.8) that

$$\|Ax\|_\varepsilon \leqq a_1 + \tfrac{1}{2}[1 + b\rho(K)] \|x\|_\varepsilon \qquad (x \in L_2), \tag{4.10}$$

where

$$a_1 = a \left\| \int_\Omega K(t, s)\, ds \right\|_\varepsilon.$$

Let T_R be the ball $\|x\|_\varepsilon \leqq R$, where

$$R = \frac{2a_1}{1 - b\rho(K)}.$$

By (4.10) the operator A maps T_R into itself.

Now assume that $R(t, s, x)$ satisfies a Lipschitz condition

$$|R(t, s, u) - R(t, s, v)| \leqq K_1(t, s)|u - v|$$

$$(t, s \in \Omega ; \; -\infty < u, v < \infty),$$

where $K_1(t, s)$ defines a continuous linear integral operator K_1 in L_2. If $\rho(K_1) < 1$, there is an equivalent norm in which A is a contraction operator. Equation (4.7) has a unique solution in L_2, which is the limit (in the norm of L_2) of the successive approximations

$$x_{n+1}(t) = \int_\Omega R[t, s, x_n(s)]\, ds.$$

Now let $\rho(K_1) = 1$. We may then try to construct an equivalent norm in which A is a nonexpansive operator. We confine ourselves to the case

$\|K_1\| = 1$ (e.g., the kernel $K_1(t, s)$ is symmetric and $\rho(K_1) = 1$). Then equation (4.7) may have more than one solution. It follows from Theorem 4.1 that for any initial approximation $x_0 \in T_R$ the successive approximations

$$x_{n+1}(t) = \tfrac{1}{2}\left[x_n + \int_\Omega R[t, s, x_n(s)]\, ds\right]$$

converge to a solution of equation (4.7).

4.3. *Selfadjoint nonexpansive operators*. We now consider the equation

$$x = Ax + f \tag{4.11}$$

where A is a selfadjoint linear operator on a Hilbert space H. We shall assume that $\|A\| = 1$. Then equation (4.11) need not be solvable for all $f \in H$.

Theorem 4.2.* *If* -1 *is not an eigenvalue of the operator A and equation* (4.11) *has a solution* (*not necessarily unique*) *for a given f, then the successive approximations*

$$x_{n+1} = Ax_n + f \qquad (n = 0, 1, 2, \ldots) \tag{4.12}$$

converge to a solution of equation (4.11) *for any* $x_0 \in H$.

Proof. Let H_1 be the subspace of eigenvectors of A corresponding to the eigenvalue 1, and P_1 the orthogonal projection on H_1. If equation (4.11) is solvable, there exists a solution x^* orthogonal to H_1 (if $\tilde{x}$ is any solution, then $x^* = \tilde{x} - P_1\tilde{x}$ is a solution orthogonal to H_1).

Let x_0 be any initial approximation. We claim that the successive approximations (4.12) converge to the solution $x^* + P_1x_0$ of equation (4.11). To prove this, it suffices to show that for any $\varepsilon > 0$ there exists $N = N(\varepsilon)$ such that

$$\|x_n - (x^* + P_1x_0)\| < \varepsilon$$

for $n > N$.

Let E_λ be the partition of unity associated with the selfadjoint operator A:

$$A = \int_{-1}^{1} \lambda\, dE_\lambda.$$

* Krasnosel'skii [10].

Let η be any positive number. Then the space H is the direct sum of three subspaces H_1, H_2 and H_3, where H_2 and H_3 are the ranges of the projection operators

$$P_2 x = \int_{-1+\eta}^{1-\eta} dE_\lambda x \tag{4.13}$$

and

$$P_3 x = \int_{-1}^{-1+\eta} dE_\lambda x + \int_{1-\eta}^{1-0} dE_\lambda x .$$

Since

$$\|P_3 x\|^2 = \int_{-1}^{-1+\eta} d(E_\lambda x, x) + \int_{1-\eta}^{1-0} d(E_\lambda x, x)$$

and -1 is not a discontinuity point of the function $(E_\lambda x, x)$, it follows that for any fixed $x \in H$

$$\lim_{\eta \to 0} \|P_3 x\| = 0 .$$

Choose $\eta > 0$ such that

$$\|P(x_0 - x^*)\| < \varepsilon/2 . \tag{4.14}$$

The projections P_1, P_2 and P_3 commute with A. Equality (4.12) is therefore equivalent to the three equalities

$$P_1 x_n = A P_1 x_{n-1} + P_1 f \quad (n = 1, 2, \ldots), \tag{4.15}$$

$$P_2 x_n = A P_2 x_{n-1} + P_2 f \quad (n = 1, 2, \ldots), \tag{4.16}$$

$$P_3 x_n = A P_3 x_{n-1} + P_3 f \quad (n = 1, 2, \ldots). \tag{4.17}$$

Since equation (4.11) is solvable, f is orthogonal to H_1, and so equality (4.15) becomes

$$P_1 x_n = P_1 x_0 \quad (n = 1, 2, \ldots). \tag{4.18}$$

It follows from (4.13) that the norm of the operator A in the subspace H_2 is at most $1 - \eta$. Thus the successive approximations (4.16) converge to a solution $P_2 x^*$ of the equality $y = Ay + P_2 f$. Consequently, there exists N such that

$$\|P_2 x_n - P_2 x^*\| < \varepsilon/2 \tag{4.19}$$

for $n > N$.

Finally, by (4.17) and (4.14),

$$\|P_3(x_n - x^*)\| = \|A(P_3 x_{n-1} - P_3 x^*)\| \leq$$

$$\leq \|P_3(x_{n-1} - x^*)\| \leq \ldots \leq \|P_3(x_0 - x^*)\| < \varepsilon/2 \tag{4.20}$$

for all n.

Combining (4.18), (4.20) and (4.19), we get

$$\|x_n - (x^* + P_1 x_0)\| \leq$$

$$\leq \|P_1(x_n - x^* - x_0)\| + \|P_2(x_n - x^*)\| + \|P_3(x_n - x^*)\| < \varepsilon$$

for $n > N$. ∎

Any equation involving a bounded linear operator in a Hilbert space and having a solution may be reduced to an equation of type (4.11) with an operator A satisfying the assumptions of Theorem 4.2. In fact, consider the equation

$$Bx = f \tag{4.21}$$

where the operator B is bounded. It is equivalent to the equation

$$B_1 x = f_1, \tag{4.22}$$

where $B_1 = B^*B$, $f_1 = B^*f$, B^* being the adjoint of B. To verify this we need only show that each solution of equation (4.22) is also a solution of equation (4.21). But this is obvious, for, on the one hand, $B_1 x^* = f_1$ means that the element $Bx^* - f$ belongs to the null space of B^*, which is orthogonal to the range of B; on the other hand, $Bx^* - f$ is in the range of B since, by assumption, equation (4.21) is solvable.

Equation (4.22) is equivalent to the equation

$$x = (I - kB_1)x + kf_1 \tag{4.23}$$

for any nonzero k. When

$$0 < k < \frac{2}{\|B_1\|}, \tag{4.24}$$

equation (4.23) is of type (4.11), with an operator

$$A = I - kB_1 \qquad (4.25)$$

which satisfies the assumptions of Theorem 4.2, since the operator B_1 is selfadjoint and nonnegative.

If the operator B in equation (4.21) is selfadjoint and nonnegative, the intermediate step via equation (4.22) is of course superfluous.

Together with Theorem 4.2, these arguments yield

Theorem 4.3. Given a solution of equation (4.21), where B is a bounded linear operator in a Hilbert space, one can construct successive approximations of type (4.12) which converge to this solution.

If the operator B_1 in equation (4.22) is positive definite, and m and M are the bounds of its spectrum, then the operator in equation (4.23) has norm smaller than 1 (equal to $\max\{|1 - km|, |1 - kM|\}$). Theorem 4.2 is then not needed to prove that the successive approximations converge — the contracting mapping principle is sufficient.

The reduction to equation (4.23) was indicated for certain cases by Natanson [1]. It was used previously by Viarda [1] for Fredholm integral equations of the second kind. Essentially, this method is used by Fridman [1] in his study of Fredholm integral equations of the first kind.

Linear equations

§ 5. Bounds for the spectral radius of a linear operator

5.1. *Spectral radius.** The concept of the spectral radius has already been used in Chapter 1, in estimating the rate of convergence of successive approximations, in estimating Lipschitz constants for nonlinear operators, etc.

The spectral radius $\rho(A)$ of a linear operator A (in both real and complex spaces) is defined by

$$\rho(A) = \lim_{n \to \infty} \sqrt[n]{\|A^n\|}. \tag{5.1}$$

Exercise 5.1. Let A be defined in a real space E. Let $\bar{A}$ denote the natural extension of A to the complex extension of the space E ($\bar{A}(x + iy) = Ax + iAy$). Prove that $\rho(\bar{A}) = \rho(A)$.

Exercise 5.2. Let E_1 and E_2 be two Banach spaces such that $E_1 \subset E_2$ and $\|x\|_{E_2} = \|x\|_{E_1}$ ($x \in E_1$). Let A be a continuous linear operator mapping E_2 into E_1. Show that the spectral radius of A as an operator on E_2 coincides with its spectral radius as an operator on E_1.

The spectral radius of many types of operators is equal to the maximum absolute value of the eigenvalues (for example, compact operators; in particular, finite-dimensional operators).

It follows directly from (5.1) that $\rho(\alpha A) = |\alpha| \rho(A)$. If the operators A and B commute, then

$$\rho(A + B) \leqq \rho(A) + \rho(B). \tag{5.2}$$

---

* For a detailed treatment see, e.g., Dunford–Schwartz [1].

It should be borne in mind that (5.2) is false for non-commuting opera-tors*; for a simple example, consider the matrices

$$A = \begin{pmatrix} 0 & 1 \\ 0 & 0 \end{pmatrix}, \qquad B = \begin{pmatrix} 0 & 0 \\ 1 & 0 \end{pmatrix}.$$

Later we shall need the formulas

$$\rho(A^*) = \rho(A), \tag{5.3}$$

where A^* is the adjoint of A (defined in the dual space E^*), and

$$\rho(A^n) = [\rho(A)]^n \qquad (n = 2, 3, \ldots). \tag{5.4}$$

The role of the spectral radius in the theory of linear equations

$$\lambda x = Ax + f \tag{5.5}$$

is determined essentially by the following theorems:

Theorem 5.1. If

$$|\lambda| > \rho(A), \tag{5.6}$$

then equation (5.5) has a unique solution

$$x^* = (\lambda I - A)^{-1} f, \tag{5.7}$$

which is the limit of successive approximations

$$\lambda x_{n+1} = Ax_n + f \qquad (n = 0, 1, 2, \ldots) \tag{5.8}$$

for any $x_0 \in E$.

The approximations (5.8) converge if and only if the solution (5.7) has a Neumann expansion

$$x^* = \sum_{n=0}^{\infty} \lambda^{-(n+1)} A^n f \tag{5.9}$$

which converges in norm.

Theorem 5.2. If the series (5.9) converges for any $f \in E$ when $|\lambda| > a$, then $\rho(A) \leq a$.

---

* V.Ya. Stetsenko and A.R. Esayan [1], A.R. Esayan [2], V.Ya. Stetsenko [5] describe classes of operators which satisfy (5.2) under less restrictive assumptions; viz., com-mutability is replaced by semicommutability: $ABx \preccurlyeq BAx$ for all x in some cone K.

5.2. *Positive linear operators.* To apply Theorem 5.1 one needs either $\rho(A)$ itself or sufficiently good estimates thereof. In this section we describe how to establish estimates for positive operators in Banach spaces with a semi-order relation generated by a cone K. The basic concepts were introduced in § 3.

We shall use the following types of cones: *solid* cones (containing interior points), *reproducing* cones (whose linear span coincides with the entire space), *normal* cones ($\theta \leqslant x \leqslant y$ implies that $\|x\| \leq N \|y\|$, where N is a constant). The cones of nonnegative functions and nonnegative vector-functions in the standard spaces (C, L_p, etc.) are reproducing and normal. Every cone in a finite-dimensional space is normal.

We shall study various classes of positive linear operators, i.e., operators which leave some cone invariant.

Later we shall need the following important proposition (Karlin [1], Bonsall [1]):

Lemma 5.1. Let A be an operator which is positive with respect to a reproducing and normal cone K. Then the spectral radius $\rho(A)$ is in the spectrum of A.

We present a brief proof of this lemma. Retain the notation A for the natural extension of A to the complex extension $\tilde{E}$ of the space E. Suppose that $\rho(A)$ is not a point of the spectrum. Then the operator-valued function $(\lambda I - A)^{-1}$ is analytic (see Hille and Phillips [1]) in a domain G containing all λ such that $|\lambda| > \rho(A)$ and a certain disk S of radius r_0 with center $\rho(A)$ (Fig. 5.1). Fix $f \in K$, $l \in K^*$ (where K^* is the cone of nonnegative functionals); then the scalar function

$$\phi(\lambda) = l[(\lambda I - A)^{-1}f]$$

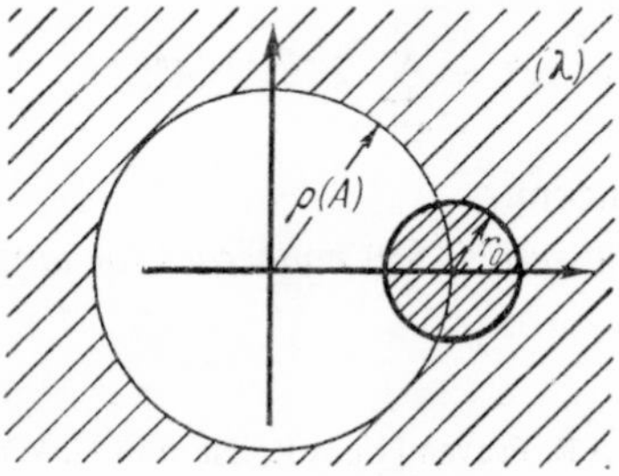

Fig. 5.1.

is also analytic in G and the coefficients of its Laurent series

$$\phi(\lambda) = \sum_{n=0}^{\infty} l(A^n f) \lambda^{-(n+1)} \tag{5.10}$$

are nonnegative. It follows from a well-known theorem of Pringsheim (see Markushevich [1]) that the Laurent series converges for $|\lambda| > \rho(A) - r_0$. Since K is reproducing, the Laurent series converges for $|\lambda| > \rho(A) - r_0$ for all $f \in \tilde{E}$. Since K is normal, it follows (Krein and Rutman [1]) that every functional $l \in \tilde{E}^*$ is the difference of two functionals in K^*. The series (5.10) therefore converge for all $f \in \tilde{E}$ and all $l \in \tilde{E}^*$ in the domain $|\lambda| > \rho(A) - r_0$. In other words, the operator-valued function $(\lambda I - A)^{-1}$ is weakly analytic for $|\lambda| > \rho(A) - r_0$. Hence (Hille and Phillips [1]) it is also strongly analytic for $|\lambda| > \rho(A) - r_0$. But this means that the series (5.9) converge in norm for $|\lambda| > \rho(A) - r_0$, and this contradicts the definition of $\rho(A)$. ∎

A linear operator A is called *quasi-compact* if there exists n_0 such that $A^{n_0} = B + C$, where B is compact and $\|C\| < [\rho(A)]^{n_0}$. The following supplement to Lemma 5.1 is important (Dunford and Schwartz [1], Karlin [1]).

Lemma 5.2. Let A be quasi-compact and positive with respect to a reproducing normal cone K. Then $\rho(A)$ is an eigenvalue of both A and A^*. The eigenvalue $\rho(A)$ has at least one eigenelement $x_0 \in K$ (as an eigenvalue of A) and an eigenelement $l_0 \in K^*$ (as an eigenvalue of A^*).

An operator A is called *strongly positive* if K is solid and A maps every nonzero point of K onto an interior point of K. When K is the cone of vectors

$$x = (\xi_1, \ldots, \xi_n)$$

with nonnegative components in n-space R^n, the matrix

$$A = (a_{ij}) \qquad (i, j = 1, \ldots, n) \tag{5.11}$$

defines a strongly positive operator if and only if all its elements are positive. The integral operator

$$Ax(t) = \int_{\Omega} K(t, s) x(s) \, ds \tag{5.12}$$

in the space C of continuous real functions on Ω is positive with respect to the cone K of nonnegative functions if, e.g., the kernel $K(t, s)$ is positive.

An operator A is called u_0-*positive* if there exists a nonzero element $u_0 \in K$ such that, for any nonzero $x \in K$, there are numbers $\alpha(x), \beta(x) > 0$ for which

$$\alpha(x) u_0 \leqslant Ax \leqslant \beta(x) u_0 . \tag{5.13}$$

We limit ourselves to a single example. Assume that the kernel of the operator (5.12) satisfies the inequalities

$$u_0(t) b_1(s) \leqq K(t, s) \leqq u_0(t) b_2(s) \qquad (t, s \in \Omega), \tag{5.14}$$

where $u_0(t)$, $b_1(t)$, $b_2(t)$ are nonnegative continuous functions, not identically zero. Then (5.13) is valid if

$$\alpha(x) = \int_\Omega b_1(t) x(t) \, dt , \qquad \beta(x) = \int_\Omega b_2(t) x(t) \, dt .$$

An operator A is called u_0-*bounded from above* if, for every nonzero $x \in K$, there exists $\beta(x) \geqq 0$ such that

$$Ax \leqslant \beta(x) u_0 . \tag{5.15}$$

M. A. Krasnosel'skii, in his book [11] (see also Krein and Rutman [1], Krasnosel'skii and Ladyzhenskii [1]), introduces and studies more general classes of strongly positive, u_0-positive and u_0-bounded operators.

Exercise 5.3. Let $K(t, s)$ be a nonnegative continuous kernel ($a \leqq t, s \leqq b$) which is positive for $|t - s| < \varepsilon_0$ ($\varepsilon_0 > 0$). Show that a certain power of the operator (5.12) is strongly positive in the space C, relative to the cone K of nonnegative functions.

Exercise 5.4. Prove that the operator (5.12) is u_0-positive if its kernel is the Green's function of the problem $\ddot{x} = \lambda x, x(a) = x(b) = 0$.

5.3. *Indecomposable operators*. Indecomposable operators, which constitute an immediate generalization of indecomposable matrices, will now be studied in greater detail. A complete exposition of the theory of indecomposable matrices may be found in Gantmakher [1] (see also Krasnosel'skii [11]). Indecomposable operators were introduced independently by Schaefer [1], Niiro [1, 2] and Stetsenko [4]. We shall follow Stetsenko's exposition.

An element x is said to be a quasi-interior element of a cone K if $l(x) > 0$ for any nonzero functional $l \in K^*$. Recall that K^* is the set of linear functionals with nonnegative values on K. In the sequel we shall use the notation $x \gg \theta$ to indicate that x is a quasi-interior element of K.

If the cone is solid, any quasi-interior element is an interior element, and vice versa.

Exercise 5.5. Let K be a reproducing cone in a separable Banach space. Show that K has quasi-interior elements.

Exercise 5.6. Construct a Banach space with a reproducing cone which has no quasi-interior elements.

Exercise 5.7. Show that any nonnegative function in L_p whose set of zeros has measure zero is a quasi-interior element with respect to the cone of nonnegative functions in L_p $(1 \leqq p < \infty)$.

A positive linear operator A is said to be *indecomposable* if the inequality

$$x \not\geqslant \alpha Ax, \qquad (5.16)$$

where $\alpha > 0$, $x \in K$, $x \neq \theta$, implies that x is a quasi-interior element of K. It is easily seen that every strongly positive operator is indecomposable.

Exercise 5.8. a) Show that a nonnegative matrix A is indecomposable if and only if the operator A is indecomposable with respect to the cone of vectors with nonnegative components.

b) A sequence $a_{i_1 i_2}, a_{i_2 i_3}, \ldots, a_{i_m i_1}$ of nonzero elements of the matrix (5.11) is called a nondegeneracy path if the indexes $i_1, \ldots, i_m$ include all numbers $1, 2, \ldots, n$ and $i_1 \neq i_2, i_2 \neq i_3, \ldots$ $i_m \neq i_1$. Prove that a matrix with nonnegative elements which has at least one nondegeneracy path is indecomposable.

Exercise 5.9. Let E be a reflexive Banach space. Show that an operator A which is positive with respect to a cone K is indecomposable if and only if its adjoint A^* is indecomposable.

As an example, consider the integral operator (5.12), where Ω is a bounded closed set in a finite-dimensional euclidean space on which some continuous measure (such as Lebesgue measure) is defined. Assume that the kernel $K(t, s)$ is nonnegative. Under what conditions is the operator (5.12) indecomposable with respect to the cone K of nonnegative functions in the relevant Banach space?

First assume that the kernel $K(t, s)$ is jointly continuous in t, $s \in \Omega$, where Ω is a bounded closed connected set in R^n. Then the operator (5.12) is defined (and compact) in the space C. It can be shown that it is indecomposable if and only if for every nonempty closed subset $\Omega_0 \subset \Omega$ $(\Omega_0 \neq \Omega)$ there exist points $t_0 \in \Omega_0$, $s_0 \in \Omega/\Omega_0$ such that $K(t_0, s_0) > 0$.

Now assume that the operator (5.12) is defined and continuous in L_p (see Krasnosel'skii, Zabreiko, Pustyl'nik, Sobolevskii [1]). The operator (5.12) is indecomposable if and only if, for every measurable set $\Omega_1 \subset \Omega$ ($0 < \text{mes } \Omega_1 < \text{mes } \Omega$), there are two closed sets F_1 and F_2 ($F_1 \subset \Omega_1$, $F_2 \subset \Omega/\Omega_1$, mes $F_1 > 0$, mes $F_2 > 0$) such that $K(t, s) > 0$ for $t \in F_1$, $s \in F_2$. We leave the proof of both indecomposability criteria to the reader.

5.4. Comparison of the spectral radii of different operators.

Theorem 5.3.[*] *Let A be a linear operator which is positive with respect to a normal reproducing cone K. Let B be a linear operator such that*

$$- Ax \leqslant Bx \leqslant Ax \qquad (x \in K).$$ (5.17)

Then

$$\rho(B) \leq \rho(A).$$ (5.18)

Proof. We first prove by induction that

$$- A^n x \leqslant B^n x \leqslant A^n x \qquad (x \in K ; n = 1, 2, \ldots).$$ (5.19)

Indeed, the inequality is valid for $n = 1$. Assume that it holds for $n = k$:

$$- A^k x \leqslant B^k x \leqslant A^k x \qquad (x \in K).$$

Applying the positive (therefore also monotone) operators $A - B$ and $A + B$ to the above inequality, we get

$$- A^{k+1}x + BA^k x \leqslant AB^k x - B^{k+1}x \leqslant A^{k+1}x - BA^k x,$$

$$- A^{k+1}x - BA^k x \leqslant AB^k x + B^{k+1}x \leqslant A^{k+1}x + BA^k x,$$

and (5.19) follows for $n = k + 1$ (since, by (5.17), $- A^{k+1}x \leqslant BA^k x \leqslant A^{k+1}x$).

Inequality (5.19) and the fact that the cone is normal imply that

$$\|B^n x\| - \|A^n x\| \leq \|B^n x + A^n x\| \leq 2N \|A^n x\| \qquad (x \in K),$$

whence

$$\|B^n x\| \leq (2N + 1) \|A^n x\| \qquad (x \in K ; n = 1, 2, \ldots).$$ (5.20)

* Stetsenko [1].

Since K is reproducing, it is *non-flattened* (Krasnosel'skii [11], p. 102): every element $f \in E$ may be expressed in the form $f = x_1 - x_2$, where $x_1, x_2 \in K$ and $\|x_1\|, \|x_2\| \leq M\|f\|$, M a constant. Thus (5.20) implies the inequality

$$\|B^n f\| \leq (2N + 1)\, 2M\, \|A^n\| \cdot \|f\| \qquad (f \in E\,;\ n = 1, 2, \ldots),$$

i.e.,

$$\|B^n\| \leq 2(2N + 1)\, M\, \|A^n\| \qquad (n = 1, 2, \ldots).$$

The proof is completed by applying formula (5.1). ∎

In many cases, Theorem 5.3 enables one to reduce estimation of the spectral radius of an arbitrary linear operator to that of the spectral radius of a positive operator. It turns out that, under very broad conditions, bounds for the spectral radius of a positive operator may be derived from its values on a fixed element of the cone. Bounds of this type for matrices were obtained by Perron and Collatz, and for integral operators by Urysohn. Bounds for compact and other operators (sometimes in implicit form) have been derived by Krein and Rutman [1], Krasnosel'skii [11], Stetsenko [1–3, 5], Zabreiko, Krasnosel'skii, Stetsenko [1], Esayan [1, 2], and others. Bounds for the spectral radius for certain classes of operators are implied by the so-called theorems on incompatible inequalities (Krasnosel'skii [11], Esayan and Sabirov [1], Stetsenko [6]).

5.5. Lower bounds for the spectral radius.
Theorem 5.4.* *If*

$$Ax_0 \succcurlyeq \gamma x_0, \tag{5.21}$$

where A is a positive linear operator, $- x_0 \notin K$, and K is a reproducing cone, then

$$\rho(A) \geq \gamma. \tag{5.22}$$

Proof. Let f be an element of K such that $x_0 \preccurlyeq f$. It follows from Theorem 5.1 that, for any $\varepsilon > 0$, the equation

$$[\rho(A) + \varepsilon]\, x = Ax + f$$

---

* Stetsenko [4].

has a solution x_ε which has a Neumann expansion (5.9). Since all the elements $A^n f\,(n = 1, 2, \ldots)$ are positive, it follows that $x_\varepsilon \in K$. The solution x_ε clearly satisfies the inequalities

$$[\rho(A) + \varepsilon]\, x_\varepsilon \succcurlyeq A x_\varepsilon, \qquad [\rho(A) + \varepsilon]\, x_\varepsilon \succcurlyeq f \succcurlyeq x_0\,.$$

These in turn imply (by induction) that

$$x_\varepsilon \succcurlyeq \gamma^{n-1}\, [\rho(A) + \varepsilon]^{-n} x_0 \qquad (n = 1, 2, \ldots)\,. \tag{5.23}$$

Now (5.23) implies that for any fixed $\varepsilon > 0$ the sequence

$$\gamma^{n-1}\, [\rho(A) + \varepsilon]^{-n} \qquad (n = 1, 2, \ldots)$$

is bounded, and this implies (5.22). ∎

It is clear from the proof that the cone K need not be reproducing; it suffices that x_0 belong to the linear span of the cone.

Another modification, following from (5.4), is to replace the condition (5.21) by the inequality

$$A^k x_0 \succcurlyeq \gamma^k x_0\,. \tag{5.24}$$

As an example, consider the integral operator (5.12) with a continuous positive kernel $K(t, s)$. Setting $x_0(t) \equiv 1$, we get

$$\rho(A) \geqq \min_{t \in \Omega} \int_\Omega K(t, s)\, ds\,. \tag{5.25}$$

But if we set

$$x_0(t) = \int_\Omega K(t, s)\, ds\,,$$

then (5.22) gives the estimate

$$\rho(A) \geqq \min_{t \in \Omega} \frac{\displaystyle\int_\Omega \int_\Omega K(t, s)\, K(s, \tau)\, ds\, d\tau}{\displaystyle\int_\Omega K(t, s)\, ds}$$

Obviously,

$$\min_{t\in\Omega} \frac{\int_{\Omega}\int_{\Omega} K(t,s)\,K(s,\tau)\,ds\,d\tau}{\int_{\Omega} K(t,s)\,ds} \geqq \min_{t\in\Omega} \int_{\Omega} K(t,s)\,ds\,.$$

Exercise 5.10. Show that the spectral radius of the operator (5.12) satisfies the inequality

$$\rho(A) \geqq \sqrt{\min_{t\in\Omega} \int_{\Omega}\int_{\Omega} K(t,s)\,K(s,\tau)\,ds\,d\tau}\,.$$

Exercise 5.11. Find a lower bound for the spectral radius of the operator (5.12) when its kernel is the Green's function of the problem in Exercise 5.4, where $x_0(t) = (t-a)(b-t)$.

5.6. *Upper bounds for the spectral radius.* The main purpose of this subsection is to describe various sufficient conditions under which the inequality

$$Ax_0 \leqslant \delta x_0 \qquad (x_0 \in K,\ x_0 \neq 0) \tag{5.26}$$

implies the following estimate for the spectral radius $\rho(A)$ of a positive linear operator A:

$$\rho(A) \leqq \delta\,. \tag{5.27}$$

Note that (5.27) does not follow from (5.26) without additional assumption. For example, the diagonal matrix

$$A = \begin{pmatrix} \frac{1}{2} & 0 \\ 0 & 1 \end{pmatrix}$$

satisfies (5.26) with $x_0 = \{1, 0\}$, $\delta = \frac{1}{2}$, but $\rho(A) = 1$.

We first assume that the cone K is reproducing (e.g., solid) and normal, and the operator A is x_0-bounded from above. It follows from (5.26) that the operator

$$Bx = \frac{1}{\delta} Ax \tag{5.28}$$

maps the *conic segment* $\langle -x_0, x_0 \rangle$ (the set of elements x such that $-x_0 \leqslant x \leqslant x_0$) into itself. Therefore, for any $f \in \langle -x_0, x_0 \rangle$, the norms of the elements $B^n f$ are uniformly bounded, so that the Neumann series (5.9) converge for $|\lambda| > \delta$. Since A is x_0-bounded above, the elements $B^n f$

are bounded in norm for any $f \in E$. Hence the series (5.9) converge for $|\lambda| > \delta$ and all $f \in E$, and inequality (5.27) follows from Theorem 5.2.

If the operator A is compact, the only assumption needed to prove (5.27) is that x_0 is a quasi-interior element of the cone. We need only consider the case $\rho(A) > 0$. It follows from the general theory of compact linear operators that $\rho(A)$ is the maximum absolute value of the eigenvalues of both A and its adjoint A^*. By Lemma 5.2, for a positive operator A we have a more precise statement: $\rho(A)$ is an eigenvalue of both A and A^*, having eigenvectors in K and K^*, respectively. The following lemma may therefore be used to prove (5.27).

Lemma 5.3. If (5.26) *holds and there exists a functional* $l_0 \in K^*$, $l_0 \neq \theta$, *such that*

$$A^* l_0 = \rho(A) l_0, \tag{5.29}$$

and $l_0(x_0) \neq 0$, *then inequality* (5.27) *holds.*

The proof follows from the chain of relations

$$\rho(A) = \frac{A^* l_0(x_0)}{l_0(x_0)} = \frac{l_0(A x_0)}{l_0(x_0)} \leqq \frac{l_0(\delta x_0)}{l_0(x_0)} = \delta.$$

Now consider a more complicated case: K is reproducing and normal, x_0 is a quasi-interior element of K, and the operator A is known to be u_0-bounded above, where $u_0 \in K$ and u_0 may differ from x_0.

We assume again that $\rho(A) > 0$. Let E_{u_0} denote the set of all $x \in E$ with a finite u_0-norm

$$\|x\|_{u_0} = \inf_{-t u_0 \leqslant x \leqslant t u_0} t. \tag{5.30}$$

Since the cone K is normal, it follows (see Krasnosel'skii [11], p. 22) that E_{u_0} is complete in the u_0-norm. The cone $K_{u_0} = K \cap E_{u_0}$ is solid and normal in E_{u_0}. It is easily verified that the spectral radii of the operator A in E and E_{u_0} are identical (to prove this, use the fact that the usual norm is dominated by the norm (5.30), and Theorem 5.2).

Let G denote the set of all elements $y \in E$ such that $y = Ax - \rho(A)x$, $x \in E_{u_0}$. Since A is u_0-bounded from above, $G \subset E_{u_0}$. We claim that the linear set G contains no interior elements of the cone K_{u_0}. Otherwise, there is a point $x_0 \in E_{u_0}$ such that

$$v_0 = Ax_0 - \rho(A) x_0 \gg \theta,$$

and thus, for sufficiently small $\varepsilon > 0$,

$$Ax_0 \succcurlyeq [\rho(A) + \varepsilon] x_0, \qquad Ax_0 \succcurlyeq [\rho(A) - \varepsilon] x_0. \qquad (5.31)$$

If $-x_0 \notin K_{u_0}$, the first inequality of (5.31) contradicts Theorem 5.4. If $-x_0 \in K_{u_0}$, then $A(-x_0) \in K_{u_0}$, and so

$$-x_0 = \frac{1}{\rho(A)} [v_0 + A(-x_0)] \succcurlyeq \frac{1}{\rho(A)} v_0 \gg \theta.$$

The second inequality of (5.31) may be rewritten as

$$A(-x_0) \preccurlyeq [\rho(A) - \varepsilon](-x_0),$$

but this is a contradiction since, even for the first case considered in this subsection, it implies that $\rho(A) \leqq \rho(A) - \varepsilon$.

By Krein's theorem on the extension of positive linear functionals (see Krein and Rutman [1]), we can define a linear functional l_1 on E_{u_0}, which is positive on K_{u_0}, such that $l_1(y) = 0$ for $y \in G$. Set

$$l_0(x) = \frac{1}{\rho(A)} l_1(Ax) \qquad (x \in E). \qquad (5.32)$$

Obviously,

$$A^* l_0(x) = l_0(Ax) = \frac{1}{\rho(A)} l_1(A^2 x) =$$

$$= \frac{1}{\rho(A)} l_1 [A^2 x - \rho(A) Ax] + l_1(Ax) = \rho(A) l_0(x) \qquad (x \in E).$$

Thus inequality (5.27) follows from Lemma 5.3.

Now assume that the positive operator A has the representation

$$A = B + C, \qquad (5.33)$$

where B is compact, $\|C\| < \gamma$, where γ is a number such that, for some $z_0 (-z_0 \notin K)$,

$$Az_0 \succcurlyeq \gamma z_0. \qquad (5.34)$$

It then follows from Theorem 5.4 that $\rho(A) \geqq \gamma$, and therefore $\|C\| < \rho(A)$. Thus the operator A is quasi-compact (see subsection 5.2). It follows from Lemma 5.2 that Lemma 5.3 is applicable to the operator

(5.33). In other words, inequality (5.27) follows from (5.26) for the operator (5.33).

The results proved in this subsection may be summarized as follows.

Theorem 5.5.* *Let A be a positive linear operator satisfying inequality* (5.26). *Assume that any one of the following conditions holds:*

a) A is compact and x_0 is a quasi-interior element of the cone K.

b) The cone K is solid and normal and x_0 is an interior element of K.

c) A is x_0-bounded from above, the cone K is reproducing and normal.

d) A is u_0-bounded from above, the cone K is reproducing and normal, and x_0 is a quasi-interior element of K.

e) A has a representation

$$A = B + C,$$

where C is compact, $\|C\| < \gamma$, K is reproducing and normal, x_0 is a quasi-interior element of K; there exists $z_0 (- z_0 \notin K)$ such that $Az_0 \gtrapprox \gamma z_0$. Then $\rho(A) \leqq \delta$.

Improved bounds may be derived by using (5.4). If some power A^{n_0} satisfies the assumptions of Theorem 5.5, and we know that

$$A^{n_0}x_0 \preccurlyeq \beta x_0 \tag{5.35}$$

then

$$\rho(A) \leqq \sqrt[n_0]{\beta} . \tag{5.36}$$

In conclusion, we wish to emphasize that the normality of the cone K is essential in b) of Theorem 5.5, as the following example will show. Let E be the space of functions $f(z)$ analytic in the disk $|z| < 1$ and continuous in the disk $|z| \leqq 1$, whose values for real z are real. E becomes a real Banach space when the norm is defined by

$$\|f\| = \max_{|z| \leqq 1} |f(z)| .$$

Let K denote the set of all functions $f(z) \in E$ which are nonnegative for real $z \in [-1, -\frac{1}{2}]$. It is easily seen that K is a solid cone which is not normal. Define a positive linear operator on E by

$$Af(z) = -(z + \tfrac{1}{2})f(z).$$

---

* Zabreiko, Krasnosel'skii and Stetsenko [1].

A simple calculation shows that the spectrum of this operator is exactly the disk $|\lambda + \frac{1}{2}| \leq 1$, and so $\rho(A) = \frac{3}{2}$. On the other hand, the function $v_0(z) \equiv 1$ is an interior element of the cone K, and $Av_0 \leq \frac{1}{2} v_0$. Thus (5.26) does not imply (5.27).

Exercise 5.12. Show that Theorem 5.5 remains valid if (5.26) is replaced by the weaker condition

$$\sum_{n=0}^{2k-1} (-1)^n \delta^{-n} A^n x_0 \geq \theta.$$

5.7. *Strict inequalities.*

Theorem 5.6. *If, under the conditions of Theorem 5.5, $\delta x_0 - Ax_0$ is a quasi-interior element of the cone K, then we have the strict inequality*

$$\rho(A) < \delta. \tag{5.37}$$

Proof. The assumptions of Theorem 5.5 imply that there exists a non-zero linear functional $l_0 \in K^*$ such that $A^* l_0 = \rho(A) l_0$. Applying this functional to the element $\delta x_0 - Ax_0$, we get the inequality

$$l_0(\delta x_0 - Ax_0) = \delta l_0(x_0) - \rho(A) l_0(x_0) > 0,$$

which implies (5.37). ∎

It is clear from the proof that the element $\delta x_0 - Ax_0$ need only be nonzero and positive, not necessarily quasi-interior, provided some other arguments show that the eigenelement l_0 assumes positive values on $\delta x_0 - Ax_0$ (or on all nonzero elements of K).

Assume that (5.26) holds, and let $\delta x_0 \neq Ax_0$. By Theorem 5.1, the equation

$$[\rho(A) + 1] x = Ax + \delta x_0 - Ax_0$$

has a solution

$$x = \sum_{n=0}^{\infty} [\rho(A) + 1]^{-(n+1)} A^n (\delta x_0 - Ax_0). \tag{5.38}$$

This solution obviously satisfies the inequality

$$x \geq \frac{1}{\rho(A) + 1} Ax. \tag{5.39}$$

Assume that A is indecomposable (subsection 5.3). It then follows from (5.39) that x is a quasi-interior element of K. Thus, for any nonzero $l_0 \in K^*$,

$$l_0(x) = \sum_{n=0}^{\infty} [\rho(A) + 1]^{-(n+1)} l_0 [A^n(\delta x_0 - Ax_0)] > 0. \qquad (5.40)$$

Let l_0 be an eigenelement of the operator A^* associated with the eigenvalue $\rho(A) > 0$. Then

$$l_0 [A^n(\delta x_0 - Ax_0)] = (A^*)^n l_0(\delta x_0 - Ax_0) = [\rho(A)]^n l_0(\delta x_0 - Ax_0)$$

and it follows from (5.40) that

$$[\delta - \rho(A)] l_0(x_0) = l_0(\delta x_0 - Ax_0) > 0.$$

Consequently, $\rho(A) < \delta$, and we have proved

Theorem 5.7. *If under the conditions of Theorem 5.5, A is an indecomposable operator such that $Ax_0 \neq \delta x_0$, then the strict inequality (5.37) is valid.*

Exercise 5.13. Prove that a positive linear operator A is indecomposable if and only if, for every nonzero $x \in K$ and every nonzero $l \in K^*$, there exists $n = n(x, l)$ such that $l(A^n x) > 0$.

Another assertion about strict inequalities for the spectral radius is

Theorem 5.8. *Assume that the cone K is reproducing and normal, and the linear operator A is u_0-positive; let*

$$Ay_0 \leqslant \delta y_0, \quad Ay_0 \neq \delta y_0, \qquad (5.41)$$

where y_0 is a nonzero element of K. Then

$$\rho(A) < \delta. \qquad (5.42)$$

Proof. Since A is u_0-positive, there exist positive numbers α, β such that

$$A(\delta y_0 - Ay_0) \geqslant \alpha u_0, \quad Ay_0 \leqslant \beta u_0.$$

Therefore

$$A(Ay_0) = \delta Ay_0 - A(\delta y_0 - Ay_0) \leqslant \delta Ay_0 - \alpha u_0$$

and, further,

$$A(Ay_0) \leqslant \left(\delta - \frac{\alpha}{\beta}\right) Ay_0. \qquad (5.43)$$

Since A is u_0-positive, it is obviously x_0-positive, where $x_0 = Ay_0$. Inequality (5.43) plays the role of (5.26) for Theorem 5.5. Applying this theorem (with condition c), we get

$$\rho(A) \leqq \delta - \frac{\alpha}{\beta} . \quad \blacksquare$$

5.8. *Examples*. Our first example is the linear integral operator

$$Ax(t) = \int_{\Omega} K(t, s) x(s) \, ds , \qquad (5.44)$$

where Ω is a bounded closed set in a finite-dimensional euclidean space. Assume that this operator is defined and continuous in the space $C = C(\Omega)$. It follows from Theorem 5.3 that the spectral radius of the operator (5.44) is no greater than that of the integral operator with nonnegative kernel $| K(t, s) |$ (provided, of course, that the latter operator is defined in C). We restrict ourselves to nonnegative kernels.

Let K be the cone of nonnegative functions; it is solid and normal. It follows from Theorem 5.5 (case b) that

$$\rho(A) \leqq \max_{t \in \Omega} \frac{1}{x_0(t)} \int_{\Omega} K(t, s) x_0(s) \, ds , \qquad (5.45)$$

where $x_0(t)$ is an arbitrary function which is positive on Ω. The quality of the estimate (5.45) depends, of course, on "successful" choice of the function $x_0(t)$. For example, if $x_0(t) \equiv 1$, we get the very coarse estimate

$$\rho(A) \leqq \max_{t \in \Omega} \int_{\Omega} K(t, s) \, ds$$

whose right-hand side is simply the norm $\| A \|$ of the operator A in C. It is easily verified that the operator (5.44) is u_0-bounded above, where

$$u_0(t) = \int_{\Omega} K(t, s) \, ds . \qquad (5.46)$$

Theorem 5.5 (case c) therefore implies that

$$\rho(A) \leqq \sup_{t \in \Omega} \frac{\displaystyle\int_{\Omega} \int_{\Omega} K(t, s) K(s, \tau) \, ds \, d\tau}{\displaystyle\int_{\Omega} K(t, s) \, ds} . \qquad (5.47)$$

Exercise 5.14. Prove the inequality

$$\rho(A) \leqq \max_{t \in \Omega} \left\{ \frac{1}{x_0(t)} \int_\Omega K_m(t, s) x_0(s) \, ds \right\}^{1/m},$$

where

$$K_m(t, s) = \int_\Omega \ldots \int_\Omega K(t, \tau_1) K(\tau_1, \tau_2) \ldots K(\tau_{m-1}, s) \, d\tau_1 \, d\tau_2 \ldots d\tau_{m-1}, \qquad (5.48)$$

and $x_0(t)$ is any positive continuous function.

Exercise 5.15. Prove the inequality

$$\rho(A) \leqq \sup_{t \in \Omega} \left\{ \frac{\int_\Omega K_m(t, s) \, ds}{\int_\Omega K(t, s) \, ds} \right\}^{1/(m-1)}. \qquad (5.49)$$

We devote more detailed attention to the case in which $K(t, s)$ is the Green's function of the problem $- \ddot{x} = \lambda x$, $x(0) = x(1) = 0$:

$$K(t, s) = \begin{cases} t(1 - s), & \text{if } \quad t \leqq s, \\ s(1 - t), & \text{if } \quad s \leqq t. \end{cases} \qquad (5.50)$$

This operator is u_0-positive for $u_0(t) = t(1 - t)$. A simple calculation shows that

$$Au_0(t) \leqq 0.1041 \, u_0(t), \quad A^2 u_0(t) \leqq 0.1016 \, Au_0(t).$$

Thus Theorem 5.5 implies a series of estimates for the spectral radius:

$$\rho(A) \leqq 0.1041 \, ; \, \rho(A) \leqq 0.1016.$$

The exact value of the spectral radius is $1/\pi^2 \approx 0.10132$.

If the kernel of the operator (5.44) has essential singularities, it should be considered in spaces other than C.

As a second example, consider the infinite matrix

$$A = (a_{ij}) \qquad (5.51)$$

with nonnegative elements, satisfying the condition

$$\sum_{i=1}^{\infty} \left(\sum_{j=1}^{\infty} a_{ij}^{p/p-1} \right)^{p-1} < \infty, \qquad (5.52)$$

where $p > 1$. It is well known (see, e.g., Kantorovich and Akilov [1]) that the matrix A generates a compact linear operator in the space l_p.

This operator is positive with respect to the cone K of elements x with nonnegative components, and is u_0-bounded above, where

$$u_0 = \left\{ \left(\sum_{j=1}^{\infty} a_{1j}^{p/p-1} \right)^{p-1/p}, \ldots, \left(\sum_{j=1}^{\infty} a_{nj}^{p/p-1} \right)^{p-1/p}, \ldots \right\}.$$

It follows from Theorem 5.5 (case c) that

$$\rho(A) \leqq \sup_{i=1,2,\ldots} \frac{\sum_{j=1}^{\infty} a_{ij} \left(\sum_{k=1}^{\infty} a_{jk}^{p/p-1} \right)^{p-1/p}}{\left(\sum_{j=1}^{\infty} a_{ij}^{p/p-1} \right)^{p-1/p}}. \tag{5.53}$$

Theorem 5.6 shows that this is a strict inequality if the supremum in the right-hand side is not attained for any i. By Theorem 5.7, we know that it is also a strict inequality if the matrix A is indecomposable and, for some $i = i_1, i_2$,

$$\frac{\sum_{j=1}^{\infty} a_{i_1j} \left(\sum_{k=1}^{\infty} a_{jk}^{p/p-1} \right)^{p-1/p}}{\left(\sum_{j=1}^{\infty} a_{i_1j}^{p/p-1} \right)^{p-1/p}} < \frac{\sum_{j=1}^{\infty} a_{i_2j} \left(\sum_{k=1}^{\infty} a_{jk}^{p/p-1} \right)^{p-1/p}}{\left(\sum_{j=1}^{\infty} a_{i_2j}^{p/p-1} \right)^{p-1/p}}.$$

Exercise 5.16. Show that the operator A defined by the matrix (5.51) is indecomposable if and only if, for any partition of the natural number sequence into two disjoint subsequences n_i, k_j, there exist n_{i_0} and k_{j_0} such that $a_{n_{i_0}k_{j_0}} > 0$.

§ 6. The block method for estimating the spectral radius

6.1. *Fundamental theorem.* Let E_1 and E_2 be Banach spaces with cones K_1 and K_2, respectively. Throughout this section we shall assume that K_1 is reproducing and K_2 normal.

Let T be a mapping of K_1 into K_2 such that

$$\|Tx\|_{E_2} \geqq c \|x\|_{E_1} \qquad (x \in K_1), \tag{6.1}$$

where c is a positive constant. T is not assumed to be linear or even continuous.

Theorem 6.1.* Let A and B be positive linear operators defined in E_1 and E_2, respectively. If

$$TAx \leqslant BTx \qquad (x \in K_1), \tag{6.2}$$

then

$$\rho(A) \leqq \rho(B). \tag{6.3}$$

Proof. Let $f \in K_1$. By (6.2),

$$TA^nf \leqslant B^nTf \qquad (n = 1, 2, \ldots), \tag{6.4}$$

and, since the cone K_2 is normal,

$$\|TA^nf\|_{E_2} \leqq N \|B^nTf\|_{E_2} \qquad (n = 1, 2, \ldots).$$

Inequality (6.1) implies the inequalities

$$c \|A^nf\|_{E_1} \leqq \|TA^nf\|_{E_2} \qquad (n = 1, 2, \ldots).$$

Therefore,

$$\overline{\lim_{n \to \infty}} \|A^nf\|_{E_1}^{1/n} \leqq \overline{\lim_{n \to \infty}} \|B^n\|^{1/n} \left\{ \frac{N}{c} \|Tf\|_{E_2} \right\}^{1/n} = \overline{\lim_{n \to \infty}} \|B^n\|^{1/n}.$$

This inequality, together with (5.1), implies that

$$\overline{\lim_{n \to \infty}} \|A^nf\|_{E_1}^{1/n} \leqq \rho(B) \qquad (f \in K_1). \tag{6.5}$$

Any element $g \in E_1$ may be expressed in the form $g = f_1 - f_2$, where $f_1, f_2 \in K_1$. Obviously,

$$\overline{\lim_{n \to \infty}} \|A^n(f_1 - f_2)\|_{E_1}^{1/n} \leqq \overline{\lim_{n \to \infty}} (\|A^nf_1\|_{E_1} + \|A^nf_2\|_{E_1})^{1/n} \leqq$$

$$\leqq \overline{\lim_{n \to \infty}} \{2^{1/n} \max [\|A^nf_1\|_{E_1}^{1/n}, \|A^nf_2\|_{E_1}^{1/n}]\}.$$

Thus (6.5) implies the inequality

$$\overline{\lim_{n \to \infty}} \|A^ng\|_{E_1}^{1/n} \leqq \rho(B) \qquad (g \in E_1),$$

---

* Zabreiko, Krasnosel'skii and Stetsenko [1]; similar assertions for spaces with mini-hedral cones may be found in Kantorovich, Vulikh and Pinsker [1].

and this in turn implies that the linear equation

$$\lambda x = Ax + g$$

has a solution representable by a Neumann series, for $|\lambda| > \rho(B)$ and arbitrary free term $g \in E_1$. Now use Theorem 5.2. ∎

6.2. *Examples.* In applications, it is convenient to use operators T with finite-dimensional ranges, since then the operator B is a matrix whose spectral radius coincides with its greatest nonnegative eigenvalue.

Let E_1 be the space C of functions continuous on a bounded closed set Ω. Let $\omega_1, \omega_2, \ldots, \omega_m$ be a closed covering of Ω. We define a mapping T of C into m-space R^m by

$$Tx = \{\xi_1(x),\ \xi_2(x),\ldots,\xi_m(x)\}, \tag{6.6}$$

where

$$\xi_i(x) = \max_{t \in \omega_i} |x(t)| \qquad (i = 1,\ldots,m). \tag{6.7}$$

Condition (6.1) is obviously satisfied. Consider the integral operator

$$Ax(t) = \int_\Omega K(t,s)\,x(s)\,ds \tag{6.8}$$

with continuous nonnegative kernel. It is easily seen that the operator (6.8) satisfies the assumptions of Theorem 6.1 if the elements of the matrix B are defined by

$$b_{ij} = \max_{t \in \omega_i} \int_{\omega_j} K(t,s)\,ds \qquad (i,j = 1,\ldots,m). \tag{6.9}$$

Exercise 6.1. Let A be a positive operator in $C(\Omega)$ and T the mapping defined by (6.6). Show that the matrix B satisfies the condition (6.2) if and only if its elements b_{ij} satisfy the inequalities

$$b_{ij} \geqq \lim_{\delta \to 0}\ \sup_{x(t)\,\in\,\mathfrak{N}_\delta^{(j)},\ \|x(t)\| \leqq 1}\ \max_{t \in \omega_i} Ax(t),$$

where $\mathfrak{N}_\delta^{(j)}$ is the set of continuous functions with support in a δ-neighborhood of the set ω_j.

Now consider $L_2(\Omega)$. As before, we assume that Ω is covered by sets $\omega_1, \omega_2, \ldots, \omega_m$; the covering sets are assumed to have nonzero measure.

Define the mapping T by (6.6), setting

$$\xi_i(x) = \left[\int_{\omega_i} x^2(t) \, dt \right]^{1/2} \qquad (i = 1, \ldots, m).\tag{6.10}$$

We can then use Theorem 6.1 to estimate the spectral radius of the operator (6.8) defined in L_2, if the elements of the matrix B are defined by

$$b_{ij}^2 = \int_{\omega_i} \int_{\omega_j} K^2(t, s) \, ds \, dt \, .$$

Our last example concerns the simplest case, in which A is an operator in an n-dimensional space E_1. The space E_2 is then constructed as follows. Regard E_1 as the direct sum of n subspaces F_i. Then every element $x \in E_1$ has a unique representation

$$x = P_1 x + P_2 x + \ldots + P_m x \, ,$$

where P_i is the projection on F_i, which vanishes on all $F_j, j \neq i$. Consider some norm $\| \cdot \|_i$ in each subspace F_i, and define the operator T by

$$Tx = \{ \|P_1 x\|_1 \, , \, \|P_2 x\|_2 \, , \, \ldots \, , \, \|P_m x\|_m \} \, .\tag{6.11}$$

The matrix B then satisfies condition (6.2) if its elements are defined by

$$b_{ij} = \max_{\|P_j x\|_j \leq 1} \|P_i A P_j x\|_i \, .$$

If the basis of E_1 is the union of bases of the subspaces F_i, the matrices of the operators $P_i A P_j$ are blocks along the principal diagonal of the matrix A. This is the motivation for the heading of this section.

We shall not consider specific examples here. Several examples may be found in Kantorovich, Vulikh and Pinsker [1]. See also Stetsenko [2].

6.3. *Conic norm.* Theorem 6.1 may be interpreted as a theorem on the use of special generalized metrics.* This and the following subsection will be devoted to these generalized metrics.

Let K be a normal cone in a Banach space E.

R is said to be a *generalized metric space* if, for every pair $x, y \in R$,

---

* Generalized metrics have been studied and employed by various authors. Apparently, the first results are due to Kurepa [1]; important results were proved by Kantorovich (Kantorovich, Vulikh and Pinsker [1]). Metrics of this type have been systematically

there is an element $\rho(x, y) \in K$ satisfying the usual axioms: $\rho(x, y) = \theta$ if and only if $x = y$; $\rho(y, x) = \rho(x, y)$; and the triangle inequality B.

Let R be a real linear system; R is said to be a *generalized normed space* if there is a function $p(x)$ defined on R with values in K satisfying $p(\theta) = \theta$ and $p(x) \neq \theta$ for $x \neq \theta$; $p(\alpha x) = |\alpha| p(x)$; $p(x + y) \leqslant p(x) + p(y)$.

Any generalized normed space may be made into a generalized metric space by setting

$$\rho(x, y) = p(x - y). \tag{6.12}$$

Examples of generalized norms are the values of the operators (6.6) and (6.11) considered in subsection 6.2.

Let $l(u)$ be a nonnegative sublinear functional on the cone K ($u, v \succcurlyeq \theta$ implies $0 \leq l(u + v) \leq l(u) + l(v)$). Assume, moreover, that $l(u)$ does not vanish on nonzero values of $\rho(x, y)$ or $p(x)$. Then the formulas

$$\rho_*(x, y) = l\left[\rho(x, y)\right] \tag{6.13}$$

and (if $l(tu) \equiv tl(u)$ for $t \geq 0$)

$$\|x\|_* = l\left[p(x)\right] \tag{6.14}$$

made the generalized metric space and the generalized normed space into an ordinary metric space and an ordinary normed space, respectively.

$l(u)$ might be, say, the norm of u or a linear functional with positive values on the nonzero elements of K (if there is a functional with these properties).

6.4. *Generalized contracting mapping principle.* Consider an operator A defined in a generalized metric space R. A is said to satisfy a generalized Lipschitz condition if

$$\rho(Ax, Ay) \leqslant B\rho(x, y) \qquad (x, y \in R), \tag{6.15}$$

where B is a positive linear operator in E.

applied to iterative processes by Collatz [1]. Other papers worthy of mention are Perov [1], Perov and Kibenko [1], Mukhamediev and Stetsenko [1]; some of their constructions will be used below.

For a new approach to the theory of generalized normed spaces, see Antonovskii, Boltyanskii and Sarymsakov [1–4].

[*Translator's note:* Some authors (including Kurepa and Collatz) use the term *pseudometric*; in English, however, this term is usually used in a slightly different sense.]

Theorem 6.2. Let R be complete in the metric (6.13), with $l(u) = \|u\|$. If A maps R into itself and satisfies condition (6.15) with an operator B whose spectral radius is smaller than unity, then the equation

$$x = Ax \tag{6.16}$$

has a unique solution in R, which is the limit of the successive approximations

$$x_{n+1} = Ax_n \qquad (n = 0, 1, 2, \dots) \tag{6.17}$$

for any initial approximation $x_0 \in R$.

Proof. By Theorem 5.1, the equation

$$u = Bu + \rho(Ax_0, x_0) \tag{6.18}$$

has a solution u_0 in E, with the series expansion

$$u_0 = \sum_{m=0}^{\infty} B^m \rho(Ax_0, x_0). \tag{6.19}$$

By (6.15), for any $n, m \geq 0$,

$$\rho(A^{n+m}x_0, A^nx_0) \leqslant \rho(A^{n+m}x_0, A^{n+m-1}x_0) + \dots + \rho(A^{n+1}x_0, A^nx_0) \leqslant$$

$$\leqslant B^{n+m-1}\rho(Ax_0, x_0) + \dots + B^n\rho(Ax_0, x_0)$$

and it follows from (6.19) that

$$\rho(A^{n+m}x_0, A^nx_0) \leqslant B^n u_0. \tag{6.20}$$

Hence, since K is normal,

$$\|\rho(A^{n+m}x_0, A^nx_0)\| \leqq N \|B^n u_0\|.$$

The right-hand side tends to zero as $n \to \infty$, and so $A^n x_0$ is a Cauchy sequence. Its limit x^* satisfies the inequality

$$\rho(A^*x^*, x^*) \leqslant \rho(Ax^*, A^nx_0) + \rho(A^nx_0, x^*) \leqslant$$

$$\leqslant B\rho(x^*, A^{n-1}x_0) + \rho(A^nx_0, x^*),$$

whence

$$\|\rho(Ax^*, x^*)\| \leqq N(\|B\| \cdot \|\rho(x^*, A^{n-1}x_0)\| + \|\rho(A^nx_0, x^*)\|),$$

and, since the right-hand side tends to zero, $Ax^* = x^*$. ∎

Exercise 6.2. Show that, under the assumptions of Theorem 6.2, the original norm of E may be replaced by an equivalent norm $\| \cdot \|_1$ such that A satisfies the ordinary contracting mapping principle with respect to the metric $\rho_1(x, y) = \|\rho(x, y)\|_1$.

Let $l(u)$ be a sublinear functional with positive values on the nonzero elements of the cone K. A positive linear operator B is said to be *l-nonexpansive* if

$$l(Bx) \leqq l(x) \qquad (x \in K). \tag{6.21}$$

Theorem 6.3. *Let R be complete in the metric* (6.13). *If a) A maps R into a subset of itself that is compact in the metric* (6.13) *and b) A satisfies condition* (6.15) *with an l-nonexpansive operator B and equality holds in* (6.15) *only when $x = y$, then the assertions of Theorem 6.2 are valid.*

For the proof it suffices to note that the operator A satisfies the condition $\rho_*(Ax, Ay) < \rho_*(x, y)$ for $x \neq y$ (in the metric (6.13)), and Theorem 1.4 is then applicable.

Exercise 6.3. Show that Theorem 6.3 remains valid if the assumption that AR is compact is replaced by the assumption that every sequence $A^n x_0 (x_0 \in R)$ is compact.

Exercise 6.4. Construct a complete metric space R and a continuous operator A on R such that AR is not compact, the sequence $A^n y_0$ is dense in R for some y_0, and every sequence $A^n x_0 (x_0 \in R)$ contains a convergent subsequence.

Tests for the existence of a functional l satisfying (6.21) are of independent interest. For example, assume that to the eigenvalue $\rho(B)$ of the adjoint operator B^* there corresponds a functional l_0 which is positive on the nonzero elements of K. This condition is obviously fulfilled if K is the cone of nonnegative vectors in a finite-dimensional space and B an indecomposable matrix. It holds if B is an integral operator with positive continuous kernel, in the space C with the cone of nonnegative functions. Other cases were mentioned in the proof of Theorem 5.5; see also Krein and Rutman [1], Krasnosel'skii [11], Perov [1], Mukhamediev and Stetsenko [1]. If $\rho(B) \leqq 1$, then B is l_0-nonexpansive with respect to the functional l_0.

§ 7. Transformation of linear equations

7.1. *General scheme.* In this section we shall consider various transformations of the linear equation

$$Bx = b \tag{7.1}$$

into an equivalent equation

$$x = Sx + f \tag{7.2}$$

where S is a linear operator whose spectral radius satisfies the condition

$$\rho(S) < 1. \tag{7.3}$$

Once an equation (7.2) has been found, the solution of equation (7.1) may be sought by successive approximations.

We shall assume that the operator B has a continuous inverse B^{-1}. Equation (7.1) is clearly equivalent to the equation

$$Bx + B_1 x = B_1 x + b, \tag{7.4}$$

where B_1 is any linear operator. Assume that $B + B_1$ has a bounded inverse. Then equation (7.4) (and therefore also equation (7.1)) is equivalent to the equation

$$x = (B + B_1)^{-1} B_1 x + (B + B_1)^{-1} b. \tag{7.5}$$

Equation (7.5) is an equation of type (7.2). If the operator B_1 can be chosen so that

$$\rho \left[(B + B_1)^{-1} B_1 \right] < 1, \tag{7.6}$$

the problem formulated above will be solved.

Despite its obvious and rather trivial nature, the above idea is quite general. Indeed, assume that equation (7.2) is equivalent to equation (7.1), and that $(I - S)^{-1}$ exists and is continuous. Set

$$B_1 = BS(I - S)^{-1}.$$

Then equation (7.5) becomes

$$x = \left[B + BS(I - S)^{-1} \right]^{-1} BS(I - S)^{-1} x + \left[B + BS(I - S)^{-1} \right]^{-1} b$$

or, since $B + BS(I - S)^{-1} = B(I - S)^{-1}$,

$$x = Sx + (I - S)B^{-1} b. \tag{7.7}$$

Equations (7.7) and (7.2) are equivalent, and therefore have the same solutions. The element f is thus given by the formula

$$f = (I - S)B^{-1} b = (B + B_1)^{-1} b.$$

Naturally, the choice of the operator B_1 is a complicated problem. The main requirements from B_1 are that $(B + B_1)^{-1}$ be sufficiently easy to compute.

7.2. Chebyshev polynomials.* Below we shall need the Chebyshev polynomials

$$T_n(t) = \cos(n \arccos t) \qquad (n = 0, 1, 2, \ldots). \tag{7.8}$$

In particular,

$$T_0(t) = 1, \, T_1(t) = t, \, T_2(t) = 2t^2 - 1, \, T_3(t) = 4t^3 - 3t, \ldots \tag{7.9}$$

The Chebyshev polynomial $T_n(t)$ is of degree n; all its roots are real and lie in the interval $[-1, 1]$; moreover,

$$|T_n(t)| > 1 \qquad (t \notin [-1, 1]) \tag{7.10}$$

and

$$\max_{-1 \le t \le 1} T_n(t) = 1, \qquad \min_{-1 \le t \le 1} T_n(t) = -1. \tag{7.11}$$

An important property of the Chebyshev polynomials is that any polynomial $P_n(t)$ of degree n whose value at some point $t_0 \notin [-1, 1]$ is $T_n(t_0)$ satisfies the inequality

$$\max_{-1 \le t \le 1} |P_n(t)| \ge \max_{-1 \le t \le 1} |T_n(t)|. \tag{7.12}$$

Equality holds only when $P_n(t) = T_n(t)$. Performing the substitution

$$\lambda = \frac{M + m}{2} + \frac{M - m}{2} t \qquad (0 < m < M < \infty)$$

in (7.12), we get

Lemma 7.1. Of all polynomials $P_n(\lambda)$ of degree n such that $P_n(0) = 1$, the polymonial

$$R_n(\lambda; m, M) = \frac{T_n\left(\dfrac{2\lambda - M - m}{M - m}\right)}{T_n\left(-\dfrac{M + m}{M - m}\right)} \tag{7.13}$$

---

* See Goncharov [1].

is the one with least deviation from zero over the interval $[m, M]$:

$$\max_{m \leq \lambda \leq M} |P_n(\lambda)| \geq \max_{m \leq \lambda \leq M} |R_n(\lambda; m, M)|. \tag{7.14}$$

It follows from (7.10) and (7.11) that

$$\max_{m \leq \lambda \leq M} |R_n(\lambda; m, M)| = \frac{1}{T_n\left(-\dfrac{M + m}{M - m}\right)} < 1. \tag{7.15}$$

7.3. *Equations with selfadjoint operators in Hilbert spaces.** We now return to investigation of equation (7.1), assuming that B is a positive definite selfadjoint operator:

$$B = \int_m^M \lambda dE_\lambda. \tag{7.16}$$

Let $Q_n(\lambda)$ be a polynomial

$$Q_n(\lambda) = 1 + a_1\lambda + \ldots + a_n\lambda^n.$$

As usual, $Q_n(B)$ denotes the operator

$$Q_n(B) = I + a_1 B + \ldots + a_n B^n.$$

Obviously,

$$Q_n(B) = \int_m^M Q_n(\lambda) \, dE_\lambda.$$

If $\rho[Q_n(B)] < 1$, equation (7.1) is equivalent to the equation

$$x = Q_n(B)x + f, \tag{7.17}$$

where

$$f = -a_1 b - a_2 Bb - \ldots - a_n B^{n-1} b \tag{7.18}$$

(to prove this, it suffices to note that every solution of equation (7.1) is a solution of (7.17) and the latter equation has a unique solution).

---

* Abramov [1], Birman [1, 2], Gavurin [1], Viarda [1], Natanson [1] and Krasnosel'skii [10].

Exercise 7.1. Find an operator B_1 such that equation (7.5) coincides with equation (7.17).

The spectral radius of a selfadjoint operator coincides with its norm, and (see Akhiezer and Glazman [1]) is defined by

$$\rho[Q_n(B)] = \|Q_n(B)\| = \max_{\lambda \in \sigma} |Q_n(\lambda)|,$$

where σ is the spectrum of B. Therefore

$$\rho[Q_n(B)] \leq \max_{m \leq \lambda \leq M} |Q_n(\lambda)|.$$

This inequality and Lemma 7.1 imply that the best choice for $Q_n(\lambda)$ is the polynomial (7.13).

In particular, we have proved

Theorem 7.1. *If* (7.16) *is a positive definite operator, the solution of equation* (7.1) *is the limit of successive approximations*

$$x_{k+1} = R_n(B; m, M)x_k + f \qquad (k = 1, 2, \ldots), \tag{7.19}$$

where f is defined by (7.18) *(a_i being the coefficients of the polynomial* (7.13)). *The rate of convergence of the successive approximations is that of a geometric progression with quotient*

$$\|R_n(B; m, M)\| \leq \frac{1}{\left| T_n\left(-\dfrac{M+m}{M-m}\right)\right|}. \tag{7.20}$$

It follows from (7.9) that

$$R_1(B) = I - \frac{2}{M+m}B,$$

$$R_2(B) = I - \frac{8(M+m)}{M^2+6Mm+m^2}B + \frac{8}{M^2+6Mm+m^2}B^2.$$

Thus, successive approximations solving equation (7.1) may be defined by the equalities

$$x_{k+1} = x_k = \frac{2}{M+m}Bx_k + \frac{2}{M+m}f \tag{7.21}$$

or

$$x_{k+1} = x_k - \frac{8(M+m)}{M^2+6Mm+m^2}Bx_k + \frac{8}{M^2+6Mm+m^2}B^2x_k -$$

$$- \frac{8(M+m)}{M^2 + 6Mm + m^2} f + \frac{8}{M^2 + 6Mm + m^2} f . \qquad (7.22)$$

Exercise 7.2. Show that in general the convergence rate of (7.22) is greater than that of (7.21).

Exercise 7.3. Consider a new method of solution, each step of which consists of two successive approximations by (7.21). Show that in general the convergence of this method is slower than that of (7.22).

In conclusion, we mention that the method described above may be used to solve certain types of equations in Banach spaces (such as equations with oscillatory operators (in the sense of Krein) all of whose eigenvalues are real),* and certain equations with nonselfadjoint operators in Hilbert spaces. Sometimes one has more information on the distribution of the spectrum of B than that used above. If the spectrum is concentrated in a set F different from the interval $[m, M]$, equation (7.1) is conveniently replaced by equation (7.2) with an operator S which is a polynomial $P_n(B)$ such that $P_n(0) = 1$ and $P_n(\lambda)$ has least deviation from zero over F. Unfortunately, the theory of polynomials of least deviation from zero on sets other than intervals is far more complicated than the theory of Chebyshev polynomials.

7.4. *Equations with compact operators.* Assume that equation (7.1) has the form

$$x = Ax + f_0 . \qquad (7.23)$$

If the operator $(I - A_1)^{-1}$ exists and is continuous, then (7.23) is equivalent to equation (7.2) with

$$S = (I - A_1)^{-1} (A - A_1) \qquad (7.24)$$

and

$$f = (I - A_1)^{-1} f_0 . \qquad (7.25)$$

Exercise 7.4. Construct an operator B_1 for which the transformation of (7.23) into (7.24) is the same as that of (7.1) into (7.5).

---

* See Gantmakher and Krein [1].

If the operator A_1 can be so chosen that $\rho(S) < 1$, then equation (7.23) has a unique solution, which can be obtained by successive approximations.

We shall consider an operator

$$A_1 = PA, \qquad (7.26)$$

where P is the projection onto some subspace $E_0 \subset E$. To determine the values of the operator $(I - A_1)^{-1}$ one must solve equations in the subspace E_0. For the element

$$u = (I - A_1)^{-1} f = (I - PA)^{-1} f \qquad (7.27)$$

is a solution of the equation

$$u = PAu + f,$$

whence the equalities

$$(I - P)u = (I - P)f, \quad Pu = PAPu + PA(I - P)f + Pf. \qquad (7.28)$$

This observation is extremely useful if P is the projection onto a finite-dimensional subspace. For then construction of the operator $(I - PA)^{-1}$ involves solution of a finite system of linear algebraic equations (or, equivalently, inversion of the matrix of the operator $I - PA$ in E_0).

If A_1 is defined by (7.26), equation (7.2) becomes

$$x = (I - PA)^{-1}(I - P)Ax + (I - PA)^{-1} f. \qquad (7.29)$$

If A_1 is defined as $A_1 = AP$, equation (7.2) is

$$x = (I - AP)^{-1} A(I - P)x + (I - AP)^{-1} f. \qquad (7.30)$$

Exercise 7.5. Show that construction of the operator $(I - AP)^{-1}$ is equivalent to matrix inversion if P is the projection on a finite-dimensional subspace.

If A is compact and the space has a basis, A may be approximated to any desired accuracy (in norm) by operators PA, where P is the projection on a finite-dimensional subspace. If 1 is not an eigenvalue of A, the operators $(I - PA)^{-1}$ exist and approximate the operator $(I - A)^{-1}$. Therefore, if PA is close to A, the operator $(I - PA)^{-1} \cdot (A - PA)$ has a small norm.

Much attention has been devoted by Sokolov and his co-workers (see the monographs of Sokolov [1], Luchka [1], Kurpel' [1] and the references cited there) to an approximate method which is, in essence, an ordinary iterative method based on equations (7.29) or (7.30):

$$x_{n+1} = (I - PA)^{-1}(A - PA)x_n + (I - PA)^{-1}f \qquad (7.31)$$

or

$$x_{n+1} = (I - PA)^{-1}(A - AP)x_n + (I - AP)^{-1}f \qquad (7.32)$$

with a special choice of the operator P (which is usually one-dimensional).

Exercise 7.6. Let P be a projection in a Banach space E, and A a bounded linear operator. Show that the operators AP, PA and PAP have equal spectral radii.

Exercise 7.7. Show that the spectral radius of the operator $(I - AP)^{-1}A(I - P)$ in (7.30) is equal to that of the operator $(I - P)(I - AP)^{-1}A(I - P)$.

Exercise 7.8. Let A be a selfadjoint operator defined on a Hilbert space H, and P an orthogonal projection. Show that the operator

$$L = (I - P)(I - AP)^{-1}A(I - P)$$

is also selfadjoint.

Exercise 7.9. Show that if any one of the operators $I - AP$, $I - PA$, $I - PAP$ is invertible, so are the other two. Find formulas expressing each of the operators $(I - AP)^{-1}$, $(I - PA)^{-1}$, $(I - PAP)^{-1}$ in terms of any one of the other two.

Exercise 7.10. Show that determination of the successive approximations (7.32) is equivalent to determination of Px_n and $Qx_n = x_n - Px_n$ from the recurrence relations

$$\left. \begin{array}{l} Px_n = PAPx_n + PAQx_{n-1} + Pf, \\ Qx_n = QAPx_n + QAQx_{n-1} + Qf. \end{array} \right\} \qquad (7.33)$$

The above arguments concerning the passage from equation (7.2) to (7.29) or (7.30) contain no specific recommendations for the choice of the operator P.

Let λ_0 be an eigenvalue of the operator A, of greatest absolute value, with corresponding eigenvector e_0. Let l_0 be an eigenelement of the adjoint A^* corresponding to the same eigenvalue λ_0. Let ρ_0 denote the radius of a disk containing the entire spectrum of A with the exception of λ_0. A simple calculation shows that the spectral radii of the operators $(I - PA)^{-1}(A - PA)$ and $(I - AP)^{-1}(A - AP)$ are at most ρ_0, where P is defined by

$$Px = l_0(x)e_0 \qquad (x \in E) \qquad (7.34)$$

(in this case these two operators coincide). Therefore, if $\rho_0 < 1$, a suitable projection is either (7.34) or any sufficiently close approximation to it.

Similar reasoning holds when several eigenvalues lie outside a disk of radius $\rho_0 < 1$; here P should project the space onto a subspace whose dimension is at least the sum of multiplicities of those eigenvalues of A outside the disk $|\lambda| \leqq \rho_0$.

Exericse 7.11. Assume that the sum of multiplicities of the eigenvalues of A outside the disk $|\lambda| \leqq \rho_0$ is greater than n. Let P be the projection on an n-dimensional subspace. Show that $\rho(A - PA) > \rho_0$.

The arguments of this section are also applicable when the equation is amenable to solution by successive approximations without any preliminary modifications; the result is simply a higher rate of convergence.

7.5. *Seidel's method.* For approximate solution of a system of linear equations

$$\sum_{k=1}^{m} b_{ik}\xi_k = b_i \qquad (i = 1, \ldots, m)$$

one often uses Seidel's method.* The system is rewritten as

$$\sum_{k=1}^{i} b_{ik}\xi_k = - \sum_{k=i+1}^{m} b_{ik}\xi_k + b_i \qquad (i = 1, \ldots, m),$$

and successive approximations $x_n = \{\xi_1^{(n)}, \ldots, \xi_m^{(n)}\}$ are determined by the formulas

$$\sum_{k=1}^{i} b_{ik}\xi_k^{(n)} = - \sum_{k=i+1}^{m} b_{ik}\xi_k^{(n-1)} + b_i \qquad (i = 1, \ldots, m; n = 1, 2, \ldots).$$

It is known that these approximations converge if the matrix (b_{ik}) is symmetric and positive definite.

Seidel's method may be generalized to operator equations.** Consider the equation

$$Bx = b, \qquad (7.35)$$

where B is a bounded selfadjoint operator defined in a Hilbert space H. Let

$$B = B_1 + B_2 + B_2^*. \qquad (7.36)$$

---

* See, e.g., Faddeev and Faddeeva [1].
** Krein and Prozorovskaya [1].

Then equation (7.35) becomes

$$(B_1 + B_2)x = - B_2^*x + b$$

or, if $(B_1 + B_2)^{-1}$ exists,

$$x = - (B_1 + B_2)^{-1} B_2^*x + (B_1 + B_2)^{-1} b. \tag{7.37}$$

Equations (7.37) and (7.35) are equivalent. If the spectral radius of the operator

$$C = (B_1 + B_2)^{-1} B_2^* \tag{7.38}$$

is smaller than unity, equation (7.37) may be solved by successive approximations.

Set

$$\gamma_0 = \inf_{||x||=1} \left\{ \left| ((B_1 + B_2)x, x) \right| - \left| (B_2^*x, x) \right| \right\}. \tag{7.39}$$

We claim that this number is positive if the operators B and B_1 are positive definite, i.e.,

$$\left. \begin{array}{c} (Bx, x) \geqq m(x, x), \\[4pt] (B_1x, x) \geqq m_1(x, x) \\[4pt] (x \in H; m, m_1 > 0). \end{array} \right\} \tag{7.40}$$

First note that the obvious inequality

$$(B_1x, x)\,((B_1 + B_2 + B_2^*)x, x) \geqq mm_1(x, x)^2$$

may be rewritten as

$$((B_1 + B_2^*)x, x)\,((B_1 + B_2)x, x) \geqq (B_2x, x)(B_2^*x, x) + mm_1(x, x)^2$$

or, what is the same,

$$\left| ((B_1 + B_2)x, x) \right|^2 - \left| (B_2^*x, x) \right|^2 \geqq mm_1(x, x)^2. \tag{7.41}$$

Let

$$\sup_{||x||=1} \left| ((B_1 + B_2)x, x) \right| = \beta_0.$$

Then

$$\gamma_0 \geqq \inf\{\beta - \alpha : \beta^2 - \alpha^2 \geqq mm_1, \beta \leqq \beta_0\}.$$

Therefore (see Fig. 7.1)

$$\gamma_0 \geqq \beta_0 - \sqrt{\beta_0^2 - mm_1} .$$

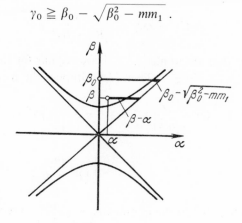

Fig. 7.1.

*Theorem 7.2.** If B and B_1 are positive definite, the spectral radius $\rho(C)$ of the operator (7.38) satisfies the inequality*

$$\rho(C) \leqq \rho_0 = 1 - \frac{\gamma_0}{\|B_1 + B_2\|}, \tag{7.42}$$

where γ_0 is given by (7.39).

Proof. Since

$$Cx - \lambda x = (B_1 + B_2)^{-1} [B_2^* x - \lambda(B_1 + B_2) x],$$

it follows that, for $\|x\| \neq 0$,

$$\|Cx - \lambda x\| \geqq \frac{1}{\|B_1 + B_2\|} \|B_2^* x - \lambda(B_1 + B_2) x\| \geqq$$

$$\geqq \frac{1}{\|B_1 + B_2\| \cdot \|x\|} |(B_2^* x - \lambda(B_1 + B_2)x, x)| \geqq$$

$$\geqq \frac{1}{\|B_1 + B_2\| \cdot \|x\|} \{|\lambda| \cdot |((B_1 + B_2)x, x)| - |(B_2^* x, x)|\}$$

* Krasnosel'skii and Stetsenko [1]. In the cited work of Krein and Prozorovskaya it is proved in another way that under the conditions of Theorem 7.2 $\rho(C) < 1$.

and for $|\lambda| \geqq 1$, by (7.39),

$$\|Cx - \lambda x\| \geqq \frac{\gamma_0}{\|B_1 + B_2\|} \|x\|. \tag{7.43}$$

This means that all points $|\lambda| \geqq 1$ are regular for the operator C. Now let $\rho_0 < |\lambda| \leqq 1$. Then (7.43) implies the inequality

$$\|Cx - \lambda x\| \geqq \left\| Cx - \frac{\lambda}{|\lambda|} x \right\| - \left| \lambda - \frac{\lambda}{|\lambda|} \right| \cdot \|x\| \geqq$$

$$\geqq \left\{ \frac{\gamma_0}{\|B_1 + B_2\|} - (1 - |\lambda|) \right\} \|x\| > 0.$$

Consequently, all λ in the annulus $\rho_0 < |\lambda| \leqq 1$ are also regular. Thus, the range $\mathfrak{N}_\lambda$ of the operator $C - \lambda I$ for $|\lambda| > \rho_0$ is closed, and the inverse operator $(C - \lambda I)^{-1}$ is defined on $\mathfrak{N}_\lambda$. The proof will be completed if we show that $\mathfrak{N}_\lambda$ coincides with H for $|\lambda| > \rho_0$. Let d_λ denote the dimension of the orthogonal complement of $\mathfrak{N}_\lambda$. This number is known as the *deficiency index* of the operator C. The basic property of the deficiency index is its invariance: If G is a connected domain (of the complex plane) consisting of regular points of the operator C, then the deficiency indices d_λ are the same for all $\lambda \in G$ (Krasnosel'skii [1]; see also Akhiezer and Glazman [1]). To complete the proof, therefore, it suffices to show that $d_\lambda = 0$ for large λ. But this is obvious, since for large λ the operator $C - \lambda I$ has continuous inverse defined on all of H:

$$(C - \lambda I)^{-1} = -\frac{1}{\lambda} \sum_{n=0}^{\infty} \lambda^{-n} C^n. \quad \blacksquare$$

7.6. *Remark on convergence in other norms.* Let A be an operator defined in E such that $\rho(A) < 1$. Moreover, suppose that some power A^k is known to be a continuous operator mapping E into a (proper) subspace E_1. It turns out that the successive approximations

$$x_{n+1} = Ax_n + f \qquad (f, x_0 \in E) \tag{7.44}$$

converge to a solution $x^* \in E$ of the equation

$$x = Ax + f \tag{7.45}$$

not only in the norm of the space E, but also in the norm of E_1 (for large n, the difference $x_n - x^*$ belongs to E_1 and $\|x_n - x^*\|_{E_1} \to 0$). To prove this, one need only note that

$$x^* - x_n = \sum_{k=n}^{\infty} A^k f - A^n x_0 = A^n (x^* - x_0),$$

whence

$$\|x^* - x_n\|_{E_1} \leqq \|A^k\|_{E \to E_1} \|A^{n-k}(x^* - x_0)\| = O[\rho^n(A)].$$

This observation is useful if equation (7.45) is to be solved in C or L_p, and it is known that A^k is defined in some space of differentiable functions (for example, if A is a potential-type integral operator). It can then be shown that the successive approximations converge uniformly together with their derivatives up to a certain order.

Exercise 7.12. Let B be a linear operator which commutes with A in E and maps E continuously into E_1. Let $\rho(A) < 1$. Show that the successive approximations $y_{n+1} = Ay_n + Bf$ converge in the norm of E_1 to a solution of the equation $y = Ay + Bf$ if the initial approximation has the form $y_0 = Bx_0$, $x_0 \in E$.

Exercise 7.13. Let P be a linear projection mapping a Banach space E onto a subspace E_0, A and A_0 linear operators defined in E and E_0, respectively. Assume that $I - \lambda A$ has an inverse $R(\lambda)$. Let $\|PAx_0 - Ax_0\| \leqq \varepsilon \|x_0\|$ $(x_0 \in E_0)$ and assume that for every $x \in E$ there is an element $x_0 \in E_0$ such that $\|Ax - x_0\| \leqq \varepsilon_1 \|x\|$. Show that, for any $y \in E$, the equation $x - \lambda A_0 x = Py$ has a unique solution $x \in E$, if

$$q = |\lambda| \cdot \|R(\lambda)\| \, (\|P\| \cdot \|I - \lambda A\| \varepsilon_1 + \varepsilon + |\lambda| \varepsilon \varepsilon_1) < 1. \tag{7.46}$$

When this condition is satisfied, the solution satisfies the inequality

$$\|x\| \leqq \frac{1}{1-q}(1 + |\lambda| \varepsilon_1) \|R(\lambda)\| \cdot \|P\| \cdot \|Py\|$$

(Kantorovich [2]).

Exercise 7.14. Show that the assertion of Exercise 7.13 remains valid if (7.46) is replaced by the condition

$$r = |\lambda| \cdot \|R(\lambda)\| \qquad (\|P\|\varepsilon_1 + \varepsilon_1 + \varepsilon) < 1.$$

The equation $x - \lambda A_0 x = Py$ then has a unique solution x for any y, and

$$\|x\| \leqq \frac{1}{1-r} \|R(\lambda)\| \cdot \|Py\|$$

(Krasnosel'skii and Chechik [1]).

§ 8. Method of minimal residuals*

8.1. *Statement of the problem.* Most of this section is devoted to approximate solution of the equation

$$Bx = b \qquad (8.1)$$

with a bounded positive definite operator

$$Bx = \int_m^M \lambda dE_\lambda x \qquad (0 < m < M < \infty) \qquad (8.2)$$

in a real Hilbert space H. Let x^* be a solution of the equation. If x_0 is an approximate solution, we denote the residual $Bx_0 - b$ by Δ_0, and the error $x_0 - x^*$ by δ_0. Obviously,

$$\Delta_0 = B\delta_0 . \qquad (8.3)$$

Beginning with the approximation x_0, let us find a new approximation x_1 by the formula

$$x_1 = x_0 - c_1 \Delta_0 , \qquad (8.4)$$

where c_1 is an as yet undefined factor, to be determined by solution of an appropriate extremum problem (e.g., minimization of the norm of the residual $\Delta_1 = Bx_1 - b$ or the norm of the error $\delta_1 = x_1 - x^*$, etc.). x_2 is then determined from x_1 in the same way, and so on. The result is an iterative process

$$x_{n+1} = x_n - c_{n+1}\Delta_n \qquad (n = 0, 1, 2, \ldots), \qquad (8.5)$$

where

$$\Delta_n = Bx_n - b = B\delta_n , \qquad \delta_n = x_n - x^* . \qquad (8.6)$$

If we set $c_{n+1} = k$ in (8.5), where k is some nonzero number, the recurrence relation becomes

$$x_{n+1} = (I - kB)x_n + kb \qquad (n = 0, 1, 2, \ldots). \qquad (8.7)$$

---

* Most of this section is based on a paper of Krasnosel'skii and Krein [1]. The ideas of the method of minimal residuals have been extensively applied and developed by Marchuk, Samarskii and their students.

Of course, k should be so chosen that the norm of the operator $I - kB$ is as small as possible. It is not difficult to see (see subsection 7.3) that when $k = 2/(M + m)$,

$$\| I - kB \| = \left\| \int_m^M \left(1 - \frac{2\lambda}{M + m} \right) dE_\lambda \right\| \leqq$$

$$\leqq \max \left\{ \left| 1 - \frac{2M}{M + m} \right|, \left| 1 - \frac{2m}{M + m} \right| \right\} = \frac{M - m}{M + m}. \qquad (8.8)$$

It follows that, for successive approximations (8.7) (with $k = 2/(M + m)$), the successive errors and residuals satisfy the inequalities

$$\| \Delta_{n+1} \| \leqq \frac{M - m}{M + m} \| \Delta_n \| \qquad (n = 0, 1, 2, \ldots) \qquad (8.9)$$

and

$$\| \delta_{n+1} \| \leqq \frac{M - m}{M + m} \| \delta_n \| \qquad (n = 0, 1, 2, \ldots). \qquad (8.10)$$

To prove these inequalities it suffices to note that

$$\delta_{n+1} = (I - kB)\delta_n, \qquad \Delta_{n+1} = (I - kB)\Delta_n. \qquad (8.11)$$

8.2. *Convergence of the method of minimal residuals.* The approximation x_1 is sought in the form (8.4), the factor c_1 being chosen so as to minimize the norm of $\Delta_1 = Bx_1 - b$. Obviously,

$$\Delta_1 = Bx_0 - b - c_1 B\Delta_0 = \Delta_0 - c_1 B\Delta_0,$$

and hence

$$\| \Delta_1 \|^2 = (\Delta_1, \Delta_1) = \| \Delta_0 \|^2 - 2c_1 (B\Delta_0, \Delta_0) + c_1^2 (B\Delta_0, B\Delta_0). \qquad (8.12)$$

This quadratic trinomial is a minimum when

$$c_1 = \frac{(B\Delta_0, \Delta_0)}{(B\Delta_0, B\Delta_0)}. \qquad (8.13)$$

Since the operator B is positive definite, so are both numerator and denominator in the right-hand side of (8.13) (provided, of course, that $\Delta_0 \neq 0$, i.e., $x_0 \neq x^*$).

The same method is then used to find the second approximation x_2 from the first x_1, and so on. The result is an iterative process

$$x_{n+1} = x_n - \frac{(\Delta_n, B\Delta_n)}{(B\Delta_n, B\Delta_n)} \Delta_n \qquad (n = 0, 1, 2, \ldots), \qquad (8.14)$$

where $\Delta_n = Bx_n - b$. We call this process the *method of minimal residuals*.

Theorem 8.1. The successive approximations (8.14) *of the method of minimal residuals converge to a solution* x^* *of equation* (8.1). *The rate of convergence is described by the inequality*

$$\|x_n - x^*\| \le \frac{1}{m} \left(\frac{M - m}{M + m} \right)^n \|\Delta_0\|. \qquad (8.15)$$

Proof. It follows directly from the definition that

$$\|\Delta_{n+1}\| \le \|\Delta_n - cB\Delta_n\|$$

for any c. Set $c = 2/(M + m)$; then, by (8.8), we get the inequality

$$\|\Delta_{n+1}\| \le \frac{M - m}{M + m} \|\Delta_n\| \qquad (n = 0, 1, 2, \ldots). \qquad (8.16)$$

Therefore

$$\|\Delta_n\| \le \left(\frac{M - m}{M + m} \right)^n \|\Delta_0\|. \qquad (8.17)$$

Inequality (8.17) implies (8.15), since $\|x_n - x^*\| \le (1/m) \|\Delta_n\|$. ∎

Set

$$Ly = y - \frac{(By, y)}{(By, By)} By \qquad (y \in H, y \ne \theta). \qquad (8.18)$$

It follows from (8.6) and (8.14) that

$$\Delta_{n+1} = L\Delta_n \qquad (n = 0, 1, 2, \ldots). \qquad (8.19)$$

A simple calculation shows that

$$L\left(\sqrt{M}\, e_m \pm \sqrt{m}\, e_M \right) = \frac{M - m}{M + m} \left(\sqrt{M}\, e_m \pm \sqrt{m}\, e_M \right),$$

where e_m, e_M are unit eigenvectors of the operator B corresponding to the eigenvalues m and M, respectively. Therefore, if

$$\Delta_0 = \sqrt{M}\, e_m + \sqrt{m}\, e_M$$

the residual Δ_n will be

$$\Delta_n = L^n \Delta_0 = \left(\frac{M - m}{M + m}\right)^n \left[\sqrt{M}\, e_m + (-1)^{n+1} \sqrt{m}\, e_M \right],$$

and so

$$\|\Delta_n\| = \left(\frac{M - m}{M + m}\right)^n \|\Delta_0\|.$$

This equality shows that the estimate (8.17) cannot be sharpened if m and M are eigenvalues of B.

Exercise 8.1. Prove that (8.17) cannot be sharpened even if m and M are not eigenvalues of B but infimum and supremum of its spectrum.

8.3. *The moment inequality.** The numbers

$$b_s = (B^s x, x) = \int_m^M \lambda^s d(E_\lambda x, x)$$

are known as the *moments* of the operator (8.2).

Lemma 8.1. *Let s be an arbitrary number, p and q positive numbers. Then*

$$b_s^{p+q} \leqq b_{s+q}^p b_{s-p}^q. \tag{8.20}$$

Proof. Let $p_1 > 1$, $p_1' = p_1/(p_1 - 1)$; then, any functions $\alpha(\lambda)$ and $\beta(\lambda)$ satisfy the Hölder inequality

$$\int_m^M \alpha(\lambda) \beta(\lambda) d(E_\lambda x, x) \leqq$$

$$\leqq \left\{ \int_m^M |\alpha(\lambda)|^{p_1} d(E_\lambda x, x) \right\}^{1/p_1} \left\{ \int_m^M |\beta(\lambda)|^{p_1'} d(E_\lambda x, x) \right\}^{1/p_1'}.$$

In this inequality, set

$$\alpha(\lambda) = \lambda^{p(s+q)/(p+q)}, \quad \beta(\lambda) = \lambda^{q(s-p)/(p+q)} \qquad (m \leqq \lambda \leqq M).$$

---

* The inequalities in this subsection were first established in connection with certain problems in the theory of nonlinear integral equations (see Krasnosel'skii [8]).

Since

$$\frac{p(s + q)}{p + q} + \frac{q(s - p)}{p + q} = s,$$

it follows that

$$\int_m^M \lambda^s \, d(E_\lambda x, x) \leqq$$

$$\leqq \left\{ \int_m^M \lambda^{pp_1(s+q)/(p+q)} \, d(E_\lambda x, x) \right\}^{1/p_1} \times$$

$$\times \left\{ \int_m^M \lambda^{qp_1'(s-p)/(p+q)} \, d(E_\lambda x, x) \right\}^{1/p_1'}.$$

Now set $p_1 = (p + q)/p$; the result is

$$b_s \leqq \left\{ \int_m^M \lambda^{s+q} \, d(E_\lambda x, x) \right\}^{p/(p+q)} \left\{ \int_m^M \lambda^{s-p} \, d(E_\lambda x, x) \right\}^{q/(p+q)}.$$

Raising both sides of this inequality to the $(p + q)$-th power, we get (8.20). ∎

Introduce the notation

$$b_{s_1, \ldots, s_k}^{\alpha_1, \ldots, \alpha_k} = b_{s_1}^{\alpha_1} b_{s_2}^{\alpha_2} \ldots b_{s_k}^{\alpha_k}, \tag{8.21}$$

where $s_1, \ldots, s_k$ are arbitrary numbers, $\alpha_1, \ldots, \alpha_k$ positive numbers. The pair of numbers $\{\omega, \tau\}$,

$$\omega = \alpha_1 + \ldots + \alpha_k, \qquad \tau = \alpha_1 s_1 + \ldots + \alpha_k s_k,$$

is called the *weight* of $b_{s_1, \ldots, s_k}^{\alpha_1, \ldots, \alpha_k}$.

In this notation, inequality (8.20) becomes

$$b_s^{p+q} \leqq b_{s+q, s-p}^{p, q}.$$

It is clear from homogeneity arguments that general moment inequalities can exist only for expressions (8.21) of the same weight.

Lemma 8.2. Let $s_1 < s < s_2$. Then, for any $\beta > 0$, there exist positive β_1 and β_2 such that

$$b_s^\beta \leqq b_{s_1}^{\beta_1} b_{s_2}^{\beta_2}, \tag{8.22}$$

where $\beta_1 + \beta_2 = \beta$, $\beta_1 s_1 + \beta_2 s_2 = \beta s$.

The proof follows from Lemma 8.1, with

$$p = s - s_1, \quad q = s_2 - s, \quad \beta_1 = \beta\,\frac{s_2 - s}{s_2 - s_1}, \quad \beta_2 = \beta\,\frac{s - s_1}{s_2 - s_1}\,.$$

Theorem 8.2.

$$b^{\alpha_1, \ldots, \alpha_k}_{p_1, \ldots, p_k} \leqq b^{\beta_1, \beta_2}_{s_1, s_2} = b^{\beta_1}_{s_1} b^{\beta_2}_{s_2}, \tag{8.23}$$

provided both sides of the inequality have the same weight and

$$s_1 < p_1, \ldots, p_k < s_2\,.$$

Proof. By Lemma 8.2, each factor $b^{\alpha_i}_{p_i}$ appearing in the left-hand side of (8.23) satisfies the inequality

$$b^{\alpha_i}_{p_i} \leqq b^{\beta_{1i}}_{s_1} b^{\beta_{2i}}_{s_2} \qquad (\beta_{1i}, \; \beta_{2i} > 0)\,.$$

Multiplying these inequalities together, we get

$$b^{\alpha_1, \ldots, \alpha_k}_{p_1, \ldots, p_k} \leqq b^{\beta_{11} + \ldots + \beta_{1k}}_{s_1} b^{\beta_{21} + \ldots + \beta_{2k}}_{s_2}\,. \tag{8.24}$$

Since both sides have the same weight

$$(\beta_{11} + \ldots + \beta_{1k}) + (\beta_{21} + \ldots + \beta_{2k}) = \beta_1 + \beta_2\,,$$

$$s_1(\beta_{11} + \ldots + \beta_{1k}) + s_2(\beta_{21} + \ldots + \beta_{2k}) = s_1\beta_1 + s_2\beta_2\,,$$

which implies that

$$\beta_{11} + \ldots + \beta_{1k} = \beta_1\,, \qquad \beta_{21} + \ldots + \beta_{2k} = \beta_2\,,$$

so that (8.24) is the same as (8.23). ∎

8.4. *α-processes.* Let α be any fixed real number. Define a sequence

$$x^{(\alpha)}_{n+1} = x^{(\alpha)}_n - \frac{(B^\alpha \Delta_n, \Delta_n)}{(B^{\alpha+1} \Delta_n, \Delta_n)}\,\Delta_n\,, \tag{8.25}$$

where $x^{(\alpha)}_0$ is any element of H and

$$\Delta_n = Bx^{(\alpha)}_n - b\,. \tag{8.26}$$

The iterations defined by (8.25) are known as an *α-process.*

For $\alpha = 1$ this is simply the method of minimal residuals. For $\alpha = 0$

(8.25) gives the well-known *method of steepest descent*:

$$x_{n+1}^{(0)} = x_n^{(0)} - \frac{(\Delta_n, \Delta_n)}{(B\Delta_n, \Delta_n)} \Delta_n \tag{8.27}$$

(see, e.g., Kantorovich and Akilov [1]).

Of special interest is the (-1)-process

$$x_{n+1}^{(-1)} = x_n^{(-1)} - \frac{(B^{-1}\Delta_n, \Delta_n)}{(\Delta_n, \Delta_n)} \Delta_n \tag{8.28}$$

which we shall call the *method of minimal errors*. It is clear from (8.28) that this method is not always feasible—the calculations involve application of the operator B^{-1}. However, we shall show in subsection 8.6 that, strange as it may seem, there are equations for which the method of minimal errors is quite effective. For the moment we indicate some general properties of α-processes.

Theorem 8.3. For $\alpha \geqq -1$, *all α-processes are monotone, in the sense that*

$$\|x_{n+1}^{(\alpha)} - x^*\| \leqq \|x_n^{(\alpha)} - x^*\|. \tag{8.29}$$

Proof. It follows from

$$\Delta_n = Bx_n^{(\alpha)} - b = Bx_n^{(\alpha)} - Bx^* = B(x_n^{(\alpha)} - x^*)$$

that

$$x_n^{(\alpha)} - x^* = B^{-1}\Delta_n.$$

Hence the equality

$$\|x_{n+1}^{(\alpha)} - x^*\|^2 =$$

$$= \|x_n^{(\alpha)} - x^*\|^2 - \frac{2(B^\alpha \Delta_n, \Delta_n)}{(B^{\alpha+1}\Delta_n, \Delta_n)} (x_n^{(\alpha)} - x^*, \Delta_n) + \frac{(B^\alpha \Delta_n, \Delta_n)^2}{(B^{\alpha+1}\Delta_n, \Delta_n)^2} \|\Delta_n\|^2 =$$

$$= \|x_n^{(\alpha)} - x^*\|^2 - \frac{(B^\alpha \Delta_n, \Delta_n)}{(B^{\alpha+1}\Delta_n, \Delta_n)} \left[2(B^{-1}\Delta_n, \Delta_n) - \right.$$

$$\left. - \frac{(B^\alpha \Delta_n, \Delta_n)(\Delta_n, \Delta_n)}{(B^{\alpha+1}\Delta_n, \Delta_n)} \right]. \tag{8.30}$$

By Theorem 8.2,

$$(B^\alpha \Delta_n, \Delta_n)(\Delta_n, \Delta_n) \leqq (B^{\alpha+1}\Delta_n, \Delta_n)(B^{-1}\Delta_n, \Delta_n)$$

and, *a fortiori*,

$$\frac{(B^\alpha \Delta_n, \Delta_n)(\Delta_n, \Delta_n)}{(B^{\alpha+1}\Delta_n, \Delta_n)} \leqq 2(B^{-1}\Delta_n, \Delta_n).$$

Thus (8.30) implies (8.29). ∎

Let γ be some real number. Define a new inner product and norm in H by

$$(x, y)_\gamma = (B^\gamma x, y), \qquad \|x\|_\gamma = \sqrt{(x, y)_\gamma}. \qquad (8.31)$$

The new norm $\|x\|_\gamma$ is equivalent to the original norm in H: for $\gamma \geqq 0$,

$$m^{\gamma/2}\|x\| \leqq \|x\|_\gamma \leqq M^{\gamma/2}\|x\| \qquad (x \in H)$$

and for $\gamma < 0$,

$$M^{-\gamma/2}\|x\| \leqq \|x\|_\gamma \leqq m^{-\gamma/2}\|x\| \qquad (x \in H).$$

It is easily seen that every α-process is a β-process in the space H with the inner product (8.31), where $\gamma = \alpha - \beta$; this follows from

$$\frac{(B^\alpha \Delta, \Delta)}{(B^{\alpha+1}\Delta, \Delta)} = \frac{(B^{\alpha-\beta}B^\beta \Delta, \Delta)}{(B^{\alpha-\beta}B^{\beta+1}\Delta, \Delta)} = \frac{(B^\beta \Delta, \Delta)_{\alpha-\beta}}{(B^{\beta+1}\Delta, \Delta)_{\alpha-\beta}}.$$

In particular, every α-process coincides with the method of minimal residuals for the inner product $(x, y)_{\alpha-1}$. Hence, by Theorem 8.1,

Theorem 8.4. Every α-process is convergent, with rate of convergence equal to that of a geometric progression with quotient $(M - m)/(M + m)$.

Exercise 8.2. Show that the factor $c_\alpha = \dfrac{(B^\alpha \Delta, \Delta)}{(B^{\alpha+1}\Delta, \Delta)}$ in (8.25) is a nonincreasing function of α for fixed Δ.

Exercise 8.3. Let $x_n^{(\alpha)} = x_n^{(\beta)}$, where $-1 \leqq \alpha < \beta$. Prove the inequality

$$\|x_{n+1}^{(\alpha)} - x^*\| \leqq \|x_{n+1}^{(\beta)} - x^*\|. \qquad (8.32)$$

8.5. *Computation scheme.* It follows from (8.25) and (8.26) that for any α-process

$$\Delta_{n+1} = L_\alpha \Delta_n \qquad (n = 0, 1, 2, \ldots), \qquad (8.33)$$

where

$$L_\alpha \Delta = \Delta - \frac{(B^\alpha \Delta, \Delta)}{(B^{\alpha+1}\Delta, \Delta)} B\Delta .$$ (8.34)

The properties of the operator L_α have been studied in detail by Krasnosel'skii and Krein [1] (pp. 321–328).

Using formula (8.33), one can compute the sequence Δ_i without having to compute x_i at each step of the process. To compute x_n, therefore, one can use the formula

$$x_n = x_0 - c_1 \Delta_0 - \ldots - c_n \Delta_{n-1} ,$$ (8.35)

where

$$c_i = \frac{(B^\alpha \Delta_{i-1}, \Delta_{i-1})}{(B^{\alpha+1}\Delta_{i-1}, \Delta_{i-1})} .$$

This computation scheme has a serious drawback: in computations according to (8.35) errors in determination of $c_i \Delta_{i-1}$ accumulate, while this is not so in iterations according to (8.25).

8.6. *Application to equations with nonselfadjoint operators.* Consider the equation

$$Ax = a ;$$ (8.36)

where A and A^{-1} are continuous. Equation (8.36) is equivalent to the equation

$$A^*Ax = A^*a .$$ (8.37)

The operator $B = A^*A$ is selfadjoint and positive definite:

$$(A^*Ax, y) = (Ax, Ay) = (x, A^*Ay)$$

and

$$(A^*Ax, x) = (Ax, Ax) = \|Ax\|^2 \geq \frac{1}{\|A^{-1}\|^2} (x, x) .$$

The solution of equation (8.37) may therefore be approximated by the method of minimal residuals and other α-processes.

As indicated above, the most convenient α-processes for the solution of equation (8.1) are those with integral nonnegative α. It turns out that equation (8.37) is amenable to treatment by a (-1)-process (method of minimal errors). In fact, in this case formulas (8.28) become

$$x_{n+1} = x_n - \frac{((A^*A)^{-1}A^*(Ax_n - a), A^*(Ax_n - a))}{(A^*(Ax_n - a), A^*(Ax_n - a))} A^*(Ax_n - a) ;$$

alternatively, in view of the identity

$$((A^*A)^{-1}A^*z, A^*z) \equiv (z, z) \qquad (z \in H),$$

we get*

$$x_{n+1} = x_n - \frac{(Ax_n - a, Ax_n - a)}{(A^*(Ax_n - a), A^*(Ax_n - a))} A^*(Ax_n - a). \qquad (8.38)$$

§ 9. Approximate computation of the spectral radius

9.1. *Use of powers of an operator.* In §5 and §6 we indicated a few methods for estimating the spectral radius $\rho(A)$ of a linear operator A in a Banach space E. The resulting bounds often prove to be exaggerated. In applications it is important to have more precise estimates of $\rho(A)$.

Gel'fand's formula

$$\rho(A) = \lim_{n \to \infty} \rho_n(A), \qquad (9.1)$$

where

$$\rho_n(A) = \|A^n\|^{1/n} \qquad (n = 1, 2, \ldots), \qquad (9.2)$$

means that, for fixed and sufficiently large n, the numbers $\rho_n(A)$ may be regarded as approximations to $\rho(A)$. However, only rarely is actual computation of the numbers (9.2) possible, and other methods are therefore needed.

Quite often, approximations to $\rho(A)$ may be defined by a formula simpler than (9.1):

$$\rho(A) = \lim_{n \to \infty} \|A^n f\|^{1/n}, \qquad (9.3)$$

---

* Fridman [3–5].

where f is an element of E, usually chosen by "trial and error." Formula (9.3) is of course not valid for every $f \in E$. Indeed, if f belongs to an A-invariant subspace E_0, then, by Gel'fand's formula,

$$\varlimsup_{n \to \infty} \|A^n f\|^{1/n} \leq \rho_0(A),$$

where $\rho_0(A)$ is the spectral radius of the restriction A_0 of A to E_0; $\rho_0(A)$ may of course be smaller than $\rho(A)$. The words "trial and error" are therefore essential.

First let A be a selfadjoint operator

$$A = \int_a^b \lambda \, dE_\lambda,$$

where a, b are the infimum and supremum, respectively, of the spectrum. To simplify matters, we shall assume that $b = \|A\|$ and $|a| < b$. Then (9.3) is valid, provided $f - E_{b-\varepsilon} f \neq 0$ for any $\varepsilon > 0$, since then

$$b^{2n}\|f\|^2 = \|A^n\|^2 \|f\|^2 \geq \|A^n f\|^2 = \int_a^b \lambda^{2n} \, d(E_\lambda f, f) \geq$$

$$\geq \int_{b-\varepsilon}^b \lambda^{2n} d(E_\lambda f, f) \geq (b - \varepsilon)^{2n} \int_{b-\varepsilon}^b d(E_\lambda f, f) = (b - \varepsilon)^{2n}\|f - E_{b-\varepsilon} f\|^2$$

whence

$$b - \varepsilon \leq \varliminf_{n \to \infty} \|A^n f\|^{1/n} \leq \varlimsup_{n \to \infty} \|A^n f\|^{1/n} \leq b,$$

and the assertion follows from the fact that ε is arbitrary.

The elements f for which the difference $f - E_{b-\varepsilon} f$ vanishes for small ε constitute a linear set which is not dense in the entire space; only for these elements is (9.3) false.

Another important type of operators are compact operators. If A is compact and its spectral radius is not zero, then $\rho(A)$ coincides with the greatest absolute value of the eigenvalues. Let $\lambda_1, \lambda_2, \ldots, \lambda_k$ be the eigenvalues of A whose absolute values coincide with $\rho(A)$. Let the corresponding root subspaces be $E_1, E_2, \ldots, E_k$. Then E is the direct sum

$$E = E_1 \oplus E_2 \oplus \ldots \oplus E_k \oplus E_0, \tag{9.4}$$

where E_0 an A-invariant subspace, the restriction of A to which has spectral radius ρ smaller than $\rho(A)$. Corresponding to this direct decomposition of E we have a representation of each $f \in E$ as a sum

$$f = P_1 f + P_2 f + \ldots + P_k f + P_0 f \qquad (P_i f \in E_i), \tag{9.5}$$

where P_i is the projection onto the corresponding subspace. Let us define a new norm in E,

$$\|f\|_* = \|P_1 f\| + \ldots + \|P_k f\| + \|P_0 f\| \qquad (f \in E), \tag{9.6}$$

which is equivalent to the original norms,

$$\alpha \|f\| \leq \|f\|_* \leq \beta \|f\| \qquad (f \in E; \alpha, \beta > 0). \tag{9.7}$$

We claim that (9.3) holds if $f \neq P_0 f$, or, equivalently, if one of the projections $P_1 f, \ldots, P_k f$ does not vanish. For simplicity we assume that $P_1 f \neq 0$.

Let $e_1, e_2, \ldots, e_s$ be a basis of E_1 relative to which the restriction A_1 of A to E_1 is represented by a matrix in Jordan normal form. Then one of the nonzero components ξ_{i_0} of the vector

$$P_1 f = \xi_1 e_1 + \ldots + \xi_s e_s \tag{9.8}$$

relative to the above basis is such that all vectors $P_1 A^n P_1 f = A_1^n P_1 f$ have the representation

$$A_1^n P_1 f = \xi_1^{(n)} e_1 + \ldots + \xi_{i_0-1}^{(n)} e_{i_0-1} + \lambda_1^n \xi_{i_0} e_{i_0} + \xi_{i_0+1}^{(n)} e_{i_0+1} + \ldots$$
$$\ldots + \xi_s^{(n)} e_s \tag{9.9}$$

(the proof is left to the reader). Each coefficient in (9.8) is a continuous linear functional on E_1, and so

$$|\xi_{i_0}| \rho^n(A) = |\xi_{i_0} \lambda_1^n| \leq c \|A^n P_1 f\| \qquad (n = 1, 2, \ldots). \tag{9.10}$$

We are now ready to prove (9.3). It follows from (9.6) and (9.10) that

$$\|A^n f\|_* \geq \|P_1 A^n f\| = \|A^n P_1 f\| \geq \frac{1}{c} |\xi_{i_0}| [\rho(A)]^n = c_1 [\rho(A)]^n,$$

whence, by (9.7),

$$\|A^n f\| \geq \alpha c_1 [\rho(A)]^n$$

and thus

$$\varlimsup_{n \to \infty} \left\| A^n f \right\|^{1/n} \geqq \rho(A).$$

On the other hand, Gel'fand's formula implies the inequality

$$\varlimsup_{n \to \infty} \left\| A^n f \right\|^{1/n} \leqq \rho(A).$$

This implies (9.3).

Exercise 9.1. Let A be quasi-compact. Show that (9.3) holds for all f outside a certain subspace.

It is easy to describe other types of operators satisfying (9.3) for "almost all" f.

9.2. *Positive operators.** If A is a positive operator (in a space E with a cone K), a class of elements f satisfying (9.3) is easily determined.

Let the cone K be solid and f one of its nonzero elements. Recall that the f-norm $\| \cdot \|_f$ is defined in E by

$$\left\| x \right\|_f = \min \left\{ t: - tf \leqslant x \leqslant tf \right\}. \tag{9.11}$$

For positive x, $\| x \|_f$ is precisely the minimal t such that $x \leqslant tf$.

Theorem 9.1. *If the cone K is solid and normal, and f is an interior element of K, then formula (9.3) is valid.*

Proof. By Theorem 5.5, it follows from

$$A^n f \leqslant \left\| A^n f \right\|_f \cdot f$$

that

$$\left[\rho(A) \right]^n \leqq \left\| A^n f \right\|_f, \tag{9.12}$$

whence

$$\varlimsup_{n \to \infty} \left\| A^n f \right\|_f^{1/n} \geqq \rho(A).$$

---

* Stetsenko [3]; only the simplest theorems will be given.

On the other hand, by Gel'fand's formula,

$$\varlimsup_{n \to \infty} \|A^n f\|_f^{1/n} \leqq \rho(A).$$

Thus

$$\lim_{n \to \infty} \|A^n f\|_f^{1/n} = \rho(A). \tag{9.13}$$

Now (9.13) implies (9.3), since the norms $\|x\|$ and $\|x\|_f$ are equivalent. ∎

Inequality (9.12) implies the following important supplement to Theorem 9.1.

Theorem 9.2. *Under the assumptions of Theorem 9.1, the numbers* $\|A^n f\|_f^{1/n}$ *converge to* $\rho(A)$ *from above:*

$$\|A^n f\|_f^{1/n} \geqq \rho(A) \qquad (n = 1, 2, \ldots).$$

Exercise 9.2. Show that the sequence $\|A^n f\|_{A^{n-1} f}$ is monotone.

Exercise 9.3. Show that the sequence $\{\|A^{2^k} f\|_f\}^{2^{-k}}$ is monotone.

Theorem 9.2 remains valid when the cone K is not solid, provided the operator A satisfies additional assumptions.

Theorem 9.2 provides approximations to $\rho(A)$ from above. Similar arguments lead to approximations from below for fairly extensive classes of operators.

Theorem 9.3. *Let the cone K be solid and normal and let f be any interior element of K. If the operator A has an eigenvector u_0 which is an interior point of K and corresponds to the eigenvalue $\rho(A)$, then*

$$\rho(A) = \lim_{n \to \infty} \sqrt[n]{\alpha_n}, \tag{9.14}$$

where

$$\alpha_n = \sup \{\alpha : A^n f \geqslant \alpha f\}. \tag{9.15}$$

Moreover, $\sqrt[n]{\alpha_n}$ *converges to* $\rho(A)$ *from below.*

Proof. Since f is an interior element of K, there exist positive c_1 and c_2 such that

$$c_1 u_0 \leqslant f \leqslant c_2 u_0.$$

Therefore

$$c_1 [\rho(A)]^n u_0 \leqslant A^n f \leqslant c_2 [\rho(A)]^n u_0$$

and

$$A^n f \succcurlyeq \frac{c_1}{c_2} [\rho(A)]^n f .$$

This implies the inequalities

$$\alpha_n \geqq \frac{c_1}{c_2} [\rho(A)]^n ,$$

so that

$$\lim_{n \to \infty} \sqrt[n]{\alpha_n} \geqq \rho(A) . \tag{9.16}$$

On the other hand, by Theorem 5.4,

$$\sqrt[n]{\alpha_n} \leqq \rho(A) \qquad (n = 1, 2, \ldots) . \tag{9.17}$$

Inequalities (9.16) and (9.17) imply (9.14). The second assertion of the theorem follows from (9.17). ∎

The assumptions of Theorem 9.3 hold, for example, for strongly positive compact operators, and in the more general case of indecomposable quasi-compact operators.

9.3. *Computation of the greatest eigenvalue.* We have repeatedly mentioned that, for many operators (such as compact operators) the spectral radius coincides with the greatest absolute value of the eigenvalues. Thus any approximate method for computation of eigenvalues is at the same time a method for computation of the spectral radius.

The Galerkin and Galerkin-Petrov methods are often convenient in applications. These methods have been discussed in detail, for example, in Mikhlin's monographs [3, 8]. Several aspects of these methods in their general formulation will be considered in Chapter 4.

Neither shall we discuss Kantorovich's method of steepest descent as applied to the problem of the eigenvalues of selfadjoint operators. The basic results have been presented by Kantorovich and Akilov [1].

Finally, we wish to mention the fundamental monograph of Faddeev and Faddeeva [1] on numerical methods of linear algebra. Many of the methods described therein carry over directly to operators in Banach and Hilbert spaces.

9.4. *Approximate computation of the eigenvalues of selfadjoint operators.*
Till the end of this section, A denotes a positive definite selfadjoint
operator in a real Hilbert space H. We shall describe several approximate
methods for finding the infimum m and the supremum M of the spectrum
of A.

To derive suitable approximate formulas,* we first assume that H is
n-dimensional and that all eigenvalues of A are simple and distinct.
To simplify the notation, let

$$M = \lambda_1 > \lambda_2 > \ldots > \lambda_n = m. \tag{9.18}$$

Let $e_1, \ldots, e_n$ be unit eigenvectors of the matrix A, corresponding to the
eigenvalues $\lambda_1, \ldots, \lambda_n$.

The simplest methods giving the approximation x_{k+1} in terms of the
previous approximation x_k are based on representation of the vector
x_{k+1} as a linear combination of x_k and Ax_k, with suitable choice of numerical coefficients. In other words, the next approximation x_{k+1} is
assumed to be a point of the plane Π spanned by the vectors x_k and Ax_k.
Fig. 9.1 illustrates the ellipsoid $(Ax, x) = 1$; its smallest semi-axis is

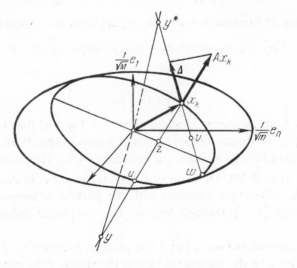

Fig. 9.1.

---

* Here we are following Krasnosel'skii [7].

$e_1/\sqrt{M}$, its greatest semi-axis is $e_n/\sqrt{m}$; the dashed line indicates the semi-axes of the ellipse in which the ellipsoid cuts the plane Π.

We first try to approximate the vector e_1 (if $x \approx e_1$, then the number $(Ax, x)/(x, x)$ approximates M).

Let y denote the point at which the straight line through x_k in the direction of Ax_k intersects the semiminor axis of the ellipse (or its continuation). To approximate x_{k+1}, we normalize y in such a way that the point x_{k+1} lies on the ellipsoid $(Ax, x) = 1$, to this end it suffices to set

$$x_{k+1} = \frac{1}{\sqrt{(Ay, y)}} y. \tag{9.19}$$

Express the point y as

$$y = x_k - \varepsilon Ax_k. \tag{9.20}$$

To determine ε we use the fact that y is an eigenvalue of the operator PA, where P is the orthogonal projection on the plane Π. Thus $PAy = \lambda^* y$, so that

$$(Ay - \lambda^* y, x_k) = 0, \qquad (Ay - \lambda^* y, Ax_k) = 0.$$

Eliminating λ^* from these equations, we get a quadratic equation for ε:

$$\varepsilon^2(a_2^2 - a_1 a_3) - \varepsilon(a_1 a_2 - a_0 a_3) + (a_1^2 - a_0 a_2) = 0, \tag{9.21}$$

where

$$a_i = (A^i x_k, x_k).$$

It is clear from geometric considerations (see Fig. 9.1) that both roots ε_1 and ε_2 of this equation are real and positive. This is also easy to prove analytically. To fix ideas, assume that $\varepsilon_1 < \varepsilon_2$. Then, to determine y, set $\varepsilon = \varepsilon_2$ in (9.20). The sequence x_k then converges to $e_1/\sqrt{M}$. The method described here coincides with the method of steepest descent (Kantorovich [1, 2]) (though here we have employed different arguments).

Had we considered not y but z, the point of intersection of the same straight line with the semimajor axis of the ellipse, determination of z would have reduced to the formula.

$$z = x_k - \varepsilon Ax_k, \tag{9.22}$$

which is analogous to (9.20); in this case ε must be the smaller root ε_1 of the quadratic equation (9.21). The approximation x_{k+1} is then defined by

$$x_{k+1} = \frac{z}{\sqrt{(Az, z)}}. \tag{9.23}$$

The sequence (9.23) converges to $e_n/\sqrt{m}$. Thus the sequence of numbers

$$\frac{(Ax_k, x_k)}{(x_k, x_k)} \qquad (k = 1, 2, \dots) \tag{9.24}$$

converges to m.

Exercise 9.4. Show that the discriminant of equation (9.21) is nonnegative.

Exercise 9.5. Show that the sequence (9.24) also converges to m when A is an arbitrary positive definite selfadjoint operator in a Hilbert space, provided the initial approximation x_0 does not belong to a certain fixed subspace H_0. Describe the subspace H_0.

We now consider iterative processes which do not involve solution of a quadratic equation at each step. First consider approximation of the vector e_1. Geometric considerations (Fig. 9.1) show that, with a "small error," the point y may be replaced by the point u at which the straight line through x_k in the direction of Ax_k intersects the ellipse. Then $x_{k+1} = x_k - \varepsilon Ax_k$, and ε is determined by

$$(Ax_{k+1}, x_{k+1}) = 1$$

or, equivalently, by the equation

$$(Ax_k - \varepsilon A^2x_k, x_k - \varepsilon Ax_k) = (Ax_k, Ax_k).$$

This equation has one nonzero root. After finding it we obtain the recurrence relation

$$x_{k+1} = x_k - 2\frac{(A^2x_k, x_k)}{(A^3x_k, x_k)}Ax_k \qquad (k = 0, 1, 2, \dots). \tag{9.25}$$

This method is due to Kostarchuk [1], who called it the *method of normal chords*.

If the space is provided with an auxiliary inner product

$$(x, y)_\gamma = (A^\gamma x, y)$$

and the normal chord formula derived in terms of this inner product, we get the recurrence relation

$$x_{k+1} = x_k - 2\frac{(A^{2+\gamma}x_k, x_k)}{(A^{3+\gamma}x_k, x_k)}Ax_k \qquad (k = 0, 1, 2, \ldots). \qquad (9.26)$$

The method of (9.26) is most convenient for $\gamma = -2$:

$$x_{k+1} = x_k - 2\frac{(x_k, x_k)}{(Ax_k, x_k)}Ax_k \qquad (k = 0, 1, 2, \ldots). \qquad (9.27)$$

Let h_k denote the projection of the vector Ax_k on the line perpendicular to x_k in the plane Π (see Fig. 9.1). Obviously,

$$h_k = Ax_k - \frac{(Ax_k, x_k)}{(x_k, x_k)}x_k, \qquad (k = 0, 1, 2, \ldots). \qquad (9.28)$$

Let w be the point at which the straight line through x_k in the direction of h_k intersects the ellipse. This point $w = x_k - \varepsilon h_k$ may be determined from the condition $(Aw, w) = 1$. The point w may be regarded as the next approximation x_{k+1} if the aim is to find the eigenvector e_n. The recurrence relation is

$$x_{k+1} = x_k - 2\frac{(h_k, h_k)}{(Ah_k, h_k)}h_k \qquad (k = 0, 1, 2, \ldots), \qquad (9.29)$$

where h_k is determined by (9.28). The convergence of (9.29) to e_n is clear from geometric considerations.

The main idea of the last process we describe in this subsection is to go from x_k to the midpoint v of the segment joining x_k and w. Here the iterative process for approximation of e_n is

$$x_{k+1} = x_k - \frac{(h_k, h_k)}{(Ah_k, h_k)}h_k \qquad (k = 0, 1, 2, \ldots). \qquad (9.30)$$

Note that the approximations (9.30) no longer lie on an ellipsoid $(Ax, x) = c$. The norms of the approximations do not decrease, for passage from x_k to the next approximation involves addition of an orthogonal vector.

9.5. *The method of normal chords.* The method of normal chords has been studied in detail by Kostarchuk [2] and Pugachev [3].

Formulas (9.25), (9.26) and (9.27), which describe different variants of the method of normal chords, are meaningful for arbitrary positive definite selfadjoint operators, not merely positive definite matrices. Of course, it may be impossible to find the eigenvectors of an arbitrary positive definite operator A by these formulas—A may not have any eigenvectors. Nevertheless, in any case one can expect the numbers

$$c_k = \frac{(A^{2+\gamma}x_k, x_k)}{(A^{3+\gamma}x_k, x_k)} \qquad (k = 0, 1, 2, \ldots) \qquad (9.31)$$

to converge to $1/M$ for "almost all" x_0. This is indeed the case, as we shall now prove.

Lemma 9.1. *Let x_0 be arbitrary and x_k defined by (9.26); then, for any κ, the sequence*

$$(A^\kappa x_0, x_0), \ (A^\kappa x_1, x_1), \ldots, (A^\kappa x_k, x_k), \ldots, \qquad (9.32)$$

has a finite limit.

Proof. A simple calculation shows that

$$(A^\kappa x_{k+1}, x_{k+1}) = (A^\kappa x_k, x_k) - \frac{4(A^{2+\gamma}x_k, x_k)}{(A^{3+\gamma}x_k, x_k)}(A^{\kappa+1}x_k, x_k) +$$

$$+ \frac{4(A^{2+\gamma}x_k, x_k)^2}{(A^{3+\gamma}x_k, x_k)^2}(A^{\kappa+2}x_k, x_k). \qquad (9.33)$$

It follows from (9.33) that the sequence (9.32) is stationary when $\kappa = \gamma + 1$:

$$(A^\kappa x_k, x_k) = (A^\kappa x_0, x_0) \qquad (k = 1, 2, \ldots).$$

Let $\kappa > \gamma + 1$. Then Theorem 8.2 implies the inequality

$$(A^{\kappa+1}x_k, x_k)(A^{3+\gamma}x_k, x_k) \leqq (A^{2+\gamma}x_k, x_k)(A^{2+\kappa}x_k, x_k).$$

It therefore follows from (9.33) that the sequence (9.32) is nondecreasing. It is obviously bounded above.

Finally, let $\kappa < \gamma + 1$. Again by Theorem 8.2,

$$(A^{2+\gamma}x_k, x_k)(A^{\kappa+2}x_k, x_k) \leqq (A^{\kappa+1}x_k, x_k)(A^{3+\gamma}x_k, x_k),$$

and therefore the sequence (9.32) is nonincreasing. ∎

It follows immediately from this lemma that the sequence (9.31) has a limit c^*.

The following lemma is useful when (9.31) is to be used to approximate c^*:

Lemma 9.2. The sequence (9.31) *is nonincreasing.*

Proof. By (9.33), we have the following expression for the difference $c_k - c_{k+1}$:

$$c_k - c_{k+1} = \frac{a_{2+\gamma}}{a_{3+\gamma}} - \frac{a_{2+\gamma} - \dfrac{4a_{2+\gamma}}{a_{3+\gamma}^2}(a_{3+\gamma}^2 - a_{2+\gamma}a_{4+\gamma})}{a_{3+\gamma} - \dfrac{4a_{2+\gamma}}{a_{3+\gamma}^2}(a_{4+\gamma}a_{3+\gamma} - a_{2+\gamma}a_{5+\gamma})},$$

where $a_i = (A^i x_k, x_k)$. The sign of the difference is the same as that of the numerator:

$$\text{sign}\,(c_k - c_{k+1}) =$$

$$= \text{sign}\left\{a_{2+\gamma}a_{5+\gamma} - a_{4+\gamma}a_{3+\gamma} + \frac{a_{3+\gamma}}{a_{2+\gamma}}(a_{3+\gamma}^2 - a_{2+\gamma}a_{4+\gamma})\right\}.$$

Since, by Theorem 8.2,

$$a_{3+\gamma}^2 \leqq a_{2+\gamma}a_{4+\gamma},$$

it follows that

$$a_{2+\gamma}a_{5+\gamma} - a_{4+\gamma}a_{3+\gamma} + \frac{a_{3+\gamma}}{a_{2+\gamma}}(a_{3+\gamma}^2 - a_{2+\gamma}a_{4+\gamma}) \geqq$$

$$\geqq a_{2+\gamma}a_{5+\gamma} - a_{4+\gamma}a_{3+\gamma} + \frac{a_{4+\gamma}}{a_{3+\gamma}}(a_{3+\gamma}^2 - a_{2+\gamma}a_{4+\gamma}) =$$

$$= a_{2+\gamma}a_{5+\gamma} - \frac{a_{4+\gamma}^2 a_{2+\gamma}}{a_{3+\gamma}}.$$

The right-hand side (again by Theorem 8.2) is nonnegative; thus

$$a_{2+\gamma}a_{5+\gamma} - a_{4+\gamma}a_{3+\gamma} + \frac{a_{3+\gamma}}{a_{2+\gamma}}(a_{3+\gamma}^2 - a_{2+\gamma}a_{4+\gamma}) \geqq 0,$$

and so $c_k \geqq c_{k+1}$. ∎

The fundamental theorem is

Theorem 9.4. Let

$$A = \int_m^M \lambda\, dE_\lambda.$$

If the initial approximation x_0 satisfies the condition $x_0 \neq E_\lambda x_0$ for $\lambda < M$, then $c^* = 1/M$, where c^* is the limit of the sequence (9.31).

Proof. If the assertion is false, there exists λ_0 such that $1/c^* < \lambda_0 < M$. By Lemma 9.2,

$$|1 - 2c_n\lambda| = 2c_n\lambda - 1 \geq 2c^*\lambda_0 - 1 = 1 + \varepsilon \qquad (\lambda_0 \leq \lambda \leq M), \quad (9.34)$$

for all n, where $\varepsilon > 0$. Apply the operator $I - E_{\lambda_0}$ to (9.26). Since this operator commutes with A,

$$(I - E_{\lambda_0}) x_{k+1} = (I - 2c_k A)(I - E_{\lambda_0}) x_k = \int_{\lambda_0}^{M} (1 - 2c_k\lambda) \, dE_\lambda x_k.$$

It now follows from (9.34) that

$$\left\| (I - E_{\lambda_0}) x_{k+1} \right\| \geq (1 + \varepsilon) \left\| \int_{\lambda_0}^{M} dE_\lambda x_k \right\| = (1 + \varepsilon) \left\| (I - E_{\lambda_0}) x_k \right\|.$$

Consequently,

$$\left\| x_k \right\| \geq \left\| (I - E_{\lambda_0}) x_k \right\| \geq (1 + \varepsilon)^k \left\| x_0 - E_{\lambda_0} x_0 \right\| \qquad (k = 1, 2, \dots).$$

By assumption, $x_0 \neq E_{\lambda_0} x_0$, and so $\left\| x_k \right\| \to \infty$. But this contradicts the equality $(Ax, x) = 1$. ∎

Exercise 9.6. Show that if the operator A is compact the even-indexed (or odd-indexed) terms of the sequence (9.26) converge to one of the eigenvectors of A. Show that this eigenvector corresponds to the greatest eigenvalue of the restriction of A to the minimal invariant subspace containing the initial approximation x_0.

9.6. *The method of orthogonal chords.* We now return to the iterative processes (9.29) and (9.30), the first of which we call the *method of orthogonal chords.* Recall that in this process the successive approximations are given by

$$x_{k+1} = x_k - 2 \frac{(h_k, h_k)}{(Ah_k, h_k)} h_k \qquad (k = 0, 1, 2, \dots), \quad (9.35)$$

where

$$h_k = Ax_k - \frac{(Ax_k, x_k)}{(x_k, x_k)} x_k \qquad (k = 0, 1, 2, \dots). \quad (9.36)$$

The second method, following the recurrence relation

$$x_{k+1} = x_k - \frac{(h_k, h_k)}{(Ah_k, h_k)} h_k, \tag{9.37}$$

is known as the *method of orthogonal semichords*.

The methods of orthogonal chords and semichords* have been studied in detail by Pugachev [1, 2] and Bezsmertnykh [1–3]. Here the fundamental fact is that for "almost all" initial approximations x_0 the number sequence

$$d_k = \frac{(h_k, h_k)}{(Ah_k, h_k)} \qquad (k = 0, 1, 2, \dots) \tag{9.38}$$

converges to $1/m$. We shall not discuss the proof of the relevant theorems.

Fridman has applied the methods (9.35) and (9.37) to compute the so-called resonance detuning.

A minute analysis of the distribution of the normalized vectors (9.36) in H has been carried out by Bezsmertnykh. Under fairly general assumptions, these vectors converge to an eigenvector corresponding either to the eigenvalue M or to the second (in order of magnitude) eigenvalue λ_2.

9.7. *Simultaneous computation of several iterations.* We again consider the case in which A is a positive definite matrix (though the methods described are valid when H is not finite-dimensional, and the formulas are applicable in the general case). We shall try to base the construction of the approximation x_{k+1} from x_k on several iterations:

$$x_k, Ax_k, \dots, A^s x_k. \tag{9.39}$$

Let Π_k denote the linear span of the vectors (9.39), and P_k the orthogonal projection on Π_k. The approximation x_{k+1} will be the unit eigenvector of the operator $P_k A$ corresponding to its greatest eigenvalue v (see Lanczos [1], Lyusternik [2], Birman [2], Vorob'ev [1–3]). Let

$$x_{k+1} = \alpha_0 x_k + \alpha_1 A x_k + \dots + \alpha_s A^s x_k. \tag{9.40}$$

Obviously, the equality

$$P_k(A x_{k+1} - v x_{k+1}) = 0$$

---

* These methods were suggested by Krasnosel'skii [7].

must hold, and so

$$(Ax_{k+1} - vx_{k+1},\ A^i x_k) = 0 \qquad (i = 0, 1, \ldots, s)$$

or, equivalently,

$$(x_{k+1}, A^{i+1} x_k - v A^i x_k) = 0 \qquad (i = 0, 1, \ldots, s).$$

Written out in full, these equalities are

$$\alpha_0(a_1 - va_0) + \alpha_1(a_2 - va_1) + \ldots + \alpha_s(a_{s+1} - va_s) = 0,$$

$$\alpha_0(a_2 - va_1) + \alpha_1(a_3 - va_2) + \ldots + \alpha_s(a_{s+2} - va_{s+1}) = 0,$$

$$\cdot \quad \cdot \quad \cdot \quad \cdot \quad \cdot \quad \cdot \quad \cdot \quad \cdot \quad \cdot \quad \cdot \quad \cdot \quad \cdot \quad \cdot \quad \cdot \quad \cdot \quad \cdot$$

$$\alpha_0(a_{s+1} - va_s) + \alpha_1(a_{s+2} - va_{s+1}) + \ldots + \alpha_s(a_{2s+1} - va_{2s}) = 0. \tag{9.41}$$

The number v is the greatest root of the characteristic equation

$$\begin{vmatrix} a_1 - va_0 & a_2 - va_1 & \ldots \ldots & a_{s+1} - va_s \\ a_2 - va_1 & a_3 - va_2 & \ldots \ldots & a_{s+2} - va_{s+1} \\ \cdot & \cdot & \cdot & \cdot \\ a_{s+1} - va_s & a_{s+2} - va_{s+1} & \ldots \ldots & a_{2s+1} - va_{2s} \end{vmatrix} = 0. \tag{9.42}$$

Once v is found, the numbers $\alpha_0, \alpha_1, \ldots, \alpha_s$ my be determined, up to a multiplicative factor which may be determined from the normalization condition.

No modification is needed to construct successive approximations to the vector e_n. To this end, we again need formula (9.40), the numbers α_1 being determined from the system (9.41); but here the number v is the *smallest* root of the characteristic equation (9.42). Geometric considerations show that the subspace Π_k "approaches" the required eigenvectors "fairly rapidly."

Solution of the characteristic equation is a troublesome stage in computation of the new approximation. It is natural to try to avoid this stage by "coarsening" the above scheme.

Let Π^* be a hyperplane through x_k and orthogonal to it. Let R_k be the intersection of Π^* with the $(s + 1)$-dimensional linear span of the vectors (9.39). Let P^* denote the orthogonal projection onto Π^*, and denote $P^*Ax_k, \ldots, P^*A^s x_k$ by $h_1, \ldots, h_s$. Obviously,

$$h_i = A^i x_k - \frac{(A^i x_k, x_k)}{(x_k, x_k)} x_k \qquad (i = 1, \ldots, s). \qquad (9.43)$$

As the next approximation we take the center of the s-dimensional ellipsoid formed by the intersection of R_k and the ellipsoid $(Ax, x) = 1$ (Fig. 9.2). Formulas for the vectors

$$x_{k+1} = x_k - \beta_1 h_1 - \ldots - \beta_s h_s \qquad (9.44)$$

may be derived as follows. The vector Ax_{k+1} must be orthogonal to R_k.

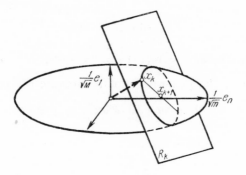

Fig. 9.2.

The numbers $\beta_1, \ldots, \beta_s$ are therefore determined by the system

$$(Ax_{k+1}, h_i) = 0 \qquad (i = 1, \ldots, s),$$

which is conveniently represented as

$$\left.\begin{array}{l}
\beta_1(h_1, h_1) + \beta_2(h_2, h_1) + \ldots + \beta_s(h_s, h_1) = (x_k, h_1), \\[2mm]
\beta_1(h_1, h_2) + \beta_2(h_2, h_2) + \ldots + \beta_s(h_s, h_2) = (x_k, h_2), \\[2mm]
\cdot \;\; \cdot \;\; \cdot \;\; \cdot \;\; \cdot \;\; \cdot \;\; \cdot \;\; \cdot \;\; \cdot \;\; \cdot \;\; \cdot \;\; \cdot \;\; \cdot \;\; \cdot \;\; \cdot \;\; \cdot \\[2mm]
\beta_1(h_1, h_s) + \beta_2(h_2, h_s) + \ldots + \beta_s(h_s, h_s) = (x_k, h_s).
\end{array}\right\} \qquad (9.45)$$

This process is a direct generalization of the method of orthogonal semichords. The analogue of the method of orthogonal chords is the iterative process

$$x_{k+1} = x_k - 2\beta_1 h_1 - \ldots - 2\beta_s h_s,$$ (9.46)

where $\beta_1, \ldots, \beta_s$ are again determined by the system (9.45).

No serious investigation of the processes (9.44) and (9.45) has as yet been undertaken.

§ 10. Monotone iterative processes*

10.1. *Statement of the problem.* In § 3 we described certain classes of equations with positive (linear and nonlinear) operators, for which the relevant iterative solution processes led to monotone successive approximations.

Here we shall dwell in greater detail on the linear equation

$$x = Ax + f,$$ (10.1)

where A is a positive linear operator relative to a solid normal cone K in a Banach space $E, f \in E$. We shall assume that $\rho(A) < 1$. Equation (10.1) therefore has a unique solution x^*, which is the limit of the successive approximations

$$x_{n+1} = Ax_n + f \qquad (n = 0, 1, 2, \ldots)$$ (10.2)

for any initial approximation $x_0 \in E$. Suppose that $x_0 = u_0$ is so chosen that

$$u_0 \leqslant Au_0 + f.$$ (10.3)

Then the successive approximations (10.3) (which, for convenience, we denote by u_n) satisfy the inequalities

$$u_0 \leqslant u_1 \leqslant \ldots \leqslant u_n \leqslant \ldots \leqslant x^*.$$ (10.4)

Similarly, if the initial approximation $x_0 = v_0$ is such that

$$v_0 \geqslant Av_0 + f,$$ (10.5)

the successive approximations v_n satisfy the inequalities

$$v_0 \geqslant v_1 \geqslant \ldots \geqslant v_n \geqslant \ldots \geqslant x^*.$$ (10.6)

* Stetsenko [7].

Thus, if we can find elements u_0 and v_0 satisfying (10.3) and (10.5), the result will be monotone approximations to x^*. The availability of two-sided monotone approximations also yields an error estimate: if we approximate the solution x^* by the element $\frac{1}{2}(u_n + v_n)$, then (10.4) and (10.6) imply the estimate

$$-\tfrac{1}{2}(v_n - u_n) \leqslant \tfrac{1}{2}(u_n + v_n) - x^* \leqslant \tfrac{1}{2}(v_n - u_n). \tag{10.7}$$

The principal difficulty is determination of the initial approximations u_0 and v_0. Below we shall indicate various means to this end.

In this section we shall also investigate the construction of monotone approximations to x^* which converge more rapidly than the sequences (10.4) and (10.6). Finally, we shall consider the contruction of monotone approximations for certain nonpositive operators A.

10.2. *Choice of initial approximations.* Let z_0 be an interior point of the cone K. It follows from Theorem 9.2 that, for some r,

$$A^r z_0 \leqslant \kappa z_0, \tag{10.8}$$

where $\kappa < 1$. Then the element

$$z_1 = \kappa^{1 - 1/r} z_0 + \kappa^{1 - 2/r} A z_0 + \ldots + A^{r-1} z_0, \tag{10.9}$$

satisfies the inequality

$$A z_1 \leqslant \kappa^{1/r} z_1. \tag{10.10}$$

The element (10.9) is an interior point of K, and so there exists a $b > 0$ such that

$$-b(1 - \kappa^{1/r}) z_1 \leqslant f \leqslant b(1 - \kappa^{1/r}) z_1. \tag{10.11}$$

Now set

$$u_0 = -bz_1, \qquad v_0 = bz_1. \tag{10.12}$$

These elements satisfy (10.3) and (10.5), respectively. Indeed, by (10.10),

$$Au_0 + f = -bAz_1 + f \geqslant -b\kappa^{1/r} z_1 + f,$$

and (10.3) follows from the left member of inequality (10.11). Similarly,

$$Av_0 + f = bAz_1 + f \leqslant b\kappa^{1/r} z_1 + f \leqslant b\kappa^{1/r} z_1 + b(1 - \kappa^{1/r}) z_1 = v_0.$$

Of course, there are various special cases in which u_0 and v_0 can be chosen more simply. For example, if $f \in K$, we may obviously take u_0 equal to f.

Exercise 10.1. Let $A = (a_{ij})$ $(i, j = 1, \ldots, n)$ be a matrix with nonnegative elements which satisfy the inequalities

$$a_{i1} + a_{i2} + \ldots + a_{in} \leqq \alpha_i < 1 \qquad (i = 1, \ldots, n).$$

Let f be a vector with nonnegative components $f_1, \ldots, f_n$. Show that the elements

$$u_0 = w_0 \cdot \min_i \frac{f_i}{1 - \alpha_i}, \quad v_0 = w_0 \cdot \max_i \frac{f_i}{1 - \alpha_i},$$

where $w_0 = \{1, \ldots, 1\}$, satisfy (10.3) and (10.5).

Exercise 10.2. Let

$$Ax(t) = \int_\Omega K(t, s) x(s) \, ds \tag{10.13}$$

where $K(t, s)$ is a nonnegative continuous kernel. Assume that the function

$$\alpha(t) = \int_\Omega K(t, s) \, ds$$

satisfies the inequality $\alpha(t) < 1$ $(t \in \Omega)$. Finally, let $f(t) \geqq 0$. Show that, when a semi-order relation is defined by the cone of nonnegative functions, conditions (10.3) and (10.5) hold for the (constant) functions

$$u_0(t) \equiv \min_{s \in \Omega} \frac{f(s)}{1 - \alpha(s)}, \quad v_0(t) \equiv \max_{s \in \Omega} \frac{f(s)}{1 - \alpha(s)}.$$

10.3. *Acceleration of convergence.* Let u_0 and v_0 satisfy (10.3) and (10.5), respectively. Moreover, assume that

$$Au_0 + f - u_0 \geqq p_1(v_0 - Av_0 - f), \tag{10.14}$$

$$v_0 - Av_0 - f \geqq q_1(Au_0 + f - u_0), \tag{10.15}$$

where one of the numbers p_1, q_1 is positive (when $p_1 = q_1 = 0$ these inequalities coincide with (10.3) and (10.5)). Define elements

$$u_1^* = \frac{Au_0 + f + p_1(Av_0 + f)}{1 + p_1}, \tag{10.16}$$

$$v_1^* = \frac{Av_0 + f + q_1(Au_0 + f)}{1 + q_1}. \tag{10.17}$$

Theorem 10.1.

$$Au_0 + f \leqslant u_1^* \leqslant x^* \leqslant v_1^* \leqslant Av_0 + f \,. \qquad (10.18)$$

Proof. The first and last inequalities in (10.18) follow immediately from (10.16), (10.17) and the inequality $Au_0 + f \leqslant Av_0 + f$. Obviously,

$$Au_1^* + f - u_1^* = \frac{1}{1 + p_1}[A(Au_0 + f - u_0) + p_1 A(Av_0 + f - v_0)]$$

and, by (10.14),

$$Au_1^* + f - u_1^* \geqslant \theta \,.$$

This means that u_1^* satisfies (10.3). As we already know, the latter condition implies that $u_1^* \leqslant x^*$. Similarly,

$$v_1^* - (Av_1^* + f) = \frac{1}{1 + q_1}[A(v_0 - Av_0 - f) + q_1(u_0 - Au_0 - f)] \geqslant \theta \,.$$

Thus v_1^* satisfies (10.5), and so $v_1^* \geqslant x^*$. ∎

Formulas (10.16) and (10.17) may be regarded as recurrence relations defining sequences u_n^*, v_n^*. Theorem 10.1 means that these sequences converge at least as rapidly as (10.4) and (10.6) and that the successive approximations are still monotone.

Examples show that, under fairly broad conditions, the sequences (10.6) and (10.7) converge much more rapidly than the usual method.

For example, consider the system

$$\xi = 0.68\xi + 0.40\eta + 1 \,, \qquad \eta = 0.30\xi + 0.60\eta + 1.5 \,.$$

The exact solution is $\xi = 125$, $\eta = 97.5$. The spectral radius of the matrix $\begin{pmatrix} 0.68 & 0.40 \\ 0.30 & 0.60 \end{pmatrix}$ is 0.9887. Let us first use the successive approximations (10.2) with $\xi_0 = \eta_0 = 10$. The result is

$$\xi_{307} = 121.20 \,, \qquad \eta_{307} = 94.59 \,,$$
$$\xi_{616} = 124.88 \,, \qquad \eta_{616} = 97.41 \,,$$
$$\xi_{993} = 124.99 \,, \qquad \eta_{998} = 97.49 \,.$$

But if we use (10.16) and (10.17), setting $u_0 = \{10, 10\}$, $v_0 = \{260, 200\}$, six steps are sufficient to yield the following estimates for ξ^*, η^*:

$$124.99 \leqq \xi^* \leqq 125.01 \,, \qquad 97.499 \leqq \eta^* \leqq 97.506 \,.$$

In a computation for the system (10.1), with

$$A = \begin{bmatrix} 0.40 & 0.16 & 0.12 & 0.07 & 0.03 \\ 0.16 & 0.26 & 0.31 & 0.19 & 0.07 \\ 0.12 & 0.31 & 0.34 & 0.31 & 0.12 \\ 0.07 & 0.19 & 0.31 & 0.26 & 0.16 \\ 0.03 & 0.07 & 0.12 & 0.16 & 0.04 \end{bmatrix}$$

and a free term $f = \{1, 0.1, 0.1, 0.1, 1\}$, $u_0 = f$ and $v_0 = \{10, 14, 17, 14, 10\}$, five iterations of (10.16) gave the same result as 167 iterations of (10.2).

In conclusion, note that the fact that the cone K is solid was used only in constructing elements u_0 and v_0 satisfying (10.3) and (10.5). All our arguments remain valid when the cone K is not solid, provided u_0 and v_0 can be found. The fact that the cone is normal will be used only in subsection 10.4.

10.4. *Equations with nonpositive operators.* Consider the equation

$$x = A_1 x - A_2 x + f, \tag{10.19}$$

where A_1 and A_2 are positive operators. Let us assume that

$$\rho(A_1 + A_2) < 1. \tag{10.20}$$

An example is the integral equation

$$x(t) = \int_\Omega K(t, s) x(s) \, ds + f(t)$$

with a continuous kernel $K(t, s)$ such that the spectral radius of the operator with kernel $|K(t, s)|$ is smaller than 1. A suitable operator A_1 is then the integral operator with kernel

$$K_1(t, s) = \max \{K(t, s), 0\},$$

and A_2 is the operator with kernel

$$K_2(t, s) = |K(t, s)| - K_1(t, s).$$

We claim that the solution of equation (10.19) can be approximated above and below by monotone sequences:

$$u_{n+1} = A_1 u_n - A_2 v_n + f \qquad (n = 0, 1, 2, \ldots), \tag{10.21}$$

$$v_{n+1} = A_1 v_n - A_2 u_n + f \qquad (n = 0, 1, 2, \dots), \qquad (10.22)$$

if u_0 and v_0 are suitably chosen.

Repeating the reasoning of subsection 10.2, one can construct an element z_1 which is an interior point of the cone K and satisfies the inequality

$$(A_1 + A_2) z_1 \leqslant q z_1, \qquad (10.23)$$

with $q < 1$. One can then choose $b > 0$ such that

$$- b(1 - q) z_1 \leqslant f \leqslant b(1 - q) z_1. \qquad (10.24)$$

Define elements u_0 and v_0 by (10.12):

$$u_0 = - b z_1, \qquad v_0 = b z_1.$$

A simple calculation shows that in this case the sequences (10.21) and (10.22) satisfy the inequalities

$$u_0 \leqslant u_1 \leqslant \dots \leqslant u_n \leqslant \dots \leqslant v_n \leqslant \dots \leqslant v_1 \leqslant v_0. \qquad (10.25)$$

Since

$$v_{n+1} - u_{n+1} = (A_1 + A_2)(v_n - u_n),$$

it follows from (10.20) that

$$\lim_{n \to \infty} \|v_n - u_n\| = 0. \qquad (10.26)$$

It follows from (10.25) that, for any $m > 0$,

$$\theta \leqslant u_{n+m} - u_n \leqslant v_n - u_n$$

and, since the cone K is normal,

$$\|u_{n+m} - u_n\| \leqq N \|v_n - u_n\|.$$

Thus (10.26) implies that u_n is a Cauchy sequence. Its limit x^* is (again by (10.26)) also the limit of the sequence v_n. One can thus extend the inequalities (10.25), adding the inequalities $u_n \leqslant x^* \leqslant v_n$.

Letting $n \to \infty$ in (10.21), we see that x^* is a solution of equation (10.19). We have thus proved

Theorem 10.2. *The successive approximations* (10.21) *and* (10.22) *converge monotonically to a solution x^* of equation* (10.19).

Exercise 10.3. Show that Theorem 10.2 remains valid if u_0 and v_0 are arbitrary elements such that $u_0 \leqslant v_0$ and

$$u_0 \leqslant A_1 u_0 - A_2 v_0 + f, \tag{10.27}$$

$$v_0 \geqslant A_1 v_0 - A_2 u_0 + f. \tag{10.28}$$

Exercise 10.4. Let u_0 and v_0 satisfy (10.27) and (10.28). Moreover, let

$$v_0 - A_1 v_0 + A_2 u_0 - f \geqslant m(A_1 u_0 - A_2 v_0 + f - u_0),$$

$$A_1 u_0 - A_2 v_0 + f - u_0 \geqslant m(v_0 - A_1 v_0 + A_2 u_0 + f),$$

where $m = m(u_0, v_0) > 0$. Define elements

$$u_1^* = \frac{1}{1+m} [A_1 u_0 - A_2 v_0 + f + m(A_1 v_0 - A_2 u_0 + f)], \tag{10.29}$$

$$v_1^* = \frac{1}{1+m} [A_1 v_0 - A_2 u_0 + f + m(A_1 u_0 - A_2 v_0 + f)]. \tag{10.30}$$

Show that

$$u_1 \leqslant u_1^* \leqslant x^* \leqslant v_1^* \leqslant v_1, \tag{10.31}$$

where u_1 and v_1 are defined by (10.21) and (10.22).

Exercise 10.5. Formulas (10.29) and (10.30) may be regarded as recurrence relations defining sequences u_n^* and v_n^*. Show that the resulting approximations converge monotonically (from above and below) to x^*.

Equations with smooth operators

§ 11. The Newton–Kantorovich method

11.1. *Linearization of equations.* Consider the equation

$$Fx = 0, \qquad (11.1)$$

where F is a Fréchet-differentiable nonlinear operator mapping a subset $\mathfrak{M}$ of a Banach space E_1 into a Banach space E_2. We shall find it convenient to assume that $\mathfrak{M}$ is a ball. The principal method for constructing successive approximations x_n to a solution x^* (if it exists) of equation (11.1) is based on successive linearization of the equation.

Suppose that the approximation x_n has been found. To determine the next approximation x_{n+1}, we replace equation (11.1) by an equation which is linearized at x_n:

$$Fx_n + F'(x_n)(x - x_n) = 0. \qquad (11.2)$$

If the linear operator $[F'(x_n)]^{-1}$, which maps a subset of E_2 into E_1, is defined, the approximation x_{n+1} follows immediately:

$$x_{n+1} = x_n - [F'(x_n)]^{-1}Fx_n \qquad (n = 0, 1, 2, \ldots). \qquad (11.3)$$

The method defined by (11.3) is known as the *Newton–Kantorovich method*, and is the subject of an extensive literature; we mention Kantorovich [2], Kantorovich and Akilov [1], Kantorovich [3–7], Mysovskikh [1, 2], Vertgeim [1–3], Gavurin [3], Kivistik [1–3], Polyak [1].

There are various aspects to the Newton–Kantorovich method. First, one is interested in effectively verifiable conditions for its applicability. Second, both estimates of convergence rate and *a priori* error estimates are important. Third, one needs methods for determining initial

138

approximations x_0 for which the method converges. A final important factor is the degree of "stability" of the method with respect to both random computational errors and errors incurred by replacing (11.3) by other approximate methods for solution of (11.2).

11.2. *Convergence.* Consider the operator

$$Ax = x - \Gamma(x) Fx, \tag{11.4}$$

where

$$\Gamma(x) = [F'(x)]^{-1}. \tag{11.5}$$

The Newton–Kantorovich method may then be regarded as the usual iterative method

$$x_{n+1} = Ax_n \quad (n = 0, 1, 2, \dots) \tag{11.6}$$

for approximate solution of the equation

$$x = Ax. \tag{11.7}$$

All the results of Chapter 1 are therefore applicable to investigations of convergence and convergence rates in the present case. We shall see below that, under natural assumptions, the derivative $A'(x^*)$ in the solution x^* of equation (11.7) vanishes, and we may therefore use the estimates of subsection 2.5.

Assume that the linear operator $F'(x^*)$ has a bounded inverse, and

$$\lim_{\|x - x^*\| \to 0} \| F'(x) - F'(x^*) \| = 0. \tag{11.8}$$

Then it is easily seen that, at points x close to x^*, the linear operators $F'(x)$ also have bounded inverses $\Gamma(x)$, and

$$\lim_{\|x - x^*\| \to 0} \| \Gamma(x) - \Gamma(x^*) \| = 0. \tag{11.9}$$

It follows from the identity

$$Ax - Ax^* = \Gamma(x) [F'(x^*)(x - x^*) - Fx + Fx^*] +$$

$$+ \Gamma(x) [F'(x) - F'(x^*)](x - x^*) \tag{11.10}$$

that

$$\frac{\|Ax - Ax^*\|}{\|x - x^*\|} \leq \|\Gamma(x)\| \left[\frac{\|Fx - Fx^* - F'(x^*)(x - x^*)\|}{\|x - x^*\|} + \right.$$

$$\left. + \|F'(x) - F'(x^*)\| \right].$$

Therefore

$$\lim_{\|x - x^*\| \to 0} \frac{\|Ax - Ax^*\|}{\|x - x^*\|} = 0,$$

i.e., $A'(x^*) = 0$. Theorem 2.3 now implies

Theorem 11.1. *Let x^* be a solution of equation* (11.1). *If the linear operator $F'(x^*)$ has a bounded inverse and* (11.8) *holds, then the successive approximations* (11.3) *converge to x^*, provided the initial approximation x_0 is sufficiently close to x^*. The rate of convergence is described by the inequality*

$$\|x_n - x^*\| \leq c(x_0; \varepsilon) \varepsilon^n, \tag{11.11}$$

where ε is any positive number.

In the general case, this theorem cannot be strengthened, in the sense that for every sequence of positive numbers α_n such that

$$\lim_{n \to \infty} \frac{\alpha_{n+1}}{\alpha_n} = 0, \tag{11.12}$$

there is an equation for which the Newton–Kantorovich method converges less rapidly than α_n. The following example is from a paper of Levin and Strygin [1].

Set

$$t_n = \begin{cases} \alpha_{n/2}, & \text{if } n \text{ is even,} \\ \sqrt{\alpha_{(n-1)/2} \, \alpha_{(n+1)/2}}, & \text{if } n \text{ is odd.} \end{cases}$$

Clearly, $t_n \to 0$, $t_{n+1}/t_n \to 0$, and

$$\lim_{n \to \infty} \frac{\alpha_n}{t_{n+k}} = 0 \qquad (k = 1, 2, \dots). \tag{11.13}$$

Define a scalar function $F(t)$ by

$$F(t) = \begin{cases} 0, & \text{if } t = 0, \\ \\ t_{n+1} - t_{n+2} + \left(1 + \dfrac{t_{n+1} - t_{n+2}}{t_n - t_{n+1}}\right)(t - t_{n+1}) + \\ \\ \quad + \dfrac{t_{n+1} - t_{n+2}}{2\pi} \sin\left(2\pi \dfrac{t - t_{n+1}}{t_n - t_{n+1}}\right), & \text{if } t_{n+1} \leqq t \leqq t_n, \\ \\ -F(-t), & \text{if } t < 0. \end{cases}$$

It is easily verified that $F(t)$ is continuous and continuously differentiable in the interval $[-t_1, t_1]$, and $F'(0) = 1$. Construct successive approximations x_n by (11.3), with the initial approximation $x_0 = t_k$, k being fixed. Direct calculation shows that $x_n = t_{n+k}$ (using the fact that $F'(t_n) = 1$). It follows from (11.13) that, for any m,

$$\lim_{n \to \infty} \frac{x_{n+m}}{\alpha_n} = \infty .$$

Thus the approximations x_n converge to the zero solution of the equation $F(t) = 0$, less rapidly than the sequence α_n.

11.3. *Further investigation of convergence rate.* Theorem 11.1 can be considerably sharpened if some index of the smoothness of $F'(x)$ as a function of x (with values in the space of linear operators) is available. We confine ourselves to the simplest case.

Assume that $F'(x)$ satisfies a Lipschitz condition in some neighborhood of a solution x^* of equation (11.1):

$$\|F'(x) - F'(y)\| \leqq L \|x - y\| . \tag{11.14}$$

Then the identity

$$Fx - Fy - F'(y)(x - y) = \int_0^1 \{F'[y + t(x - y)] - F'(y)\}(x - y)\, dt$$

implies the inequality

$$\|Fx - Fy - F'(y)(x - y)\| \leqq \frac{L}{2} \|x - y\|^2 \tag{11.15}$$

Exercise 11.1. Assume that the operator F satisfies a Hölder condition

$$\|F'(x) - F'(y)\| \leqq L \|x - y\|^{\alpha},$$ (11.16)

with $0 < \alpha < 1$, on some convex set. Show that for any x, y in this set

$$\|Fx - Fy - F'(y)(x - y)\| \leqq \frac{L}{1 + \alpha} \|x - y\|^{1 + \alpha}.$$ (11.17)

Setting $y = x_n$, $x = x^*$, we can rewrite (11.15) as

$$\|F'(x_n)(x_{n+1} - x^*)\| \leqq \frac{L}{2} \|x_n - x^*\|^2.$$ (11.18)

Therefore (see (2.25)):

Theorem 11.2. *If under the assumptions of Theorem* 11.1, *condition* (11.14) *is satisfied in some neighborhood of a solution* x^* *of equation* (11.1), *then*

$$\|x_n - x^*\| \leqq c(\delta; x_0)\delta^{2^{n - n_0(\delta)}},$$ (11.19)

where δ *is an arbitrarily small positive number.*

It is clear that for fixed δ the number $n_0(\delta)$ may be assumed equal to zero if x_0 is sufficiently close to x^*.

Exercise 11.2. Under the assumptions of Theorem 11.1, assume that (11.16) is satisfied in a neighborhood of x^*. Show that

$$\|x_n - x^*\| \leqq c(\delta; x_0)\delta^{(1 + \alpha)^{n - n_0(\delta)}}$$ (11.20)

Stirling's formula and (11.19) imply that, for large n,

$$\|x_n - x^*\| \leqq (n!)^{-2^n n^{-2}}.$$ (11.21)

Exericse 11.3. Show that, under the assumptions of Exercise 11.2,

$$\|x_n - x^*\| \leqq (n!)^{-(1 + \alpha)^n n^{-2}}$$ (11.22)

for large n.

11.4. *Global convergence condition.* Let $S(x, r)$ denote the ball of radius r centered at x. We shall assume that the operator F is defined in a ball $S(x_0, R)$, where x_0 is an initial approximation for the sequence (11.3).

Let us assume that the operator F is differentiable in $S(x_0, R)$ and its derivative $F'(x)$ satisfies a condition of type (11.14) there. Throughout the sequel we shall assume that the operator $\Gamma_0 = \Gamma(x_0)$ exists (where $\Gamma(x)$ is defined by (11.5)).

Theorem 11.3. *Assume that*

$$\|\Gamma_0\| \leqq b_0, \|\Gamma_0 F x_0\| \leqq \eta_0, \qquad h_0 = b_0 L \eta_0 \leqq \tfrac{1}{2}. \tag{11.23}$$

If

$$R \geqq r_0 = \frac{1 - \sqrt{1 - 2h_0}}{h_0} \eta_0, \tag{11.24}$$

then the successive approximations (11.3) *converge to a solution* x^* *of equation* (11.1) *in the ball* $S(x_0, r_0)$.

There are several proofs of this theorem; we present one due to Kantorovich.

Define a number sequence

$$b_{n+1} = \frac{b_n}{1 - h_n}, \qquad \eta_{n+1} = \frac{h_n}{2(1 - h_n)} \eta_n, \qquad h_{n+1} = b_{n+1} L \eta_{n+1},$$

$$r_{n+1} = \frac{1 - \sqrt{1 - 2h_{n+1}}}{h_{n+1}} \eta_{n+1}.$$

We claim that under the assumptions (11.23), (11.24) the successive approximations (11.3) exist; moreover

$$\|\Gamma(x_n)\| \leqq b_n, \qquad \|\Gamma(x_n) F x_n\| \leqq \eta_n, \qquad h_n \leqq \tfrac{1}{2} \tag{11.25}$$

and

$$S(x_n, r_n) \subset S(x_{n-1}, r_{n-1}). \tag{11.26}$$

The proof is by induction. Assume that (11.25) and (11.26) hold for $n = m$. Since $\|x_{m+1} - x_m\| = \|\Gamma(x_m) F x_m\| \leqq \eta_m$, it follows from the definition of r_m that $x_{m+1} \in S(x_m, r_m)$; a fortiori, $x_{m+1} \in S(x_0, R)$. The derivative $F'(x_{m+1})$ therefore exists. By (11.14),

$$\|\Gamma(x_m) [F'(x_{m+1}) - F'(x_m)]\| \leqq b_m L \|x_{m+1} - x_m\| \leqq h_m \leqq \tfrac{1}{2};$$

the operator $\Gamma(x_{m+1})$ therefore exists, and has the representation

$$\Gamma(x_{m+1}) = \{I + \Gamma(x_m) [F'(x_{m+1}) - F'(x_m)]\}^{-1} \Gamma(x_m) =$$

$$= \sum_{i=0}^{\infty} (-1)^i \{\Gamma(x_m) [F'(x_{m+1}) - F'(x_m)]\}^i \Gamma(x_m).$$

Hence

$$\left\|\Gamma(x_{m+1})\right\| \leq \sum_{i=0}^{\infty} \left\|\Gamma(x_m)\left[F'(x_{m+1}) - F'(x_m)\right]\right\|^i b_m \leq \frac{b_m}{1 - h_m} = b_{m1}.$$
(11.27)

Now consider the second inequality of (11.25) (for $n = m + 1$). It follows from the identity

$$Fx_{m+1} = Fx_{m+1} - Fx_m - F'(x_m)(x_{m+1} - x_m)$$

and from (11.15) that

$$\left\|Fx_{m+1}\right\| \leq \frac{L}{2}\left\|x_{m+1} - x_m\right\|^2 \leq \frac{L}{2}\eta_m^2,$$
(11.28)

and, by (11.27),

$$\left\|\Gamma(x_{m+1})Fx_{m+1}\right\| \leq \frac{b_m L \eta_m^2}{2(1 - h_m)} = \frac{h_m}{2(1 - h_m)}\eta_m = \eta_{m+1}.$$

The third inequality of (11.25) is easily proved; by definition,

$$h_{m+1} = b_{m+1}L\eta_{m+1} = \frac{b_m}{1 - h_m}L\frac{h_m}{2(1 - h_m)}\eta_m = \frac{h_m^2}{2(1 - h_m)^2} \leq \frac{1}{2}.$$

To prove the inclusion (11.26), it suffices to note that if $\|x - x_{k+1}\| \leq r_{k+1}$ then

$$\|x - x_k\| \leq \|x - x_{k+1}\| + \|x_{k+1} - x_k\| \leq r_{k+1} + \eta_k,$$

since the right-hand side is identically equal to r_k (the simple computation is left to the reader).

Thus the successive approximations (11.3) are all well defined.

The third inequality of (11.25) implies that $\eta_{m+1} \leq \frac{1}{2}\eta_m$; therefore $r_n \to 0$ as $n \to \infty$. Thus the successive approximations converge to some point $x^* \in S(x_0, r_0)$. To complete the proof, it suffices to let $m \to \infty$ in (11.28). ∎

It is clear from the proof that, under the assumptions of Theorem 11.3,

$$\|x_n - x^*\| \leq r_n \qquad (n = 1, 2, \ldots).$$

Exercise 11.4. Under the assumptions of Theorem 11.3, prove that

$$\|x_n - x^*\| \leq \frac{1}{2^n}(2h_0)^{2^n - 1}\eta_0.$$

Exercise 11.5. Assume that the derivative $F'(x)$ satisfies a Hölder condition (11.16) in the ball $S(x_0, R)$. Let $h_0 = b_0 L \eta_0^\alpha \leqq \beta_0$, where β_0 is a root of the equation

$$\left(\frac{\beta}{1+\alpha}\right)^\alpha = (1-\beta)^{1+\alpha} \qquad (0 \leqq \beta \leqq 1), \qquad (11.29)$$

and let $R \geqq r_0 = \eta_0/(1 - \gamma_0)$, where $\gamma_0 = h_0/(1 + \alpha)(1 - h_0)$. Prove that the successive approximations (11.3) converge to a solution x^* of equation (11.1) in the ball $S(x_0, r_0)$.

11.5. *Simple zeros.* It is natural to call a solution x^* of equation (11.1) a *simple zero* of the operator F if the operator $\Gamma(x^*)$ exists and is continuous.

Theorem 11.4. *If, under the assumptions of Theorem* 11.3, $h_0 < \frac{1}{2}$, *then the zero* x^* *of* F *to which the successive approximations* (11.3) *converge is simple.*

To prove this it suffices to note that $r_0 < (Lb_0)^{-1}$ for $h_0 < \frac{1}{2}$, and that $\|x - x_0\| < (Lb_0)^{-1}$ implies $\|\Gamma_0 F'(x) - I\| < 1$. Thus both operators $\Gamma_0 F'(x)$ and $F'(x)$ are invertible.

Note that when $h_0 = \frac{1}{2}$ the successive approximations may converge to a "multiple" zero. An example is the scalar equation $x^2 = 0$ for any $x_0 \neq 0$.

§ 12. Modified Newton–Kantorovich method

12.1. *The modified method.* The basic defect of the method (11.3) is that each step involves the solution of a linear equation with a different linear operator $F'(x_n)$. For this reason, one often constructs successive approximations which employ linear equations other than (11.2), though similar to them.

The most frequently used substitute for (11.2) is the equation

$$Fx_n + F'(x_0)(x - x_n) = 0, \qquad (12.1)$$

where x_0 is the initial approximation. The successive approximations are then defined by the recurrence relation

$$x_{n+1} = x_n - \Gamma_0 Fx_n \qquad (n = 0, 1, 2, \dots). \qquad (12.2)$$

Recall that $\Gamma_0 = [F'(x_0)]^{-1}$.

Formula (12.2) describes the *modified Newton–Kantorovich method* for approximate solution of the equation

$$Fx = 0. \tag{12.3}$$

The method (12.2) coincides with the usual iterative method

$$x_{n+1} = Ax_n \qquad (n = 0, 1, 2, \ldots) \tag{12.4}$$

for approximate solution of the equation

$$x = Ax, \tag{12.5}$$

where

$$Ax = x - \Gamma_0 Fx. \tag{12.6}$$

12.2. *Fundamental theorem.* We shall assume that the operator F is defined and Fréchet-differentiable in a ball $S(x_0, R) = \{x : \|x - x_0\| < R\}$, in which the derivative $F'(x)$ satisfies a Lipschitz condition

$$\|F'(x) - F'(y)\| \leqq L \|x - y\|. \tag{12.7}$$

Moreover, assume that $\Gamma_0 = [F'(x_0)]^{-1}$ exists, and let

$$\|\Gamma_0\| \leqq b_0, \qquad \|\Gamma_0 F x_0\| \leqq \eta_0 \tag{12.8}$$

Theorem 12.1. If

$$h_0 = b_0 L \eta_0 < \tfrac{1}{2} \tag{12.9}$$

and

$$r_0 = \frac{1 - \sqrt{1 - 2h_0}}{h_0} \eta_0 \leqq R, \tag{12.10}$$

then the successive approximations (12.2) *converge to a solution* $x^* \in S(x_0, r_0)$ *of equation* (12.3).

Proof. First note that equation (12.3) indeed has a solution in the ball $S(x_0, r_0)$—this follows from Theorem 11.3. Below we shall prove that the operator (12.6) satisfies the assumptions of the contracting mapping principle in the ball $S(x_0, r_0)$. This will imply that the solution x^* in the ball $S(x_0, r_0)$ is unique, and that the approximations (12.2) converge.

Obviously, for any x, $y \in S(x_0, r)$ $(r \leq R)$,

$$Ax - Ay = x - y - \Gamma_0(Fx - Fy) =$$

$$= \Gamma_0 \int_0^1 \{F'(x_0) - F'[y + t(x - y)]\}(x - y)\, dt. \qquad (12.11)$$

This identity, together with (12.7), implies the estimate

$$\|Ax - Ay\| \leq b_0 L r \|x - y\|. \qquad (12.12)$$

Consequently, A is a contraction operator in the ball $S(x_0, r_0)$. To complete the proof, it remains to show that

$$AS(x_0, r_0) \subset S(x_0, r_0).$$

Let $\|x - x_0\| \leq R$. Then, by (12.11),

$$\|Ax - x_0\| \leq \|Ax - Ax_0\| + \|Ax_0 - x_0\| \leq$$

$$\leq \left\| \Gamma_0 \int_0^1 \{F'(x_0) - F'[x_0 + t(x - x_0)]\}(x - x_0)\, dt \right\| + \eta_0$$

and, again by (12.7),

$$\|Ax - x_0\| \leq b_0 L \int_0^1 t\, dt \|x - x_0\|^2 + \eta_0 = \frac{b_0 L \|x - x_0\|^2}{2} + \eta_0.$$

Therefore, when $\|x - x_0\| \leq r_0$,

$$\|Ax - x_0\| \leq \frac{b_0 L r_0^2}{2} + \eta_0 = r_0. \quad \blacksquare$$

Note that, by (12.12), the operator A satisfies a Lipschitz condition with constant $q = 1 - \sqrt{1 - 2h_0}$ in the ball $S(x_0, r_0)$. The modified method thus converges at the rate of a geometric progression with quotient $1 - \sqrt{1 - 2h_0}$.

The above analysis of the modified Newton–Kantorovich method relates to the simplest case. More subtle arguments (see, e.g., Kantorovich and Akilov [1]) show that Theorem 12.1 remains valid if the sign $<$ in (12.9) is replaced by $\leq$.

12.3. *Uniqueness ball.* Consider the operator A defined by (12.6). Assume that the conditions of Theorem 12.1 hold, and set

$$\alpha(r) = \sup_{\|x - x_0\| \leq r} \|Ax - x_0\| . \tag{12.13}$$

The function $\alpha(r)$ is obviously continuous and nondecreasing. It was shown in the proof of Theorem 12.1 that

$$\|Ax - x_0\| \leq \frac{b_0 L \|x - x_0\|^2}{2} + \eta_0 \qquad (\|x - x_0\| \leq R). \tag{12.14}$$

Hence follows

Lemma 12.1. *The function $\alpha(r)$ satisfies the inequality*

$$\alpha(r) \leq \frac{b_0 L r^2}{2} + \eta_0 \qquad (r_0 \leq r \leq R). \tag{12.15}$$

Theorem 12.2. *If, under the assumptions of Theorem* 11.1,

$$r_0 = \frac{1 - \sqrt{1 - 2h_0}}{h_0} \eta_0 \leq R < \frac{1 + \sqrt{1 - 2h_0}}{h_0} \eta_0 , \tag{12.16}$$

equation (12.3) *has a unique solution in the ball* $S(x_0, R)$.

To prove this, note that when $r < \|x - x_0\| < R$, Lemma 12.1 gives

$$\|Ax - x_0\| \leq \alpha(\|x - x_0\|) \leq \frac{b_0 L \|x - x_0\|^2}{2} + \eta_0 < \|x - x_0\| , \tag{12.17}$$

as the quadratic trinomial $\frac{1}{2} b_0 L r^2 - r + \eta_0$ is negative in the interval

$$\left(r_0, \frac{1 + \sqrt{1 - 2h_0}}{h_0} \eta_0 \right).$$

Exercise 12.1. Show that, under the assumptions of Theorem 12.2, the successive approximations $\tilde{x}_{n+1} = A\tilde{x}_n$ converge to a solution of equation (12.3) for any initial approximation $\tilde{x}_0 \in S(x_0, R)$.

12.4. *Modified method with perturbations.* Consider the equation

$$x = Ax + A_1 x , \tag{12.18}$$

where A is the operator (12.6) and A_1 is any nonlinear operator defined in the ball $S(x_0, R)$ $(R > r_0)$ such that

$$\|A_1 x\| \leq a_0 \qquad (x \in S(x_0, R)) \tag{12.19}$$

and

$$\|A_1x - A_1y\| \leqq q_0 \|x - y\| \qquad (x, y \in S(x_0, R)), \qquad (12.20)$$

where $0 < q_0 < 1$. If a_0 and q_0 are sufficiently small, equation (12.18) approximates equation (12.5).

Theorem 12.3. *If the numbers a_0 and q_0 satisfy the inequality*

$$2b_0La_0 + q_0^2 < 1 - 2b_0L\eta_0 \qquad (12.21)$$

and

$$r^{**} = \frac{1 - \sqrt{1 - 2b_0L(\eta_0 + a_0)}}{b_0L} \leqq R, \qquad (12.22)$$

equation (12.18) has a solution x^{**} in the ball $S(x_0, r^{**})$; this solution is the limit of successive approximations

$$x_{n+1} = Ax_n + A_1x_n \qquad (n = 0, 1, 2, \ldots). \qquad (12.23)$$

The approximations (12.23) converge at the rate of a geometric progression with quotient

$$q = q_0 + b_0Lr^{**}. \qquad (12.24)$$

Proof. By (12.21), the number r^{**} is smaller than the real root of the quadratic equation

$$\frac{b_0Lr^2}{2} - r + (\eta_0 + a_0) = 0.$$

Therefore, when $\|x - x_0\| \leqq r^{**}$ it follows from Lemma 12.1 that

$$\|(A + A_1)x - x_0\| \leqq \|Ax - x_0\| + \|A_1x\| \leqq \frac{b_0L(r^{**})^2}{2} +$$

$$+ \eta_0 + a_0 = r^{**}.$$

This means that the operator $D = A + A_1$ maps the ball $S(x_0, r^{**})$ into itself.

It follows from (12.12) and (12.20) that

$$\|Dx - Dy\| \leqq (b_0Lr^{**} + q_0) \|x - y\| \qquad (x, y \in S(x_0, r^{**})).$$

Hence D is a contraction operator in the ball $S(x_0, r^{**})$, since, by (12.21),

$$b_0Lr^{**} + q_0 = 1 - \sqrt{1 - 2b_0L(\eta_0 + a_0)} + q_0 < 1.$$

Thus D satisfies the assumptions of the contracting mapping principle in the ball $S(x_0, r^{**})$. ∎

Exercise 12.2. Under the assumptions of Theorem 12.3, show that the solution x^{**} of equation (12.8) is unique in any ball $S(x, r)$, where

$$r \leqq R \text{ and } r < \frac{1 + \sqrt{1 - 2b_0 L(\eta_0 + a_0)}}{b_0 L}.$$

Exercise 12.3. Show that if r satisfies the assumptions of Exercise 12.2 the successive approximations $\tilde{x}_{n+1} = (A + A_1)\tilde{x}_n$ converge to a solution x^{**} of equation (12.8) for any initial approximation $\tilde{x}_0 \in S(x_0, r)$.

12.5. *Equations with compact operators.* Assume that equation (12.3) has the form

$$x = Bx, \tag{12.25}$$

where B is a compact operator in a Banach space E. In this case formulas (12.21) are

$$x_{n+1} = x_n - [I - B'(x_0)]^{-1}(x_n - Bx_n) \qquad (n = 0, 1, 2, \ldots). \tag{12.26}$$

It is sometimes convenient to write them as

$$x_{n+1} = [I - B'(x_0)]^{-1}[Bx_n - B'(x_0)x_n] \qquad (n = 0, 1, 2, \ldots). \tag{12.27}$$

Since the operator B is compact, so is the linear operator $B'(x_0)$ (see Krasnosel'skii [8]). In "well-behaved" Banach spaces, a linear compact operator may be approximated up to any desired accuracy by a finite-dimensional linear operator. In other words, for any $\delta > 0$ there is a finite-dimensional linear operator C_δ such that

$$\|B'(x_0) - C_\delta\| < \delta. \tag{12.28}$$

Assume that the space E has a basis $e_1, \ldots, e_n, \ldots$ Then any element $x \in E$ has a representation

$$x = \sum_{i=1}^{\infty} \xi_i(x) e_i.$$

Set

$$P_m x = \sum_{i=1}^{m} \xi_i(x) e_i \qquad (x \in E). \tag{12.29}$$

It is well known (see, e.g., Lyusternik and Sobolev [1]) that

$$\lim_{m \to \infty} \|B'(x_0) - P_m B'(x_0)\| = 0 \ ;$$

thus the role of the operator C_δ may be fulfilled by $C_\delta = P_m B'(x_0)$ if m is sufficiently large.

Replacing the operator $[I - B'(x_0)]^{-1}$ in (12.26) and (12.27) by $(I - C_\delta)^{-1}$, we get the iterative processes

$$x_{n+1} = x_n - (I - C_\delta)^{-1}(x_n - Bx_n) \qquad (n = 0, 1, 2, \dots) \qquad (12.30)$$

and

$$x_{n+1} = (I - C_\delta)^{-1}[Bx_n - B'(x_0)x_n] \qquad (n = 0, 1, 2, \dots). \qquad (12.31)$$

It is natural to expect the processes (12.30) and (12.31) to converge to a solution x^* of equation (12.25) for small δ. We shall consider (12.30).

It is evident that computations using formulas (12.30) are much simpler than those using formulas (12.26) (the modified Newton–Kantorovich method). The reason is that determination of the operator $(I - C_\delta)^{-1}$ is equivalent to the inversion of a certain matrix of finite order; in other words, to find the values of the operator $(I - C_\delta)^{-1}$ one need only solve a finite system of linear algebraic equations.

In fact, let the operator C_δ be given by

$$C_\delta x = \sum_{i=1}^{m} \eta_i(x) e_i, \qquad (12.32)$$

where $e_1, \dots, e_m$ is a basis of the range of the operator C_δ, while $\eta_1(x), \dots, \eta_m(x)$ are suitable linear functionals. Let $g_1, \dots, g_m$ be elements in E such that $C_\delta g_i = e_i$ $(i = 1, \dots, m)$, and set

$$Px = \sum_{i=1}^{m} \eta_i(x) g_i, \qquad Qx = x - Px. \qquad (12.33)$$

Obviously,

$$C_\delta Px = C_\delta x \qquad (x \in E), \qquad (12.34)$$

and therefore

$$C_\delta Qx = 0 \qquad (x \in E). \qquad (12.35)$$

Assume that 1 is not an eigenvalue of C_δ. Then, for any $f \in E$, the element

$$z = (I - C_\delta)^{-1} f,$$

which is a solution of the equation

$$z = C_\delta z + f \tag{12.36}$$

or, equivalently (by (12.34)), of the equation

$$z = C_\delta P z + f \tag{12.37}$$

is well defined. It follows from the last equation that Pz may be determined from the equation

$$Pz = PC_\delta Pz + Pf . \tag{12.38}$$

Equation (12.38) is a finite system of linear algebraic equations: if

$$Pz = \xi_1 g_1 + \ldots + \xi_m g_m ,$$

the coefficients $\xi_1, \ldots, \xi_m$ are determined by the equations

$$\left.\begin{array}{l} \xi_1 = \eta_1(e_1)\,\xi_1 + \eta_1(e_2)\,\xi_2 + \ldots + \eta_1(e_m)\,\xi_m + \eta_1(f), \\[2mm] \xi_2 = \eta_2(e_1)\,\xi_1 + \eta_2(e_2)\,\xi_2 + \ldots + \eta_2(e_m)\,\xi_m + \eta_2(f), \\[2mm] \cdot\ \cdot \\[2mm] \xi_m = \eta_m(e_1)\,\xi_1 + \eta_m(e_2)\,\xi_2 + \ldots + \eta_m(e_m)\,\xi_m + \eta_m(f). \end{array}\right\} \tag{12.39}$$

The system (12.39) has a unique solution, since it is solvable for any free terms $\eta_i = \eta_i(f)$. Once Pz is known, z may be determined from (12.37).

In many cases the operator C_δ is, as we have already said, determined by the formula

$$C_\delta = P_m B'(x_0), \tag{12.40}$$

where P_m is the projection on a certain finite-dimensional subspace $E_m \subset E$. Then $(I - P_m)C_\delta = 0$, and it therefore follows from (12.36) that $z - P_m z = f - P_m f$. As for the element $P_m z$, it is determined from the equation

$$P_m z = C_\delta P_m z + C_\delta(f - P_m f) + P_m f ,$$

which is conveniently written as

$$(P_m - C_\delta P_m)(P_m z - P_m f) = C_\delta f . \tag{12.41}$$

Now this is an equation in a finite-dimensional space; its solution reduces to solution of a system of linear algebraic equations or determination of the inverse $(I_m - P_m C_\delta P_m)^{-1}_{E_m}$ of the operator $(I_m - P_m C_\delta P_m)^{-1}_{E_m}$ in the subspace E_m. By (12.41),

$$P_m z = (I_m - P_m C_\delta P_m)^{-1}_{E_m} C_\delta f + P_m f,$$

and the element $(I - C_\delta)^{-1} f = z = P_m z + f - P_m f$ is therefore determined by the simple formula

$$z = f + (I_m - P_m C_\delta P_m)^{-1}_{E_m} C_\delta f . \tag{12.42}$$

The iterations defined by (12.30) are precisely ordinary iterations for the equation $x = Dx$, where

$$Dx = x - (I - C_\delta)^{-1}(x - Bx).\qquad(12.43)$$

To study the process (12.30), we shall treat the equation $x = Dx$ as an equation of type (12.18). It is immediate that the solutions of the equation $x = Dx$ coincide with those of (12.25).

Writing the equation $x = Dx$ as (12.18), we have

$$A_1 x = \{[I - B'(x_0)]^{-1} - (I - C_\delta)^{-1}\}(x - Bx).$$

Assume, as usual, that $\|[I - B'(x_0)]^{-1}\| \leqq b_0$. It then follows from (12.28) that, for $b_0\delta < 1$, the operator $(I - C_\delta)^{-1}$ is defined, and

$$\|(I - C_\delta)^{-1}\| \leqq \frac{b_0}{1 - b_0\delta}.\qquad(12.44)$$

The identity

$$A_1 x = (I - C_\delta)^{-1}[B'(x_0) - C_\delta][I - B'(x_0)]^{-1}(x - Bx)$$

implies the estimate

$$\|A_1 x\| \leqq \frac{b_0\delta}{1 - b_0\delta}\|[I - B'(x_0)]^{-1}(x - Bx)\| \qquad (x \in S(x_0, R)),$$

and since

$$\|[I - B'(x_0)]^{-1}(x - Bx)\| \leqq \frac{b_0 L\|x - x_0\|^2}{2} + \|x - x_0\| + \eta_0,$$

it follows that

$$\|A_1 x\| \leqq a_0 = \frac{b_0\delta}{1 - b_0\delta}\left(\frac{b_0 L R^2}{2} + R + \eta_0\right).\qquad(12.45)$$

It is easy to verify that

$$\|A_1 x - A_1 y\| \leqq q_0\|x - y\|,$$

where

$$q_0 = \frac{b_0 + b_0^2 L R}{1 - b_0\delta}\delta.$$

Thus, for sufficiently small δ, the numbers a_0 and q_0 satisfy inequality (12.21) (provided $2b_0 L\eta_0 < 1$). Theorem 12.3 implies

Theorem 12.4. *If δ is so small that condition* (12.21) *of Theorem* 12.3 *is fulfilled, the successive approximations* (12.30) *converge to a solution x^* of equation* (12.25).

Exercise 12.4. Show that, under the assumptions of Theorem 12.4, the successive approximations converge at the rate of a geometric progression.

Exercise 12.5. Study the convergence of the method (12.31).

Exercise 12.6. Let P be the projection on a finite-dimensional subspace, $u = Px$, $v = x - Px (x \in E)$. Then equation (12.25) may be rewritten as

$$u = PB(u + v), \qquad v = B(u + v) - PB(u + v).$$

Show that if the operator PB is a "good" approximation to B, and u_{n+1} and v_{n+1} are defined by the recurrence relations

$$\left. \begin{aligned} u_{n+1} &= u_n - \left[I - PB'(x_0) \right]^{-1} \left[u_n - PB(u_n + v_n) \right] \quad (n = 0, 1, 2, \ldots), \\ v_{n+1} &= B(u_n + v_n) - PB(u_n + v_n) \quad (n = 0, 1, 2, \ldots), \end{aligned} \right\} \qquad (12.46)$$

then the successive approximations $x_{n+1} = u_{n+1} + v_{n+1}$ converge to a solution x^* of equation (12.25). State the analogue of Theorem 12.4 for this method.

12.6. *Equations with nondifferentiable operators.* Consider the equation

$$Gx = 0 \qquad (12.47)$$

where G is a nondifferentiable operator. In many cases, this equation may be solved approximately by the iterative process

$$x_{n+1} = x_n - \left[F'(x_0) \right]^{-1} Gx_n \qquad (n = 0, 1, 2, \ldots), \qquad (12.48)$$

where F is a differentiable operator which approximates G sufficiently well in the uniform norm.

Theorem 12.5. *Assume that the operator F is differentiable in the ball $S(x_0, R)$ and that its derivative $F'(x)$ satisfies a Lipschitz condition* (12.7). *Let $G = F + F_1$, where F_1 satisfies a Lipschitz condition in $S(x_0, R)$:*

$$\left\| F_1 x - P_1 y \right\| \leq L_1 \left\| x - y \right\|. \qquad (12.49)$$

Assume that $\Gamma_0 = \left[F'(x_0) \right]^{-1}$ exists and $\left\| \Gamma_0 \right\| \leq b_0$, $\left\| \Gamma_0 G x_0 \right\| \leq \eta_0$. Finally, let

$$L_1 < \frac{1 - \sqrt{2b_0 L\eta_0}}{b_0} \qquad (12.50)$$

and

$$R \geq r^{**} = \frac{1 - b_0 L_1 - \sqrt{(1 - b_0 L_1)^2 - 2 b_0 L \eta_0}}{b_0 L}. \tag{12.51}$$

Then equation (12.47) *has a unique solution* x^{**} *in the ball* $S(x_0, r^{**})$; *this solution is the limit of the successive approximations* (12.48).

This statement follows from Theorem 12.3, by setting

$$A_1 x = \Gamma_0 (Gx - Fx).$$

Since formula (12.48) involves only the derivative $F'(x_0)$ of F, there is no loss of generality in assuming that $Fx_0 = Gx_0$. A simple verification shows that conditions (12.19) and (12.20) hold with constants

$$a_0 = b_0 L_1 r^{**}, \qquad q_0 = b_0 L_1.$$

Thus we need only verify the inequality (12.21); this we leave to the reader.

This method of constructing successive approximations (12.48) is due to Krasnosel'skii. It has been studied by Zinchenko and Kusakin.

Note that Theorem 12.5 remains valid if the sign $<$ in (12.50) is replaced by $\leq$.

12.7. *Remark on the Newton–Kantorovich method.* Our construction of the iterative processes (12.30) (12.31), (12.46) and (12.48) was based on the modified Newton–Kantorovich method. Analogous processes may be based on the Newton–Kantorovich method (11.3) itself. We consider a single example.

Let P be a projection in E such that $PF'(x)$ is a good approximation to $F'(x)$ for all $x \in S(x_0, R)$. Approximate solution of the equation $x = Bx$ may then be based on the iterative process

$$x_{n+1} = x_n - [I - PB'(x_n)]^{-1}(x_n - Bx_n) \qquad (n = 0, 1, 2, \ldots). \tag{12.52}$$

The advantage of this method over the usual Newton–Kantorovich method is that inversion of the operators $I - PB'(x_n)$ reduces to matrix inversion.

§ 13. Approximate solution of linearized equations

13.1. *Statement of the problem.* We continue our investigation of the Newton–Kantorovich method for approximate solution of the equation

$$Fx = 0, \tag{13.1}$$

where F is a differentiable operator. Recall that the method is to construct successive approximations x_n, each of which is an exact solution of the linearized equation

$$Fx_{n-1} + F'(x_{n-1})(x - x_{n-1}) = 0. \tag{13.2}$$

Exact solution of this equation is usually impossible, and one must resort to an approximate solution. In effect, therefore, the successive approximations actually employed are not those of the Newton–Kantorovich method.

Suppose that the equation

$$Bx = b \tag{13.3}$$

is to be solved approximately by some fixed method, which may be one or more steps of the method of minimal residuals, the ordinary iterative methods, the method of steepest descent, Seidel's method, etc. (see §7 and §8). In all these methods, one derives a "better" approximation x_1 from an approximation x_0 via some nonlinear operator

$$x_1 = V(x_0; B, b). \tag{13.4}$$

Applying (13.4) successively to the linearized equation (13.2), we get an iterative process:

$$x_{n+1} = V[x_n; F'(x_n); F'(x_n)x_n - Fx_n]. \tag{13.5}$$

Similarly, for approximate solution of the equation

$$Fx_{n-1} + F'(x_0)(x - x_{n-1}) = 0 \tag{13.6}$$

in the modified Newton–Kantorovich method, we get the iterative process

$$x_{n+1} = V[x_n; F'(x_0); F'(x_0)x_n - Fx_n]. \tag{13.7}$$

Let $\tilde{x}$ be an exact solution of equation (13.3). Throughout the sequel we shall assume that

$$\|V(x; B, b) - \tilde{x}\| \leq q \|x - \tilde{x}\|, \tag{13.8}$$

where $0 < q < 1$. More precisely, we assume that (13.8) is valid for all B in some class of linear operators containing the operator $F'(x)$. It is clear from the constructions of the previous chapter that the operator (13.4) may always be constructed in such a way that the number q is sufficiently small.

We shall limit ourselves to a single theorem on the convergence of (13.5). Several other theorems on the method (13.5), on the convergence of the method (13.7) and the effect of round-off errors on convergence are given in papers of Krasnosel'skii and Rutitskii [2–4].

13.2. *Convergence theorem.* We recall the basic notation of the last two sections.

The operator F is assumed to be defined and differentiable in the ball $S(x_0, R)$. We assume that the operator $\Gamma(x_0)$ exists, where $\Gamma(x) = [F'(x)]^{-1}$ and $\|\Gamma(x_0)\| \leq b_0$, $\|\Gamma(x_0) Fx_0\| \leq \eta_0$. Finally, in the ball $S(x_0, R)$ we assume the condition

$$\|F'(x) - F'(y)\| \leq L \|x - y\|. \tag{13.9}$$

The subsequent constructions will use the smaller root

$$\beta(q) = \frac{5 + q - \sqrt{9 + 26q + q^2}}{4} \tag{13.10}$$

of the quadratic equation

$$q + \frac{\beta(1 + q)}{2} = (1 - \beta)^2. \tag{13.11}$$

Theorem 13.1. *If*

$$h_0 = b_0 L \eta_0 \leq \frac{\beta(q)}{1 + q}, \tag{13.12}$$

and

$$r^* = \frac{1 + q}{\beta(q)} \eta_0 \leq R, \tag{13.13}$$

the successive approximations (13.5) *converge to a solution* x^* *of equation* (13.1).

Proof. Set

$$b_{n+1} = \frac{b_n}{1 - \beta(q)}, \qquad \eta_{n+1} = [1 - \beta(q)]\,\eta_n \qquad (n = 0, 1, 2, \ldots). \quad (13.14)$$

Under the assumptions of the theorem the successive approximations (13.5) are defined (all operators $F'(x_n)$ have continuous inverses Γ_n), and

$$\|\Gamma_n\| \leqq b_n, \qquad \|\Gamma_n F x_n\| \leqq \eta_n, \qquad h_n = b_n L \eta_n \leqq \frac{\beta(q)}{1 + q}. \quad (13.15)$$

The last equality (13.15) is obvious, since $h_n = h_0$ for all n. We prove the first two by induction, assuming them true for all $n \leqq m$.

It follows from (13.8) that

$$\|x_{m+1} - x_m\| \leqq \|x_{m+1} - \tilde{x}_{m+1}\| +$$

$$+ \|\tilde{x}_{m+1} - x_m\| \leqq (1 + q)\|\tilde{x}_{m+1} - x_m\|,$$

where $\tilde{x}_{m+1} = x_m - \Gamma_m F x_m$. Therefore

$$\|x_{m+1} - x_m\| \leqq (1 + q)\,\eta_m = (1 + q)\,[1 - \beta(q)]^m \eta_0. \quad (13.16)$$

Consequently,

$$\|x_{m+1} - x_0\| \leqq \|x_{m+1} - x_m\| + \|x_m - x_{m-1}\| + \ldots + \|x_1 - x_0\| \leqq$$

$$\leqq (1 + q)\,\{[1 - \beta(q)]^m + [1 - \beta(q)]^{m-1} + \ldots + 1\}\,\eta_0 \leqq$$

$$\leqq \frac{1 + q}{\beta(q)}\,\eta_0 = r^*.$$

Hence the operator $F'(x_{m+1})$ is defined.

The identity

$$F'(x_{m+1}) = F'(x_m)\,\{I - \Gamma_m[F'(x_m) - F'(x_{m+1})]\}$$

implies that Γ_{m+1} exists and that

$$\|\Gamma_{m+1}\| \leqq \frac{b_m}{1 - b_m L \|x_{m+1} - x_m\|} \leqq \frac{b_m}{1 - b_m L \eta_m (1 + q)} \leqq$$

$$\leqq \frac{b^m}{1 - \beta(q)} = b_{m+1}. \quad (13.17)$$

This proves the first inequality of (13.15).

We now proceed to the second inequality. Obviously,

$$\Gamma_{m+1} F x_{m+1} = \{I - \Gamma_m [F'(x_m) - F'(x_{m+1})]\}^{-1} \Gamma_m F x_{m+1}.$$

Therefore

$$\|\Gamma_{m+1} F x_{m+1}\| \leqq \frac{1}{1 - \beta(q)} \|\Gamma_m F x_{m+1}\|.$$

But

$$\Gamma_m F x_{m+1} =$$

$$= \Gamma_m F x_m + (x_{m+1} - x_m) + \Gamma_m [F x_{m+1} - F x_m - F'(x_m)(x_{m+1} - x_m)]$$

and, by (11.15),

$$\|\Gamma_m F x_{m+1}\| \leqq \|x_{m+1} - \tilde{x}_{m+1}\| + \frac{b_m L \|x_{m+1} - x_m\|^2}{2} \leqq$$

$$\leqq q \eta_m + \frac{b_m L (1 + q)^2 \eta_m^2}{2} \leqq q \eta_m + \frac{\beta(q)(1 + q)}{2} \eta_m = [1 - \beta(q)]^2 \eta_m.$$

Consequently,

$$\|\Gamma_{m+1} F x_{m+1}\| \leqq [1 - \beta(q)] \eta_m = \eta_{m+1}.$$

It follows from (13.16) that x_n is a Cauchy sequence, which converges to some point x^*. The chain of inequalities

$$\|\tilde{x}_{n+1} - x^*\| \leqq \|\tilde{x}_{n+1} - x_{n+1}\| + \|x_{n+1} - x^*\| \leqq q \|\tilde{x}_{n+1} - x_n\| +$$

$$+ \|x_{n+1} - x^*\| \leqq q \|\tilde{x}_{n+1} - x^*\| + q \|x_n - x^*\| + \|x_{n+1} - x^*\|$$

implies that

$$\|\tilde{x}_{n+1} - x^*\| \leq \frac{q}{1-q} \|x_n - x^*\| + \frac{1}{1-q} \|x_{n+1} - x^*\|.$$

Thus the sequence $\tilde{x}_n$ also converges to x^*. Letting $n \to \infty$ in the identity

$$F'(x_n)(\tilde{x}_{n+1} - x_n) + Fx_n = 0,$$

we get $Fx^* = 0$. ∎

For $q = 0$ Theorem 13.1 implies Theorem 11.3 (convergence of the Newton–Kantorovich method).

Exercise 13.1. Assume the conditions of Theorem 13.1, except that (13.12) is a strict inequality. Show that

$$\|x_n - x^*\| \leq (q + \varepsilon_n) \|x_{n-1} - x^*\|, \tag{13.18}$$

when $\varepsilon_n \to 0$.

Exercise 13.2. Assume that the derivative $F'(x)$ satisfies a Hölder condition

$$\|F'(x) - F'(y)\| \leq L \|x - y\|^\alpha,$$

where $0 < \alpha < 1$, in the ball $S(x_0, R)$. Assume that $\Gamma_0 = [F'(x_0)]^{-1}$ is defined, and $\|\Gamma_0\| \leq \leq b_0$, $\|\Gamma_0 F x_0\| \leq \eta_0$,

$$h_0 = b_0 L \eta_0^\alpha \leq \frac{\beta(q;\alpha)}{(1+q)^\alpha}, \tag{13.19}$$

where $\beta(q;\alpha)$ is the smallest positive root of the equation

$$\frac{1}{(1-\beta)^{1+\alpha}} \left[q + \frac{\beta(1+q)}{1+\alpha} \right]^\alpha = 1.$$

Finally, let

$$r^* = \frac{1+q}{1 - [1 - \beta(q;\alpha)]^{1/\alpha}} \eta_0 \leq R.$$

Show that the successive approximations (13.5) converge to a solution x^* of equation (13.1).

Exercise 13.3. Assume the conditions of Exercise 13.2, except that (13.19) is a strict inequality. Show that (13.18) is valid.

13.3. *Application of the method of minimal residuals.* In this subsection we shall describe several iterative processes of type (13.5) for equations in a Hilbert space H. If F is a potential operator (i.e., it is the gradient of some functional), a well-known theorem of Lyusternik implies that its derivative $F'(x)$ is a selfadjoint operator. Moreover, if $F(x)$ is the gradient of a strongly convex functional $\Phi(x)$, the operators $F'(x)$ are positive definite. In this situation, equation (13.2) may be solved approximately by one of the α-processes studied in §8.

For example, suppose that the operator (13.4) is equivalent to one iteration of the method of minimal residuals. The iterative process (13.5) is then

$$x_{n+1} = x_n - \frac{(F'(x_n) Fx_n, Fx_n)}{\|F'(x_n) Fx_n\|^2} Fx_n \qquad (n = 0, 1, 2, \ldots). \qquad (13.20)$$

In the general case (F is not a potential operator), equation (13.2) may be solved approximately by the method of subsection 8.6 (see (8.38)). If the operator (13.4) represents one iteration of this method, we get the iterative process

$$x_{n+1} = x_n - \frac{\|Fx_n\|^2}{\|[F'(x_n)]^* Fx_n\|^2} [F'(x_n)]^* Fx_n, \qquad (13.21)$$

where $[F'(x_n)]^*$ is a linear operator—the adjoint of $F'(x_n)$. The process (13.20) may also be considered for nonselfadjoint $F'(x_n)$. These and similar processes have been studied by various authors (Altman, Kivistik, Pugachev, Polyak, Kusakin and others). They usually base their constructions on variational arguments.

Application of Theorem 13.1 to the processes just described is not very fruitful, and it is more convenient to investigate them directly. As an example, we present a theorem due to Kivistik.

As before, we shall assume that the derivative $F'(x)$ satisfies a Lipschitz condition (13.9) in some ball $S(x_0, R)$.

Theorem 13.2. Let

$$\|Fx_0\| \leqq \delta_0 ; \qquad (13.22)$$

assume that, for all $x \in S(x_0, R)$,

$$|F'(x) h, h)| \geqq \frac{1}{c} \|h\|^2 \qquad (h \in H; c > 0), \qquad \|F'(x)\| \leqq \alpha; \qquad (13.23)$$

also let

$$q = \sqrt{1 - \frac{1}{a^2 c^2} + \frac{c^2 L \delta_0}{2}} < 1 \qquad (13.24)$$

and

$$r_0 = \frac{c\delta_0}{1-q} \leqq R.$$ (13.25)

Under the above conditions equation (13.1) *has a solution in the ball* $S(x_0, r_0)$ *which is the limit of the successive approximations* (13.20).

Proof. We first prove by induction that all the approximations (13.20) are defined (lie in the ball $S(x_0, r_0)$) and satisfy the inequalities

$$\|Fx_n\| \leqq q^n\delta_0 \qquad (n = 0, 1, 2, \ldots).$$ (13.26)

This is obvious for $n = 0$. Assuming the assertion true for $n \leqq m$, we shall prove it for $n = m + 1$.

By the first inequality of (13.23)

$$\|F'(x_m) Fx_m\| \geqq \frac{1}{c} \|Fx_m\|.$$

It therefore follows from (13.20) (with $n = m$) that

$$\|x_{m+1} - x_m\| = \frac{|(F'(x_m) Fx_m, Fx_m)|}{\|F'(x_m) Fx_m\|^2} \|Fx_m\| \leqq$$

$$\leqq \frac{\|Fx_m\|^2}{\|F'(x_m) Fx_m\|} \leqq c \|Fx_m\| \leqq cq^m\delta_0.$$ (13.27)

Hence

$$\|x_{m+1} - x_0\| \leqq \|x_{m+1} - x_m\| + \|x_m - x_{m-1}\| + \ldots + \|x_1 - x_0\| \leqq$$

$$\leqq c\delta_0(q^m + q^{m-1} + \ldots + 1) < \frac{c\delta_0}{1-q},$$

i.e., $x_{m+1} \in S(x_0, r_0)$.

It follows from the identity

$$Fx_{m+1} =$$

$$= Fx_{m+1} - Fx_m - F'(x_m)(x_{m+1} - x_m) + Fx_m + F'(x_m)(x_{m+1} - x_m)$$

and from inequalities (11.5), (13.27) that

$$\|Fx_{m+1}\| \leqq \frac{L\|x_{m+1} - x_m\|^2}{2} + \|Fx_m + F'(x_m)(x_{m+1} - x_m)\| \leqq$$

$$\leqq \frac{c^2 L}{2}\|Fx_m\|^2 + \|Fx_m + F'(x_m)(x_{m+1} - x_m)\|. \tag{13.28}$$

Now consider the second term in the right-hand side of (13.28). Since

$$Fx_m + F'(x_m)(x_{m+1} - x_m) = Fx_m - \frac{(F'(x_m)Fx_m, Fx_m)}{\|F'(x_m)Fx_m\|^2} F'(x_m)Fx_m,$$

it follows from (13.23) that

$$\|Fx_m + F'(x_m)(x_{m+1} - x_m)\|^2 =$$

$$= \|Fx_m\|^2 - \frac{(F'(x_m)Fx_m, Fx_m)^2}{\|F'(x_m)Fx_m\|^2} \leqq \left(1 - \frac{1}{a^2 c^2}\right)\|Fx_m\|^2.$$

Together with (13.28), this implies (13.26):

$$\|Fx_{m+1}\| \leqq \left(\frac{c^2 L}{2}\|Fx_m\| + \sqrt{1 - \frac{1}{a^2 c^2}}\right)\|Fx_m\| \leqq$$

$$\leqq \left(\frac{c^2 L}{2} q^m \delta_0 + \sqrt{1 - \frac{1}{a^2 c^2}}\right) q^m \delta_0 \leqq q^{m+1}\delta_0.$$

It follows from (13.27) that x_n is a Cauchy sequence. By (13.26), its limit x^* is a solution of equation (13.1). ∎

Exercise 13.4. Prove that under the assumptions of Theorem 13.2 the solution x^* is unique in the ball $S(x_0, r_0)$.

Exercise 13.5. Prove that under the assumptions of Theorem 13.2 the rate of convergence of the successive approximations x_n is determined by the inequality

$$\|x_n - x^*\| \leqq \frac{c\delta_0}{1 - q} q_n.$$

Exercise 13.6. Assume that conditions (13.22), (13.23) and (13.25) of Theorem 13.2 are satisfied. Let

$$q = \sqrt{c^2(a^2 + L\delta_0) - 1} < 1. \tag{13.29}$$

Show that the successive approximations (13.21) converge to a solution of equation (13.1).

Exercise 13.7. Under the assumptions of Exercise 13.6, show that the successive approximations

$$x_{n+1} = x_n - \frac{\|Fx_n\|^2}{(F'(x_n)Fx_n, Fx_n)} Fx_n \tag{13.30}$$

converge to a solution of equation (13.1).

13.4. *Choice of initial approximations.* Approximate solution of equation (13.1) by one of the methods described above involves a complex problem: choice of an initial approximation x_0 sufficiently close to the true solution. The method of random choice is often successful. Another frequently used method is to replace (13.1) by a "similar" equation and to regard the exact solution of the latter as the initial approximation x_0. Of course, there are no general "prescriptions" for admissible initial approximations. Nevertheless, one can describe various devices suitable for extensive classes of equations (in this respect, see, e.g., Kantorovich and Akilov [1]).

As usual, let F map E_1 into E_2. To simplify the exposition, we shall assume that F is defined throughout E_1. Assume that $G(x; \lambda)$ ($x \in E_1$; $0 \leqq \lambda \leqq 1$) is an operator with values in E_2 such that

$$G(x; 1) \equiv Fx \qquad (x \in E_1) \tag{13.31}$$

and the equation

$$G(x; 0) = 0 \tag{13.32}$$

has an obvious solution x^0. For example, the operator $G(x; \lambda)$ might be defined by

$$G(x; \lambda) = Fx - (1 - \lambda)Fx^0. \tag{13.33}$$

Consider the equation

$$G(x; \lambda) = 0. \tag{13.34}$$

Suppose that equation (13.34) has a continuous solution $x = x(\lambda)$, defined for $0 \leqq \lambda \leqq 1$ and satisfying the condition

$$x(0) = x^0. \tag{13.35}$$

Were the solution $x(\lambda)$ known,

$$x^* = x(1) \tag{13.36}$$

would be a solution of equation (13.1). One can thus find a point x_0 close to x^* by approximating $x(\lambda)$.

Our problem is thus to approximate the implicit function defined by (13.34) and the initial conditions (13.35). Global propositions are relevant here—theorems on implicit functions defined on the entire interval $[0, 1]$. The theory of implicit functions of this type is at present insufficiently developed.

The idea of extending solutions with respect to a parameter is due to S.N. Bernshtein; it has found extensive application in various theoretical and applied problems.

Assume that the operator $G(x; \lambda)$ is differentiable with respect to both x and λ, in the sense that there exist linear operators $G'_x(x; \lambda)$ mapping E_1 into E_2 and elements $G'_\lambda(x; \lambda) \in E_2$ such that

$$\lim_{\|h\| + |\Delta\lambda| \to 0} \frac{\|G(x + h; \lambda + \Delta\lambda) - G(x; \lambda) - G'_x(x; \lambda)h - G'_\lambda(x, \lambda)\Delta\lambda\|}{\|h\| + |\Delta\lambda|} = 0.$$

The implicit function $x(\lambda)$ is then a solution of the differential equation

$$G'_x(x; \lambda)\frac{dx}{d\lambda} + G'_\lambda(x; \lambda) = 0, \qquad (13.37)$$

satisfying the initial condition (13.35). Conditions for existence of a solution of this Cauchy problem defined on $[0, 1]$ are precisely conditions for existence of the implicit function. Assuming the existence of a continuous operator

$$\Gamma(x; \lambda) = [G'_x(x; \lambda)]^{-1}, \qquad (13.38)$$

we can rewrite equation (13.37) as

$$\frac{dx}{d\lambda} = -\Gamma(x; \lambda)G'_\lambda(x; \lambda). \qquad (13.39)$$

One must bear in mind that Peano's Theorem is false for ordinary differential equations in Banach spaces (see Bourbaki [1]). Therefore, even in the local existence theorem for equation (13.39) with condition (13.35) one must assume that the right-hand side of the equation satisfies certain smoothness conditions. However, there are no sufficient smoothness conditions for the existence of a global extension of the solution to the entire interval $0 \leqq \lambda \leqq 1$. Even formulation of the various global existence theorems for differential equations in Banach spaces is beyond the scope of this book (see, e.g., Krasnosel'skii [12]). We shall only mention a trivial fact: if the equation

$$\frac{dx}{d\lambda} = f(x; \lambda) \qquad (13.40)$$

in a Banach space satisfies the local existence theorem for some initial condition, and

$$\|f(x; \lambda)\| \leqq a + b \|x\| \qquad (0 \leqq \lambda \leqq 1; x \in E_1), \qquad (13.41)$$

then every solution of equation (13.40) can be extended to the entire interval $0 \leqq \lambda \leqq 1$. Thus, if

$$\|\Gamma(x; \lambda) G'_\lambda(x; \lambda)\| \leqq a + b \|x\|, \qquad (13.42)$$

then equation (13.34) defines an implicit function which satisfies (13.35) and is defined for $0 \leqq \lambda \leqq 1$. Consequently, condition (13.42) guarantees that equation (13.1) is solvable and its solution can be constructed by integrating the differential equation (13.39).

To approximate a solution $x(\lambda)$ of equation (13.39), one can use, for example, Euler's method. To this end, divide the interval $[0, 1]$ into m subintervals by points

$$\lambda_0 = 0 < \lambda_1 < \ldots < \lambda_m = 1. \qquad (13.43)$$

The approximate values $x(\lambda_i)$ of the implicit function $x(\lambda)$ are then determined by the equalities $x(\lambda_0) = x^0$ and

$$x(\lambda_{i+1}) = x(\lambda_i) - \Gamma[x(\lambda_i); \lambda_i] G'_\lambda[x(\lambda_i); \lambda_i] (\lambda_{i+1} - \lambda_i). \qquad (13.44)$$

The element $x(\lambda_m)$ is in general close to the solution x^* of equation (13.1), and one may therefore expect it to fulfill the demands imposed on initial approximations for iterative solution of equation (13.1). We emphasize that (13.44) does *not* describe an iterative process: it only yields a finite sequence of operations, whose result is an element which may (?!) be a suitable initial approximation for iterative solution of equation (13.1).

Other constructions may be used to approximate the values of the implicit function $x(\lambda)$. Partition the interval $[0, 1]$ by the points (13.43).

The point $x(\lambda_1)$ is a solution of the equation $G(x; \lambda_1) = 0$. Now $x(\lambda_0) = x^0$ is a suitable initial approximation to $x(\lambda_1)$. Approximate $x(\lambda_1)$ by performing a fixed number of steps of some iterative process. The result is an element x_1 which should be fairly close to $x(\lambda_1)$. x_1 is obtained from x^0 by a certain operator

$$x_1 = W\left[x^0; G(x; \lambda_1)\right].$$

Now regard x_1 as an initial approximation to the solution $x(\lambda_2)$ of the equation $G(x; \lambda_2)$, and proceed as before. The result is an element x_2:

$$x_2 = W\left[x_1; G(x; \lambda_2)\right].$$

Continuing in this way, we obtain a finite set of points

$$x_{i+1} = W\left[x_i; G(x; \lambda_i)\right] \qquad (i = 0, 1, \ldots, m-1), \qquad (13.45)$$

the last of which, x_m, may be regarded as an initial approximation for iterative solution of equation (13.1).

If the operator W represents one iteration of the method (11.3), formula (13.45) is

$$x_{i+1} = x_i - \left[G'_x(x_i; \lambda_{i+1})\right]^{-1} G(x_i; \lambda_{i+1}) \qquad (i = 1, \ldots, m-1). \quad (13.46)$$

If W is an operator of type (13.4), formula (13.45) becomes

$$x_{i+1} = V\left[x_i; G'_x(x_i; \lambda_{i+1}); G'_x(x_i; \lambda_{i+1})x_i - G(x_i; \lambda_{i+1})\right]. \quad (13.47)$$

Several authors have employed this device of introducing a parameter and studying its variations to construct approximate methods. The pioneer work is apparently due to Davidenko.

§ 14. A posteriori error estimates*

14.1. *Error estimates and existence theorems.* Consider the equation

$$Gx = 0, \qquad (14.1)$$

where G is a sufficiently smooth operator mapping a Banach space E into another space F. Suppose that we have somehow found an approxi-

---

* Krasnosel'skii [13].

mate solution x_0 of equation (14.1). The point x_0 may be determined by computations according to one of the iterative processes described in this chapter, by a projection method (see Chapter 4), and so on. Another widely used method is to compute the value of G for a certain given set of points, and to regard the point x_0 at which the operator G is "minimal" (say, in norm) as an approximate solution. To find a "minimum" point of the operator G one sometimes sets up a random process (in unscientific parlance—a "scientific probe"). In all these cases one is interested in *a posteriori* error estimates, which should be as accurate as possible. Error estimates are sometimes obtained by linearizing the equations— these cases are the simplest. In other cases the problem is quite complicated.

Since we are dealing with nonlinear equations, which usually have several solutions, the term "error estimate" must be properly defined.

We define an *error estimate* for the approximate solution x_0 of equation (14.1) to be any domain $\Omega \subset E$ which contains x_0 and at least one solution x^* of equation (14.1). Let $\mathfrak{M}$ be the set of solutions of equation (14.1). We define the *error* of the approximate solution x_0 *in the norm* $\| \cdot \|_0$ to be the number

$$\delta(x_0) = \inf_{x \in \mathfrak{M}} \|x_0 - x\|_0, \qquad (14.2)$$

where $\| \cdot \|_0$ is some norm in E equivalent to the original norm $\| \cdot \|$. Of course, the error depends essentially on the choice of the norm $\| \cdot \|_0$. If equation (14.1) has no solutions in E, it is natural to define $\delta(x_0) = \infty$ for any $x_0 \in E$ and any norm $\| \cdot \|_0$.

The exact value of the error may be determined only when exact solutions of equation (14.1) are known. We must therefore confine ourselves to estimates. The definition implies that $\delta(x_0) \leqq \delta_0$, if the ball $\|x - x_0\|_0 \leqq \delta_0$ is an error estimate, or, equivalently, if the ball $\|x - x_0\|_0 \leqq \delta_0$ contains at least one solution x^* of equation (14.1). Thus, *any estimate* $\delta(x_0) \leqq \delta_0$ *for the error* $\delta(x_0)$ *of an approximate solution* x_0 *is equivalent to the existence of at least one solution* x^* *of equation* (14.1) *in the ball* $\|x - x_0\|_0 \leqq \delta_0$. This principle converts any method for proving existence theorems (see, e.g., Kantorovich and Akilov [1], Krasnosel'skii [8, 11]) into a method for error estimation.

To estimate the error of a solution x_0 of equation (14.1) is precisely the same as estimating the error of the trivial solution of the equation $G(x_0 + x) = 0$. In the sequel, therefore, we shall simplify matters by assuming that $x_0 = \theta$ and deriving estimates for $\delta(\theta)$.

14.2. *Linearization of the equation.* First consider the particular case

$$x = Ax \qquad (14.3)$$

of equation (14.1), where A is a compact operator in E. We assume that A is differentiable at θ, and the derivative $A'(\theta) = A_1$ is also a compact operator. Write equation (14.3) in the form

$$x = A_1 x + Tx + A\theta ; \qquad (14.4)$$

$A\theta$ is the residual of the zero approximate solution; the compact operator T satisfies the inequality

$$\|Tx\| \leqq a_1(\|x\|) \qquad (\|x\| \leqq r_0), \qquad (14.5)$$

where r_0 is some positive number, and the function $a_1(r)$ is continuous for $0 \leqq r \leqq r_0$, $a_1(r) = o(r)$. Inequality (14.5) is awlays true, with $a_1(r) = ar^2$, if the operator A is twice differentiable in a neighborhood of θ.

The basic case here is that in which 1 is not an eigenvalue of A_1, i.e., the operator $\Gamma = (I - A_1)^{-1}$ is defined. Equation (14.4) is then equivalent to the equation $x = \Pi x$, where

$$\Pi x = \Gamma Tx + \Gamma A\theta .$$

The compact operator Π satisfies the obvious inequality

$$\|\Pi x\| \leqq \|\Gamma\| a_1(\|x\|) + \|\Gamma A\theta\| .$$

Therefore, if δ_0 is a number in the interval $[0, r_0]$ such that

$$\|\Gamma\| a_1(\delta_0) + \|\Gamma A\theta\| \leqq \delta_0 ,$$

the operator Π maps the boundary $\|x\| = \delta_0$ of the ball $\|x\| \leqq \delta_0$ into the interior of the ball; it follows from Schauder's principle that Π has at least one fixed point. We have proved

Theorem 14.1. Let A be a compact operator having the representation (14.4), where 1 is not an eigenvalue of the operator $A_1 = A'(\theta)$ and T

satisfies condition (14.5). *Let the residual $A\theta$ satisfy the inequality*

$$\|\Gamma A\theta\| \leqq \delta_0 - \|\Gamma\| \, a_1(\delta_0), \qquad (14.6)$$

where $\delta_0 \in [0, r_0]$.
Then

$$\delta(\theta) \leqq \delta_0.$$

Exercise 14.1. Replace (14.5) by the inequality

$$\|\Gamma T x\| \leqq a_1^*(\|x\|) \qquad (0 \leqq \|x\| \leqq r_0).$$

Show that Theorem 14.1 remains valid if (14.6) is replaced by the inequality

$$\|\Gamma A\theta\| \leqq \delta_0 - a_1^*(\delta_0).$$

Fixed point principles other than Schauder's are often convenient in proving assertions like that of Theorem 14.1. As an example we present a proposition in which norm inequalities are replaced by less restrictive one-sided inequalities.

Let E be a Hilbert space H. Compact operators in Hilbert spaces satisfy the following fixed point principle (Krasnosel'skii [8], p. 314):

Lemma 14.1. *If A is compact in the ball* $\|x\| \leqq r$ *and*

$$(Ax, x) \leqq r^2 \qquad (\|x\| = r), \qquad (14.7)$$

then A has at least one fixed point in the ball $\|x\| \leqq r$.

This lemma directly implies

Theorem 14.2. *Let A be an operator in a Hilbert space H which is compact in a ball* $\|x\| \leqq r$ *and has the representation* (14.4), *where $A_1 = A'(\theta)$ satisfies the inequality*

$$(A_1 x, x) \leqq b(x, x) \qquad (x \in H),$$

where $b < 1$, and

$$(Tx, x) \leqq a_2(\|x\|) \qquad (\|x\| \leqq r_0).$$

Let the residual $A\theta$ satisfy the inequality

$$\|A\theta\| \leqq (1 - b)\,\delta_0 - \delta_0^{-1}\, a_2(\delta_0),$$

where $\delta_0 \in [0, r_0]$.
Then

$$\delta(\theta) \leqq \delta_0.$$

The proof reduces to verification of (14.7) for $r = \delta_0$.

Exercise 14.2. Find explicit values of δ_0 in Theorems 14.1 and 14.2, for the cases $r_0 = \infty$, $a_1(r) = ar^2, a_2(r) = ar^{1+v}$.

Returning to the general equation (14.1), assume that G has the representation

$$Gx = G\theta + G_1 x + T_1 x, \qquad (14.8)$$

where $G_1 = G'(\theta)$ and T satisfies the conditions

$$\|T_1 x\| \leq a_3(\|x\|) \qquad (\|x\| \leq r_0) \qquad (14.9)$$

and

$$\|T_1 x - T_1 y\| \leq q(r) \|x - y\| \qquad (\|x\|, \|y\| \leq r; 0 \leq r \leq r_0). \quad (14.10)$$

Since $T_1\theta = \theta$, it follows from (14.10) that $\|T_1 x\| \leq q(\|x\|) \|x\|$. However, this estimate may be coarser than that of (14.9).

Theorem 14.3. *Assume that G has the representation* (14.8). *Let G_1 have a continuous inverse Γ. Let δ_0 be a number such that $0 \leq \delta_0 \leq r_0$ and*

$$\|\Gamma G\theta\| \leq \delta_0 - \|\Gamma\| a_3(\delta_0), \qquad \|\Gamma\| q(\delta_0) < 1. \qquad (14.11)$$

Then

$$\delta(\theta) \leq \delta_0.$$

Proof. Replace (14.1) by the equivalent equation $x = \Pi x$, where

$$\Pi x = -\Gamma T_1 x - \Gamma G\theta.$$

It follows from (14.9) and the first condition of (14.11) that Π maps the ball $\|x\| \leq \delta_0$ into itself:

$$\|\Pi x\| \leq \|\Gamma\| \cdot \|T_1 x\| + \|\Gamma G\theta\| \leq \|\Gamma\| a_3(\|x\|) + \|\Gamma G\theta\| \leq \delta_0 (\|x\| \leq \delta_0).$$

The second condition of (14.11) implies that Π is a contraction operator in the ball $\|x\| \leq \delta_0$, since, by (14.10),

$$\|\Pi x - \Pi y\| \leq \|\Gamma\| \cdot \|T_1 x - T_1 y\| \leq \|\Gamma\| q(\delta_0) \|x - y\|.$$

The contracting mapping principle therefore shows that Π has a unique fixed point in the ball $\|x\| \leq \delta_0$. ∎

Exercise 14.3. Write out the error estimates following from the theorems on the modified Newton–Kantorovich method presented in §12.

The theorems of this subsection are well known. As a rule, they yield quite satisfactory error estimates, provided the residuals ($A\theta$ for equation (14.3) and $G\theta$ for equation (14.1)) are small and the norm of the operator Γ (the inverse of $I - A'(\theta)$ for (14.3) and the inverse of $G'(\theta)$ for (14.1)) is not too large. If Γ does not exist, or if its norm is large, new arguments are necessary—these are the subject of the rest of this section.

14.3. *Rotation of a finite-dimensional vector field.* The rotation is an integer which is a topological invariant of a vector field on the boundary of a bounded domain. To make the exposition complete, we present the formulation of the relevant propositions in full. A complete exposition may be found, e.g., in Aleksandrov [1] and Pontryagin [1]. The monograph of Krasnosel'skii, Perov, Povolotskii and Zabreiko [1] contains a detailed study of the rotation of plane vector fields. Krasnosel'skii [8] devotes much attention to the rotation of finite-dimensional vector fields.

Let Ω be a bounded domain in n-space R^n, S the boundary of Ω. Let Φx be a continuous vector field defined on S, without zeros. One can then define a continuous mapping $\|\Phi x\|^{-1} \Phi x$ of the boundary S onto the unit sphere $\|x\| = 1$ of R^n. The degree of this mapping is known as the *rotation of the field* Φ on S, denoted by $\gamma(\Phi; S)$.

The rotation may be any integer $0, \pm 1, \pm 2, \dots$. The following two properties are important here:

Lemma 14.2. *Let the continuous vector field* Φ *be defined on a closed domain* $\bar{\Omega} = \Omega \cup S$. *If* $\gamma(\Phi; S) \neq 0$, *the field* Φ *vanishes for at least one point of the domain* Ω.

Two vector fields Φ_0 and Φ_1 are said to be *homotopic on* S if there exists a jointly continuous vector-function $\Phi(t; x)$ ($0 \leq t \leq 1$; $x \in S$) with values in R^n such that

$$\Phi(0; x) \equiv \Phi_0 x, \qquad \Phi(1; x) \equiv \Phi_1 x \qquad (x \in S)$$

and

$$\Phi(t; x) \neq 0 \qquad (0 \leq t \leq 1; x \in S).$$

For example, the fields Φ_0 and Φ_1 are homotopic on S if the vectors $\Phi_0 x$ and $\Phi_1 x$ are not zero and not in opposite directions, for any $x \in S$.

Lemma 14.3. *Homotopic vector fields have the same rotation,* $\gamma(\Phi_0; S) = \gamma(\Phi_1; S)$.

Lemma 14.3 indicates the basic method for determining rotations: computation of the rotation of complicated fields reduces to a computation for a small number of simple (standard) fields. The rotation of these simple fields is easily computed.

We begin with one-dimensional spaces; the only domains considered will be intervals. The rotation of a one-dimensional field Φ is determined by the directions of the vectors at the ends of the interval. Regard the line R^1 as the set of points $te\,(-\infty < t < \infty)$, where $e \in R^1$ and $e \neq 0$. Consider the field Φ on an interval $a \leqq t \leqq b$; let $\Phi(ae) = \alpha e$, $\Phi(be) = \beta e$.

Lemma 14.4. *If* $\alpha\beta > 0$, *then* $\gamma(\Phi; S) = 0$. *If* $\alpha < 0$ *and* $\beta > 0$, *then* $\gamma(\Phi; S) = 1$. *If* $\alpha > 0$ *and* $\beta < 0$, *then* $\gamma(\Phi; S) = -1$.

The statement of this lemma is illustrated graphically in Figure 14.1.

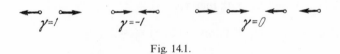

$$\gamma = 1 \qquad\qquad \gamma = -1 \qquad\qquad\qquad \gamma = 0$$

Fig. 14.1.

Now consider two-dimensional vector fields. We limit ourselves to homogeneous fields.

Assume that R^2 is provided with a vector coordinate system $\{\xi, \eta\}$. Let the field Φ be linear:

$$\Phi\{\xi, \eta\} = \{\alpha\xi + \beta\eta, \gamma\xi + \delta\eta\}, \tag{14.12}$$

where $\alpha\delta - \beta\gamma \neq 0$. Then Φx vanishes only at the zero point. The rotation of the field (14.12) on the boundary S of any bounded domain containing the zero point is

$$\gamma(\Phi; S) = \text{sign}\,(\alpha\delta - \beta\gamma). \tag{14.13}$$

Thus, the rotation of a linear field (14.12) on the circles $\xi^2 + \eta^2 = r^2$ is either 1 or -1.

Now let the field Φ be quadratic:

$$\Phi\{\xi,\eta\} = \{a_1\xi^2 + 2a_2\xi\eta + a_3\eta^2, b_1\xi^2 + 2b_2\xi\eta + b_3\eta^2\}. \quad (14.14)$$

Define three determinants:

$$D_1 = \begin{vmatrix} a_1 & a_2 \\ b_1 & b_1 \end{vmatrix}, \quad D_2 = \begin{vmatrix} a_2 & a_3 \\ b_2 & b_3 \end{vmatrix}, \quad D_3 = \begin{vmatrix} a_1 & a_3 \\ b_1 & b_3 \end{vmatrix}. \quad (14.15)$$

We assume that

$$D = 4D_1D_2 - D_3^2 \neq 0. \quad (14.16)$$

Exercise 14.4. Show that inequality (14.16) holds if and only if the field (14.14) vanishes only at the zero point $\{0, 0\}$.

Lemma 14.5. *The rotation* $\gamma(\Phi; S)$ *of the field* (14.14) *on the circles* $\xi^2 + \eta^2 = r^2$ *is given by the following rules:*

a) *if* $D < 0$, *then* $\gamma(\Phi; S) = 0$;

b) *if* $D > 0$ *and* $D_1 > 0$, *then* $\gamma(\Phi; S) = 2$;

c) *if* $D > 0$ *and* $D_1 < 0$, *then* $\gamma(\Phi; S) = -2$.

Now consider a plane field Φ of a more general type:

$$\Phi\{\xi,\eta\} = \{a_0\xi^k + a_1\xi^{k-1}\eta + \ldots + a_k\eta^k, b_0\xi^k + b_1\xi^{k-1}\eta + \ldots$$

$$\ldots + b_k\eta^k\}, \quad (14.17)$$

where $k \geqq 3$. To simplify matters we assume that $a_0 \neq 0$ (this may always be ensured by suitable choice of the coordinate system). Introduce the notation

$$T_0(\mu) = a_0 + a_1\mu + \ldots + a_k\mu^k,$$

$$T_1(\mu) = b_0 + b_1\mu + \ldots + b_k\mu^k.$$

Now use Euclid's algorithm to construct polynomials

$$T_0(\mu), T_1(\mu), \ldots, T_l(\mu), \quad (14.18)$$

such that

$$T_0(\mu) = \varepsilon_1(\mu)\, T_1(\mu) - T_2(\mu),$$

$$T_1(\mu) = \varepsilon_2(\mu)\, T_2(\mu) - T_3(\mu),$$

$$\cdot \quad \cdot \quad \cdot \quad \cdot \quad \cdot \quad \cdot \quad \cdot \quad \cdot \quad \cdot \quad \cdot \quad \cdot \quad \cdot \quad \cdot$$

$$T_{i-1}(\mu) = \varepsilon_i(\mu)\, T_i(\mu) - T_{i+1}(\mu),$$

$$\cdot \quad \cdot \quad \cdot \quad \cdot \quad \cdot \quad \cdot \quad \cdot \quad \cdot \quad \cdot \quad \cdot \quad \cdot \quad \cdot \quad \cdot$$

$$T_{l-1}(\mu) = \varepsilon_l(\mu)\, T_l(\mu).$$

The polynomials (14.18) form a so-called generalized Sturm sequence. Let $s(\infty)$ and $s(-\infty)$ denote the number of changes of sign in the sequence (14.18) for large positive μ and negative μ with large absolute value, respectively.

Lemma 14.6. *The vector field* (14.17) *does not vanish at zero points of the plane if and only if the equation*

$$T_l(\mu) = 0 \qquad\qquad (14.19)$$

has no real roots.

Lemma 14.7. *If equation* (14.19) *has no real roots, then the rotation* $\gamma(\Phi; S)$ *of the field* (14.17) *on the circles* $\xi^2 + \eta^2 = r^2$ *is*

$$\gamma(\Phi; S) = s(\infty) - s(-\infty).$$

Now consider a vector field in R^n, $n \geqq 3$. We confine ourselves to the simplest cases.

Let

$$\Phi x = Bx, \qquad\qquad (14.20)$$

where B is a matrix. If B is nonsingular, this field does not vanish on any sphere $\|x\| = r$. Its rotation $\gamma(\Phi; S)$ on these spheres coincides with the sign of det B (det B is the determinant of B). The sign of det B, in turn, is $(-1)^\beta$, where β is the sum of multiplicities of the real negative eigenvalues of B. Thus, the rotation of the field (14.20) is

$$\gamma(\Phi; S) = (-1)^\beta. \qquad\qquad (14.21)$$

Formula (14.21) is convenient, in that it easily generalizes to compact vector fields (see subsection 14.4).

Note, moreover, that *the rotation* $\gamma(\Phi; S)$ *of an odd vector field* Φ $(\Phi(-x) = -\Phi(x))$ *without zero vectors on the sphere S in an odd number.*

The rotation $\gamma(\Phi; S)$ *of an even field on S is even, and* $\gamma(\Phi; S) = 0$ *if the space is odd-dimensional* $(n = 2k + 1)$.

14.4. *Rotation of a compact vector field.** A vector field

$$\Phi x = x - Ax \qquad (14.22)$$

is said to be *compact* if A is a compact operator in a Banach space E.

Let Ω be a bounded domain in E, and S the boundary of Ω. Assume that a compact vector field (14.22) is defined on S and has no zeros there. The field (14.22) is then said to be *nondegenerate* on S.

Exercise 14.5. Assume that the compact vector field (14.22) is nondegenerate on S. Show that

$$\inf_{x \in S} \|x - Ax\| = \alpha > 0. \qquad (14.23)$$

Let E_ε denote a finite-dimensional subspace of E containing some ε-net $y_1, \ldots, y_k$ of the (sequentially) compact set AS. Following Schauder, set

$$P_\varepsilon y = \frac{\sum\limits_{i=1}^{k} \mu_i(y)\, y_i}{\sum\limits_{i=1}^{k} \mu_i(y)} \qquad (y \in AS),$$

where

$$\mu_i(y) = \begin{cases} \varepsilon - \|y - y_i\|, & \text{if} \quad \|y - y_i\| \leq \varepsilon, \\ 0, & \text{if} \quad \|y - y_i\| \geq \varepsilon. \end{cases}$$

The nonlinear operator P_ε is defined in AS and projects it onto E_ε. It is easy to see that

$$\|Ax - P_\varepsilon Ax\| \leq \varepsilon \qquad (x \in S).$$

Let S_ε denote the boundary of the intersection $\Omega \cap E_\varepsilon$, and γ_ε the rotation of the finite-dimensional (in E_ε) field $x - P_\varepsilon Ax$ on S_ε. If the intersection of E_ε and Ω is empty, we set $\gamma_\varepsilon = 0$.

---

* [*Translator's note:* Literal translation—"completely continuous." As for operators, we use the term "compact."]

It turns out that, for sufficiently small ε ($\varepsilon < \alpha$, where α is defined by (14.23)), the number γ_ε depends neither on ε, on the choice of the ε-net, nor on the choice of the subspace containing the net. The common value of the numbers γ_ε is known as the rotation* of the compact field (14.22) on S, denoted (as in the finite-dimensional case) by $\gamma(\Phi; S)$.

The rotation of a compact field has the same properties as that of a continuous field in a finite-dimensional space. We present the basic propositions.

Lemma 14.8. *Let the compact vector field* (14.22) *be defined on* $\Omega \cup S$. *If* $\gamma(\Phi; S) \neq 0$, *then the operator* A *has at least one fixed point in the domain* Ω.

An operator $A(t, x)$ ($0 \leq t \leq 1$, $x \in S$) with values in E is said to be *compact* if its range is (sequentially) compact in E and it is jointly continuous. Two compact vector fields $x - A_0 x$ and $x - A_1 x$ are *homotopic* on S if there exists a compact operator ($0 \leq t \leq 1$, $x \in S$) such that

$$A(0, x) \equiv A_0 x, \qquad A(1, x) \equiv A_1 x \qquad (x \in S)$$

and

$$A(t, x) \neq x \qquad (0 \leq t \leq 1, \ x \in S).$$

Lemma 14.9. *Fields homotopic on* S *have the same rotation on* S.

Lemma 14.9 reduces the computation of rotations to the simplest case. The so-called *Rouché Theorem* follows directly from Lemma 14.9: *If the compact vector field* (14.22) *is nondegenerate on* S *and the compact operator* A_0 *satisfies the condition*

$$\|A_0 x - Ax\| < \|x - Ax\| \qquad (x \in S), \qquad (14.24)$$

then the field $\Phi_0 x = x - A_0 x$ *is also nondegenerate on* S *and*

$$\gamma(\Phi_0; S) = \gamma(\Phi; S).$$

The following more general proposition is often useful: Two compact fields Φ_0 and Φ_1 are homotopic on S if, at every point $x \in S$, the vectors $\Phi_0 x$ and $\Phi_1 x$ are not opposite in direction.

---

* Krasnosel'skii [8]. The concept of the rotation of a compact vector field is equivalent to that of the Leray–Schauder degree of a mapping relative to zero (see Leray–Schauder [1]).

The field (14.22) is said to be *linear* if the compact operator A is linear. Assume that 1 is not an eigenvalue of A. Then the field Φ vanishes only at the zero point and

$$\|x - Ax\| > \|(I - A)^{-1}\|^{-1}\|x\| \qquad (x \in E). \qquad (14.25)$$

The rotation $\gamma(I - A; S)$ of a nondegenerate linear field on the sphere $\|x\| = r$ is given by the following analogue of (14.21):

$$\gamma(I - A; S) = (-1)^{\beta}, \qquad (14.26)$$

where β is the sum of multiplicities of the real eigenvalues of A exceeding 1.

The rotation of any odd compact vector field on a sphere is an odd number.

Exercise 14.6. Show that there are no even compact vector fields in an infinite-dimensional Banach space.

14.5. *Index of a fixed point.* Let A be a compact operator defined in a neighborhood of a point $x_0 \in E$, where x_0 is an isolated fixed point of A. Then the field (14.22) is nondegenerate on all spheres $\|x - x_0\| = r$ for sufficiently small r. It turns out that the field (14.22) has the same rotation on all these spheres. This common rotation is known as the *index* of the fixed point x_0, denoted by $\gamma(x_0; A)$. The index $\gamma(x_0; A)$ is the same as the rotation of the field (14.22) on the boundary S of any bounded domain in which x_0 is the only fixed point of A.

Besides the index, another important characteristic of an isolated fixed point x_0 is the scalar function

$$\alpha(r; x_0) = \inf_{\|x\| = r} \|x_0 + x - A(x_0 + x)\|. \qquad (14.27)$$

Since the function $\alpha(r; x_0)$ can only rarely be determined explicitly, we shall often employ *minorants* of the function (14.27), i.e., functions $\alpha_0(r; x_0)$ such that

$$\|x - Ax\| \geqq \alpha_0(\|x - x_0\|, x_0) \qquad (\|x - x\| \leqq r_0). \qquad (14.28)$$

We shall assume henceforth that $x_0 = \theta$. The second argument in the function (14.27) and its minorants $\alpha_0(r; x_0)$ may therefore be omitted, and we shall write $\alpha(r)$ and $\alpha_0(r)$.

Theorem 14.4. *Let A be a compact operator in a neighborhood of the point θ, such that*

$$Ax = A_1 x + Tx \qquad (\|x\| \leqq r_0), \qquad (14.29)$$

where $A_1 = A'(\theta)$, $T\theta = \theta$, and T satisfies the inequality

$$\|Tx\| \leqq a_1(\|x\|) \qquad (\|x\| \leqq r_0), \qquad (14.30)$$

$a_1(r) = o(r)$. *Assume that 1 is not an eigenvalue of the operator A_1.*

Then θ is an isolated fixed point of A; its index is given by (14.26), *where β is the sum of multiplicities of the real eigenvalues of A_1 exceeding 1. The function* (14.27) *has the minorant*

$$\alpha_0(r) = \frac{r}{\|(I - A_1)^{-1}\|} - a_1(r) \qquad (0 \leqq r \leqq r_0). \qquad (14.31)$$

Proof. The minorant (14.31) follows directly from (14.30):

$$\|x - Ax\| \geqq \|x - A_1 x\| - \|Tx\| \geqq \frac{\|x\|}{\|(I - A_1)^{-1}\|} - a_1(\|x\|).$$

Since $a_1(r) = o(r)$, it follows from (14.31) that, for small nonzero $\|x\|$, $\|x - Ax\| > 0$. The point θ is therefore an isolated fixed point of A.

Since the rotation of the linear field $\Phi_0 = I - A_1$ on the spheres $\|x\| = r$ is $(-1)^\beta$ (see (14.26)), the equality

$$\gamma(\theta; A) = (-1)^\beta$$

will follow if we can show that, for small nonzero x, $\|Ax - A_1 x\| < \|x - A_1 x\|$, or, equivalently, $\|Tx\| < \|x - A_1 x\|$. But this is obvious, since $\|Tx\| = o(\|x\|)$, while $\|x - A_1 x\| \geqq \|(I - A_1)^{-1}\|^{-1}\|x\|$. ∎

The function $a_1(r)$ in (14.30) is generally of the form $a_1(r) = ar^2$.

If 1 is an eigenvalue of the operator A_1, computation of the index and determination of the functions $\alpha_0(r)$ become quite laborious.

Let 1 be an eigenvalue of A_1, whose eigensubspace E_1 consists only of eigenvectors. There is no loss of generality in the latter assumption; see Zabreiko and Krasnosel'skii [1, 2]. Since A_1 is compact, E_1 is finite-dimensional; the dimension m of the subspace E_1 coincides with the multiplicity of the eigenvalue 1.

It follows from the general Riesz–Schauder theory of compact linear operators (see, e.g., Riesz and Nagy [1] or Kantorovich and Akilov [1]) that the space E is a direct sum

$$E = E_1 \oplus E^1$$

of A_1-invariant subspaces. Each element $x \in E$ then has a unique representation

$$x = u + v \qquad (u \in E_1, v \in E^1).$$

This representation defines two linear operators

$$P_1 x = u, \qquad P^1 x = v \tag{14.32}$$

which commute with A_1 and project E onto the subspaces E_1 and E^1, respectively; $P^1 = I - P_1$, $P_1 P^1 = P^1 P_1 = 0$.

Denote the restriction of A_1 to E^1 by A_1^0. Since 1 is not an eigenvalue of A_1^0, the operator

$$\Gamma^0 = (I - A_1^0)^{-1} \tag{14.33}$$

is defined on E^1. Let β_0 denote the sum of multiplicities of the real eigenvalues of A_1^0 exceeding 1 (β_0 is also the sum of multiplicities of the real eigenvalues of A_1 exceeding 1).

Suppose that in a neighborhood of θ the operator A may be expressed in the form

$$Ax = A_1 x + A_k x + Tx, \tag{14.34}$$

where $A_1 = A'(\theta)$, A_k is a homogeneous form* of degree k:

$$A_k(tx) \equiv t^k A_k x,$$

$T\theta = \theta$, and

$$\|Tx\| \leqq a_2(\|x\|) \qquad (\|x\| \leqq r_0), \tag{14.35}$$

where $a_2(r) = o(r^k)$.

Consider the vector field

$$\Phi_0 u = - P_1 A_k u \tag{14.36}$$

in the space E_1. If a coordinate system is chosen in E_1, the components of the field (14.36) are homogeneous polynomials of degree k in the co-

---

* See Lyusternik and Sobolev [1]; the basic propositions of the theory of multilinear and homogeneous operators will be stated in Chapter 5.

ordinates of the point. In the general case, the homogeneous field (14.36) vanishes only at the zero point—this we shall now assume.

We first present a theorem on the index (Krasnosel'skii [8], p. 219), without proof.

Theorem 14.5. *Let A be a compact operator having the representation* (14.34). *Let* 1 *be an eigenvalue of* A_1 *whose eigensubspace* E_1 *consists only of eigenvectors,* β_0 *the sum of multiplicities of the real eigenvalues exceeding* 1. *Finally, assume that the finite-dimensional vector field* (14.36) *vanishes only at the zero point.*

Then θ *is an isolated fixed point of the operator A, and its index is*

$$\gamma(\theta; A) = (-1)^{\beta_0} \gamma_0 , \tag{14.37}$$

where γ_0 *is the rotation of the field* (14.36) *on any shpere* $\|u\| = r$ *of the finite-dimensional space* E_1.

This theorem is effective only when one can determine or estimate γ_0. When E_1 is one- or two-dimensional, simple formulas are available for γ_0 (see subsection 14.3). If dim $E_1 \geqq 3$, we must confine ourselves, for the present, to special cases (in the main, the only applicable arguments are those relating to the oddness or evenness of the field (14.36)).

We now consider the function $\alpha_0(r)$ under the assumptions of Theorem 14.5.

First, some notation.

Since the field (14.36) is assumed to be nondegenerate on the spheres $\|x\| = r$, there exists a constant $b_1 > 0$ such that

$$\|P_1 A_k u\| \geqq b_1 \|u\|^k \qquad (u \in E_1) ; \tag{14.38}$$

for example, we can set

$$b_1 = \min_{\|u\| = 1} \|P_1 A_k u\| .$$

Since the homogeneous operator A_k is continuous (even compact),

$$\|P^1 A_k x\| \leqq b_2 \|x\|^k \qquad (x \in E), \tag{14.39}$$

where b_2 is a constant; one possibility is

$$b_2 = \sup_{\|x\| = 1} \|P^1 A_k x\| .$$

Finally, the operator A_k satisfies a Lipschitz condition in every ball $\|x\| \leqq r$, with constant of order r^{k-1}. We shall only need the estimate

$$\|P_1 A_k x - P_1 A_k P_1 x\| \leqq q_0 \|x\|^{k-1} \|P^1 x\| \qquad (x \in E) \qquad (14.40)$$

(recall that $P^1 = I - P_1$).

Set

$$\Phi_1 x = x - (A_1 x + A_k x) \qquad (x \in E). \qquad (14.41)$$

By assumption, $P_1 A_1 x = A_1 P_1 x = P_1 x$. Therefore,

$$P_1 \Phi_1 x = - P_1 A_k x = - P_1 A_k P_1 x - (P_1 A_k x - P_1 A_k P_1 x)$$

and it follows from (14.38) and (14.40) that

$$\|P_1 \Phi_1 x\| \geqq b_1 \|P_1 x\|^k - q_0 \|x\|^{k-1} \|P^1 x\| \qquad (x \in E). \qquad (14.42)$$

Obviously,

$$\|P^1 x - P^1 A_1 x\| = \|(I - A_1) P^1 x\| \geqq \frac{1}{\kappa_0} \|P^1 x\|,$$

where

$$\kappa_0 = \|\Gamma^0\| = \sup_{\|v\|=1, v \in E^1} \|(I - A_1)^{-1} v\| ; \qquad (14.43)$$

thus (14.39) implies the estimate

$$\|P^1 \Phi_1 x\| \geqq \frac{1}{\kappa_0} \|P^1 x\| - b_2 \|x\|^k \qquad (x \in E). \qquad (14.44)$$

Let $\phi(r)$ be any function such that

$$0 \leqq \phi(r) \leqq 1 \qquad (0 \leqq r \leqq r_0). \qquad (14.45)$$

Then (14.42) and (14.44) (together with the obvious inequality $\|P_1 x\| \geqq \|x\| - \|P^1 x\|$) imply the estimates

$$\|P_1 \Phi_1 x\| \geqq b_1 [r - r\phi(r)]^k - q_0 r^k \phi(r)$$
$$(\|P^1 x\| \leqq r\phi(r), \|x\| = r) \qquad (14.46)$$

and

$$\|P^1 \Phi_1 x\| \geqq \frac{1}{\kappa_0} r\phi(r) - b_2 r^k \qquad (\|P^1 x\| \geqq r\phi(r), \|x\| = r). \qquad (14.47)$$

Therefore

$$\max \left\{ \|P_1 \Phi x\| \; ; \; \|P^1 \Phi_1 x\| \right\} \geq$$

$$\geq \min \left\{ b_1 \left[r - r\phi(r) \right]^k - q_0 r^k \phi(r) \; ; \; \frac{1}{\kappa_0} r\phi(r) - b_2 r^k \right\} \qquad (\|x\| = r)$$

for any function $\phi(r)$ satisfying (14.45). Consequently,

$$\max \left\{ \|P_1 \Phi_1 x\|, \|P^1 \Phi_1 x\| \right\} \geq \psi(\|x\|) \qquad (x \in E), \qquad (14.48)$$

where

$$\psi(r) = \max_{0 \leq \phi \leq 1} \min \left\{ b_1 (r - r\phi)^k - q_0 r^k \phi; \frac{1}{\kappa_0} r\phi - b_2 r^k \right\}. \qquad (14.49)$$

Analysis of the function (14.49) (and analogous functions arising in this connection) is quite simple, if one notes that the first expression in the braces decreases with increasing ϕ, while the second increases. The value of ϕ at which the maximum is attained is therefore a root of the equation

$$b_1 (r - r\phi)^k - q_0 r^k \phi = \frac{1}{\kappa_0} r\phi - b_2 r^k, \qquad (14.50)$$

provided, of course, that the root belongs to the interval $[0, 1]$. Having determined the root ϕ and substituting it in one of the expressions in braces, we get an explicit formula for $\psi(r)$.

Exercise 14.7. Find an explicit formula for $\psi(r)$ when $k = 2$.

In the general case, equation (14.50) cannot be solved explicitly. We shall therefore coarsen the estimate (14.48).

An elementary proof shows that

$$(r - r\phi)^k > r^k - k r^k \phi \qquad (0 < \phi < 1; r > 0).$$

Therefore

$$\psi(r) \geq \psi_0(r) = \max_{0 \leq \phi \leq 1} \min \left\{ b_1 r^k - k b_1 r^k \phi - q_0 r^k \phi; \; \frac{1}{\kappa_0} r\phi - b_2 r^k \right\}.$$

$$(14.51)$$

To determine $\psi_0(r)$, solve the equation

$$b_1 r^k - (k b_1 + q_0) r^k \phi = \frac{1}{\kappa_0} r\phi - b_2 r^k$$

in ϕ. The solution ϕ_0 is

$$\phi_0 = \frac{(b_1 + b_2)\, r^{k-1}}{\dfrac{1}{\kappa_0} + (kb_1 + q_0)\, r^{k-1}}.$$

We shall consider only values of r for which $\phi_0 \leqq 1$, i.e.,

$$(b_1 + b_2)\, r^{k-1} \leqq \frac{1}{\kappa_0} + (kb_1 + q_0)\, r^{k-1}$$

or, equivalently,

$$[b_2 - (k-1)\, b_1 - q_0]\, r^{k-1} \leqq \frac{1}{\kappa_0}.$$

If $b_2 > (k-1)\, b_1 + q_0$, we shall assume that

$$r \leqq \left\{ \frac{1}{\kappa_0\, [b_2 - (k-1)\, b_1 - q_0]} \right\}^{1/(k-1)} = r^* ;$$

otherwise, we consider r in the interval $[0, \infty)$. Obviously,

$$\psi_0(r) = \min \left\{ b_1 r^k - kb_1 r^k \phi_0 - q_0 r^k \phi_0;\ \frac{1}{\kappa_0}\, r\phi_0 - b_2 r^k \right\},$$

which implies the formula

$$\psi_0(r) = \frac{b_1 r^k - \kappa_0\, (kb_1 + q_0)\, b_2 r^{2k-1}}{1 + \kappa_0\, (kb_1 + q_0)\, r^{k-1}}. \tag{14.52}$$

Formulas (14.48), (14.51) and (14.52) imply the fundamental estimate

$$\max \{\|P_1\Phi_1 x\|,\, \|P^1\Phi_1 x\|\} \geqq \frac{b_1 \|x\|^k - \kappa_0(kb_1 + q_0)\, b_2 \|x\|^{2k-1}}{1 + \kappa_0\, (kb_1 + q_0)\, \|x\|^{k-1}}$$

$$(\|x\| \leqq r^*). \tag{14.53}$$

The estimate (14.53) may be improved and simplified in certain special norms in E. For example, if

$$\|x\| = \|x\|_* = \max \{\|P_1 x\|,\, \|P^1 x\|\},$$

then (14.42) implies the following estimate, which is more precise than (14.46):

$$\|P_1 \Phi_1 x\| \geqq b_1 r^k - q_0 r^k \phi(r) \qquad (\|P^1 x\| \leqq r \phi(r); \|x\| = r).$$

This estimate, together with (14.47), implies the inequality

$$\|\Phi_1 x\|_* \geqq \max_{0 \leqq \phi \leqq 1} \min \left\{ b_1 r^k - q_0 r^k \phi; \ \frac{1}{\kappa_0} r \phi - b_2 r^k \right\} \qquad (\|x\| = r),$$

whence it follows that

$$\|\Phi_1 x\|_* \geqq \frac{b_1 \|x\|_*^k - \kappa_0 q_0 b_2 \|x\|_*^{2k-1}}{1 + \kappa_0 q_0 \|x\|_*^{k-1}} \qquad (\|x\|_* \leqq r_*)$$

for

$$r_* = \left\{ \frac{1}{\kappa_0 (b_1 + b_2 - q_0)} \right\}^{1/(k-1)}.$$

Inequalities (14.53) and (14.35) imply

Theorem 14.6. *Under the assumptions of Theorem* 14.5,

$$\|x - Ax\|_* \geqq \frac{b_1 \|x\|^k - \kappa_0 (kb_1 + q_0) b_2 \|x\|^{2k-1}}{1 + \kappa_0 (kb_1 + q_0) \|x\|^{k-1}} - a_2(\|x\|)$$

$$(\kappa_0 (b_1 + b_2 - kb_1 - q_0) \|x\|^{k-1} \leqq 1 \ ; \ \|x\| \leqq r_0), \qquad (14.54)$$

where $\|x\|_* = \max \{ \|P_1 x\|, \|P^1 x\| \}$.

One can often establish the estimate $a_2(r) \leqq b_3 r^{k+1}$. The estimate (14.54) then becomes

$$\|x - Ax\|_* \geqq \frac{b_1 \|x\|^k - \kappa_0 (kb_1 + q_0) b_2 \|x\|^{2k-1}}{1 + \kappa_0 (kb_1 + q_0) \|x\|^{k-1}} - b_3 \|x\|^{k+1}. \quad (14.55)$$

The main restriction in the assumptions of Theorems 14.5 and 14.6 is that the field (14.36) is nondegenerate at nonzero points of the subspace E_1. If this homogeneous field vanishes on certain lines or cones, additional constructions are necessary before fixed points can be studied. At this point we wish to warn the reader of a serious error.

Assume that

$$Ax = A_1 x + A_k x + A_{k+1} x + \ldots + A_l x + Tx,$$

where A_i are homogeneous operators of degree i and T involves terms of smaller order of magnitude. Let 1 be an eigenvalue of the operator A_1, P_1 and P^1 the operators (14.32). If the field (14.36) vanishes on certain lines, one might suppose, by analogy with Theorem 14.5, that the index $\gamma(\theta; A)$ is defined (up to its sign) by the rotation γ_0 of the finite-dimensional field

$$\Phi_0 u = - P(A_k u + A_{k+1} u + \ldots + A_l u) \qquad (u \in E_1)$$

on spheres $\|u\| = r$ for small r (if the field is not degenerate there). This assumption is erroneous. Even more: the field Φ_0 may vanish only at $u = \theta$ even though θ is a non-isolated fixed point of A!

Zabreiko and Krasnosel'skii have evolved a general algorithm for the study of fixed points and computation of their indexes when the field (14.36) vanishes on certain lines or cones (Zabreiko and Krasnosel'skii [2]). Their constructions yield the functions $\alpha_0(r)$ in many cases.

14.6. *Index of a fixed point and* a posteriori *error estimates.* Consider the equation

$$x = Bx \tag{14.56}$$

where B is a compact operator. Let θ be an approximate solution of equation (14.56). θ is obviously a fixed point of the operator

$$Ax = Bx - B\theta. \tag{14.57}$$

Let

$$\|Ax - x\|_1 = \|Bx - x - B\theta\|_1 \geqq \alpha_0(\|x\|_2) \qquad (\|x\|_2 \leqq r_0), \tag{14.58}$$

where $\| \cdot \|_1$, $\| \cdot \|_2$ are norms in E, which may coincide with the usual norm $\| \cdot \|$. Suppose that the residual $B\theta$ satisfies the inequality

$$\|B\theta\|_1 \leqq \alpha_0(\delta_0), \tag{14.59}$$

where $\delta_0 \leqq r_0$. Then the vector field $x - Bx$ (if it is nondegenerate) is homotopic to the field $x - Ax$ on the sphere $\|x\|_2 = \delta_0$. To prove this, it suffices to note that, by (14.58) and (14.59), the vectors $x - Ax$ and $x - Bx$ are not opposite in direction for $\|x\|_2 = \delta_0$. It then follows from Lemma 14.9 that the rotation of the field $x - Bx$ on the sphere $\|x\|_2 = \delta_0$ coincides with that of the field $x - Ax$, or, equivalently, with the index $\gamma(\theta; A)$ of the fixed point θ of A. Consequently, if $\gamma(\theta; A) \neq 0$, the operator B has at least one fixed point in the ball $\|x\|_2 \leqq \delta_0$. We have proved

Theorem 14.7. *Assume that inequalities* (14.58) *and* (14.59) *are satisfied. Let* $\gamma(\theta; A) \neq 0$.
Then

$$\delta(\theta) \leqq \delta_0 .$$

This general principle and Theorem 14.6 imply the following theorem, which is convenient for applications:

Theorem 14.8. *Assume that the operator* (14.57) *satisfies the conditions of Theorem* 14.6, $\gamma_0 \neq 0$, *and the residual* $B\theta$ *satisfies the inequality*

$$\|B\theta\|_* \leqq \frac{b_1 \delta_0^k - \kappa_0(kb_1 + q_0) b_2 \delta_0^{2k-1}}{1 + \kappa_0(kb_1 + q_0) \delta_0^{k-1}} - a_2(\delta_0), \qquad (14.60)$$

where $\delta_0 \in [0, r_0]$.
Then

$$\delta(\theta) \leqq \delta_0 .$$

In the main, our interest is concentrated on approximate solutions x_0 whose errors $\delta(x_0)$ are small. As a first approximation for the estimate, therefore, one can determine δ_0 from the equation

$$\|B\theta\|_* = b_1 \delta_0^k ,$$

or, equivalently, for small residuals one can assume that

$$\delta(\theta) \lesssim \left\{ \frac{\|B\theta\|_*}{b_1} \right\}^{1/k} . \qquad (14.61)$$

Of course, the estimate (14.61) is applicable only when $\gamma_0 \neq 0$.

Exercise 14.8. Show that Theorem 14.7 implies Theorems 14.1 and 14.2.

The proof of the following theorem is analogous to that of Theorem 14.7.

Theorem 14.9. *Let* θ *be an approximate solution of equation* (14.56), *where* B *is a compact operator which has the representation*

$$Bx = B\theta + Ax + Dx$$

in a neighborhood of θ; *let* θ *be an isolated fixed point of the compact operator* A, $\gamma(\theta; A) \neq 0$, *and*

$$\|Ax - x\|_1 \geqq \alpha_0(\|x\|_2) \qquad (\|x\|_2 \leqq r_0),$$

where $\| \cdot \|_1$, $\| \cdot \|_2$ *are norms in E. Assume that, for some* $\delta_0 \in [0, r_0]$,

$$\| B\theta + Dx \|_1 \leqq \alpha_0(\delta_0) \qquad (\| x \|_2 = \delta_0).$$

Then

$$\delta(\theta) \leqq \delta_0 .$$

Theorem 14.9 often yields better estimates than Theorem 14.7.

14.7. Special case. Theorems 14.7 and 14.9 are no longer valid if $\gamma(\theta; A) = 0$. Nevertheless, under certain addtional assumptions, one can still derive estimates of $\delta(\theta)$. The basic difference as regards Theorems 14.7 and 14.9 is that, besides the norm of the residual $B\theta$, its direction must also be taken into consideration.

We limit ourselves to an approximate solution $x_0 = \theta$ if the equation

$$x = B\theta + Ax, \qquad (14.62)$$

where A has the representation (14.34). We shall assume that the operators A_1, A_k, T satisfy all the assumptions of Theorems 14.5 and 14.6, and use the same notation. Condition (14.35) will have the specific form

$$\| Tx \| \leqq b_3 \| x \|^{k+1} \qquad (\| x \| \leqq r_0), \qquad (14.63)$$

where r_0 is a real number.

In addition, we assume that E_1 is one-dimensional and k even. It then follows from Theorem 14.5 and Lemma 14.4 that $\gamma(\theta; A) = 0$. Theorems 14.7 and 14.9 are therefore useless for deriving estimates of the error $\delta(\theta)$.

To prove the theorem of this subsection we shall employ the following estimates, which follow from (14.42), (14.44) and (14.63):

$$\| P_1(x - Ax) \| \geqq b_1 \| P_1 x \|^k - q_0 \| x \|^{k-1} \| P^1 x \| - \| P_1 \| b_3 \| x \|^{k+1}, \quad (14.64)$$

$$\| P^1(x - Ax) \| \geqq \frac{1}{\kappa_0} \| P^1 x \| - b_2 \| x \|^k - \| P^1 \| b_3 \| x \|^{k+1}, \qquad (14.65)$$

and the obvious inequality

$$\| P_1(x - Ax) \| \leqq q_0 \| x \|^k + \| P_1 \| b_3 \| x \|^{k+1} \qquad (14.66)$$

$$(P_1 x = 0, \; \| x \| \leqq r_0).$$

Let e be a unit vector in the (one-dimensional) space E_1. Then the operator P_1 is

$$P_1 x = l(x)\, e \qquad (x \in E),$$

where l is a linear functional such that $l(e) = 1$, $\|l\| = \|P_1\|$. Since the field (14.36) is nondegenerate, $l(A_k e) \neq 0$. The numbers b_1 and b_2 in inequalities (14.38) and (14.39) are here identical:

$$b_1 = b_2 = |\,l(A_k e)\,|.$$

Theorem 14.10. *Let* ρ_0 *and* δ_0 *be positive numbers satisfying the following four inequalities:*

$$\left.\begin{aligned}
\rho_0 + \delta_0 &\leq r_0, \\[4pt]
b_1 \delta_0^k - q_0(\rho_0 + \delta_0)^{k-1}\rho_0 - \|P_1\|\,b_3(\rho_0 + \delta_0)^{k+1} &\geq \|P_1 B\theta\|, \\[4pt]
\frac{1}{\kappa_0}\rho_0 - b_2(\rho_0 + \delta_0)^k - \|P^1\|\,b_3(\rho_0 + \delta_0)^{k+1} &\geq \|P^1 B\theta\|, \\[4pt]
q_0 \rho_0^k + \|P_1\|\,b_3 \rho_0^{k+1} &\leq \|P_1 B\theta\|,
\end{aligned}\;\right\} \qquad (14.67)$$

and let

$$l(A_k e)\, l(B\theta) < 0. \tag{14.68}$$

Then

$$\delta(\theta) \leq \rho_0 + \delta_0.$$

Proof. Let Ω_1 and Ω_2 denote the domains

$$\Omega_1 = \{x \colon \|P^1 x\| \leq \rho_0;\, 0 \leq l(x) \leq \delta_0\},$$

$$\Omega_2 = \{x \colon \|P^1 x\| \leq \rho_0;\, -\delta_0 \leq l(x) \leq 0\}$$

and S_1, S_2 their boundaries. By Lemma 14.8, the theorem will be proved if we show that the rotations of the compact vector field $\Phi x = x - Ax - B\theta$ on S_1 and S_2 do not vanish.

To fix ideas, consider the field Φ on S_1 (the reasoning for S_2 is analogous).

Define an auxiliary field:

$$\Psi x = x - A_1 P^1 x - l(x)\, e - [l(x) - \tfrac{1}{2}\delta_0]\, l(A_k e)\, e.$$

This field vanishes at all points x_0 for which $P_1 \Psi x_0 = 0$ and $P^1 \Psi x_0 = 0$. The first of these equations may be written as $[l(x_0) - \frac{1}{2} \delta_0] l(A_k e) = 0$, whence it follows that $P_1 x_0 = \frac{1}{2} \delta_0 e$. The second is $P^1 x_0 - A_1 P^1 x_0 = 0$, whence it follows that $P^1 x_0 = 0$. Consequently, the field Ψ vanishes only at the point $x_0 = \frac{1}{2} \delta_0 e \in \Omega_1$. The rotation $\gamma(\Psi; S_1)$ coincides with the index γ_0 of the fixed point x_0 of the operator $x - \Psi x$. Since the field Ψ is linear, $|\gamma| = 1$. Thus $\gamma(\Psi; S_1) \neq 0$.

By Lemma 14.9, the proof will be completed if we prove that the fields Φ and Ψ are homotopic on S_1.

To prove this, we shall show that the vectors Φx and Ψx are not in opposite directions for any $x \in S_1$. Assume the converse. Then there exist a point $x_* \in S_1$ and a number $t > 0$ such that

$$\Phi x_* = - t \Psi x_* . \tag{14.69}$$

Since $x_* \in S_1$, x_* belongs to one of the three sets

$$\Gamma_1 = \{x: \|P^1 x\| = \rho_0; \ 0 \leqq l(x) \leqq \delta_0\} ,$$

$$\Gamma_2 = \{x: \|P^1 x\| < \rho_0; \ l(x) = 0\} ,$$

$$\Gamma_3 = \{x: \|P^1 x\| < \rho_0; \ l(x) = \delta_0\} .$$

First assume that $x_* \in \Gamma_1$. Apply the operator P^1 to (14.69):

$$P^1 x_* - A_1 P^1 x_* - P^1 A_k x_* - P^1 T x_* - P^1 B\theta = - t(P^1 x_* - A_1 P^1 x_*),$$

so that

$$P^1 x_* - A_1 P^1 x_* = \frac{1}{1+t} P^1 A_k x_* + \frac{1}{1+t} P^1 T x_* + \frac{1}{1+t} P^1 B\theta$$

and so

$$\|(I - A_1) P^1 x_*\| < \|P^1 A_k x_*\| + \|P^1 T x_*\| + \|P^1 B\theta\| .$$

Therefore

$$\frac{1}{\kappa_0} \|P^1 x_*\| < b_2 \|x_*\|^k + \|P^1\| \cdot b_3 \|x_*\|^{k+1} + \|P^1 B\theta\| ,$$

and this inequality contradicts the third inequality of (14.67) (since $\|x_*\| \leqq \rho_0 + \delta_0$, $\|P^1 x_*\| = \rho_0$).

Now let $x_* \in \Gamma_2$. Apply the functional l to (14.69):

$$- l(A_k x_*) - l(Tx_*) - l(B\theta) = - \frac{t\delta_0}{2} l(A_k e).$$

This equality implies, by (14.68), that

$$\|P_1 B\theta\| = |l(B\theta)| < \left| l(B\theta) - \frac{t\delta_0}{2} l(A_k e) \right| \leqq |l(A_k x_*)| + |l(Tx_*)|.$$

Hence, by (14.40) and (14.63),

$$\|P_1 B\theta\| < q_0 \|x_*\|^k + \|P_1\| \cdot b_3 \cdot \|x_*\|^{k+1}$$

and so

$$\|P_1 B\theta\| < q_0 \rho_0^k + \|P_1\| b_3 \rho_0^{k+1}.$$

This contradicts the fourth inequality (14.67).

Finally, let $x_* \in \Gamma_3$. Again, apply the functional l to (14.69); the resulting equality

$$- l(A_k x_*) - l(Tx_*) - l(B\theta) = \frac{t\delta_0}{2} l(A_k e)$$

may be rewritten as

$$\left(\delta_0^k + \frac{t\delta_0}{2} \right) l(A_k e) = l(A_k P_1 x_* - A_k x_*) - l(Tx_*) - l(B\theta),$$

so that

$$b_1 \delta_0^k = |\delta_0 l(A_k e)| < \left| \left(\delta_0^k + \frac{t\delta_0}{2} \right) l(A_k e) \right| \leqq$$

$$\leqq |l(A_k P_1 x_* - A_k x_*)| + |l(Tx_*)| + |l(B\theta)| \leqq$$

$$\leqq \|P_1 A_k P_1 x_* - P_1 A_k x_*\| + \|P_1\| \cdot \|Tx_*\| + \|P_1 B\theta\|.$$

Consequently,

$$b_1 \delta_0^k < q_0 \|x_*\|^{k-1} \|P^1 x_*\| + \|P_1\| \cdot b_3 \|x_*\|^{k+1} + \|P_1 B\theta\| <$$

$$< q_0 (\rho_0 + \delta_0)^{k-1} \rho_0 + \|P_1\| \cdot b_3 (\rho_0 + \delta_0)^{k+1} + \|P_1 B\theta\|.$$

This contradicts the second inequality of (14.67).

The first inequality of (14.67) was used in all parts of the proof that (14.69) is contradictory, since we made use of the estimate (14.63). ∎

It is not difficult to choose numbers ρ_0 and δ_0 satisfying (14.67), provided the residual $B\theta$ is small and the norms $\|P_1 B\theta\|$, $\|P^1 B\theta\|$ of the residual have the same order of magnitude.

Assume that all the coefficients b_1, q_0, $\|P_1\| b_3$, $1/\kappa_0$, b_2, $\|P^1\| b_3$ in (14.67) have the same order, and that they are all considerably greater than the residual $\|B\theta\|$. Choose the numbers δ_0, ρ_0 in such a way that ρ_0 is considerably smaller than δ_0 and considerably greater than δ_0^k, in the sense that

$$b_4 \delta_0^{k-\nu} < \rho_0 < b_5 \delta_0^{1+\nu},$$

where $\nu > 0$, and b_4, b_5 have the same order of magnitude as the coefficients in (14.67). In approximation of δ_0 and ρ_0, inequalities (14.67) may then be replaced by the simpler ones

$$b_1 \delta_0^k \geqq \|P_1 B\theta\|, \quad \frac{1}{\kappa_0} \rho_0 \geqq \|P^1 B\theta\|, \quad q_0 \rho_0^k \leqq \|P_1 B\theta\|;$$

in other words, one can set

$$\rho_0 = \kappa_0 \|P^1 B\theta\|, \quad \delta_0 = \left(\frac{\|P_1 B\theta\|}{b_1}\right)^{1/k}$$

and assume that

$$\delta(\theta) \gtreqless \left(\frac{\|P_1 B\theta\|}{b_1}\right)^{1/k}.$$

14.8. *Relaxing the compactness condition.* The theorems of the preceding subsections may be extended to non-compact operators A. One must then assume that, in the ball $\|x\| \leq r_0$, the operator T satisfies not only an inequality of type (14.35), but also a Lipschitz condition, with a constant that depends in some way on r_0. We shall limit ourselves to a generalization of Theorem 14.8. The method applied below is actually based on a topological investigation of the Lyapunov–Schmidt branching equation (see § 22).

We return to the general equation

$$Gx = 0, \tag{14.70}$$

where the operator G maps a Banach space E into a space F. We shall again consider the zero approximate solution $x_0 = \theta$.

Let G have the representation

$$Gx = G\theta + G_1 x + G_k x + Tx, \tag{14.71}$$

where $G_1 = G'(\theta)$, G_k is a homogeneous operator of degree k, and T includes terms of lower order of magnitude.

Let us assume that the operator G_1 is normally solvable (i.e., $G_1 E$ is a closed subspace F^0 of the space F). Let E_0 denote the null subspace of G_1; assume that its dimension m is finite and equal to the deficiency index of the subspace F^0 in F. The space E is the direct sum of E_0 and a certain subspace E^0; each element $x \in E$ may be expressed uniquely in the form

$$x = u + v \qquad (u \in E_0,\ v \in E^0)\,;$$

this equality defines projections

$$P_1 x = u, \qquad P^1 x = v \qquad (x \in E)\,;$$

and we have the equalities $P_1 P^1 = P^1 P_1 = 0$, $G_1 P_1 = 0$, $G_1 P^1 = G_1$. The space F is the direct sum of F^0 and a certain subspace F_0 (whose dimension is the same as that of E_0). Each element $f \in F$ has a unique representation

$$f = z + w \qquad (z \in F_0, w \in F^0)\,;$$

this defines projections

$$Q_1 f = z, \qquad Q^1 f = w \qquad (f \in F)$$

in F, and we have the equalities $Q_1 Q^1 = Q^1 Q_1 = 0$, $Q_1 G_1 = 0$, $Q^1 G_1 = G_1$. We shall not dwell here on the construction of the operators P_1, P^1, Q_1, Q^1, assuming that they are known (see below, §22).

Denote the restriction of the linear operator G_1 to E^0 by G_1^0, and the inverse of G_1^0 by Γ^0 (this inverse maps F^0 into E^0); let

$$\|\Gamma^0\| = \|(G_1^0)^{-1}\| = \kappa_0\,.$$

Equation (14.70) is equivalent to the system of two equations

$$Q_1 G\theta + Q_1 G_k x + Q_1 Tx = 0, \tag{14.72}$$

$$Q^1 x = -\Gamma^0 Q^1 G\theta - \Gamma^0 Q^1 G_k x - \Gamma^0 Q^1 Tx\,. \tag{14.73}$$

It is therefore convenient to describe the properties of the operators G_k and T as inequalities in the operators $Q_1 G_k$, $Q_1 T$, $\Gamma^0 Q^1 G_k$, $\Gamma^0 Q^1 T$.

We first write down the conditions on the operators $\Gamma^0 Q^1 G_k$ and $\Gamma^0 Q^1 T$:

$$\left\| \Gamma^0 Q^1 (G_k P_1 x + T P_1 x) \right\| \leqq b_1 \left\| P_1 x \right\|^k \qquad (\left\| P_1 x \right\| \leqq r_1) \qquad (14.74)$$

and

$$\left\| \Gamma^0 Q^1 G_k (u + v_1) - \Gamma^0 Q^1 G_k (u + v_2) + \Gamma^0 Q^1 T (u + v_1) - \right.$$

$$\left. - \Gamma^0 Q^1 T (u + v_2) \right\| \leqq q_1 (\delta + \rho)^{k-1} \left\| v_1 - v_2 \right\| \qquad (14.75)$$

$$(u \in E_0, \left\| u \right\| \leqq \delta \leqq r_1 ; v_1, v_2 \in E^0 ; \left\| v_1 \right\|, \left\| v_2 \right\| \leqq \rho \leqq r_2).$$

In the sequel, these conditions will enable us to solve the equation (14.73) in $P^1 x$, that is, to express $P^1 x$ in terms of $P_1 x$. Once this has been done, solution of equation (14.70) will reduce to the study of the finite-dimensional equation (14.72), from which we must determine $P_1 x$.

Regarding $Q_1 G_k$ and $Q_1 T$, we assume that

$$\left\| Q_1 G_k x - Q_1 G_k P_1 x \right\| \leqq q_2 (\delta + \rho)^{k-1} \left\| P^1 x \right\|$$

$$(\left\| P_1 x \right\| \leqq \delta \leqq r_1 ; \left\| P_1 x \right\| \leqq \rho \leqq r_2) \qquad (14.76)$$

and

$$\left\| Q_1 T x \right\| \leqq b_2 (\delta + \rho)^{k+1} \qquad (\left\| P_1 x \right\| \leqq \delta \leqq r_1 ; \left\| P^1 x \right\| \leqq \rho \leqq r_2). \qquad (14.77)$$

In practice, the coefficients b_1, q_1, b_2, q_2 are fairly easy to compute or estimate.

Consider the operator $Q_1 G_k$, which maps E_0 into F_0. We shall assume that this operator is nondegenerate: $Q_1 G_k u \neq 0$ for $u \in E_0$ and $u \neq 0$. Then

$$\left\| Q_1 G_k u \right\|_F \geqq b_0 \left\| u \right\|_E^k \qquad (u \in E_0), \qquad (14.78)$$

where $b_0 > 0$.

The spaces E_0 and F_0 have the same dimension, and so one can construct a linear and continuously invertible operator D_0 mapping F_0 onto E_0. Define a homogeneous vector field on E_0 by

$$\Phi_0 u = D_0 Q_1 G_k u \qquad (u \in E_0). \qquad (14.79)$$

This field vanishes only at the zero point. Let γ_0 denote the rotation of the field (14.79) on the unit sphere $\|u\| = 1$ of the subspace E_0.

The operator D_0 is, of course, not uniquely defined. Let D_1 be another operator satisfying the same conditions, and consider the field

$$\Phi_1 u = D_1 Q_1 G_k u \qquad (u \in E_0)$$

on E_0. Let γ_1 be the rotation of the field Φ_1 on the sphere $\|u\| = 1$ in E_0. The field Φ_1 may be expressed as

$$\Phi_1 u = (D_1 D_0^{-1}) \Phi_0 u \qquad (u \in E_0).$$

It then follows from general theorems on vector fields (see Krasnosel'skii [8], p. 100) that $\gamma_1 = \gamma_0 \varepsilon$, where ε is the sign of the determinant of the operator $D_1 D_0^{-1}$ relative to some basis of the space E_0, i.e., $\varepsilon = (-1)^\beta$, where β is the sum of multiplicities of the negative real eigenvalues of $D_1 D_0^{-1}$. We are interested only in whether γ_0 is different from zero, and we may therefore assume that the operator D_0 has been chosen once and for all. Define a new norm in E_0, equivalent to the original norm $\| \cdot \|_E$, by

$$\|u\| = \|D_0^{-1} u\|_{F_0} \qquad (u \in E_0). \tag{14.80}$$

Theorem 14.11. Assume that conditions (14.74)–(14.78) are satisfied, and that the rotation γ_0 of the field (14.79) on the sphere $\|u\| = 1$ in E_0 is different from zero. Assume that the residual $G\theta$ satisfies the inequalities

$$\|\Gamma^0 Q^1 G\theta\| \leqq \rho_0 - b_1 \delta_0^k - q_1 (\delta_0 + \rho_0)^{k-1} \rho_0, \tag{14.81}$$

$$\|Q_1 G\theta\| \leqq b_0 \delta_0^k - q_2 (\delta_0 + \rho_0)^{k-1} \rho_0 - b_2 (\delta_0 + \rho_0)^{k+1}, \tag{14.82}$$

where $\delta_0 \leqq r_1$, $\rho_0 \leqq r_2$.
Then

$$\delta(\theta) \leqq \delta_0 + \rho_0.$$

Proof. We claim that equation (14.70) has at least one solution x^* such that $\|P_1 x^*\| \leq \delta_0$ and $\|P^1 x^*\| \leq \rho_0$. This will prove the theorem.

Let u be a point of the subspace E_0, $\|u\| \leq \delta_0$. In the space E_0, consider the equation

$$v = A(v; u), \tag{14.83}$$

where

$$A(v; u) = - \Gamma^0 Q^1 G\theta - \Gamma^0 Q^1 G_k(u + v) - \Gamma^0 Q^1 T (u + v).$$

By (14.75), this operator satisfies a Lipschitz condition in the ball $\|v\| \leq \rho_0$:

$$\|A(v_1; u) - A(v_2; u)\| \leq q_1 (\delta_0 + \rho_0)^{k-1} \|v_1 - v_2\|,$$

where, by (14.81),

$$q_1 (\delta_0 + \rho_0)^{k-1} < 1.$$

The operator $A(v; u)$ maps the ball $\|v\| \leq \rho_0$ into itself, since by (14.75), when $\|v\| \leq \rho_0$

$$\|A(v; u)\| \leq \|A(\theta; u)\| + \|A(v; u) - A(\theta; u)\| \leq$$

$$\leq \|A(\theta; u)\| + q_1 (\delta_0 + \rho_0)^{k-1} \rho_0,$$

whence, by (14.74),

$$\|A(v; u)\| \leq \|\Gamma^0 Q^1 G\theta\| + b_1 \delta_0^k + q_1 (\delta_0 + \rho_0)^{k-1} \rho_0,$$

and it follows from (14.81) that $\|A(v; u)\| \leq \rho_0$. Consequently, equation (14.83) has a unique solution $v = W(u)$ ($u \in E_0$; $\|u\| \leq \delta_0$) in the ball $\|v\| \leq \rho_0$. It is easily seen that this solution depends continuously on u.

If we set $u = P_1 x, v = P^1 x$ in equation (14.83), we get equation (14.73). Thus equation (14.73) may be solved for $P^1 x$. The solution

$$P^1 x = W(P_1 x)$$

is defined for $\|P_1 x\| \leq \delta_0$.

It follows that equation (14.70) is equivalent to equation (14.72), with $P^1 x$ replaced by $W(P_1 x)$. The proof will thus be complete if we show that the equation

$$Q_1 G\theta + Q_1 G_k [P_1 x + W(P_1 x)] + Q_1 T [P_1 x + W(P_1 x)] = 0 \quad (14.84)$$

has at least one solution in the ball $\|P_1 x\| < \delta_0$. Consider the continuous vector field

$$\Psi u = D_0 \{Q_1 G\theta + Q_1 G_k [u + W(u)] + Q_1 T [u + W(u)]\} \quad (14.85)$$

in the ball $\|u\| \leq \delta_0$ in E_0. If this field vanishes for at least one point of the ball $\|u\| \leq \delta_0$, equation (14.84) has a solution in the ball $\|P_1 x\| \leq \delta_0$.

To complete the proof, we must show that the field (14.85) has a zero in the ball $\|u\| \leq \delta_0$. By Lemma 14.2, it suffices to prove that the rotation $\gamma(\Psi; S)$ of the field (14.85) on the sphere $\|u\| = \delta_0$ is not zero. By Lemma 14.3, we need only show that the fields (14.85) and (14.79) are homotopic on the sphere $\|u\| = \delta_0$, since the rotation of the latter field is nonzero by assumption.

To prove that the fields in question are homotopic, we shall show that the vectors Ψu and $\Phi_0 u$ are not in opposite directions.

Indeed, suppose that for some $t > 0$ and $u_0 \in E_0$, $\|u_0\| = \delta_0$,

$$\Psi u_0 = - t\Phi_0 u_0,$$

i.e.,

$$D_0 \{Q_1 G\theta + Q_1 G_k [u_0 + W(u_0)] + Q_1 T [u_0 + W(u_0)] + tQ_1 G_k u_0\} = 0.$$

Rewrite this equality as

$$D_0 Q_1 G_k u_0 = - \frac{1}{1 + t} D_0 \{Q_1 G_k [u_0 + W(u_0)] -$$

$$- Q_1 G_k u_0 + Q_1 G\theta + Q_1 T [u_0 + W(u_0)]\}.$$

Computing the norms (14.80) of both sides and using (14.76) and (14.77), we get the chain of inequalities

$$\|Q_1 G_k u_0\|_F < \|Q_1 G_k [u_0 + W(u_0)] - Q_1 G_k u_0\| + \|Q_1 G\theta\| +$$

$$+ \|Q_1 T [u_0 + W(u_0)]\| \leq q_2 [\|u_0\| + \|W(u_0)\|]^{k-1} \|W(u_0)\| +$$

$$+ \|Q_1 G\theta\| + b_2 [\|u_0\| + \|W(u_0)\|]^{k+1} \leq q_2 (\delta_0 + \rho_0)^{k-1} \rho_0 +$$

$$+ b_2 (\delta_0 + \rho_0)^{k+1} + \|Q_1 G\theta\|.$$

It then follows from (14.78) that

$$b_0 \delta_0^k < q_2 (\delta_0 + \rho_0)^{k-1} \rho_0 + b_2 (\delta_0 + \rho_0)^{k+1} + \|Q_1 G\theta\|,$$

which contradicts (14.82). ∎

We conclude the chapter with one more remark on the equation $x = Bx$, where B is compact (though the remark is also relevant to the general equation $Gx = 0$). We have assumed that θ is an approximate

solution and 1 is an eigenvalue of the operator $B'(\theta)$. One often encoun-
ters situations in which $B'(\theta) = A_1 + D$, where the linear operator A_1
has the eigenvalue 1, the number (14.43) is relatively small, and D has a
small norm (D may not be known explicitly—an estimate of its norm is
sufficient). In dealing with such situations, the norms of the residuals $B\theta$
should be replaced in all estimates by $\|B\theta + Dx\|$ (as in Theorem 14.9).

4

Projection methods

§ 15. General theorems on convergence of projection methods

15.1. *Projection methods.* In the preceding chapters, we studied various iterative methods for the approximate solution of operator equations. This chapter is devoted to approximate methods based on an essentially new idea: First "approximate" the equation, then determine an exact solution of the "approximating" equation. The structure of the approximating equation usually aims at reducing its solution to that of a finite system of scalar equations.

In this chapter we shall present the general theory of the so-called projection methods, and show that this theory includes various specific methods for the numerical solution of operator, differential, integral, etc. equations.

Let E and F be (complex or real) Banach spaces. Consider the equation

$$Lu = f, \qquad (15.1)$$

where L is a (usually unbounded) linear operator with domain $D(L) \subset E$ and range $R(L) \subset F$. The *projection method* for solution of this equation is as follows. Let $\{E_n\}$ and $\{F_n\}$ be two given sequences of subspaces,

$$E_n \subset D(L) \subset E, \quad F_n \subset F \quad (n = 1, 2, \ldots),$$

and let P_n be linear projection operators mapping F onto F_n, i.e.,

$$P_n^2 = P_n, \ P_n F = F_n \quad (n = 1, 2, \ldots).$$

Now replace equation (15.1) by the approximation

$$P_n(Lu_n - f) = 0 \qquad (u_n \in E_n); \qquad (15.2)$$

the last inclusion means that the solution is sought in E_n.

When $E = F$, $E_n = F_n$ $(n = 1, 2, \dots)$, the projection method is known as the *Galerkin method*.

We now describe the best known constructions of the subspaces E_n and F_n and the operators P_n.

Let E and F be Hilbert spaces. Let $\{\phi_j\}$ and $\{\psi_j\}$ be given complete sequences,

$$\phi_j \in D(L) \subset E, \quad \psi_j \in F \qquad (j = 1, 2, \dots)$$

(the *coordinate sequences*). The approximate solution is assumed to be a linear combination

$$u_n = \sum_{j=1}^{n} \xi_j \phi_j \tag{15.3}$$

and determined by the condition that the residuals $Lu_n - f$ are orthogonal to the first n elements of the second coordinate sequence:

$$(Lu_n - f, \psi_i) = 0 \qquad (i = 1, \dots, n).$$

This gives a system of linear equations for the coefficients ξ_j:

$$\sum_{j=1}^{n} (L\phi_j, \psi_i)\, \xi_j = (f, \psi_i) \qquad (i = 1, \dots, n). \tag{15.3'}$$

It is easy to see that conditions (15.3), (15.3') are equivalent to (15.2), where E_n and F_n are the linear spans of $\phi_1, \dots, \phi_n$ and $\psi_1, \dots, \psi_n$, respectively, while P_n is the orthogonal projection onto F_n. The orthogonality of the projections P_n is essential. This is the so-called *Galerkin–Petrov method*. Note that the subspaces E_n and F_n in the Galerkin–Petrov method are finite-dimensional and

$$E_n \subset E_{n+1}, F_n \subset F_{n+1} \qquad (n = 1, 2, \dots).$$

The term "Galerkin–Petrov method" is sometimes used also for the projection method (15.2) where P_n is an orthogonal projection but the subspaces are infinite-dimensional and the above inclusions do not hold.

There are various ways of choosing the coordinate sequences $\{\phi_i\}$ and $\{\psi_i\}$ for approximate solution of equations by the Galerkin–Petrov method. For example, the elements ψ_i are often defined in terms of ϕ_i

by setting

$$\psi_i = L\phi_i \qquad (i = 1, 2, \dots).$$

This method is known as the *least-squares method.*

The variant of the Galerkin–Petrov method in which $E = F$ and the coordinate sequences coincide ($\psi_i = \phi_i$) is known as the *Bubnov–Galerkin method.*

Exercise 15.1. Prove that u_n minimizes the functional $\|Lu_n - f\|^2$ in the subspace spanned by the elements (15.3) if and only if u_n is determined by the least-squares method.

Exercise 15.2. Let L be a selfadjoint positive definite operator in a Hilbert space H, and $\{\phi_i\}$ a complete orthonormal system of eigenelements of L. Show that both the Bubnov–Galerkin and the least-squares methods are equivalent to determination of the partial sums of the Fourier expansion of the solution of equation (15.1) in the system $\{\phi_i\}$.

A sequence of elements $\{e_i\}$ is called a *basis* of the Banach space F if every element $x \in F$ has a unique representation

$$x = \sum_{i=1}^{\infty} \eta_i(x)\, e_i.$$

It is well known (see, e.g., Lyusternik and Sobolev [1]) that the coefficients $\eta_i(x)$ are continuous linear functionals with uniformly bounded norms. Moreover, the norms of the linear projections

$$P_n x = \sum_{i=1}^{n} \eta_i(x)\, e_i \qquad (x \in F) \tag{15.4}$$

are also uniformly bounded. Each P_n maps F onto a finite-dimensional subspace F_n with basis $e_1, \dots, e_n$. The projection method may be based on this construction of finite-dimensional subspaces F_n and projections P_n.

Exercise 15.3. Let F_n be an n-dimensional subspace of a Banach space F. Show that there exists a projection $P_n (P_n F = F_n)$ whose norm is at most n. A subspace F^n of a Banach space F is said to have finite defect n if the subspace of linear functionals that vanish on F^n is n-dimensional. Show that any subspace of defect n is the range of a projection P^n whose norm is at most $n + 1$.

15.2. *Fundamental convergence theorem.* We shall say that a sequence of subspaces $\{E_n\}$ is *ultimately dense* in a (Banach) space E if, for every $z \in E$,

$$\rho(z, E_n) \to 0 \qquad \text{as} \qquad n \to \infty,$$

where

$$\rho(z, E_n) = \inf_{z_n \in E_n} \|z - z_n\|.$$

Theorem 15.1. *Let the domain* $D(L)$ *of the operator* L *be dense in* E *and the range* $R(L)$ *dense in* F, *and assume that* L *is one-to-one on* $D(L)$. *Assume that the subspaces* LE_n *and* F_n *are closed in* F. *Finally, assume that the projections* P_n *are uniformly bounded*:

$$\|P_n\| \leq c \qquad (n = 1, 2, \ldots). \tag{15.5}$$

Let (A) *be the following statement: For any* $f \in F$, *for sufficiently large* n ($n \geq n_0$, *say*) *there exists a unique solution* u_n *of equation* (15.2), *and the residual* $Lu_n - f$ *converges to zero in norm as* $n \to \infty$. *A necessary and sufficient condition for* (A) *to hold is*

1) *the sequence of subspaces* LE_n *is ultimately dense in* F;
2) *for* $n \geq n_0$, *the operator* P_n *maps* LE_n *biuniquely onto* F_n;
3) $\tau \equiv \varlimsup\limits_{n \to \infty} \tau_n > 0$, *where*

$$\tau_n = \inf_{z_n \in LE_n, \|z_n\| = 1} \|P_n z_n\|.$$

When conditions 1–3 *are satisfied, the rate of convergence is described by the inequality*

$$\rho(f, LE_n) \leq \|Lu_n - f\| \leq (1 + c/\tau_n)\,\rho(f, LE_n). \tag{15.6}$$

If E_n and F_n are finite-dimensional and $\dim E_n = \dim F_n$, condition 2 follows from 3.

Proof of Theorem 15.1. Under the substitution

$$Lu_n = x_n,$$

equation (15.2) becomes

$$P_n x_n = P_n f \qquad (x_n \in LE_n). \tag{15.7}$$

Sufficiency. Let $\tilde{P}_n$ denote the restriction of P_n to the subspace LE_n. For $n \geq n_0$, it follows from 2 that the operator $\tilde{P}_n$ maps the Banach space LE_n biuniquely onto the Banach space F_n, and it thus has a bounded inverse $\tilde{P}_n^{-1}$ mapping F_n onto LE_n. Hence there exists a unique element

x_n satisfying (15.7):

$$x_n = \tilde{P}_n^{-1} P_n f .$$

But then $u_n = L^{-1} x_n$ is the unique element satisfying (15.2).

It follows from condition 3 that $\|\tilde{P}_n^{-1}\| = 1/\tau_n$. Together with (15.5), this implies that the norms $\|\tilde{P}_n^{-1} P_n\|$ are uniformly bounded:

$$\|\tilde{P}_n^{-1} P_n\| \leqq c/\tau_n, \ \overline{\lim_{n \to \infty}} \ c/\tau_n = c/\tau < \infty .$$

If $f_n \in LE_n$, then $\tilde{P}_n^{-1} P_n f_n = f_n$, and so

$$Lu_n - f = x_n - f = \tilde{P}_n^{-1} P_n f - f = \tilde{P}_n^{-1} P_n (f - f_n) - (f - f_n) \qquad (15.8)$$

and

$$\|Lu_n - f\| \leqq (c/\tau_n + 1) \|f - f_n\| .$$

Since the element $f_n \in LE_n$ is arbitrary, it follows that the residual $Lu_n - f$ tends to zero, and (15.6) holds.

Necessity. Suppose that for any $f \in F$ the approximation x_n is uniquely determined by (15.7) for $n \geqq n_0$, and $\|x_n - f\| \to 0$ as $n \to \infty$. We must prove conditions 1–3. The truth of 1 and 2 is obvious. For 3, it suffices to prove that the norms $\|\tilde{P}_n^{-1}\| = 1/\tau_n$ ($n \geqq n_0$) are uniformly bounded (the operator $\tilde{P}_n$ was defined in the sufficiency proof; that $\tilde{P}_n^{-1}$ is bounded follows from 2). We have $x_n = \tilde{P}_n^{-1} P_n f$ for $n \geqq n_0$. Thus, for any $f \in F$,

$$\tilde{P}_n^{-1} P_n f \to f \qquad \text{as} \qquad n \to \infty .$$

By the Banach–Steinhaus Theorem,* the norms $\|\tilde{P}_n^{-1} P_n\|$ are uniformly bounded:

$$\|\tilde{P}_n^{-1} P_n\| \leqq c' \qquad (n \geqq n_0) .$$

In particular, for $f_n \in F_n$,

$$\|\tilde{P}_n^{-1} f_n\| = \|\tilde{P}_n^{-1} P_n f_n\| \leqq c' \|f_n\| \qquad (n \geqq n_0) ,$$

and therefore $\|\tilde{P}_n^{-1}\| \leqq c' \ (n \geqq n_0)$. ∎

---

* This theorem (see Lyusternik and Sobolev [1]) states that if a sequence of continuous linear operators (defined in the Banach space F or in another Banach space F') converges at every element of F, its terms are uniformly bounded.

As formulated above, Theorem 15.1 was first proved by Vainikko; a similar theorem has been proved by Pol'skii. Necessity was indicated by Krasnosel'skii in a special case.

We emphasize that nothing is assumed in Theorem 15.1 concerning the solvability of equation (15.1). In the general case, therefore, the question of the convergence of the approximations u_n does not arise naturally. However, if $f \in R(L)$ and the operator L^{-1} exists and is bounded, the fact that the residuals $Lu_n - f$ converge to zero directly implies that the approximations u_n converge to a solution of equation (15.1). This follows from the obvious inequality

$$\| u_n - L^{-1}f \| \leqq \| L^{-1} \| \cdot \| Lu_n - f \|.$$

Note that (15.5) implies a considerable restriction on the applicability of Theorem 15.1. For in many cases, even when the subspaces F_n are finite-dimensional, there are no projections P_n with uniformly bounded norms. For example, if F is the space C of functions continuous on $[0, 1]$ and F_n is the set of algebraic polynomials of degree $\leqq n$, then any projection P_n mapping C onto these subspaces has norm

$$\| P_n \| \geqq c \ln n \qquad (c = \text{const} > 0)$$

(see Natanson [2]). One must thus consider projection methods for cases in which $\| P_n \| \to \infty$ as $n \to \infty$. Moreover, as will be made clear in the sequel, there are problems which lead naturally to projection methods involving unbounded projections.

Exercise 15.4. Assume condition 2 of Theorem 15.1, and suppose that for some given $f \in F$

$$\frac{\| P_n \|}{\tau_n} \rho(f, LE_n) \to 0 \qquad \text{as } n \to \infty$$

(the cases $\| P_n \| \to \infty, \tau_n \to 0$ are not excluded). Show that the residual $Lu_n - f$ tends to zero.

Exercise 15.5. Let $P_n F = F_n$ (the assumption $P_n^2 = P_n$ is dropped, so that P_n is not a projection). Show that Theorem 15.1 remains valid if condition 3 is replaced by the condition

$$3') \quad \| \tilde{P}_n^{-1} P_n \| \leqq c' \qquad (c' = \text{const}; \ n \geqq n_0).$$

Show that $\tilde{P}_n^{-1} P_n$ is a projection of F onto LE_n.

We conclude this subsection with a rather disappointing remark. The projection method is not an iterative method, in the sense that previous approximations $u_1, \ldots, u_{n-1}$ are not used to determine the next approximation u_n. In actual computations, therefore, one must usually make do with a single approximation u_{n_0}. In this respect, any theorem on the convergence of projection methods should be viewed only as a "compensating" consideration. True, it must be emphasized that compensating arguments of this type do have a considerable role to play, in convincing the computer that computations according to some plan or another are feasible. The results of these computations should always be subjected to *a posteriori* analysis.

It would be important and interesting to have *a priori* error estimates for each approximation. Unfortunately, derivation of such estimates involves unsurmountable difficulties. We shall not touch upon these estimates here. The first results of this type were obtained in the well-known work of Bogolyubov and Krylov (see Krylov [1]). Several further results may be found in Lehmann [1], Dzishkariani [2] and Vainikko [1–3].

15.3. *The aperture of subspaces of a Hilbert space.* We shall now present another formulation of Theorem 15.1 for the Galerkin–Petrov method, employing the concept of the aperture of subspaces.

Let H be a Hilbert space, H_1 and H_2 closed subspaces of H, P_1 and P_2 the corresponding orthogonal projections, $P^{(1)} = I - P_1$, $P^{(2)} = I - P_2$ (where I is the identity operator). The *aperture* of the subspaces H_1 and H_2 (Krein and Krasnosel'skii [1]) is the number

$$\Theta(H_1, H_2) = \max \left\{ \sup_{x \in H_1, \|x\| = 1} \|P^{(2)}x\|, \quad \sup_{x \in H_2, \|x\| = 1} \|P^{(1)}x\| \right\}. \quad (15.9)$$

It is clear that $0 \leq \Theta(H_1, H_2) \leq 1$, and $\Theta(H_1, H_2) = 0$ if and only if $H_1 = H_2$. Another definition of the aperture (Nagy [1]) is

$$\Theta(H_1, H_2) = \|P_1 - P_2\|. \quad (15.10)$$

Let us prove that these definitions are indeed equivalent. First rewrite (15.9) as

$$\Theta(H_1, H_2) = \max \{ \|P^{(2)}P_1\|, \quad \|P^{(1)}P_2\| \}; \quad (15.11)$$

we shall prove that (15.10) and (15.11) are equivalent.
Since

$$P_1(P_1 - P_2)x = P_1 P^{(2)}x, \quad P^{(1)}(P_1 - P_2)x = -P^{(1)}P_2 x$$

and $P_2^2 = P_2$, $(P^{(2)})^2 = P^{(2)}$, it follows that

$$(P_1 - P_2) x = (P_1 + P^{(1)})(P_1 - P_2) x = P_1 P^{(2)} P^{(2)} x - P^{(1)} P_2 P_2 x,$$

whence

$$\| (P_1 - P_2) x \|^2 \leqq \| P_1 P^{(2)} \|^2 \| P^{(2)} x \| + \| P^{(1)} P_2 \|^2 \| P_2 x \|^2 \leqq$$

$$\leqq \max \{ \| P_1 P^{(2)} \|^2, \| P^{(1)} P_2 \|^2 \} \| x \|^2 \qquad (x \in H).$$

But $\| P_1 P^{(2)} \| = \| (P_1 P^{(2)})^* \| = \| P^{(2)} P_1 \|$ (since P_1 and $P^{(2)}$ are orthogonal projections, they are selfadjoint), and so

$$\| P_1 - P_2 \| \leqq \max \{ \| P^{(2)} P_1 \|, \| P^{(1)} P_2 \| \}.$$

The converse inequality is easily verified:

$$\| P^{(1)} P_2 \| = \| (I - P_1) P_2 \| = \| (P_2 - P_1) P_2 \| \leqq \| P_2 - P_1 \|$$

and, similarly,

$$\| P^{(2)} P_1 \| \leqq \| P_1 - P_2 \|.$$

This proves that the two definitions are equivalent.

Lemma 15.1. *Let G_n and H_n be closed subspaces of a Hilbert space H, Q_n and P_n the corresponding orthogonal projections. The operator P_n maps G_n biuniquely onto H_n if and only if $\Theta(H_n, G_n) < 1$. If $\theta_n = \Theta(H_n, G_n)$ and*

$$\tau_n = \inf_{z \in G_n, \|z\| = 1} \| P_n z \| \text{ then}$$

$$\theta_n^2 + \tau_n^2 = 1. \tag{15.12}$$

Proof. Let $\theta_n < 1$. We claim that P_n maps G_n biuniquely onto H_n. By (15.10), we have $\| Q_n - P_n \| = \theta_n < 1$, and so the operator $I + (Q_n - P_n)$ maps H *onto* itself:

$$(I + Q_n - P_n) H = H.$$

Applying the operator P_n to both sides of this equality, we get $P_n Q_n H = P_n H$, or $P_n G_n = H_n$, i.e., P_n maps G_n onto H_n. It is not difficult to see that this mapping is also biunique. In fact, if $P_n x_0 = 0$ and $x_0 \in G_n$, $x_0 \neq 0$, the result is the contradictory chain of inequalities

$$\| x_0 \| = \| P^{(n)} x_0 \| \leqq \theta_n \| x_0 \| < \| x_0 \| \qquad (P^{(n)} = I - P_n).$$

This proves one direction of the theorem.

Conversely, assume that the projection P_n maps G_n biuniquely onto H_n. We must show that $\theta_n < 1$. The restriction $\tilde{P}_n$ of P_n to G_n is invertible; $\tilde{P}_n^{-1}$ maps H_n onto G_n, is bounded, and $\|\tilde{P}_n^{-1}\| = 1/\tau_n$ (compare the sufficiency proof in Theorem 15.1). Then

$$\|P_n x\| \geq \tau_n \|x\| \qquad (x \in G_n),$$

and for any $\varepsilon > 0$ there exists an element $x_\varepsilon \in G_n$ such that

$$\|P_n x_\varepsilon\| \leq (\tau_n + \varepsilon) \|x_\varepsilon\|.$$

Thus,

$$\|P^{(n)} x\|^2 = \|x\|^2 - \|P_n x\|^2 \leq (1 - \tau_n^2) \|x\|^2 \qquad (x \in G_n),$$

$$\|P^{(n)} x_\varepsilon\| \geq [1 - (\tau_n + \varepsilon)]^2 \|x_\varepsilon\|^2,$$

i.e.,

$$\sup_{x \in G_n, \|x\| = 1} \|P^{(n)} x\|^2 = 1 - \tau_n^2. \tag{15.13}$$

The adjoint of $\tilde{P}_n$ is Q_n, the restriction of the projection Q_n to H_n. In fact, for any $x \in G_n$, $y \in H_n$,

$$(\tilde{P}_n x, y)_{H_n} = (P_n x, y) = (x, P_n y) = (x, y) = (Q_n x, y) =$$
$$= (x, Q_n y) = (x, \tilde{Q}_n y)_{G_n}.$$

Consequently, the adjoint $\tilde{Q}_n$ of $\tilde{P}_n$ is also invertible:

$$\|\tilde{Q}_n^{-1}\| = \|\tilde{P}_n^{-1}\| = 1/\tau_n.$$

This is equivalent to the statement

$$\inf_{z \in H_n, \|z\| = 1} \|Q_n z\| = \tau_n.$$

Reasoning in the same way (interchanging G_n and H_n), we get the following analogue of (15.13):

$$\sup_{z \in H_n, \|z\| = 1} \|Q^{(n)} z\|^2 = 1 - \tau_n^2 \qquad (Q^{(n)} = I - Q_n).$$

Together with (15.13), this gives

$$\Theta(G_n, H_n) = \sqrt{1 - \tau_n^2} < 1.$$

This proves (15.12), and completes the proof of the lemma.

Exercise 15.6. Let $\Theta\,(G_n,\,H_n) = 1$. Show that either $P_nG_n \neq H_n$ or $Q_nH_n \neq G_n$.

Exercise 15.7. Suppose that one of the subspaces G_n and H_n contains a nonzero element orthogonal to the other subspace. Show that $\Theta\,(G_n,\,H_n) = 1$. Show by an example that the converse statement is false.

We return to equation (15.1) and the projection method (15.2). We shall now assume that $F = H$ (which contains the range of the operator L) is a Hilbert space and that P_n is the orthogonal projection onto $F_n = H_n$. In view of Lemma 15.1, we can reformulate Theorem 15.1 as follows.

Theorem 15.2. *Let the domain $D(L)$ of the operator L be dense in the Banach space E, and the range $R(L)$ dense in the Hilbert space H. Assume that the subspaces LE_n and H_n are closed in H and let Q_n and P_n be the corresponding orthogonal projections.*

Let (A) *be the same statement as in Theorem 15.1. A necessary and sufficient condition for* (A) *to hold is*

1) $Q_nf \rightarrow f$ *in norms as $n \rightarrow \infty$, for any $f \in H$;*

2) $\overline{\lim_{n \to \infty}}\ \Theta(LE_n, H_n) < 1.$

The rate of convergence is then described by the inequality

$$\|Q^{(n)}f\| \leq \|Lu_n - f\| \leq \frac{1}{\sqrt{1 - \theta_n^2}}\,\|Q^{(n)}f\|, \qquad (15.14)$$

where $Q^{(n)} = I - Q_n$, $\theta_n = \Theta(LE_n, H_n)$.

We need only prove the estimate (15.14), which does not follow from the coarser estimate (15.6). We start from (15.8), setting $f_n = Q_nf$:

$$Lu_n - f = \tilde{P}_n^{-1}P_nQ^{(n)}f - Q^{(n)}f\,.$$

The elements $\tilde{P}_n^{-1}P_nQ^{(n)}f \in LE_n$ and $Q^{(n)}f \in H \ominus LE_n$ are orthogonal, and so

$$\|Lu_n - f\|^2 =$$

$$= \|\tilde{P}_n^{-1}P_nQ^{(n)}f\|^2 + \|Q^{(n)}f\|^2 \leq \left(\frac{1}{\tau_n^2}\,\|P_nQ^{(n)}\|^2 + 1\right)\|Q^{(n)}f\|^2\,,$$

but since $\|P_nQ^{(n)}\| = \|Q^{(n)}P_n\| \leq \theta_n$ (see (15.11)) and $\tau_n^2 = 1 - \theta_n^2$ (see (15.12)), this implies (15.14).

Note that condition 2 of Theorem 15.2 is satisfied trivially for the least-squares method, for then $H_n = LE_n$ and $\Theta(LE_n, H_n) = 0$ ($n = 1$, 2, ...).

Exercise 15.8. Let E and H be Hilbert spaces, and the operator L continuous and continuously invertible (L maps E onto H). Let $\{E_n\}$ be a sequence of subspaces ultimately dense in E. Show that the following two conditions are then equivalent:

$$\text{a) } \overline{\lim_{n \to \infty}} \; \Theta(LE_n, H_n) < 1; \qquad \text{b) } \overline{\lim_{n \to \infty}} \; \Theta(E_n, L^*H_n) < 1.$$

Exercise 15.9. It does not follow from conditions 1 and 2 of Theorem 15.2 that $P_n f \to f$ as $n \to \infty$ for any $f \in H$. Verify this by the following example: LE_n and H_n are the linear spans of the elements $\phi_1, \phi_2, \ldots, \phi_n$ and $\phi_1 + \phi_{n+1}, \phi_2, \ldots, \phi_n$, respectively, where $\{\phi_i\}$ is a complete orthonormal system in H.

The essential point in this example is that $\{H_n\}$ is not an ascending sequence.

Exercise 15.10. Let $\{\chi_i\}$ and $\{\Psi_i\}$ be two systems of elements in H, LE_n and H_n the linear spans of the elements $\chi_1, \ldots, \chi_n$ and $\psi_1, \ldots, \psi_n$, respectively. Show that conditions 1 and 2 of Theorem 15.2 then imply that $P_n f \to f$ ($f \in H$).

Exercise 15.11. Let $\{x_i\}$ and $\{y_i\}$ be complete biorthogonal systems in H. Show that they form a basis of H if and only if $\overline{\lim_{n \to \infty}} \; \Theta(G_n, H_n) < 1$ G_n and H_n are the linear spans of the elements $x_1, \ldots, x_n$ and $y_1, \ldots, y_n$, respectively.

In particular, it follows from Lemma 15.1 that the subspaces H_1 and H_2 of the Hilbert space H coincide if $\Theta(H_1, H_2) < 1$. The concept of aperture may be generalized (Krein, Krasnosel'skii and Mil'man [1]) to subspaces E_1 and E_2 of a Banach space E. Formula (15.9) is then replaced by the analogous equality

$$\Theta(E_1, E_2) = \max \left\{ \sup_{x \in E_1, \|x\|=1} \rho(x, E_2), \; \sup_{x \in E_2, \|x\|=1} \rho(x, E_1) \right\},$$

where, as usual,

$$\rho(x, E_i) = \inf_{z \in E_i} \|x - z\| \qquad (i = 1, 2).$$

It is known that in this case the dimensions of E_1 and E_2 coincide if $\Theta(E_1, E_2) < \frac{1}{2}$. It is not known whether this remains true for $\Theta(E_1, E_2) < 1$.

15.4. The Galerkin method for equations of the second kind.

When studying projection methods for the approximate solution of linear equations, one is usually able to reduce the discussion to that of some other projection method for an equation of the second kind

$$x = Tx + f, \qquad (15.15)$$

where T is a continuous linear operator in some Banach space E.

The special form of equation (15.15) permits formulation of simple convergence tests for projection methods when the projections P_n are not bounded.

Let $\{E_n\}$ be a sequence of closed subspaces of E, and assume that for every n we have a (generally unbounded) linear operator P_n which projects its domain $D(P_n) \subset E$ onto E_n ($E_n \subset D(P_n)$). Throughout the sequel we shall assume that $f \in D(P_n)$ and $TE \subset D(P_n)$; then the solutions of equation (15.15), if they exist, also belong to $D(P_n)$. Every operator $P_n T$ is assumed to be bounded in E.

Since $P_n x_n = x_n$ for $x_n \in E_n$, the Galerkin method

$$P_n(x_n - Tx_n - f) = 0 \qquad (x_n \in E_n)$$

gives the following equation in E:

$$x_n = P_n T x_n + P_n f ; \qquad (15.16)$$

the condition $x_n \in E$ in (15.16) becomes superfluous, since the solutions of (15.16) (if they exist) necessarily belong to E_n.

As usual, we shall use the notation $P^{(n)} = I - P_n$, where I is the identity operator.

Theorem 15.3. Let the operator $I - T$ be continuously invertible, and $\|P_n^{(n)} T\| \to 0$ as $n \to \infty$.

Then for sufficiently large n, equation (15.16) has a unique solution x_n. The sequence x_n converges to a solution x_0 of equation (15.15) if and only if $P^{(n)} f \to 0$ as $n \to \infty$. Moreover,

$$c_1 \|P^{(n)} x_0\| \leq \|x_n - x_0\| \leq c_2 \|P^{(n)} x_0\| \qquad (c_1, c_2 = \text{const} > 0), \quad (15.17)$$

and, if the projections P_n are bounded,

$$\|x_n - x_0\| \leq c \|P_n\| \rho(x_0, E_n) \qquad (c = \text{const}), \qquad (15.18)$$

where

$$\rho(z, E_n) = \inf_{z_n \in E_n} \|z - z_n\| .$$

The proof of this theorem involves the following simple result (used repeatedly in the sequel), whose proof we leave to the reader.

Lemma 15.2. Let A and B be bounded linear operators in a Banach space F, where A is invertible and $\|B\| \cdot \|A^{-1}\| < 1$. Then the operator $A + B$ is also invertible, and

$$\|(A + B)^{-1}\| \leq \frac{\|A^{-1}\|}{1 - \|B\| \cdot \|A^{-1}\|} .$$

Proof of Theorem 15.3. For sufficiently large n ($n \geq n_0$, say), $\|P^{(n)}T\| \cdot \|(I - T)^{-1}\| \leq q < 1$. The operators $I - P_n T$ are therefore invertible for $n \geq n_0$, and

$$\|(I - P_n T)^{-1}\| \leq \frac{\|(I - T)^{-1}\|}{1 - q} = c_2 \qquad (n \geq n_0). \tag{15.19}$$

We have proved that equation (15.16) has a unique solution for $n \geq n_0$. The solution x_0 of equation (15.15) and the solution x_n of equation (15.16) satisfy the relation

$$(I - P_n T)(x_0 - x_n) = P^{(n)}x_0.$$

This implies (15.7), in view of (15.19) and the analogous inequalities for the operators $I - P_n T$ themselves:

$$\|I - P_n T\| \leq 1/c_1 \qquad (c_1 = \text{const}; \ n \geq n_0).$$

Since $P^{(n)}x_0 = P^{(n)}Tx_0 + P^{(n)}f$, and by assumption

$$\|P^{(n)}Tx_0\| \leq \|P^{(n)}T\| \cdot \|x_0\| \to 0 \qquad \text{as} \qquad n \to \infty,$$

it follows that $P^{(n)}x_0 \to 0$ if and only if $P^{(n)}f \to 0$. Hence, using (15.17), we get the statement of the theorem that $x_n \to x_0$.

Finally, the estimate (15.18) is a consequence of (15.17). In fact, $P^{(n)}z_n = 0$ for $z_n \in E_n$, and for any $z \in E$

$$\|P^{(n)}z\| = \|P^{(n)}(z - z_n)\| \leq (1 + \|P_n\|) \|z - z_n\|,$$

so that (since $z_n \in E_n$)

$$\|P^{(n)}z\| \leq (1 + \|P_n\|) \rho(z, E_n) \leq 2\|P_n\| \rho(z, E_n). \tag{15.20}$$

Estimating the norm $\|P^{(n)}x_0\|$ according to (15.20), we get (15.18) from (15.17). $\blacksquare$

Exercise 15.12. Prove that the estimates (15.17) and (15.18) remain valid when the norms of the operators $P^{(n)}T$ do not converge to zero, but for $n \geq n_0$

$$\|P^{(n)}T\| \cdot \|(I - T)^{-1}\| \leq q < 1$$

(it is now no longer true that $x_n \to x_0$ if and only if $P^{(n)}f \to 0$).

Exercise 15.13. Under the assumptions of Theorem 15.3, let $\|TP^{(n)}\| \to 0$ as $n \to \infty$. Prove the following estimate, which is sharper than (15.17):

$$(1 - \varepsilon_n)\|P^{(n)}x_0\| \leq \|x_n - x_0\| \leq (1 + \varepsilon_n) \|P^{(n)}x_0\|,$$

where $\varepsilon_n \to 0$ as $n \to \infty$ (N.N. Bogolyubov, N.I. Pol'skii).

Exercise 15.14. Under the assumptions of Theorem 15.3, let the subspaces E_n be finite-dimensional. Let B be an unbounded linear operator mapping E into a Banach space E', such that $x_0 \in D(B)$, $E_n \subset D(B)$. Show that anyone of the following conditions is sufficient for $\|B(x_n - x_0)\| \to 0$:

a) $\|BP_n T\| \cdot \|P^{(n)} x_0\| + \|BP^{(n)} x_0\| \to 0$;

b) $\beta_n \|P^{(n)} x_0\| + \|BP^{(n)} x_0\| \to 0$,

where $\beta_n = \sup\limits_{x \in E_n, \|x\|=1} \|Bx\|$;

c) $\|BQ_n\| \cdot \|P^{(n)} x_0\| + \|BQ^{(n)} x_0\| \to 0$,

where $Q^{(n)} = I - Q_n$, Q_n being any projection onto E_n.
Find the rate of convergence in each case.

15.5. Supplements to Theorem 15.3. To facilitate the application of Theorem 15.3, we shall indicate a few sufficient conditions for the relation $\|P^{(n)} T\| \to 0$ to hold.

Lemma 15.3. *Assume that, for any $x \in E$,*

$$\rho(Tx, E_n) \leq \eta_n \|x\| \qquad (n = 1, 2, \dots) \tag{15.21}$$

(η_n is a constant depending on n).

Then, if the operators P_n are bounded in E and

$$\|P_n\| \eta_n \to 0 \qquad as \qquad n \to \infty ,$$

we have $\|P^{(n)} T\| \to 0$ as $n \to \infty$.

In fact, by inequalities (15.20) and (15.21),

$$\|P^{(n)} Tx\| \leq 2\|P_n\| \rho(Tx, E_n) \leq 2\|P_n\| \eta_n \|x\| \qquad (x \in E) .$$

Lemma 15.4. *Let the operator T be compact in E, and the projections P_n bounded in E. Let $P_n \to I$ strongly, i.e., for any $x \in E$,*

$$\|P_n x - x\| \to 0 \qquad as \qquad n \to \infty .$$

Then $\|P^{(n)} T\| \to 0$ as $n \to \infty$.

To prove this it suffices to note that any strongly convergent sequence of continuous operators converges uniformly on any compact set. The details are left to the reader.

Exercise 15.15. The assumptions of Lemma 15.4 do not necessarily imply that $\|T P^{(n)}\| \to 0$ as $n \to \infty$. Discuss the following example. Let E be the space of functions integrable on

[0, 1], with norm

$$\|x\| = \int_0^1 |x(s)| \, ds,$$

E_n the space of functions vanishing for $0 \leqq t \leqq 1/n$,

$$P_n x(t) = \begin{cases} 0 & \text{if} \quad 0 \leqq t \leqq 1/n, \\ x(t) & \text{if} \quad 1/n < t \leqq 1, \end{cases}$$

$Tx = f(x) x_0$, where $f(x) = \int_0^1 x(s) \, ds$, and $x_0(t)$ is a fixed function in E.

The following analog of Lemma 15.4 is proved similarly.

Lemma 15.5. *Let E' be a Banach space which is continuously embeddable in E, i.e.,*

$$E' \subset E, \quad \|x\|_E \leqq c\|x\|_{E'} \quad (x \in E').$$

Let T be a compact operator mapping E into E', and the projections P_n bounded as operators mapping E' into E; assume that $P_n \to P$ strongly, where P is the operator embedding E' in E.

Then $\|P^{(n)}T\| \to 0$ as $n \to \infty$.

Let P_n be a sequence of bounded projections converging strongly to the identity operator I. It was remarked in Exercise 15.15 that compact operators T do not always satisfy the relation $\|TP^{(n)}\| \to 0$. However, if the projections P_n are constructed relative to a basis $\{e_k\}$ by formulas analogous to (15.4), and the space E is reflexive, then $\|TP^{(n)}\| \to 0$. In fact, the adjoints P_n^* then converge strongly to the identity of E^*, and the operator T^* is compact in E^*. By Lemma 15.4,

$$\|P_n^*T^* - T^*\| \to 0 \qquad \text{as} \qquad n \to \infty,$$

and our assertion follows from the equality

$$\|P_n^*T^* - T^*\| = \|TP_n - T\|.$$

Exercise 15.16. Let P_n be a sequence of projections (15.4) in a space E whose dual is not separable. Construct a compact linear operator T such that $\|TP^{(n)}\| \geqq 1 \ (n = 1, 2, \ldots)$.

We conclude with one more important remark on the approximate solution of equation (15.15) when T is compact.

The estimate (15.17) means that the more rapid the convergence of the number sequence $\|P^{(n)}x_0\|$ to zero, the "better" the convergence of the Galerkin method. In this connection, one is naturally interested in transforming equation (15.15) to the form

$$y = T_1 y + f_1 \tag{15.22}$$

for application of the Galerkin method, in such a way that, if y_0 is a solution of the new equation, the number sequence $\|P^{(n)} y_0\|$ converges to zero much more rapidly than the sequence $\|P^{(n)} x_0\|$. In many cases, this may be achieved by the following simple device. Look for a solution of equation (15.15) in the form

$$x = y + f + Tf + \ldots + T^{m-1} f \qquad (m \geqq 1). \tag{15.23}$$

Then y is determined from the equation

$$y = Ty + T^m f, \tag{15.24}$$

whose solution y_0 is related to the solution x_0 of equation (15.15) by

$$y_0 = T^m x_0. \tag{15.25}$$

Equation (15.24) is suitable if it is somehow known that the sequence $\|P^{(n)} T^m x_0\|$ converges to zero more rapidly than the sequence $\|P^{(n)} x_0\|$.

Exercise 15.17. Let $\|P^{(n)} T\| = o\left(\|P^{(n)} f\|\right)$. Show that for any $m (m \geqq 1)$

$$\|P^{(n)} T^m x_0\| = o\left(\|P^{(n)} x_0\|\right).$$

15.6. *Derivation of an equation of the second kind.* The projection method (15.2) may also be applied to equation (15.1) when the projections P_n are not bounded (provided, of course, that $LE_n \subset D(P_n)$ and $f \in D(P_n)$). We consider the case in which equation (15.1) has the form

$$Lu \equiv Au - Ku = f. \tag{15.26}$$

Equation (15.2) is then

$$P_n(Au_n - Ku_n - f) = 0 \qquad (u_n \in E_n). \tag{15.27}$$

Let us assume that the operator A is continuously invertible on F and $D(A) \subset D(K)$. We can then perform the substitution

$$x = Au \tag{15.28}$$

in equation (15.26). x must satisfy the following equation in F:

$$x = KA^{-1} x + f. \tag{15.29}$$

Equation (15.27) becomes

$$P_n(x_n - KA^{-1}x_n - f) = 0 \qquad (x_n \in AE_n).$$ (15.30)

Now assume that the subspaces E_n and F_n are such that

$$F_n = AE_n \qquad (n = 1, 2, \ldots).$$ (15.31)

Then equation (15.30) assumes the following simple form:

$$x_n = P_n KA^{-1}x_n + P_n f.$$ (15.32)

Theorem 15.3 is now applicable, and it implies

Theorem 15.4. Let the operator KA^{-1} be continuous, the operator $L = A - K$ continuously invertible, and $\|P^{(n)}KA^{-1}\| \to 0$ as $n \to \infty$. Assume that the subspaces E_n and F_n satisfy (15.31).

Then, for sufficiently large n, equation (15.27) has a unique solution u_n. The sequence Au_n converges to Au_0 (where u_0 is a solution of equation (15.26)) if and only if $P^{(n)}f \to 0$ as $n \to \infty$. Moreover,

$$c_1 \|P^{(n)}Au_0\| \leqq \|A(u_n - u_0)\| \leqq c_2 \|P^{(n)}Au_0\|.$$ (15.33)

Recall that the assumption $\|P^{(n)}KA^{-1}\| \to 0$ of Theorem 15.4 holds automatically if $P_n \to I$ strongly and the operator KA^{-1} is compact in F.

Note that in practice Theorem 15.3 is usually more convenient than Theorem 15.4; the above transformation of variables is then carried out separately for each application.

It follows from (15.33) that $\|u_n - u_0\| \leqq c_2 \|A^{-1}\| \cdot \|P^{(n)}Au_0\|$. This estimate is often very coarse.

Exercise 15.18. Let the projections P_n be bounded and assume the conditions of Theorem 15.4. Let B be a linear operator mapping E into a Banach space E', $E_n \supset D(B)$ ($n = 1, 2, \ldots$), and BA^{-1} bounded. Show that the solutions u_0 and u_n of equations (15.26) and (15.27) satisfy the inequality

$$\|B(u_n - u_0)\| \leqq \|BL^{-1}P^{(n)}Au_0\| + \varepsilon_n\|BL^{-1}P^{(n)}\| \cdot \|P^{(n)}Au_0\| \leqq$$

$$\leqq (1 + \varepsilon_n)\|BL^{-1}P^{(n)}\| \cdot \|P^{(n)}Au_0\|,$$ (15.34)

where

$$\varepsilon_n = c_2 \|P^{(n)}KA^{-1}\| \to 0 \qquad \text{as} \qquad n \to \infty.$$ (15.35)

Exercise 15.19. Let A^{-1} be continuous on F. Assume that $D(B) \supset D(A)$ and the restriction of the operator B to $D(A)$ has a closure (e.g., B is closed). Show that the operator BA^{-1} is continuous.

15.7. *The collocation method.** We now demonstrate the application of Theorem 15.3 to analysis of a specific approximate method for solution of boundary-value problems for ordinary differential equations. Consider the linear equation

$$Lu \equiv u^{(m)} - \sum_{k=0}^{m-1} p_k(t) u^{(k)} = f(t) \qquad (15.36)$$

with the homogeneous linear boundary conditions

$$\sum_{k=0}^{m-1} \left[\alpha_{ik} u^{(k)}(a) + \beta_{ik} u^{(k)}(b) \right] = 0 \qquad (i = 1, \ldots, m). \qquad (15.37)$$

Suppose that under the boundary conditions (15.37) the only solution of the homogeneous equation $u^{(m)} = 0$ is the trivial solution $u(t) \equiv 0$. Then, as is easily seen, there exists a sequence of polynomials

$$\phi_j(t) = \sum_{k=0}^{m-1} c_{jk} t^k + t^{m+j} \qquad (j = 0, 1, 2, \ldots)$$

of degrees m, $m + 1$, $m + 2$, ... which satisfy the boundary conditions (15.37). We shall look for an approximate solution of the boundary problem (15.36)–(15.37) in the form

$$u_n(t) = \sum_{j=0}^{n} \xi_j \phi_j(t). \qquad (15.38)$$

The basic idea of the collocation method is to determine the coefficients ξ_j by the condition that the residuals $Lu_n - f$ must vanish at $n + 1$ fixed points $t_{0n}, t_{1n}, \ldots, t_{nn}$ (the *interpolation points*), i.e.,

$$u_n^{(m)}(t_{in}) - \sum_{k=0}^{m-1} p_k(t_{in}) u_n^{(k)}(t_{in}) - f(t_{in}) = 0 \qquad (i = 0, \ldots, n). \qquad (15.39)$$

---

* The *collocation method* (or *coincidence method*) was proposed by Kantorovich for the approximate solution of integral and differential equations. Kantorovich investigates cases in which the coordinate sequence consists of trigonometric functions. The case of algebraic polynomials has been studied by Kaprilovskaya [1, 2] (see also Kantorovich and Akilov [1]), Kis [1, 2], Vainikko [6, 9]. Petersen [1] has considered various modifications of the collocation method.

In other words, the coefficients ξ_j are found by solving the system of equations

$$\sum_{j=0}^{n} \left[\phi_j^{(n)}(t_{in}) - \sum_{k=0}^{m-1} p_k(t_{in}) \phi_j^{(k)}(t_{in}) \right] \xi_j = f(t_{in}) \quad (i = 0, \dots, n). \quad (15.40)$$

To determine a solution $u(t)$ of the problem (15.36)–(15.37), one can first try to determine its derivative $x(t) = u^{(m)}(t)$. Then $u(t)$ is given by

$$u(t) = \int_a^b G(t, s) \, x(s) \, ds, \quad (15.41)$$

where $G(t, s)$ is the Green's function of the differential operator $Au = u^{(m)}$ with the boundary conditions (15.37). To determine the function $x(t)$ we clearly obtain the equation

$$x = Tx + f \quad (15.42)$$

with the integral operator

$$Tx = \int_a^b K(t, s) \, x(s) \, ds,$$

where

$$K(t, s) = \sum_{j=0}^{m-1} p_j(t) \frac{\partial^j G(t, s)}{\partial t^j}.$$

It is easy to see that equation (15.42) is entirely equivalent to the boundary-value problem (15.36)–(15.37).

Approximation of the solution of the problem (15.36)–(15.37) by (15.38) is equivalent to approximation of the solution of equation (15.42) by

$$x_n(t) = \sum_{j=0}^{n} \xi_j \phi_j^{(m)}(t).$$

Equations (15.39) of the collocation method for the coefficient ξ_j may be written in the form

$$x_n = P_n T x_n + P_n f, \quad (15.43)$$

where P_n is a projection mapping every continuous function onto its n-th degree interpolation polynomial at the points $t_{0_n}, t_{1_n}, \dots, t_{nn}$. In fact, $P_n z = 0$ for any function $z(t)$ such that $z(t_{in}) = 0$ $(i = 0, \dots, n)$;

thus conditions (15.39) mean that $P_n(x_n - Tx_n - f) = 0$. But $x_n = u_n^{(m)}$ is a polynomial of degree $\leq n$, and so $P_n x_n = x_n$ and we get equation (15.43).

Whether the collocation method converges or diverges depends essentially on the choice of the interpolation points. We shall assume henceforth that these points are the roots of certain orthogonal polynomials. More precisely, let $\rho(t)$ be a given nonnegative integrable function on $[a, b]$ such that

$$\int_a^b \frac{ds}{\rho(s)} < \infty \,,$$

and let $\omega_n(t)$ ($n = 0, 1, 2, \ldots$) be a system of polynomials obtained by orthogonalizing the sequence 1, t, t^2, $\ldots$ with respect to the scalar product

$$(x, y) = \int_a^b \rho(s) \, x(s) \, y(s) \, ds \,.$$

The interpolation points $t_{0n}, t_{1n}, \ldots, t_{nn}$ will be the roots of the polynomial $\omega_{n+1}(t)$. It is known (see Szegö [1]) that all these roots are simple and lie in $[a, b]$.

In actual computations one usually uses the Chebyshev points. When $a = -1$, $b = 1$, the Chebyshev points are given by

$$t_{in} = \cos \frac{2i + 1}{2(n + 1)} \pi \qquad (i = 0, 1, \ldots, n) \,;$$

the corresponding weighting function is $\rho(t) = \dfrac{1}{\sqrt{1 - t^2}}$. (In our subsequent theoretical investigation of the convergence of the collocation method, we shall assume $\rho(t)$ arbitrary).

The space E in which we consider equations (15.42) and (15.43) may be chosen in various ways. The following result pertains to the space $L_{2,\rho}$ of functions $x(t)$ which are square-integrable with weighting function $\rho(t)$, with the norm

$$\|x\|_{L_2,\rho}^2 = \int_a^b \rho(s) \, |x(s)|^2 \, ds \,.$$

Theorem 15.5. *Let the coefficients $p_j(t)$ ($j = 0, \ldots, m - 1$) and the free term $f(t)$ in equation (15.36) be continuous on $[a, b]$, and assume that the*

boundary-value problem (15.36)–(15.37) *has a unique solution* $u_0(t)$.
Then, for sufficiently large n, the system (15.40) *has a unique solution,
and the sequence of approximations* $u_n(t)$ *and their first* $m - 1$ *derivatives
converge uniformly to the solution* $u_0(t)$ *and its corresponding derivatives
as* $n \to \infty$; *the sequence* $u_n^{(m)}(t)$ *converges to* $u_0^{(m)}(t)$ *in quadratic mean
with weight* $\rho(t)$. *Moreover,*

$$\left\| u_n^{(m)} - u_0^{(m)} \right\|_{L_2,\rho} \leq c e_n(u_0^{(m)}), \tag{15.44}$$

$$\max_{a \leq t \leq b} \left| u_n^{(k)}(t) - u_0^{(k)}(t) \right| \leq c e_n(u_0^{(m)}) \quad (k = 0, \ldots, m - 1), \tag{15.45}$$

where

$$e_n(z) = \min_{\gamma_0, \ldots, \gamma_n} \ \max_{a \leq t \leq b} \left| z(t) - \sum_{j=0}^{n} \gamma_j t^j \right|$$

is the best uniform approximation of the function $z(t)$ *by polynomials of
degree* $\leq n$.

Proof. It is not difficult to see that T is compact as an operator mapping L_2 into C, *a fortiori* as an operator in $L_{2,\rho}$. On the other hand,
by the theorem of Erdös and Turan (see Natanson [2]), the Lagrange
interpolation polynomial based on the interpolation points adopted
here converges to the approximated function in quadratic means with
weight $\rho(t)$. In other words, $P_n \to P$ strongly, where P is the operator
embedding C in $L_{2,\rho}$. By Lemma 15.5 (with $E = L_{2,\rho}$, $E' = C$),
$\left\| P^{(n)} T \right\|_{L_{2,\rho}} \to 0$ as $n \to \infty$. The operator $I - T$ is continuously invertible
in $L_{2,\rho}$ since the homogenous equation $x = Tx$ has only the trivial
solution (by assumption—the boundary-value problem (15.36), (15.37)
has a unique solution).

By Theorem 15.3, equation (15.43) has a unique solution x_n for sufficiently large n, and this is equivalent to unique solvability of the system
(15.40) by the collocation method. Let us transform the estimate (15.17).
For any polynomial z_n of degree $\leq n$,

$$\left\| P^{(n)} x_0 \right\|_{L_2,\rho} = \left\| (x_0 - z_n) - P_n(x_0 - z_n) \right\|_{L_2,\rho} \leq \left\| x_0 - z_n \right\|_{L_2,\rho} +$$

$$+ \left\| P_n(x_0 - z_n) \right\|_{L_2,\rho} \leq \left\{ \left[\int_a^b \rho(t)\, dt \right]^{1/2} + \left\| P_n \right\|_{C \to L_2,\rho} \right\} \left\| x_0 - z_n \right\|_C.$$

Since $P_n \to P$ strongly, the norms $\|P_n\|_{C \to L_{2,\rho}}$ are uniformly bounded. Since z_n is an arbitrary polynomial,

$$\|P^{(n)} x_0\|_{L_{2,\rho}} \leqq c\, e_n(x_0) \qquad (c = \text{const}).$$

Now, in view of the fact that $x_n = u_n^{(m)}$, $x_0 = u_0^{(m)}$, we get the estimate (15.44) from (15.17); the estimate (15.45) follows from (15.44). Finally, the assertion of the theorem concerning the convergence of the approximations u_n and their derivatives follows from (15.44) and (15.45), since $e_n(z) \to 0$ for any continuous function $z(t)$. ∎

Note that the projections P_n are not bounded in $L_{2,\rho}$.

Exercise 15.20. Show that the projections P_n are not closed in $L_{2,\rho}$ and have no closure.

If equations (15.42) and (15.43) are considered in C, the projections P_n are bounded, but $\|P_n\|_C \to 0$ as $n \to \infty$, at least as rapidly as log n (see subsection 15.2). For the Chebyshev points, this slowest rate of growth of the norms $\|P_n\|_C$ is actually attained (see Natanson [2]):

$$\|P_n\|_C \leqq c_1 + c_2 \ln n \qquad (n = 1, 2, \dots). \tag{15.46}$$

Exercise 15.21. Let

$$\omega\left(\frac{1}{n}; f\right) = o\left(\frac{1}{\ln n}\right), \qquad \omega\left(\frac{1}{n}; p_k\right) = o\left(\frac{1}{\ln n}\right) \qquad (k = 0, \dots, m-1),$$

where

$$\omega(h; z) = \sup_{t_1, t_2 \in [-1, 1], |t_1 - t_2| \leqq h} |z(t_1) - z(t_2)|$$

is the *modulus of continuity* of the function $z(t)$. Prove that $u_n^{(m)}(t)$ converges uniformly to $u_0^{(m)}(t)$ in the case of Chebyshev points.

Hint. Use Lemma 15.3 ($E = C$), estimate (15.46) and the inequality (Natanson [2])

$$e_n(z) \leqq c\,\omega(1/n; z) \qquad (n = 1, 2, \dots),$$

where the constant c is independent of $z(t)$.

15.8. *The factor method.* As we have seen, projection methods replace the given equation by an approximating equation in a subspace. The factor method is based on an analogous idea—to replace the given equation by an equation in a factor space.

We recall a few concepts relative to factor spaces. Let F' be a closed subspace of a Banach space F. The elements of the factor space F/F'

are the cosets $\bar{x} = x + F'$ $(x \in F)$. The operator $px = \bar{x}$ is the *canonical mapping* of F into F/F'. If we define a norm

$$\|\bar{x}\|_{F/F'} = \inf_{x \in \bar{x}} \|x\|_F,$$

then F/F' becomes a Banach space (and $\|p\| = 1$). In the sequel we shall always assume that factor spaces are normed in this way.

Let us return to equation (15.1). Recall that the operator L in this equation maps the Banach space E into a Banach space F; we shall assume that it is invertible in $R(L)$. The *factor method* for equation (15.1) proceeds as follows. Two sequences of closed subspaces

$$E^{(n)} \subset E, \ F^{(n)} \subset F \qquad (n = 1, 2, \ldots)$$

are given, and (15.1) is replaced by equations in the factor spaces:

$$\bar{L}_n \bar{u}_n = p_n f. \tag{15.47}$$

Here $\bar{L}_n$ is a linear operator mapping the factor space $E/E^{(n)}$ into $F/F^{(n)}$, and p_n is the canonical mapping of F into $F/F^{(n)}$. We denote the canonical mapping of E into $E/E^{(n)}$ by r_n.

A natural indicator of the closeness of the operators L and $\bar{L}_n$ is the norm $\|p_n L u - \bar{L}_n r_n u\|_{F/F^{(n)}}$ $(u \in D(L))$. The operators $\bar{L}_n$ are said to approximate the operator L on $D(L)$ if, for every fixed $u \in D(L)$, we have the *approximation condition*

$$\|p_n L u - \bar{L}_n r_n u\|_{F/F^{(n)}} \to 0 \qquad \text{as} \qquad n \to \infty. \tag{A}$$

A second important characteristic of the factor method (15.47) is its stability or instability. The factor method (15.47) is said to be *stable* if, for sufficiently large n ($n \geq n_0$, say), the operators L_n are continuously invertible and the norms of their inverses are uniformly bounded:

$$\|\bar{L}_n^{-1}\|_{F/F^{(n)} \to E/E^{(n)}} \leq c \qquad (n = n_0, n_0 + 1, \ldots). \tag{B}$$

Finally, the factor method is said to *converge* for a given $f \in R(L)$ if

$$\|\bar{u}_n - r_n u_0\|_{E/E^{(n)}} \to 0 \qquad \text{as} \qquad n \to \infty, \tag{C}$$

where $\bar{u}_n$ and u_0 are solutions of equations (15.47) and (15.1), respectively.

The factor method is evidently the abstract analog of finite-difference methods.

By analogy with the theory of finite-difference methods, we have

Theorem 15.6. *Let* $f \in R(L)$ *and assume that* $u_0 = L^{-1}f$ *satisfies condition* (A). *Then* (B) *implies* (C).

The trivial proof is based on the obvious equality

$$\bar{u}_n - r_n u_0 = \bar{L}_n^{-1}(p_n L u_0 - \bar{L}_n r_n u_0).$$

Theorem 15.6 has the following converse.

Theorem 15.7. *Let* $R(L) = F$. *Assume that the operators* $\bar{L}_n$ $(n = 1, 2, \dots)$ *are bounded, and that* (C) *holds for any* $f \in F$.
Then (B) *is satisfied.*

A proof may be found in Gudovich [1] or in Krein's monograph [1]. The latter also contains applications of Theorems 15.6 and 15.7 to approximate solution of evolution equations in Banach spaces.

Gudovich has remarked that projection methods may be regarded in a natural way as particular cases of the factor method. We shall illustrate this for the projection method (15.2) for equation (15.1). Assume that the subspaces

$$E_n \subset D(L) \subset E, \quad F_n \subset F \qquad (n = 1, 2, \dots)$$

are finite-dimensional. Along with the projections P_n $(P_n F = F_n)$, we introduce certain projections R_n mapping E onto E_n. Assuming the projections R_n and P_n bounded, construct factor spaces $E/E^{(n)}$ and $F/F^{(n)}$, where

$$E^{(n)} = R^{(n)}E, \quad F^{(n)} = P^{(n)}F \quad (R^{(n)} = I - R_n, P^{(n)} = I - P_n).$$

It is not hard to see that the projection method (15.2) is equivalent to the factor method (15.47) with p_n the canonical mapping of F into $F/F^{(n)}$ and

$$\bar{L}_n = p_n L \Phi_n,$$

where the operator $\Phi_n \colon E/E^{(n)} \to E_n$ maps each coset $\bar{x} = x + E^{(n)}$ onto the element $R_n x$. The solutions of equations (15.2) and (15.47) are related by

$$u_n = \Phi_n \bar{u}_n.$$

Exercise 15.22. Let r_n be the canonical mapping of E into $E/E^{(n)}$ $(E^{(n)} = R^{(n)}E)$, and Φ_n the operator just defined. Show that

$$r_n = \Phi_n^{-1} R_n, \|\Phi_n\| \leq \|R_n\|, \|\Phi_n^{-1}\| \leq 1.$$

§ 16. The Bubnov–Galerkin and Galerkin–Petrov methods

16.1. *Convergence of the Bubnov–Galerkin method for equations of the second kind.* Consider the equation

$$x = Tx + f, \tag{16.1}$$

where T is a compact linear operator in a Hilbert space H. Let $\{H_n\}$ be an ultimately dense sequence of closed subspaces of H, and P_n the corresponding orthogonal projections (the fact that $\{H_n\}$ is ultimately dense in H then means that $P_n \to I$ strongly). The Bubnov–Galerkin method

$$P_n(x_n - Tx_n - f) = 0 \qquad (x_n \in H_n)$$

gives the equation

$$x_n = P_n Tx_n + P_n f. \tag{16.2}$$

By Lemma 15.4, $\|P^{(n)}T\| \to 0$ as $n \to \infty$, where $P^{(n)} = I - P_n$. Assume that the homogeneous equation $x = Tx$ has no nontrivial solutions. Then the operator $I - T$ is continuously invertible, and by Theorem 15.3 equation (16.2) has a unique solution x_n for sufficiently large n; $x_n \to x_0$ as $n \to \infty$, where x_0 is a solution of equation (16.1). We leave it to the reader to establish the inequality

$$\|P^{(n)}x_0\| \leqq \|x_n - x_0\| \leqq (1 + \varepsilon_n) \|P^{(n)}x_0\|,$$

where $\varepsilon_n = \|P^{(n)}T^*\| \cdot \|(I - P_nT)^{-1}\| \to 0$ as $n \to \infty$.

Just as in subsection 15.6, this result carries over to more general equations, via transformation of variables.

In subsection 16.3 we shall indicate another method for generalization, based on the introduction of new function spaces.

16.2 *Necessary and sufficient conditions for convergence of the Galerkin–Petrov method for equations of the second kind.* We shall now solve equation (16.1) by the Galerkin–Petrov method. Let $\{G_n\}$ and $\{H_n\}$ be two sequences of subspaces of H, Q_n and P_n the corresponding projections ($Q_n H = G_n$, $P_n H = H_n$). We wish to find an approximate solution from the condition

$$P_n(x_n - Tx_n - f) = 0 \qquad (x_n \in G_n). \tag{16.3}$$

The necessary and sufficient conditions given by Theorem 15.3 for convergence of the Galerkin–Petrov method involve the properties of the subspaces $(I - T) G_n$ and H_n. We shall now indicate more convenient conditions, expressed in terms of the subspaces G_n and H_n.

Theorem 16.1. Assume that the operator T is compact and the only solution of the homogeneous equation $x = Tx$ is trivial.

Let (P) be the statement: For any $f \in H$, for sufficiently large n ($n \geqq n_0$, say), there exists a unique element x_n satisfying the Galerkin–Petrov equation (16.3); as $n \to \infty$ the sequence x_n converges in norm to a solution x_0 of equation (16.1). Then a necessary and sufficient condition for (P) to hold is

1) $Q_n \to I$ *strongly;*

2) $\overline{\lim_{n \to \infty}} \; \Theta \, (G_n, H_n) < 1$,

where $\Theta \, (G_n, H_n)$ is the aperture of the subspaces G_n and H_n.

Under these conditions, the rate of convergence is described by

$$\|Q^{(n)} x_0\| \leqq \|x_n - x_0\| \leqq c \|Q^{(n)} x_0\| \quad (c = \text{const}; Q^{(n)} = I - Q_n). \quad (16.4)$$

Proof.

Sufficiency. Assume that conditions 1 and 2 hold. By 2, $\Theta(G_n, H_n) \leqq q$ for sufficiently large n, where $0 \leqq q < 1$. For these values of n, Lemma 15.1 implies that P_n maps G_n biuniquely onto H_n, and

$$\|\tilde{P}_n^{-1}\| = \frac{1}{\sqrt{1 - [\Theta(G_n, H_n)]^2}} \leqq \frac{1}{\sqrt{1 - q^2}},$$

where $\tilde{P}_n$ is the restriction of P_n to G_n.

The Galerkin–Petrov method (16.3) is equivalent to the Galerkin method

$$\bar{P}_n(x_n - Tx_n - f) = 0 \qquad (x_n \in G_n),$$

where $\bar{P}_n = \tilde{P}_n^{-1} P_n$ is a projection mapping H onto G_n ($\bar{P}_n$ is not an orthogonal projection if $G_n \neq H_n$). This last condition, in turn, can be expressed as an equation

$$x_n = \bar{P}_n Tx_n + \bar{P}_n f. \quad (16.5)$$

The Galerkin method (16.5) satisfies the assumptions of Theorem 15.3. In fact, we need only verify that $\|\bar{P}^{(n)} T\| \to 0$ as $n \to \infty$ ($\bar{P}^{(n)} = I - \bar{P}_n$).

For any $z \in H$,

$$\left\| \overline{P}^{(n)}z \right\| = \left\| (I - \tilde{P}_n^{-1}P_n)(z - Q_n z) \right\| \leq \left(1 + \frac{1}{\sqrt{1 - q^2}} \right) \left\| Q^{(n)}z \right\|, \quad (16.6)$$

whence, by condition 1, it follows that $\overline{P}_n \to I$ strongly. But then Lemma 15.4 implies that $\left\| \overline{P}^{(n)}T \right\| \to 0$.

By Theorem 15.3, equation (16.5) has a unique solution for sufficiently large n. In view of inequality (16.6), the estimate (15.17), which in this case is

$$\left\| x_n - x_0 \right\| \leq c_2 \left\| \overline{P}^{(n)}x_0 \right\|,$$

immediately yields (16.4). The fact that $x_n \to x_0$ follows from (16.4). This completes the proof of sufficiency.

Exercise 16.1. Prove the following inequality, which is sharper than (16.6):

$$\left\| \overline{P}^{(n)}z \right\| \leq \frac{1}{\sqrt{1 - [\Theta(G_n, H_n)]^2}} \left\| Q^{(n)}z \right\|.$$

To prove that the conditions of Theorem 16.1 are necessary, we need the following important proposition.

Lemma 16.1. *Let* T *be a compact operator in* H *such that* $I - T$ *is invertible. Let* $\{G_n\}$ *and* $\{H_n\}$ *be two sequences of closed subspaces of* H *such that* $\{G_n\}$ *is ultimately dense in* H *and*

$$\varlimsup_{n \to \infty} \Theta(G_n, H_n) < 1.$$

Then the sequence of subspaces $\{(I - T)G_n\}$ *is also ultimately dense in* H *and*

$$\varlimsup_{n \to \infty} \Theta((I - T)G_n, H_n) < 1.$$

In fact, since we have proved that conditions 1 and 2 of Theorem 16.1 are sufficient, the approximations x_n defined by (16.3) converge for any $f \in H$ to a solution x_0 of equation (16.1). This is true if and only if the residuals $(I - T)x_n - f$ converge to zero for any $f \in H$. Now apply Theorem 15.2.

Necessity. Assume that $x_n \to x_0$ for any $f \in H$; as already mentioned, this is true if and only if the residuals $(I - T)x_n - f$ converge to zero. By Theorem 15.2, the sequence $\{(I - T)G_n\}$ is then ultimately dense in

H and

$$\varlimsup_{n \to \infty} \Theta((I - T) G_n, H_n) < 1 .$$

Let S be a compact operator such that $I - S$ is invertible. By Lemma 16.1 (with S playing the role of T and $(I - T) G_n$ that of G_n), the sequence $\{(I - S)(I - T) G_n\}$ is ultimately dense in H, and

$$\varlimsup_{n \to \infty} \Theta((I - S)(I - T) G_n, H_n) < 1 .$$

Choose S so that $(I - S)(I - T) = I$, i.e., $S = -T(I - T)^{-1}$; it follows that conditions 1 and 2 are satisfied, and this completes the proof of Theorem 16.1. ∎

Exercise 16.2. Let T and S be compact operators such that $I - T$ and $I - S$ are invertible. Let $\{G_n\}$ and $\{H_n\}$ be sequences of closed subspaces of H such that $\{G_n\}$ is ultimately dense in H and

$$\varlimsup_{n \to \infty} \Theta(G_n, H_n) < 1 .$$

Show that

$$\varlimsup_{n \to \infty} \Theta((I - T) G_n, (I - S) H_n) < 1 .$$

16.3. *Convergence of the Bubnov–Galerkin method for equations with a positive definite principal part.* Let H be a Hilbert space; consider the equation

$$Au = Ku + f , \tag{16.7}$$

where A is a positive definite selfadjoint operator, i.e., $A^* = A$ and

$$(Au, u) \geqq \gamma^2 \|u\|^2 \qquad (\gamma = \text{const} > 0 ; u \in D(A)) .$$

For the moment, the only assumption on the operator K is that $D(A) \subset D(K)$. A solution of equation (16.7) is to be approximated by the Bubnov–Galerkin method

$$P_n(Au_n - Ku_n - f) = 0 \qquad (u_n \in H_n) , \tag{16.8}$$

where P_n are orthogonal projections of H onto subspaces $H_n \subset D(A) \subset H$ $(n = 1, 2, \dots)$.

Define a new inner product in $D(A)$:

$$(u, v)_A = (Au, v) .$$

The completion of $D(A)$ with respect to this inner product is a new Hilbert space H_A (the *Friedrichs space*). It is known that this completion may be carried out using only elements of H; the set of elements of H_A coincides with $D(A^{1/2})$, where $A^{1/2}$ is the unique positive definite self-adjoint operator such that $A^{1/2}A^{1/2} = A$. Then, for any $u, v \in H_A$,

$$(u, v)_A = (A^{1/2}u, A^{1/2}v), \quad \|u\|_A = \|A^{1/2}u\|.$$

The subspaces $H_n \subset D(A)$ $(n = 1, 2, \dots)$ are contained in H_A, and we shall assume that they are closed there. Let P_n^A denote the orthogonal (in the sense of the H_A-metric) projection of H_A onto H_n. Note that $P_n v = 0$ if and only if $P_n^A A^{-1}v = 0$. In fact, $P_n v = 0$ if and only if $(v, w_n) = 0$ for any $w_n \in H_n$. But

$$(v, w_n) = (AA^{-1}v, w_n) = (A^{-1}v, w_n)_A,$$

and so $P_n v = 0$ means that $(A^{-1}v, w_n)_A = 0$ for any $w_n \in H_n$, i.e., $P_n^A A^{-1}v = 0$.

In view of this remark, we can rewrite (16.8) as

$$P_n^A (u_n - A^{-1}Ku_n - A^{-1}f) = 0 \qquad (u_n \in H_n)$$

or, assuming that the operator $A^{-1}K$ is continuous in H_A,

$$u_n = P_n^A T u_n + P_n^A A^{-1}f, \tag{16.9}$$

where $T = \overline{A^{-1}K}$ is the closure* of the operator $A^{-1}K$ in H_A. The Bubnov–Galerkin method (16.8) for equation (16.7) reduces to the Bubnov–Galerkin method (16.9) for the equation

$$u = Tu + A^{-1}f, \tag{16.10}$$

in H_A.

Before formulating the result that follows from these arguments, we have two remarks.

First, we may relax the condition $H_n \subset D(A)$ $(n = 1, 2, \dots)$, replacing it by $H_n \subset H_A$ $(n = 1, 2, \dots)$. The approximations to the solutions of equation (16.7) will be the elements $u_n \in H_n$ determined by (16.9).

Second, equation (16.10) may not be equivalent to equation (16.7).

---

* The closure of a continuous linear operator is its extension by continuity to the closure of its domain of definition.

Any solution of (16.7) is clearly a solution of (16.10), but the converse is true only if the solution in question of (16.10) belongs to $D(A)$. In general, these solutions are only in H_A; we shall then regard them as generalized solutions of equation (16.7).

Exercise 16.3. Let $D(K) \supset H_A$. Show that the solutions of (16.10) are all in $D(A)$.

Exercise 16.4. Let the operator KA^{-1} be compact in H, and the operator $A^{-1}K$ compact in H_A; assume moreover that the homogeneous equation $u = Tu$ has no nontrivial solutions. Show that all the solutions of (16.10) are in $D(A)$.

The above arguments, together with the results of subsection 16.1, imply the following theorem.

Theorem 16.2.* *Assume that the operator $A^{-1}K$ is compact in H_A and the only solution of the homogeneous equation $u = Tu$ in H_A is trivial. Let the sequence of subspaces $\{H_n\}$ be ultimately dense in H_A.*

Then the Bubnov–Galerkin method yields a unique solution u_n for sufficienctly large n. The sequence u_n converges in the H_A-norm to a (true or generalized) solution u_0 of equation (16.7), the rate of convergence being

$$\left\| P_A^{(n)} u_0 \right\|_A \leqq \left\| u_n - u_0 \right\|_A \leqq (1 + \varepsilon_n) \left\| P_A^{(n)} u_0 \right\|_A ,$$

where $\varepsilon_n \to 0$ as $n \to \infty$, $P_A^{(n)}u = u - P_n^A u$.

Another widely applied method is Ritz's variational method for approximate solution of the equation

$$Au = f , \tag{16.11}$$

where A is a positive definite selfadjoint operator. This method is based on the observation that the solution of equation (16.11) minimizes the quadratic functional

$$\Phi(u) = (u, u)_A - (u, f) - (f, u) \tag{16.12}$$

on H_A. If the space is real, the functional $\Phi(u)$ may be written as

$$\Phi(u) = (u, u)_A - 2(u, f) .$$

Let u_0 denote a solution of equation (16.11). Then

$$\Phi(u) = \left\| u - u_0 \right\|_A^2 - \left\| u_0 \right\|_A^2 . \tag{16.13}$$

---

* This theorem is due to Mikhlin.

The Ritz method is to determine the minimum point u_n of the functional (16.12) on the closed subspace $H_n \subset H_A$. In view of (16.13),

$$u_n = P_n^A u_0 \, .$$

Thus the Ritz approximation is identical to the approximation given by the Bubnov–Galerkin method (16.9) (in this case $T = 0$). Consequently, all theorems on the applicability of the Bubnov–Galerkin method to equation (16.11) are relevant to the Ritz method.

A direct analysis of the Ritz method, investigating the quadratic functional (16.12), may be found in Mikhlin [2, 3, 5]. Mikhlin also describes numerous applications of the Ritz and Bubnov–Galerkin methods to various boundary-value problems.

16.4. *Lemma on similar operators.* Two positive definite selfadjoint operators A and B in a Hilbert space H are said to be *similar* if

$$D(A) = D(B) \, .$$

Assume that the inverse B^{-1} is compact. Let $\lambda_1 \leqq \lambda_2 \leqq \ldots \leqq \lambda_n \leqq \ldots$ be the eigenvalues of B ($\lambda_n > 0$ and $\lambda_n \to \infty$ as $n \to \infty$), and $\phi_1, \phi_2, \ldots, \phi_n, \ldots$ a suitable complete orthonormal system of eigenelements ($B\phi_n = \lambda_n \phi_n$).

Let α be a real number. Recall that the fractional power B^α of an operator B is defined by

$$B^\alpha u = \sum_{i=1}^\infty (u, \phi_i) \, \lambda_i^\alpha \phi_i \, ,$$

and B^α is defined on all elements $u \in H$ such that

$$\sum_{i=1}^\infty |(u, \phi_i)|^2 \, \lambda_i^{2\alpha} < \infty \, .$$

When $\alpha \geqq 0$ the operator B^α is selfadjoint and positive definite; when $\alpha \leqq 0$, it is selfadjoint and bounded. It is clear that

$$B^1 = B, \ B^0 = I, \ B^\alpha B^\beta \subset B^{\alpha+\beta}, \ (B^\alpha)^{-1} = B^{-\alpha} \, .$$

Exercise 16.5. Let A and B be similar operators, where B^{-1} is compact. Show that A^{-1} is also compact.

The fundamental theorems on similar operators are due to Heinz [1] (see also Krasnosel'skii, Zabreiko, Pustyl'nik and Sobolevskii [1]).

The principal theorem states that if A and B are similar so are A^α and B^α for all $\alpha \in [0, 1]$. Hence, the operators $A^\alpha B^{-\alpha}$ and $B^\alpha A^{-\alpha}$ are bounded in H.

Exercise 16.6. Let A and B be similar. Then $D(A^{1/2}) = D(B^{1/2})$, and therefore the spaces H_A and H_B have the same elements. Prove that their norms are equivalent.

Exercise 16.7. Let A and B be similar, and let $\{\phi_i\}$ be a complete orthonormal system of eigenelements of B. Show that, for any $\alpha \in [0, 1]$, the elements $\{A^\alpha \phi_i\}$ form a basis in H.

Lemma 16.2. *Let A and B be similar operators, with B^{-1} compact. Let $\{\phi_i\}$ be a complete orthonormal system of eigenelements of B, H_n the linear span of $\phi_1, \ldots, \phi_n$. Then, for any α and β, $0 \le \alpha, \beta \le 1$,*

$$\lim_{n \to \infty} \Theta(A^\alpha H_n, A^\beta H_n) < 1 \,,$$

where $\Theta(A^\alpha H_n, A^\beta H_n)$ is the aperture of the subspaces $A^\alpha H_n$ and $A^\beta H_n$.

Proof. Let $P_n^{(\alpha)}$ denote the orthogonal projection of H onto $A^\alpha H_n$ $(0 \le \alpha \le 1)$. In particular, $P_n^{(0)} = P_n$ projects H onto H_n. Let $P_\alpha^{(n)} = I - P_n^{(\alpha)}$, $P^{(n)} = I - P_n$. Obviously, $P_\alpha^{(n)} A^\alpha P_n = 0$; therefore,

$$P_\alpha^{(n)} A^\alpha = P_\alpha^{(n)} A^\alpha P^{(n)} \qquad (0 \le \alpha \le 1). \tag{16.14}$$

Now, since the subspace H_n is invariant under B^α,

$$P_n B^\alpha \subset P_n B^\alpha P_n = B^\alpha P_n, \quad P^{(n)} B^\alpha = P^{(n)} B^\alpha P^{(n)} = B^\alpha P^{(n)} \,.$$

It is also easy to see that

$$\| B^\delta P_n \| = \lambda_n^\delta, \quad \| P^{(n)} B^{-\delta} \| = \lambda_{n+1}^{-\delta} \qquad \text{for} \qquad \delta \ge 0 \,. \tag{16.15}$$

Since $\Theta(A^\alpha H_n, A^\beta H_n) = \Theta(A^\beta H_n, A^\alpha H_n)$, it suffices to prove the lemma for the case $\alpha \ge \beta$.

Consider the solution (in H) of the equation

$$A^\alpha v = f$$

by the Galerkin–Petrov method:

$$P_n^{(\beta)} (A^\alpha v_n - f) = 0 \qquad (v_n \in H_n). \tag{16.16}$$

We claim that for any $f \in H$ the residuals $A^\alpha v_n - f$ converge to zero; this will immediately imply the statement of our lemma, because of Theorem 15.2.

It is not difficult to see that $P_n^{(\beta)}z = 0$ if and only if $P_n^{(\gamma)}A^{\beta-\gamma}z = 0$ ($\beta \leq \gamma \leq 1$). This being so, condition (16.16) may be rewritten as

$$P_n^{(\gamma)}(A^\gamma v_n - A^{\gamma-\alpha}f) = 0 \qquad (v_n \in H_n),\qquad (16.17)$$

where $\gamma = \frac{1}{2}(\alpha + \beta)$. Now (16.17) is precisely the equation of the Galerkin–Petrov (least-squares) method for the equation $A^\gamma v = A^{\gamma-\alpha}f$. Since the sequence of subspaces $\{A^\alpha H_n\}$ is ultimately dense in H (see Exercise 16.7), and for every n

$$\Theta(A^\gamma H_n, A^\gamma H_n) = 0,$$

it follows from Theorem 15.2 that the residuals $A^\gamma v_n - A^{\gamma-\alpha}f$ converge to zero. Here the estimate (15.14) is

$$\left\| A^\gamma v_n - A^{\gamma-\alpha}f \right\| \leq \left\| P_\gamma^{(n)} A^{\gamma-\alpha}f \right\|.$$

Consider the right-hand side; by (16.14),

$$\left\| P_\gamma^{(n)} A^{\gamma-\alpha}f \right\| = \left\| P_\gamma^{(n)} A^\gamma P^{(n)} A^{-\alpha}f \right\| \leq \left\| A^\gamma P^{(n)} A^{-\alpha}f \right\| \leq$$

$$\leq \left\| A^\gamma B^{-\gamma} \right\| \cdot \left\| B^\gamma P^{(n)} A^{-\alpha}f \right\| = \left\| A^\gamma B^{-\gamma} \right\| \cdot \left\| P^{(n)} B^{\gamma-\alpha} P^{(n)} B^\alpha A^{-\alpha}f \right\|.$$

Thus the second equality of (16.15) implies that

$$\left\| A^\gamma v_n - A^{\gamma-\alpha}f \right\| \leq \lambda_{n+1}^{\gamma-\alpha} \left\| A^\gamma B^{-\gamma} \right\| \cdot \left\| P^{(n)} B^\alpha A^{-\alpha}f \right\|.$$

Hence

$$\left\| B^\gamma(v_n - A^{-\alpha}f) \right\| \leq \lambda_{n+1}^{\gamma-\alpha} \left\| B^\gamma A^{-\gamma} \right\| \cdot \left\| A^\gamma B^{-\gamma} \right\| \cdot \left\| P^{(n)} B^\alpha A^{-\alpha}f \right\|. \qquad (16.18)$$

We claim that $\left\| B^\alpha(v_n - A^{-\alpha}f) \right\| \to 0$ as $n \to \infty$. This will also prove that $\left\| A^\alpha v_n - f \right\| \to 0$ as $n \to \infty$, as required. Now

$$B^\alpha(v_n - A^{-\alpha}f) = B^\alpha(P_n + P^{(n)})(v_n - A^{-\alpha}f) =$$

$$= B^{\alpha-\gamma}P_n B^\gamma(v_n - A^{-\alpha}f) - P^{(n)}B^\alpha A^{-\alpha}f,$$

whence, by (16.18) and the first equality of (16.15),

$$\left\| B^\alpha(v_n - A^{-\alpha}f) \right\| \leq \left[\left\| B^\gamma A^{-\gamma} \right\| \cdot \left\| A^\gamma B^{-\gamma} \right\| \left(\frac{\lambda_n}{\lambda_{n+1}} \right)^{\alpha-\gamma} + 1 \right] \left\| P^{(n)} B^\alpha A^{-\alpha}f \right\|.$$

The right-hand side of this inequality tends to zero as $n \to \infty$. ∎

16.5. *Convergence of the Bubnov–Galerkin method.* We continue our study of the Bubnov–Galerkin method (16.8) for equation (16.7). Assume that the operator B is similar to A and B^{-1} is compact. Let $\{\phi_i\}$ be a complete system of eigenelements of B, arranged in order of increasing eigenvalues λ_i:

$$B\phi_i = \lambda_i\phi_i, \quad 0 < \lambda_1 \leqq \lambda_2 \leqq \ldots \quad (\lambda_i \to \infty \quad \text{as} \quad i \to \infty).$$

Assume that the operator P_n in the Galerkin equation (16.8) is the orthogonal projection onto the linear span H_n of the elements $\phi_1, \ldots, \phi_n$.

Suppose that there exists α_0, $0 \leqq \alpha_0 \leqq 1$, such that the operator $A^{\alpha_0-1}KA^{-\alpha_0}$ is compact in H, and let T_{α_0} be the closure of this operator. Assume that the only solution of the homogeneous equation $x = T_{\alpha_0}x$ is trivial. Then the equation

$$x = T_{\alpha_0}x + A^{\alpha_0-1}f \tag{16.19}$$

has a unique solution x_0. We call the element $u_0 = A^{-\alpha_0}x_0$ a *generalized solution* of equation (16.7); if $u_0 \in D(A)$, then u_0 is an ordinary solution.

Under these restrictions, the following theorem is true.

Theorem 16.3. For sufficiently large n, the Bubnov–Galerkin method (16.8) *yields a unique approximation* u_n. *If, moreover,* $u_0 \in D(B^{\alpha_1})$, *then*

$$\|P^{(n)}B^\alpha u_0\| \leqq \|B^\alpha(u_n - u_0)\| \leqq c\|P^{(n)}B^\alpha u_0\| \quad (\alpha_0 \leqq \alpha \leqq \alpha_1). \tag{16.20}$$

The proof is in the main similar to that of Lemma 16.1. Introducing the notation $x_n = A^{\alpha_0}u_n$, rewrite equation (16.8) as

$$P_n^{(1-\alpha_0)}(x_n - T_{\alpha_0}x_n - A^{\alpha_0-1}f) = 0 \quad (x_n \in A^{\alpha_0}H_n); \tag{16.21}$$

recall that $P_n^{(\alpha)}$ is the orthogonal projection of H onto $A^\alpha H_n$. Equation (16.21) is precisely the Galerkin–Petrov equation for (16.19). The convergence conditions (see Theorem 16.1) are satisfied, for the sequence of subspaces $\{A^{\alpha_0}H_n\}$ is ultimately dense in H, and, by Lemma 16.2,

$$\varlimsup_{n \to \infty} \Theta(A^{\alpha_0}H_n, A^{1-\alpha_0}H_n) < 1.$$

In this case, the estimate (16.4) is

$$\|x_n - x_0\| \leqq c_0\|P_{\alpha_0}^{(n)}x_0\|.$$

Using (16.14) and (16.15), we get

$$\left\| P_{\alpha_0}^{(n)} x_0 \right\| = \left\| P_{\alpha_0}^{(n)} A^{\alpha_0} u_0 \right\| \leqq \left\| A^{\alpha_0} P^{(n)} u_0 \right\| \leqq \left\| A^{\alpha_0} B^{-\alpha_0} \right\| \cdot \left\| B^{\alpha_0} P^{(n)} u_0 \right\| \leqq$$

$$\leqq \lambda_{n+1}^{\alpha_0 - \alpha} \left\| A^{\alpha_0} B^{-\alpha_0} \right\| \cdot \left\| P^{(n)} B^{\alpha} u_0 \right\| \qquad (\alpha_0 \leqq \alpha \leqq \alpha_1) .$$

Since $\left\| B^{\alpha_0}(u_n - u_0) \right\| \leqq \left\| B^{\alpha_0} A^{-\alpha_0} \right\| \cdot \left\| x_n - x_0 \right\|$, it follows that

$$\left\| B^{\alpha_0}(u_n - u_0) \right\| \leqq c_1 \lambda_{n+1}^{\alpha_0 - \alpha} \left\| P^{(n)} B^{\alpha} u_0 \right\| \qquad (\alpha_0 \leqq \alpha \leqq \alpha_1) , \qquad (16.22)$$

where $c_1 = c_0 \left\| B^{\alpha_0} A^{-\alpha_0} \right\| \cdot \left\| A^{\alpha_0} B^{-\alpha_0} \right\|$. Further,

$$B^{\alpha}(u_n - u_0) = B^{\alpha}(P_n + P^{(n)})(u_n - u_0) =$$

$$= B^{\alpha - \alpha_0} P_n B^{\alpha_0}(u_n - u_0) - P^{(n)} B^{\alpha} u_0 ,$$

so that, by (16.15) and (16.22),

$$\left\| B^{\alpha}(u_n - u_0) \right\| \leqq \left[c_1 \left(\frac{\lambda_n}{\lambda_{n+1}} \right)^{\alpha - \alpha_0} + 1 \right] \cdot \left\| P^{(n)} B^{\alpha} u_0 \right\| \leqq$$

$$\leqq (c_1 + 1) \left\| P^{(n)} B^{\alpha} u_0 \right\| .$$

We have thus proved the upper bound in (16.20) for the norm $\left\| B^{\alpha}(u_n - u_0) \right\|$. The lower bound is trivial:

$$\left\| B^{\alpha}(u_n - u_0) \right\| \geqq \left\| P^{(n)} B^{\alpha}(u_n - u_0) \right\| =$$

$$= \left\| P^{(n)} B^{\alpha} P^{(n)}(u_n - u_0) \right\| = \left\| P^{(n)} B^{\alpha} u_0 \right\| . \qquad ∎$$

Apart from the estimate (16.20), the equalities (16.15) also imply that

$$\left\| P^{(n)} B^{\alpha} u_0 \right\| = o(\lambda_{n+1}^{\alpha - \alpha_1}) \qquad (\alpha_0 \leqq \alpha \leqq \alpha_1) .$$

We note an important consequence of (16.20). If the operator KA^{-1} is compact in H and $I - KA^{-1}$ is invertible, the residual $Au_n - Ku_n - f$ converges to zero as $n \to \infty$. In fact, the estimates (16.20) apply to the case $\alpha = 1$ ($\alpha_0 = \alpha_1 = 1$), and so

$$\left\| Au_n - Ku_n - f \right\| \leqq \left\| (I - KA^{-1}) AB^{-1} \right\| \cdot \left\| B(u_n - u_0) \right\| \to 0 .$$

For an arbitrary coordinate sequence, one can only state that the approximations u_n converge in the H_A-norm to a solution (Theorem

16.2); the residuals may not converge to zero. The convergence of the residuals to zero in the case considered above was first established by Mikhlin.

Exercise 16.8. Prove that the operator $A^{-1/2}KA^{-1/2}$ is compact in H if and only if the operator $A^{-1}K$ is compact in H_A.

Exercise 16.9. Let α_0 and α'_0 be two values of the parameter α ($0 \leq \alpha \leq 1$) such that $A^{\alpha-1}KA^{-\alpha}$ is compact in H. Show that the operator $I - T_{\alpha_0}$ is invertible if and only if $I - T_{\alpha'_0}$ is invertible.

Exercise 16.10. Let the operator KA^{-1} be compact in H and $I - KA^{-1}$ invertible. Show that all generalized solutions of equation (16.7) are ordinary solutions, i.e., belong to $D(A)$ and satisfy equation (16.7).

Exercise 16.11. Let $K = 0$. Show that

$$\left\| B^\alpha(u_n - u_0) \right\| \leq (1 + \left\| B^{1/2}A^{-1/2} \right\| \cdot \left\| A^{1/2}B^{-1/2} \right\|) \left\| P^{(n)}B^\alpha u_0 \right\| \qquad (\tfrac{1}{2} \leq \alpha \leq 1),$$

$$\left\| u_n - u_0 \right\| \leq \lambda_{n+1}^{-1} \left\| BA^{-1} \right\| \cdot \left\| Au_n - f \right\|,$$

$$\left\| u_n - u_0 \right\|_A \leq \lambda_{n+1}^{-1/2} \left\| B^{1/2}A^{-1/2} \right\| \cdot \left\| Au_n - f \right\|.$$

Exercise 16.12. The boundary-value problem

$$- (p(t)u')' + q(t)u' + r(t)u = f(t),$$

$$u(0) = u(1) = 0$$

is to be solved by the Bubnov–Galerkin method, the coordinate sequence consisting of the functions $\phi_i(t) = \sin i\pi t (i = 1, 2, \ldots)$. Let $p(t) > 0$, $p(t) \in L_2^{(2)}$, $q(t), r(t), f(t) \in L_2^{(1)}$, where $L_2^{(i)}(i = 1, 2)$ are the sets of functions $x(t)$, with absolutely continuous $(i - 1)$-th derivative, such that $x^{(i)}(t) \in L_2[0, 1]$. Finally, assume that the problem has a unique solution $u_0(t)$. Establish the estimates

$$\left\| u_n - u_0 \right\| \leq cn^{-5/2}, \quad \left\| u'_n - u'_0 \right\| \leq cn^{-3/2}, \quad \left\| u''_n - u''_0 \right\| \leq cn^{-1/2}.$$

Show that these estimates are exact if $u''_0(0) \neq 0$ or $u''_0(1) \neq 0$.

Hint. Set $H = L_2[0, 1]$, $Au = - (p(t)u')'$, $Ku = q(t)u' + r(t)u$, $Bu = - u''$, and $D(A) = D(B) = L_2^{\circ(2)}$, where $L_2^{\circ(2)}$ is the set of $L_2^{(2)}$-functions satisfying the conditions $u(0) = u(1) = 0$. Prove that the operators KA^{-1} and $A^{-1}K$ are compact in $L_2[0, 1]$; use the estimates (16.20) for $\alpha = 0$, $\alpha = \tfrac{1}{2}$ and $\alpha = 1$ and find lower and upper bounds for the norm $\left\| P^{(n)}B^\alpha u_0 \right\|$ for these values of α.

The estimates (16.20) and Lemma 16.2 are due to Vainikko [2, 15].

16.6. *Regular operators.* A continuous linear operator L defined in a separable Hilbert space H is said to be *regular* if, for any sequence $\{H_n\}$ of finite-dimensional subspaces which is ultimately dense in H,

$$\varlimsup_{n \to \infty} \Theta(LH_n, H_n) < 1. \tag{16.23}$$

Any regular operator is compact (see Exercises 16.3–16.16).

Exercise 16.13. Let L be a regular operator. Show that $R(L)$ is dense in H and L^{-1} is defined on $R(L)$.

Exercise 16.14. Let L be a continuous linear operator such that L^{-1} is defined, but not bounded, on $R(L)$. Show that

$$\inf_{u \in H, u \perp H_n} \frac{|(Lu, u)|}{\|Lu\| \cdot \|u\|} = 0, \tag{16.24}$$

where H_n is any finite-dimensional subspace of H.

Exercise 16.15. Assume that (16.24) holds for any finite-dimensional subspace H_n. Show that H contains a sequence $\{v_n\}$ such that

1) $v_n \to 0$ weakly, $\|v_n\| = 1$;

2) $\dfrac{Lv_n}{\|Lv_n\|} \to 0$ weakly;

3) $\dfrac{|(Lv_n, v_n)|}{\|Lv_n\|} \to 0$.

Exercise 16.16. Assume that H contains a sequence $\{v_n\}$ with properties 1–3 of the previous exercise. Show that L is not regular.

It is clear that ultimate denseness in H is preserved by regular operators. In view of Theorem 15.2, therefore, the regular operators L are precisely those for which the residual $Lu_n - f$ (therefore also the error $u_n - L^{-1}f$) in the Bubnov–Galerkin method

$$P_n(Lu_n - f) = 0 \qquad (u_n \in H_n = P_n H) \tag{16.25}$$

for the equation $Lu = f$ converges to zero as $n \to \infty$, for any $f \in H$ and and sequence $\{H_n\}$ which is ultimately dense in H. (Recall that the projections P_n in the Bubnov–Galerkin method are orthogonal.)

Theorem 16.4. *A continuous operator is regular if and only if it has a representation*

$$L = \gamma I + S + T, \tag{16.26}$$

where $\gamma = \text{const} \neq 0$, *S is a continuous linear operator*, $\|S\| < |\gamma|$, *and T is a compact linear operator*.

Proof.

Sufficiency. Assuming that (16.26) is valid, we shall show that condition (16.23) holds for any sequence of subspaces $\{H_n\}$ which is ultimately dense in H. We first claim that

$$\varlimsup_{n \to \infty} \Theta((\gamma I + S) H_n, H_n) < 1. \tag{16.27}$$

To prove this, it suffices, by Theorem 15.2, to show that for any $f \in H$ the residual $\gamma u_n + S u_n - f$ in the Bubnov–Galerkin method

$$P_n(\gamma u_n + S u_n - f) = 0 \qquad (u_n \in H_n) \tag{16.28}$$

applied to the equation

$$\gamma u + S u = f \tag{16.29}$$

converges to zero as $n \to \infty$. Condition (16.28) holds if and only if

$$\gamma u_n + P_n S u_n = P_n f .$$

Since $\|P_n S\| \le \|S\| < |\gamma|$, the operators $\gamma I + P_n S$ and $\gamma I + S$ are invertible in H, and

$$\|(\gamma I + P_n S)^{-1}\| \le \frac{1}{|\gamma| - \|S\|} . \tag{16.30}$$

Thus, equations (16.28) and (16.29) have unique solutions u_n and u_0, respectively. Clearly,

$$(\gamma I + P_n S)(u_0 - u_n) = P^{(n)} u_0 \qquad (P^{(n)} - I - P_n),$$

so that, by (16.30) and the fact that $\{H_n\}$ is ultimately dense,

$$\|u_n - u_0\| \le \frac{1}{|\gamma| - \|S\|} \|P^{(n)} u_0\| \to 0 \qquad \text{as} \qquad n \to \infty .$$

At the same time,

$$\gamma u_n + S u_n - f = (\gamma I + S)(u_n - u_0) \to 0 \qquad \text{as} \qquad n \to \infty .$$

This proves (16.27). It is also clear that $\{(\gamma I + S) H_n\}$ is ultimately dense in H (this is in effect a corollary of Theorem 15.2). But

$$L H_n = (\gamma I + S + T) H_n = (I + T_1)(\gamma I + S) H_n ,$$

where $T_1 = T(\gamma I + S)^{-1}$ is a compact operator. It is also easily seen that if L is regular and invertible, then the operator $I + T_1$ is also invertible. Thus, by Lemma 16.1, inequality (16.27) implies that

$$\varlimsup_{n \to \infty} \Theta(L H_n, H_n) = \varlimsup_{n \to \infty} ((I + T_1)(\gamma I + S) H_n, H_n) < 1 .$$

This proves the sufficiency of the condition. For the proof of *necessity*, we refer the reader to Vainikko [8]; he may also derive necessity in an obvious way from Exercises 16.17 and 16.18.

Exercise 16.17. Let A be a continuous and continuously invertible linear operator which has no representation $\alpha A = I + S$, where $\alpha = \text{const} \neq 0$, $\|S\| < 1$. Show that $\inf_{u \in H} |(Au, u)| / \|u\|^2 = 0$.

Hint. By assumption, for any nonzero α there is an element $v = v(\alpha) \in H$ such that $\|\alpha Av - v\| \geqq 1 - |\alpha|^2 \|Av\|^2$, or what is the same,

$$\text{Im} \, \alpha \, \text{Im} \, (Av, v) - \text{Re} \, \alpha \, \text{Re} \, (Av, v) + |\alpha|^2 \|Av\|^2 \geqq 0 .$$

Using this inequality and the fact that the set

$$\Omega = \{\lambda \colon \lambda = (Av, v), \quad v \in H, \quad \|v\| = 1\}$$

is convex (see Stone [1]), prove that any straight line through the origin of the complex plane contains at least one point of the closure $\overline{\Omega}$; hence conclude that $0 \in \overline{\Omega}$.

Exercise 16.18. Let L be a continuous and continuously invertible operator which has no representation (16.26). Show that H contains a sequence $\{v_n\}$ having properties 1–3 of Exercise 16.15.

Hint. Let $\{H_n\}$ be any sequence of finite-dimensional subspaces which is ultimately dense in H, and $P^{(n)}$ the orthogonal projections onto the orthogonal complements $H^{(n)} = H \ominus H_n$. For any α and n, $\|P^{(n)}(\alpha L - I)\| \geqq 1$ (why?), i.e., the operator $A_n = P^{(n)}L$ satisfies the assumptions of Exercise 16.17 in $H^{(n)}$. One can thus choose $v_n \in H^{(n)}$ so that $\|v_n\| = 1$ and $|(A_n v_n, v_n)| < 1/n$. Show that $\{v_n\}$ is the desired sequence.

Exercise 16.19. Construct a unitary operator which is not regular.

Exercise 16.20. Let L be a regular operator. Show that the Bubnov–Galerkin method satisfies the inequality

$$\|P^{(n)}u_0\| \leqq \|u_n - u_0\| \leqq c \|P^{(n)}u_0\| ,$$

where $c = \text{const}$, $P^{(n)} = I - P_n$.

Exercise 16.21. Let L be an unbounded continuously invertible operator in H ($R(L) = H$). Show that there exists a sequence of finite-dimensional subspaces $H_n \subset D(L)$ such that $\{H_n\}$ and $\{LH_n\}$ are ultimately dense in H, but for some $f \in H$ the residuals in the Bubnov–Galerkin method (16.25) for the equation $Lu = f$ do not converge to zero as $n \to \infty$.

The concept of a regular operator was introduced, in a somewhat different formulation, by Pol'skii, who also established some simple conditions for regularity. The characterization of the class of regular operators in Theorem 16.4 was proposed by Vainikko [8]. The fact that the inverse of a regular operator is bounded was proved by Vainikko and Umanskii [1].

16.7. *Rate of convergence of the Galerkin–Petrov method.* Consider the equation

$$Lu \equiv Au - Ku = f , \tag{16.31}$$

where A and K are linear operators mapping a Banach space E into a Hilbert space H; the domains $D(A) \subset D(K)$ are assumed to be dense in E. Assume that A has a bounded inverse defined throughout H, the operator KA^{-1} is compact in H, and 1 is not an eigenvalue of KA^{-1}. It follows that L is invertible; the operator $L^{-1} = A^{-1}(I - KA^{-1})^{-1}$ is bounded and defined throughout H.

Equation (16.31) is to be solved by the Galerkin–Petrov method. Let $\{E_n\}$ and $\{H_n\}$ be two sequences of finite-dimensional subspaces, $E_n \subset D(A) \subset E$, $H_n \subset H$ ($n = 1, 2, \ldots$), and P_n the orthogonal projections of H onto H_n. The approximate solution is determined from the conditions

$$P_n(Lu_n - f) = 0 \qquad (u_n \in E_n). \tag{16.32}$$

Theorem 16.5. * *Let*

$$H_n = (I - S)AE_n \qquad (n = 1, 2, \ldots), \tag{16.33}$$

where S is a compact operator in H of which 1 is not an eigenvalue. Let the sequence of subspaces $\{AE_n\}$ be ultimately dense in H.

Then, for sufficiently large n, there exists a unique approximation u_n satisfying (16.32), and

$$\|B(u_n - u_0)\| \leqq c\|R^{(n)}(I - S)^{-1}(BL^{-1})^*\| \cdot \|R^{(n)}Au_0\|, \tag{16.34}$$

where u_0 is a solution of equation (16.31), $R^{(n)} := I - R_n$, R_n the orthogonal projection of H onto AE_n, and B any linear operator with domain $D(B) \supset D(A)$ in E and range in some Banach space E' such that BL^{-1} is a bounded operator.

Proof. It is not hard to see that the Galerkin–Petrov method (16.32) satisfies the assumptions of Theorem 15.2: the sequence $\{LE_n\}$ is ultimately dense in H, and (see Exercise 16.2)

$$\varlimsup_{n \to \infty} \Theta(LE_n, H_n) = \varlimsup_{n \to \infty} \Theta((I - KA^{-1})AE_n, (I - S)AE_n) < 1.$$

We start from the estimate (15.14) of Theorem 15.2:

$$\|Lu_n - f\| \leqq c_1\|Q^{(n)}f\|,$$

---

* Subsections 16.7 and 16.8 follow Vainikko [14].

where $Q^{(n)} = I - Q_n$ is the orthogonal projection onto LE_n. It follows from the definition of the operators $Q^{(n)}$ and R_n that $Q^{(n)}(I - KA^{-1})R_n = 0$, and so

$$\|Q^{(n)}f\| = \|Q^{(n)}(I - KA^{-1})Au_0\| = \|Q^{(n)}(I - KA^{-1})R^{(n)}Au_0\| \leqq$$

$$\leqq \|I - KA^{-1}\| \cdot \|R^{(n)}Au_0\|$$

and

$$\|Lu_n - f\| \leqq c_2 \|R^{(n)}Au_0\| \qquad (c_2 = c_1 \|I - KA^{-1}\|). \quad (16.35)$$

Further, since $P_n(Lu_n - f) = 0$ by (16.32), it follows that

$$B(u_n - u_0) = BL^{-1}(Lu_n - f) = BL^{-1}P^{(n)}(Lu_n - f),$$

where $P^{(n)} = I - P_n$. Together with (16.35), this gives the estimate

$$\|B(u_n - u_0)\| \leqq c_2 \|BL^{-1}P^{(n)}\| \cdot \|R^{(n)}Au_0\|. \quad (16.36)$$

It follows from the definitions of the operators $P^{(n)}$ and R_n and from (16.33) that $P^{(n)}(I - S)R_n = 0$. Hence

$$P^{(n)} = P^{(n)}(I - S)R^{(n)}(I - S)^{-1}$$

or, taking adjoints

$$P^{(n)} = (I - S^*)^{-1}R^{(n)}(I - S^*)P^{(n)}$$

($P^{(n)}$ and $R^{(n)}$, being orthogonal projections, are selfadjoint). Hence

$$BL^{-1}P^{(n)} = BL^{-1}(I - S^*)^{-1}R^{(n)}(I - S^*)P^{(n)},$$

or, again taking adjoints,

$$(BL^{-1}P^{(n)})^* = P^{(n)}(I - S)R^{(n)}(I - S)^{-1}(BL^{-1})^*.$$

Thus

$$\|BL^{-1}P^{(n)}\| = \|(BL^{-1}P^{(n)})^*\| \leqq \|I - S\| \cdot \|R^{(n)}(I - S)^{-1}(BL^{-1})^*\|,$$

and substitution of this expression in (16.36) yields (16.34). ∎

Exercise 16.22. Show that, under the assumptions of Theorem 16.5, the norm $\|B(u_n - u_0)\|$ converges to zero at least as rapidly as the norm $\|A(u_n - u_0)\|$.

16.8. *The method of moments.* This subsection considers an application of Theorem 16.5 to the solution of boundary-value problems.

As in subsection 15.7, consider the differential equation

$$Lu \equiv u^{(m)} - \sum_{k=0}^{m-1} p_k(t) u^{(k)} = f(t) \qquad (16.37)$$

with homogeneous linear boundary conditions

$$\sum_{k=0}^{m-1} [\alpha_{ik} u^{(k)}(a) + \beta_{ik} u^{(k)}(b)] = 0 \qquad (i = 1, \dots, m). \qquad (16.38)$$

As before, we shall assume that the equation $u^{(m)}$ has no nontrivial solution satisfying the boundary conditions (16.38).

In the method of moments, an approximate solution of the problem (16.37)–(16.38) is sought in the form

$$u_n(t) = \sum_{j=0}^{n} \xi_j \phi_j(t), \qquad (16.39)$$

where $\phi_j(t)$ is a polynomial of degree $m + j$ satisfying the boundary conditions (16.38). Besides Lu, let another differential form be given:

$$Mu \equiv u^{(m)} - \sum_{j=0}^{m-1} q_j(t) u^{(j)}.$$

The constants ξ_j are determined by the conditions

$$\int_a^b (Lu_n - f) M\phi_i dt = 0 \qquad (i = 0, \dots, n), \qquad (16.40)$$

which are equivalent to a system of linear equations

$$\sum_{j=0}^{n} \left(\int_a^b L\phi_j M\phi_i \, dt \right) \xi_j = \int_a^b f M\phi_i \, dt \qquad (i = 0, \dots, n). \qquad (16.41)$$

The most interesting cases in actual computations are $q_j(t) \equiv p_j(t)$, and $q_j(t) = 0$ $(j = 0, 1, \dots, m - 1)$. In the first case, the method of moments reduces to the least-squares method; the second case is interesting in that the system (16.41) is easily formulated, since then one can assume that $M\phi_i \equiv t^i$ $(i = 0, 1, \dots)$.*

---

* A more apt designation for (16.39)–(16.41) would be "the method of generalized moments"; the term "method of moments" should be reserved for the case $q_j(t) \equiv 0$ $(j = 0, \dots, m - 1)$.

Let $L_2^{(i)}$ ($i \geq 0$) denote the set of functions $u(t)$ with $i - 1$ absolutely continuous derivatives on $[a, b]$ such that $u^{(i)}(t) \in L_2 \, [a, b]$. $\|u(t)\|$ will denote the norm in $L_2^{(0)} = L_2 \, [a, b]$.

Theorem 16.6.* *Assume that each coefficient $p_j(t)$ is j times continuously differentiable, $f(t) \in L_2^{(i)}$, and*

$$q_j(t) \in L_2^{(h)} \qquad (j = 0, \ldots, m - 1 \, ; \, 0 \leq h \leq m).$$

Assume that neither of the homogeneous equations $Lu = 0$ and $Mu = 0$ has a nontrivial solution satisfying the boundary conditions (16.38).

Then, for sufficiently large n, the system of equations (16.41) *has a unique solution, and*

$$\|u_n^{(j)}(t) - u_0^{(j)}(t)\| \leq cn^{j-m} l_n(u_0^{(m)}) \qquad (j = m - h, \ldots, m), \qquad (16.42)$$

$$\max_{a \leq t \leq b} |u_n^{(j)}(t) - u_0^{(j)}(t)| \leq cn^{j-m+1/2} l_n(u_0^{(m)}) \qquad (j = m - h, \ldots, m - 1),$$
$$(16.43)$$

where $u_0(t)$ is a solution of the boundary-value problem (16.37), (16.38) *and*

$$l_n(z) = \min_{\gamma_0, \ldots, \gamma_n} \left\| z(t) - \sum_{j=0}^{n} \gamma_j t^j \right\|$$

is the best mean-square approximation of $z(t)$ by polynomials of degree $\leq n$.

Below we shall need the following simple inequality for the mean-square approximation of a function:

$$l_n(z) \leq c_i n^{-i} \|z^{(i)}(t)\| \qquad (z(t) \in L_2^{(i)}). \qquad (16.44)$$

Proof of Theorem 16.6. We regard the boundary-value problem (16.37), (16.38) as equation (16.31) in the space $E = H = L_2 \, [a, b]$, setting

$$Au = u^{(m)}, Ku = \sum_{k=0}^{m-1} p_k(t) \, u^{(k)}$$

and letting the domain $D(A) = D(K)$ be the set of $L_2^{(m)}$-functions satisfying the boundary conditions (16.38).

---

* A similar result was proved previously by Daugavet [3], for the case $q_j(t) = 0$ ($j = 0, \ldots, m - 1$).

It is well known that

$$A^{-1}x = \int_a^b G(t, s)\, x(s)\, ds,$$

where $G(t, s)$ is the Green's function of the differential form $u^{(m)}$ with the boundary conditions (16.38). The operators

$$KA^{-1}x = \int_a^b T(t, s)\, x(s)\, ds \left(T(t, s) = \sum_{k=0}^{m-1} p_k(t)\, \frac{\partial^k G(t, s)}{\partial t^k} \right)$$

and

$$Sx = \int_a^b S(t, s)\, x(s)\, ds \left(S(t, s) = \sum_{k=0}^{m-1} q_k(t)\, \frac{\partial^k G(t, s)}{\partial t^k} \right)$$

are compact in $L_2\,[a, b]$, as integral operators with square-intergrable kernels. The number 1 is not an eigenvalue of the operators KA^{-1} and S. In fact, if the equation $x = Sx$, say, has a nontrivial solution x, then the function $u = A^{-1}x$ is a nontrivial solution of the equation $Mu = 0$ satisfying the boundary conditions (16.38), contradicting the assumptions of the theorem.

The method of moments (16.39)–(16.40) is precisely the Galerkin–Petrov method, with E_n the linear span of the polynomials $\phi_0(t), \ldots, \phi_n(t)$ and H_n the linear span of the functions $M\phi_0, \ldots, M\phi_n$. These subspaces satisfy the relation (16.33), in which S is the integral operator defined above. The subspace AE_n consists of polynomials of degree $\leq n$, and the subsequence $\{AE_n\}$ is clearly ultimately dense in $L_2\,[a, b]$. Thus Theorem 16.5 applies.

In this case the R_n defined in Theorem 16.5 are orthogonal projections into the subspace of polynomials of degree $\leq n$, i.e., $\|R^{(n)}z\| = l_n(z)$. The estimates (16.42) follow from (16.34), via the observation that the operators

$$B_j = \frac{d^j}{dt^j}$$

in $L_2\,[a, b]$ satisfy the inequalities

$$\|R^{(n)}(I - S)^{-1}(B_jL^{-1})^*\| \leq cn^{j-m} \qquad (j = m - h, \ldots, m).$$

In view of (16.44), this will follow if we can show that the operators

$$B_{m-j}(I - S)^{-1}(B_j L^{-1})^* \qquad (j = m - h, \ldots, m) \qquad (16.45)$$

are defined and bounded throughout $L_2[a, b]$. We shall outline the proof.

Let $K(t, s)$ be the Green's function of the differential form Lu with the boundary conditions (16.38). Since the coefficients p_j ($j = 0, \ldots,$ $m - 1$) are j times continuously differentiable, it follows[†] that $K(t, s)$ has continuous mixed derivatives of order $\leq m - 2$ in the square $a \leq t, s \leq b$, while its derivatives of order $m - 1$ and m have a finite discontinuity at $t = s$. It follows easily that for $i + j \leq m$ the operators $B_i(B_j L^{-1})^*$ are bounded and defined throughout $L_2[a, b]$. This means, in particular, that $(B_j L^{-1})^*$ maps $L_2[a, b]$ into $L_2^{(m-j)}$ ($j = 0, \ldots, m$).

On the other hand, it is easily verified that, because of the condition $q_j(t) \in L_2^{(h)}$ ($j = 0, \ldots, m - 1$), the operator $(I - S)^{-1}$ maps each of the sets $L_2^{(m-j)}$ ($j = m - h, \ldots, m$) into itself, and moreover

$$\|B_{m-j}(I - S)^{-1}z\| \leq c' \sum_{i=0}^{m-j} \|B_i z\| \qquad (z \in L_2^{(m-j)} \, ; \; j = m - h, \ldots, m).$$

This implies that the operators (16.45) are bounded in $L_2[a, b]$ and establishes the estimate (16.42).

The estimates (16.43) follows from (16.42) and the easily proved inequality

$$\max_{a \leq t \leq b} |z(t)| \leq \left[\frac{\|z(t)\|^2}{b - a} + 2\|z(t)\| \cdot \|z'(t)\| \right]^{1/2} \qquad (z(t) \in L_2^{(1)}). \; \blacksquare$$

Exercise 16.23. Formulate an analog of Theorem 16.6 for the least-squares method.

[†] A detailed analysis of the differential properties of the Green's function of a differential operator L, under boundary conditions more general than (16.38), may be found in Levin's dissertation [1] and Pokornyi's paper [4].

§ 17. The Galerkin method with perturbations and the general theory of approximate methods

17.1. *Statement of the problem.* Our attention will now be focussed on approximate methods for the equation

$$x = Tx + f,$$ (17.1)

where T is a continuous linear operator in a Banach space E. Let $\{E_n\}$ be a sequence of closed subspaces of E. Our approximate solutions of equation (17.1) will be solutions of the equation

$$x_n = T_n x_n + f_n,$$ (17.2)

where T_n is a continuous linear operator in E_n, $f_n \in E_n$.

Since the operators T and T_n are defined in different spaces, we cannot use their difference to characterize the closeness of equations (17.1) and (17.2). Let P_n be a (usually unbounded) projection onto the subspace E_n:

$$P_n z \in E_n \text{ for } z \in D(P_n), \qquad P_n z_n = z_n \text{ for } z_n \in E_n \subset D(P_n).$$

In what follows, we shall assume (without explicitly mentioning this) that $f \in D(P_n)$, $TE \subset D(P_n)$ and $P_n T$ is bounded in E (therefore also in E_n). Each of the operators T and T_n may be compared with $P_n T$, regarding the latter as an operator in E or E_n, as the case may be. It is therefore natural to describe the closeness of (17.1) and (17.2) in terms of the norm of the operator $S_n = T_n - P_n T$ in E_n and the norm of $U_n = T - P_n T$ in E, as well as the norms $\|f_n - P_n f\|$ and $\|f - P_n f\|$. In other words, each of equations (17.1) and (17.2) is compared with the Galerkin equation

$$x_n = P_n T x_n + P_n f.$$ (17.3)

The latter, in contradistinction to (17.1) and (17.2), may be considered both in the original space E and in the subspace E_n.

It is often convenient to write equation (17.2) in the form

$$x_n = P_n T x_n + S_n x_n + P_n f + g_n,$$ (17.4)

where $S_n = T_n - P_n T$ and $g_n = f_n - P_n f$ measure the deviation of equation (17.2) from the Galerkin equation (17.3). All subsequent arguments pertain to the case in which these deviations are small ($\|S_n\| \to 0$,

$\|g_n\| \to 0$). A natural designation for the corresponding approximate methods is the *Galerkin method with perturbations*. An example of this method (which cannot be treated by projection methods without perturbations) is the method of mechanical quadratures, to be studied in detail in subsection 17.10.

Note that equation (17.2) does not involve the projection P_n, and one can therefore replace (17.2) by various variants of equation (17.4), depending on the choice of the projections P_n. Nevertheless, in most concrete applications the most natural projection P_n for the case at hand is easily indicated. It then often turns out that $P_n f = f_n$, i.e., $g_n = 0$. Henceforth, we shall always assume when studying the abstract equation (17.2) that the projection P_n has been chosen.

The perturbed Galerkin method often crops up in numerical realization of ordinary (unperturbed) projection methods. We illustrate this for the Bubnov–Galerkin method. Assume (temporarily) that E is a Hilbert space, E_n the linear span of the elements $\phi_1, \ldots, \phi_n$, and P_n an orthogonal projection. A solution of equation (17.3) may then be approximated by a linear combination $u_n = \sum_{j=1}^{n} \xi_j \phi_j$, where the coefficients ξ_j are determined from the system

$$\sum_{j=1}^{n} (\phi_j - T\phi_j, \phi_i) \xi_j = (f, \phi_i) \qquad (i = 1, \ldots, n).$$

However, in actual computation of this system the coefficient matrix and column of free terms always involve an error, and so the system is equivalent not to equation (17.3), but rather to equation (17.4) with certain perturbations S_n and g_n.

We shall return to questions arising in numerical implementation of projection methods in subsection 17.8.

It might seem tempting to reduce equation (17.2) to the form

$$x_n = T_n P_n x + P_n f,$$

and consider this equation in E. If the projection P_n is bounded and $\|T - T_n P_n\| \to 0$ as $n \to \infty$, it is quite easy to deduce convergence theorems for the approximate solutions. Unfortunately, in applications it is usually false that $\|T - T_n P_n\| \to 0$. It is far simpler to prove that $\|T - P_n T\| \to 0$, $\|T_n - P_n T\| \to 0$ as $n \to \infty$.

17.2. *Lemma on invertibility of the operators $I - \mu T$ and $I - \mu T_n$.* If $I - \mu T$ is invertible, is $I - \mu T_n$ also invertible, and *vice versa*? We now give various sufficient conditions for this to be true. The numerical parameter μ is introduced with an eye to §18; when $\mu = 1$ we get conditions for unique solvability of equations (17.1) and (17.2). We shall employ the following notation, already introduced in subsection 17.1:

$$S_n = T_n - P_n T \qquad \text{(operator in } E_n \text{)},$$

$$U_n = T - P_n T \qquad \text{(operator in } E \text{)}.$$

Lemma 17.1. *Let $I - \mu T$ be continuously invertible in E,*

$$\|(I - \mu T)^{-1}\| \leq \kappa,$$

and set

$$q_n \equiv |\mu| \kappa (\|S_n\| + \|U_n\|) < 1. \tag{17.5}$$

Then $I - \mu T_n$ is continuously invertible in E_n, and

$$\|(I - \mu T_n)^{-1}\| \leq \frac{\kappa}{1 - q_n}.$$

Proof. The proof proceeds by a double application of Lemma 15.2. First setting $F = E$, $A = I - \mu T$, $B = \mu U_n$, we have

$$\|B\| \cdot \|A^{-1}\| \leq |\mu| \cdot \|U_n\| \kappa \leq q_n < 1,$$

and so the operator $A + B = I - \mu P_n T$ is invertible in E and

$$\|(I - \mu P_n T)^{-1}\| \leq \frac{\kappa}{1 - |\mu| \kappa \|U_n\|}. \tag{17.6}$$

It is easy to see that $(I - \mu P_n T) x \in E_n$ if and only if $x \in E_n$. This implies that the operator $I - \mu P_n T$ is invertible in E_n. Inequality (17.6) remains valid, of course, for the inverse in E_n.

Now set $F = E_n$, $A = I - \mu P_n T$, $B = -\mu S_n$ (operators in E_n); then $A + B = I - \mu T_n$. It follows from (17.5) and (17.6) that $\|B\| \cdot \|A^{-1}\| < 1$, and another application of Lemma 15.2 completes the proof of the lemma. ∎

Lemma 17.2. *Let $I - \mu T_n$ be continuously invertible in E_n,*

$$\|(I - \mu T_n)^{-1}\| \leq \kappa_n,$$

and set

$$q'_n \equiv |\mu| (\tau_n \|U_n\| + \kappa_n \|S_n\|) < 1,$$

where

$$\tau_n = 1 + |\mu| \kappa_n \|P_n T\|.$$

Then $I - \mu T$ is continuously invertible in E, and

$$\|(I - \mu T)^{-1}\| \leqq \frac{\tau_n}{1 - q'_n}.$$

Proof. The proof is analogous to that of Lemma 17.1. One first verifies that the operator $I - \mu P_n T$ is invertible in E_n, and

$$\|(I - \mu P_n T)^{-1}\|_{E_n} \leqq \frac{\kappa_n}{1 - |\mu| \kappa_n \|S_n\|}.$$

Underlying the subsequent arguments is the identity

$$(I - \mu P_n T)^{-1} x = [I + \mu (I - \mu P_n T)^{-1} P_n T] x \qquad (x \in E_n).$$

The operator in the right of this identity is defined not only on E_n, but also on E, since the operator $(I - \mu P_n T)^{-1}$ is preceded by the projection P_n. It is left to the reader to verify directly that this operator $I + \mu (I - \mu P_n T)^{-1} P_n T$ is the two-sided inverse of $I - \mu P_n T$ in E. Therefore

$$\|(I - \mu P_n T)^{-1}\|_E \leqq 1 + |\mu| \frac{\kappa_n}{1 - |\mu| \kappa_n \|S_n\|} \|P_n T\| \leqq \frac{\tau_n}{1 - |\mu| \kappa_n \|S_n\|}.$$

Another application of Lemma 15.2 completes the proof. ∎

Exercise 17.1. Let $I - \mu T$ be continuously invertible in E, $\|(I - \mu T)^{-1}\| \leqq \kappa$. Let $|\mu| \kappa \delta_n < 1$, where

$$\delta_n = \sup_{x \in E_n, \|x\| = 1} \|Tx - T_n x\|.$$

Show that the operator $I - \mu T_n$ has a left inverse $(I - \mu T_n)_*^{-1}$ in E_n, and

$$\|(I - T_n)_*^{-1}\| \leqq \frac{\kappa}{1 - |\mu| \kappa \delta_n}.$$

Exercise 17.2. The assumptions of Exercise 17.1 do not guarantee the existence of a two-sided inverse $(I - \mu T_n)^{-1}$. Give an example.

Exercise 17.3. Let the projection P_n be bounded. Prove that Lemma 17.2 remains valid if τ_n is replaced throughout by $\tau'_n = 1 + \kappa_n (1 + |\mu| \cdot \|P_n T P^{(n)}\|)$, where $P^{(n)} = I - P_n$.

17.3. Convergence theorem.

Theorem 17.1.* Let $I - T$ be continuously invertible in E, and let

$$\|S_n\| \to 0, \quad \|U_n\| \to 0, \quad \|f_n - P_n f\| \to 0, \quad \|f - P_n f\| \to 0 \quad as \quad n \to \infty,$$

$$\tag{17.7}$$

where $S_n = T_n - P_n T$ and $U_n = T - P_n T$ are operators in E_n and E, respectively.

Then, for sufficiently large n, equation (17.2) has a unique solution x_n, and the sequence $\{x_n\}$ converges in norm to a solution x_0 of equation (17.1). Moreover,

$$\|x_n - x_0\| \leq c(\|g_n\| + \|S_n\| \cdot \|P_n x_0\| + \|P^{(n)} x_0\|), \tag{17.8}$$

$$c_1 \varepsilon_n \leq \|x_n - P_n x_0\| \leq c_2 \varepsilon_n, \tag{17.9}$$

where c, c_1, c_2 are positive constants, independent of n and f; $g_n = f_n - P_n f$, $P^{(n)} = I - P_n$,

$$\varepsilon_n = \|g_n + (T_n P_n - P_n T) x_0\|.$$

Proof. By the first two relations of (17.7), the assumptions of Lemma 17.1 ($\mu = 1$) are satisfied for sufficiently large n ($n \geq n_0$, say). It follows that the operator $I - T_n$ is invertible in E_n for $n \geq n_0$, and the norms of the inverses are uniformly bounded:

$$\|(I - T_n)^{-1}\| \leq c_2 \quad (n \geq n_0). \tag{17.10}$$

Again by (17.7), the norms of the operators $I - T_n$ themselves are uniformly bounded:

$$\|I - T_n\| \leq 1/c_1 \quad (n \geq 1). \tag{17.11}$$

Let x_0 and x_n be solutions of equations (17.1) and (17.2), respectively (their existence and uniqueness follow from the fact that $I - T$ and $I - T_n$ are continuously invertible). Then

$$(I - T_n)(x_n - P_n x_0) = f_n - P_n x_0 + T_n P_n x_0$$

---

* An analog of this theorem for nonlinear equations (see subsection 19.3) may be found in Vainikko [11].

or, since $P_n x_0 = P_n T x_0 + P_n f$,

$$(I - T_n)(x_n - P_n x_0) = g_n + (T_n P_n - P_n T) x_0 .$$

Hence the estimate (17.9) follows from (17.10) and (17.11).
The estimate (17.8) follows from (17.9). We need only note that

$$\|x_n - x_0\| \leqq \|P_n(x_n - x_0)\| + \|P^{(n)}(x_n - x_0)\| = \|x_n - P_n x_0\| + \|P^{(n)} x_0\|$$

and

$$\varepsilon_n \leqq \|g_n\| + \|(T_n P_n - P_n T) x_0\| = \|g_n\| + \|S_n P_n x_0 - P_n T P^{(n)} x_0\| \leqq$$

$$\leqq \|g_n\| + \|S_n\| \cdot \|P_n x_0\| + \|P_n T\| \cdot \|P^{(n)} x_0\| .$$

Finally, the fact that $x_n \to x_0$ follows from (17.7) and (17.8), since

$$\|P^{(n)} x_0\| \leqq \|P^{(n)} T x_0\| + \|P^{(n)} f\| \leqq \|U_n\| \cdot \|x_0\| + \|P^{(n)} f\| \to 0$$

as $n \to \infty$. ∎

Exercise 17.4. Under the assumptions of Theorem 17.1, show that the following estimate holds (c', $c'' = \text{const} > 0$):

$$c' \|g_n + S_n x_n - P^{(n)} x_0\| \leqq \|x_n - x_0\| \leqq c'' \|g_n + S_n x_n - P^{(n)} x_0\| .$$

Exercise 17.5. Replace the first relation in (17.7) by $\|U_n\| \cdot \|(I - T)^{-1}\| \leqq q < 1$ ($n \geqq n_0$), and show that the estimates (17.8) and (17.9) remain valid.

Exercise 17.6. Assume that the assumptions of Lemma 17.2 are valid, $\mu = 1$. Prove the estimate

$$\|x_n - x_0\| \leqq \frac{\tau_n}{1 - q'_n} \|x_n - T x_n - f\| .$$

17.4. *Relation to Kantorovich's general theory of approximate methods.*

Kantorovich's well-known theorem on the convergence of approximate methods is easily derived from Theorem 17.1. Let $f_n = P_n f$ in equation (17.2), i.e., the approximate solutions of equation (17.1) are the solutions of the equation

$$x_n = T_n x_n + P_n f , \tag{17.12}$$

where T_n, as before is a continuous linear operator in E_n. The projection P_n ($P_n E = E_n$) is assumed bounded. The spaces E and E_n and the operators T and T_n are related as follows.

I. For any $z_n \in E_n$,

$$\|P_n T z_n - T_n z_n\| \leqq \xi_n \|z_n\| .$$

II. For every $x \in E$ there exists $z_n \in E_n$ such that

$$\|Tx - z_n\| \leqq \eta_n \|x\|.$$

III. There exists an element $z_n \in E_n$ such that

$$\|f - z_n\| \leqq \xi_n.$$

Theorem 17.2.* *Let* $I - T$ *be invertible; assume that conditions* I, II, III *hold, and*

$$\xi_n \to 0, \quad \|P_n\|\eta_n \to 0, \quad \|P_n\|\xi_n \to 0 \qquad \text{as} \qquad n \to \infty. \quad (17.13)$$

Then, for sufficiently large n, *equation* (17.12) *has a unique solution* x_n *and the sequence* x_n *converges in norm to a solution* x_0 *of equation* (17.1). *Moreover,*

$$\|x_n - x_0\| \leqq c\,(\xi_n + \|P_n\|\eta_n + \|P_n\|\xi_n), \quad (17.14)$$

$$\|x_n - x_0\| \leqq c\,[\xi_n + \|P_n\|\,\rho(x_0, E_n)], \quad (17.15)$$

where

$$\rho(z, E_n) = \inf_{z_n \in E_n} \|z - z_n\|.$$

Proof. Condition I means that $\|S_n\| \leqq \xi_n$. It follows from conditions II and III that

$$\|U_n\| = \|P^{(n)}T\| \leqq (1 + \|P_n\|)\,\eta_n \leqq 2\eta_n\|P_n\|, \qquad \|P^{(n)}f\| \leqq 2\zeta_n\|P_n\|.$$

These inequalities, together with (17.13), imply that the assumptions of Theorem 17.1 are satisfied.

The estimate (17.15) follows from (17.8) and (15.20), while (17.14) follows from (17.8) and the inequality

$$\|P^{(n)}x_0\| \leqq \|P^{(n)}T\| \cdot \|x_0\| + \|P^{(n)}f\| \leqq 2\eta_n\|P_n\| \cdot \|x_0\| + 2\zeta_n\|P_n\|. \quad \blacksquare$$

Exercise 17.7. Show that the statement of Exercise 7.14 (p. 105) follows from Lemma 17.1.

---

* Kantorovich [2] (see also Kantorovich and Akilov [1]); in that paper, the theorem is proved under an additional assumption: If equation (17.2) is solvable for any right-hand side, the operator $I - T_n$ is continuously invertible; the space E is not assumed to be complete.

Exercise 17.8. Prove an analog of Theorem 17.2 for the case in which E is an incomplete normed linear space; the subspaces E_n are, as before, complete.

17.5. *Perturbation of the Galerkin approximation.* We shall now investigate the deviation of the solution of the Galerkin equation (17.3) from that of the perturbed equation (17.4). We consider the equation

$$x_n = P_n T x_n + S_n x_n + P_n f + g_n, \tag{17.4'}$$

where S_n is an arbitrary operator and g_n an arbitrary element.

Theorem 17.3. Let $I - T$ be continuously invertible in E, and

$$\|U_n\| \to 0 \qquad as \qquad n \to \infty \quad (U_n = T - P_n T).$$

Then there exist positive constants σ, c_1, c_2, c_1', c_2' and a natural number n_0, all independent of n and f, such that
1) *if $\|S_n\| \leq \sigma$, equation (17.4') is uniquely solvable and*

$$\|\tilde{x}_n - x_n\| \leq c_1 \|S_n\| \cdot \|P_n f\| + c_2 \|g_n\| \qquad (n \geq n_0), \tag{17.16}$$

where x_n and $\tilde{x}_n$ are solutions of equations (17.3) and (17.4'), respectively;
2) *if $g_n = 0$, then for any $S_n(\|S_n\| \leq \sigma)$ and some $f \in E$ (f depends on S_n and thereby on n)*

$$\|\tilde{x}_n - x_n\| \geq c_1' \|S_n\| \cdot \|f\| \qquad (n \geq n_0); \tag{17.17}$$

3) *if $S_n = 0$, then for any $g_n \in E_n$ and any f*

$$\|\tilde{x}_n - x_n\| \geq c_2' \|g_n\| \qquad (n \geq n_0). \tag{17.18}$$

Proof. Choose n_0 and a constant $\sigma > 0$ such that, if $\|S_n\| \leq \sigma$ and $n \geq n_0$,

$$\kappa(\|U_n\| + \|S_n\|) \leq q < 1 \qquad (\kappa = \|(I - T)^{-1}\|).$$

It then follows from Lemma 17.1 that the operators $I - P_n T$ and $I - P_n T - S_n$ are invertible, and

$$\|(I - P_n T)^{-1}\| \leq \frac{\kappa}{1 - q},$$

$$\|(I - P_n T - S_n)^{-1}\| \leq \frac{\kappa}{1 - q} \qquad (n \geq n_0 ; \|S_n\| \leq \sigma). \tag{17.19}$$

This proves that equations (17.3) and (17.4') are solvable. Now

$$(I - P_nT - S_n)\,\tilde{x}_n = P_nf + g_n, \qquad (I - P_nT - S_n)\,x_n = P_nf - S_nx_n,$$

so that

$$\tilde{x}_n - x_n = (I - P_nT - S_n)^{-1}(g_n + S_nx_n). \tag{17.20}$$

In view of (17.19), we get inequality (17.16) with $c_1 = [\kappa/(1 - q)]^2$, $c_2 = \kappa/(1 - q)$.

Set $S_n = 0$; it then follows from (17.20) that

$$\|\tilde{x}_n - x_n\| \geq \frac{1}{\|I - P_nT\|}\,\|g_n\|,$$

which implies inequality (17.18) with

$$c_2' = \inf_{n \geq n_0}\,\frac{1}{\|I - P_nT\|} > 0.$$

Conversely, let $g_n = 0$, and let $S_n\,(\|S_n\| \leq \sigma)$ be arbitrary. We see from (17.20) that

$$\|\tilde{x}_n - x_n\| \geq \frac{1}{\|I - P_nT - S_n\|}\,\|S_nx_n\|.$$

Choose an element $x_n \in E_n$ such that

$$\|S_nx_n\| \geq q'\|S_n\| \cdot \|x_n\| \qquad (q' = \text{const}, 0 < q' < 1).$$

This element x_n is a solution of equation (17.3) with $f = x_n - P_nTx_n$, and

$$\|x_n\| \geq \frac{1}{\|I - P_nT\|}\,\|f\|.$$

Let x_n be a solution of equation (17.4') for the same f. The last three inequalities imply the estimate (17.17) with

$$c_1' = q'\,\inf_{n \geq n_0}\,\frac{1}{\|I - P_nT - S_n\| \cdot \|I - P_nT\|} > 0. \qquad \blacksquare$$

Exercise 17.9. Under the assumptions of Theorem 17.3, let $x_n^{(i)}$ $(i = 1, 2)$ be solutions of equation (17.4') for perturbations $g_n = g_n^{(i)}$, $S_n = S_n^{(i)}\,(\|S_n^{(i)}\| \leq \sigma;\, i = 1, 2;\, n \geq n_0)$. Prove the following analog of inequality (17.16):

$$\left\| x_n^{(1)} - x_n^{(2)} \right\| \leq c_1 \left\| S_n^{(1)} - S_n^{(2)} \right\| \cdot \left\| P_n f + g_n^{(i)} \right\| + c_2 \left\| g_n^{(1)} - g_n^{(2)} \right\| \qquad (i = 1, 2).$$

State and prove the analog of the rest of Theorem 17.3 for this case.

The results of this exercise may be used in numerical applications of (17.2), for stability analysis.

17.6. Bases in subspaces of a Hilbert space.

Lemma 17.3.* Let G_n and H_n be n-dimensional subspaces of a Hilbert space H, with aperture

$$\Theta(G_n, H_n) < 1 .$$

Then G_n and H_n have orthogonal bases $u_1, \ldots, u_n$ and $v_1, \ldots, v_n$, respectively, such that

$$(u_i, v_j) = \beta_i \delta_{ij} \qquad (i, j = 1, \ldots, n), \tag{17.21}$$

where δ_{ij} is the Kronecker delta ($\delta_{ij} = 0$ for $i \neq j$, $\delta_{ii} = 1$), and the constants β_i satisfy the inequalities

$$\{1 - [\Theta(G_n, H_n)]^2\}^{1/2} \leq \beta_i \leq 1 \qquad (i = 1, \ldots, n). \tag{17.22}$$

Proof. Let Q_n and P_n be the orthogonal projections onto G_n and H_n, respectively. The operator

$$V_n = P_n Q_n P_n$$

is selfadjoint and positive definite in H_n. In fact, for any $x, y \in H_n$,

$$(V_n x, y) = (P_n Q_n P_n x, y) = (Q_n x, y) = (x, Q_n y) = (x, V_n y),$$

$$(V_n x, x) = (Q_n x, x) = \left\| Q_n x \right\|^2 =$$

$$= \left\| x \right\|^2 - \left\| x - Q_n x \right\|^2 \geq \{1 - [\Theta(G_n, H_n)]^2\} \left\| x \right\|^2 .$$

It is also clear that

$$\left\| V_n \right\| \leq 1 .$$

Let $\beta_1^2, \ldots, \beta_n^2$ denote the eigenvalues of the operator V_n. The last two inequalities imply (17.22).

---

* Krein, Krasnosel'skii and Mil'man [1].

Let $v_1, \ldots, v_n$ $(v_i \in H_n)$ be an orthogonal system of eigenelements of the operator V_n, associated with $\beta_1^2, \ldots, \beta_n^2$. It is clear that $v_1, \ldots, v_n$ is an orthogonal basis of H_n. Set

$$u_i = \frac{1}{\beta_i} Q_n v_i \qquad (i = 1, \ldots, n).$$

Since $v_i \in H_n$, it follows that $P_n v_i = v_i$ and

$$(u_i, u_j) = \frac{1}{\beta_i \beta_j} (Q_n v_i, Q_n v_j) = \frac{1}{\beta_i \beta_j} (Q_n P_n v_i, P_n v_j) =$$

$$= \frac{1}{\beta_i \beta_j} (V_n v_i, v_j) = \frac{\beta_i^2}{\beta_i \beta_j} (v_i, v_j) = \delta_{ij},$$

i.e., $u_1, \ldots, u_n$ is an orthonormal basis of G_n satisfying (17.21). ∎

17.7. Lemma on Gram matrices. Let $\phi_1, \ldots, \phi_n$ and $\psi_1, \ldots, \psi_n$ be two sets of linearly independent elements of a Hilbert space H. Let λ_n and μ_n denote the smallest eigenvalue of the Gram matrices

$$\left.\begin{aligned} \Lambda_n &= \begin{pmatrix} (\phi_1, \phi_1) & (\phi_2, \phi_1) & \cdots & (\phi_n, \phi_1) \\ \cdots & \cdots & \cdots & \cdots \\ (\phi_1, \phi_n) & (\phi_2, \phi_n) & \cdots & (\phi_n, \phi_n) \end{pmatrix} \\[2mm] M_n &= \begin{pmatrix} (\psi_1, \psi_1) & (\psi_2, \psi_1) & \cdots & (\psi_n, \psi_1) \\ \cdots & \cdots & \cdots & \cdots \\ (\psi_1, \psi_n) & (\psi_2, \psi_n) & \cdots & (\psi_n, \psi_n) \end{pmatrix} \end{aligned}\right\} \quad (17.23)$$

respectively. Let G_n and H_n denote the linear spans of the elements $\phi_1, \ldots, \phi_n$ and $\psi_1, \ldots, \psi_n$, respectively.

Exercise 17.10. Prove that the matrices (17.23) are positive definite, so that $\lambda_n > 0$, $\mu_n > 0$.

Along with the Gram matrices (17.23), we introduce the "mixed" matrix

$$A_n = \begin{pmatrix} (\phi_1, \psi_1) & (\phi_2, \psi_1) & \cdots & (\phi_n, \psi_1) \\ \cdots & \cdots & \cdots & \cdots \\ (\phi_1, \psi_n) & (\phi_2, \psi_n) & \cdots & (\phi_n, \psi_n) \end{pmatrix}$$

and regard it as an operator in euclidean n-space R_n with the usual norm

$$\|\zeta^{(n)}\| = \left[\sum_{j=1}^{n} |\zeta_j|^2 \right]^{1/2} \qquad (\zeta^{(n)} = (\zeta_1, \ldots, \zeta_n)).$$

Let Ψ_n be the operator, mapping G_n into R_n, defined by

$$\Psi_n z_n = ((z_n, \psi_1), \ldots, (z_n, \psi_n)) \quad \left(z_n = \sum_{j=1}^{n} \xi_j \phi_j \in G_n \right).$$

It follows from the definition of the operators A_n and Ψ_n that

$$\Psi_n z_n = A_n \zeta^{(n)} \quad \left(z_n = \sum_{j=1}^{n} \zeta_j \phi_j \, ; \, \zeta^{(n)} = (\zeta_1, \ldots, \zeta_n) \right). \quad (17.24)$$

Lemma 17.4. If $\Theta(G_n, H_n) < 1$ then the operators A_n and Ψ_n are invertible, and

$$\|A_n^{-1} \Psi_n\| = \frac{1}{\sqrt{\lambda_n}}, \quad (17.25)$$

$$\frac{1}{\sqrt{\mu_n}} \le \|\Psi_n^{-1}\| \le \frac{\alpha_n}{\sqrt{\mu_n}}, \quad \text{where} \quad \alpha_n = \frac{1}{\{1 - [\Theta(G_n, H_n)]^2\}^{1/2}}. \quad (17.26)$$

Proof. Let $u_1, \ldots, u_n$ and $v_1, \ldots, v_n$ be bases of the subspaces G_n and H_n, respectively, such that (see Lemma 17.3)

$$(u_i, u_j) = \delta_{ij}, \ (v_i, v_j) = \delta_{ij}, \ (u_i, v_j) = \beta_i \delta_{ij} \qquad (i, j = 1, \ldots, n), \quad (17.27)$$

where the constants β_i satisfy inequalities (17.22). Since $\phi_1, \ldots, \phi_n$ is also a basis of G_n, the elements u_i are linear combinations

$$u_i = \sum_{k=1}^{n} c_{ik} \phi_k \qquad (i = 1, \ldots, n),$$

and the matrix $C_n = (c_{ik})_{i,k=1}^{n}$ is nonsingular. Similarly,

$$v_j = \sum_{l=1}^{n} d_{jl} \psi_l \qquad (j = 1, \ldots, n)$$

and the matrix $D_n = (d_{jl})_{j,l=1}^{n}$ is nonsingular. Conditions (17.27) become

$$\sum_{k=1}^{n} c_{ik} \sum_{l=1}^{n} (\phi_k, \phi_l) \bar{c}_{jl} = \delta_{ij},$$

$$\sum_{k=1}^{n} d_{ik} \sum_{l=1}^{n} (\psi_k \cdot \psi_l) \, \bar{d}_{jl} = \delta_{ij},$$

$$\sum_{k=1}^{n} c_{ik} \sum_{l=1}^{n} (\phi_k, \psi_l) \, \bar{d}_{jl} = \beta_i \delta_{ij}$$

$$(i, j = 1, \ldots, n),$$

or, in matrix notation,

$$C_n \Lambda'_n C_n^* = I_n, \quad D_n M'_n D_n^* = I_n, \quad C_n A'_n D_n^* = B_n, \qquad (17.28)$$

where the prime indicates transposition and the star hermitian conjugation, I_n is the unit matrix, and B_n the diagonal matrix with $\beta_1, \ldots, \beta_n$ along the principal diagonal. Equalities (17.28) imply that

$$\|C_n\| = \frac{1}{\sqrt{\lambda_n}}, \qquad \|D_n\| = \frac{1}{\sqrt{\mu_n}}, \qquad (17.29)$$

$$\|\bar{D}_n A_n C'_n\| \leq 1, \quad \|(\bar{D}_n A_n C'_n)^{-1}\| \leq \alpha_n, \qquad (17.30)$$

where α_n is the number defined in (17.26) and the bar denotes complex conjugation.

In fact, it follows from (17.28) that

$$(\Lambda'_n)^{-1} = C_n^* C_n.$$

Remembering that λ_n is the smallest eigenvalue of Λ_n, therefore also of Λ'_n, we get

$$\|C_n\| = \|C_n^* C_n\|^{1/2} = \|(\Lambda'_n)^{-1}\|^{1/2} = \frac{1}{\sqrt{\lambda_n}}.$$

The second inequality of (17.29) is proved similarly. Seeing that $(C_n A'_n D_n^*)' = \bar{D}_n A_n C'_n$ and that by (17.22),

$$\|B_n\| \leq 1, \quad \|B_n^{-1}\| \leq \alpha_n,$$

we obtain inequalities (17.30) from (17.28). We emphasize that invertibility of B_n implies that of A_n. By (17.24), the matrix Ψ_n is also invertible.

We claim that

$$\|z_n\| = \|(C'_n)^{-1} \zeta^{(n)}\| \qquad \left(z_n = \sum_{k=1}^{n} \zeta_k \phi_k \, ; \; \zeta^{(n)} = (\zeta_1, \ldots, \zeta_n) \right). \qquad (17.31)$$

In fact, setting $\zeta^{(n)} = C'_n \eta^{(n)}$, we see that

$$z_n = \sum_{k=1}^{n} \zeta_k \phi_k = \sum_{k=1}^{n} \left(\sum_{i=1}^{n} c_{ik} \eta_i \right) \phi_k = \sum_{i=1}^{n} \eta_i \sum_{k=1}^{n} c_{ik} \phi_k = \sum_{i=1}^{n} \eta_i u_i ,$$

whence

$$\| z_n \| = \| \eta^{(n)} \| = \| (C'_n)^{-1} \zeta^{(n)} \| .$$

We now prove (17.25). By (17.24), (17.29) and (17.31),

$$\| A_n^{-1} \Psi_n z_n \| = \| \zeta^{(n)} \| = \| C'_n (C'_n)^{-1} \zeta^{(n)} \| \leqq \frac{\| z_n \|}{\sqrt{\lambda_n}} \qquad (z_n \in G_n),$$

i.e., $\| A_n^{-1} \Psi_n \| \leqq 1/\sqrt{\lambda_n}$. To make this inequality an equality (17.25), it suffices to find a vector $\zeta^{(n)}$ such that

$$\| \zeta^{(n)} \| = \frac{\| z_n \|}{\sqrt{\lambda_n}} .$$

Any eigenvector of the matrix Λ_n, associated with λ_n, will suffice. In fact, for this vector,

$$\| z_n \|^2 = \left\| \sum_{i=1}^{n} \zeta_i \phi_i \right\|^2 = \sum_{i,k=1}^{n} (\phi_k, \phi_i) \zeta_k \zeta_i = (\Lambda_n \zeta^{(n)}, \zeta^{(n)}) = \lambda_n \| \zeta^{(n)} \|^2 .$$

We now prove the estimates (17.26). By (17.24), (17.29), (17.30) and (17.31),

$$\| \Psi_n z_n \| = \| A_n \zeta^{(n)} \| = \| \bar{D}_n^{-1} \bar{D}_n A_n C_n (C'_n)^{-1} \zeta^{(n)} \| \geqq$$

$$\geqq \frac{\| (C'_n)^{-1} \zeta^{(n)} \|}{\| \bar{D}_n \| \cdot \| (\bar{D}_n A_n C'_n)^{-1} \|} \geqq \frac{\sqrt{\mu_n}}{\alpha_n} \| z_n \| \qquad (z_n \in G_n),$$

which implies the second inequality of (17.26).

Again using (17.24), (17.30) and (17.31), we find

$$\| \bar{D}_n \Psi_n z_n \| = \| \bar{D}_n A_n \zeta^{(n)} \| = \| \bar{D}_n A_n C_n (C'_n)^{-1} \zeta^{(n)} \| \leqq \| z_n \| \qquad (z_n \in G_n),$$

i.e.,

$$\| \bar{D}_n \Psi_n \| \leqq 1 .$$

On the other hand, by (17.29),

$$\|\bar{D}_n \Psi_n\| \geq \frac{\|\bar{D}_n\|}{\|\Psi_n^{-1}\|} = \frac{1}{\sqrt{\mu_n}\|\Psi_n^{-1}\|}.$$

Comparison of the last two inequalities yields the first inequality of (17.26). ∎

Exercise 17.11. Under the assumptions of Lemma 17.3, show that $\|A_n^{-1}\| \leq \dfrac{\alpha_n}{\sqrt{\lambda_n}\sqrt{\mu_n}}$.

17.8. *Stability of the Galerkin–Petrov method.* Let $\{\phi_k\}$ and $\{\psi_k\}$ be two coordinate sequences in a separable Hilbert space $E = H$. An approximate solution of equation (17.1) is to have the form

$$x_n = \sum_{j=1}^{n} \xi_j \phi_j \tag{17.32}$$

and is determined by the Galerkin–Petrov method from the conditions

$$(x_n - Tx_n - f, \psi_i) = 0 \qquad (i = 1, \ldots, n). \tag{17.33}$$

This yields a system of equations

$$\sum_{j=1}^{n} (\phi_j - T\phi_j, \psi_i)\,\xi_j = (f\,;\,\psi_i) \qquad (i = 1, \ldots, n). \tag{17.34}$$

Owing to inaccuracies involved in the computation of the scalar products, the system actually solved is not (17.34) but a perturbed system

$$\sum_{j=1}^{n} [(\phi_j - T\phi_j, \psi_i) - \gamma_{ij}]\,\tilde{\xi}_j = (f, \psi_i) + \gamma_i \qquad (i = 1, \ldots, n), \tag{17.35}$$

so that, instead of the "exact" approximation (17.32), we get a perturbed approximation

$$\tilde{x}_n = \sum_{j=1}^{n} \tilde{\xi}_j \phi_j. \tag{17.36}$$

In this subsection we investigate the stability of the approximation x_n under small perturbations γ_{ij} and γ_i.*

* The stability of projection methods has been studied by various authors; the main results are summarized in the monograph [8] of Mikhlin, the pioneer in the field. The description of the behavior of the perturbed approximation in Theorem 17.4 is due to Vainikko and is published here for the first time. See also similar results of Yaskova and Yakovlev [1], Vainikko [7, 12, 6] and Dovbysh [1].

We shall use the following notation for the matrix and column of perturbations:

$$\Gamma_n = (\gamma_{ij})_{i,j=1}^n, \qquad \gamma^{(n)} = (\gamma_1, \ldots, \gamma_n).$$

Matrices and elements will be regarded as operators and elements in the euclidean space R_n, and norms treated accordingly. We retain the notation of subsection 17.7: G_n and H_n are the linear spans of the elements $\phi_1, \ldots, \phi_n$ and $\psi_1, \ldots, \psi_n$, respectively, λ_n and μ_n are the smallest eigenvalues of the Gram matrices (17.23).

It is easily seen that the sequences λ_n and μ_n $(n = 1, 2, \ldots)$ are nonincreasing, and they therefore have nonnegative limits

$$\lambda_0 = \lim_{n \to \infty} \lambda_n, \qquad \mu_0 = \lim_{n \to \infty} \mu_n.$$

If $\lambda_0 > 0$, the sequence $\{\phi_k\}$ is said to be *strongly minimal*. In particular, every orthonormal sequence is strongly minimal.

Exercise 17.12. Call a system of elements *minimal* if none of its elements belongs to the closed linear span of the others. Show that a strongly minimal sequence is minimal.

Exercise 17.13.* Let the sequence $\{\chi_j\}$ be minimal, but not strongly minimal, in H. Show that there exists a sequence of numbers c_j such that $\{c_j\chi_j\}$ is strongly minimal.

It will be clear from the following theorem that the sequences of numbers λ_n and μ_n fully characterize the "degree of stability" of the approximation (17.36) under small perturbations of the Galerkin–Petrov equations.

Theorem 17.4. *Let the operator T be compact in H, the operator $I - T$ invertible, the coordinate sequences $\{\phi_i\}$ and $\{\psi_i\}$ complete in H, and*

$$\overline{\lim_{n \to \infty}} \, \Theta(G_n, H_n) < 1. \tag{17.37}$$

Then there exist positive constants $\sigma, c_1, c_2, c_1', c_2'$ and a natural number n_0, all independent of n and f, such that

1) *if $\|\Gamma_n\| \leq \sigma \sqrt{\lambda_n} \sqrt{\mu_n}$, the system of equations (17.35) is uniquely solvable, and*

$$\|\tilde{x}_n - x_n\| \leq \frac{c_1 \|f\|}{\sqrt{\lambda_n} \sqrt{\mu_n}} \, \|\Gamma_n\| + \frac{c_2}{\sqrt{\mu_n}} \|\gamma^{(n)}\| \qquad (n \geq n_0), \tag{17.38}$$

* Dovbysh [2].

where x_n and $\tilde{x}_n$ are the approximations (17.32) *and* (17.36), *corresponding to the unperturbed and perturbed systems* (17.34) *and* (17.35), *respectively;*

2) *if* $\gamma^{(n)} = 0$, *then for some* $\Gamma_n (\|\Gamma_n\| \leqq \sigma \sqrt{\lambda_n} \sqrt{\mu_n})$ *and some* $f \in H$ (*f depends on S_n, and thereby on n*),

$$\|\tilde{x}_n - x_n\| \geqq \frac{c_1' \|f\|}{\sqrt{\lambda_n} \sqrt{\mu_n}} \|\Gamma_n\| \qquad (n \geqq n_0) ; \qquad (17.39)$$

3) *if* $\Gamma_n = 0$, *then for some* $\gamma^{(n)}$ (*with arbitrarily small norm*) *and any* $f \in H$

$$\|\tilde{x}_n - x_n\| \geqq \frac{c_2'}{\sqrt{\mu_n}} \|\gamma^{(n)}\| \qquad (n \geqq n_0) . \qquad (17.40)$$

Proof. Let P_n denote the orthogonal projection onto H_n. In proving Theorem 16.1 we proved that the equations (17.34), or, equivalently, the equation

$$P_n(x_n - Tx_n - f) = 0 \qquad (x_n \in G_n)$$

may be written as a Galerkin equation

$$x_n = \bar{P}_n Tx_n + \bar{P}_n f , \qquad (17.41)$$

where $\bar{P}_n = \tilde{P}_n^{-1} P_n$ projects H onto G_n, $\tilde{P}_n$ being the restriction of P_n to H_n. It was shown there that $\|T - \bar{P}_n T\| \to 0$ as $n \to \infty$, i.e., the Galerkin equations (17.41) satisfy the assumptions of Theorem 17.3.

The unperturbed equation system (17.34) thus corresponds to the Galerkin equation (17.41). We claim that the perturbed equation system (17.35) corresponds to the perturbed Galerkin equation

$$\tilde{x}_n = \bar{P}_n Tx_n + S_n x_n + \bar{P}_n f + g_n , \qquad (17.42)$$

with

$$S_n = \Psi_n^{-1} \Gamma_n A_n^{-1} \Psi_n, \qquad g_n = \Psi_n^{-1} \gamma^{(n)} ; \qquad (17.43)$$

the operators A_n and Ψ_n were defined in subsection 17.7.

Let $\bar{\Psi}_n$ denote the operator mapping H into R_n defined by

$$\bar{\Psi}_n z = ((z, \psi_1), \ldots, (z, \psi_n)) \qquad (z \in H) .$$

$\bar{\Psi}_n$ is obviously the restriction of $\bar{\Psi}_n$ to G_n. We shall prove that

$$\bar{P}_n = \Psi_n^{-1} \bar{\Psi}_n . \qquad (17.44)$$

In fact,

$$\Psi_n^{-1}\bar{\Psi}_n = \Psi_n^{-1}\bar{\Psi}_n(\bar{P}_n + \bar{P}^{(n)}) = \bar{P}_n + \Psi_n^{-1}\bar{\Psi}_n\bar{P}^{(n)},$$

where $\bar{P}^{(n)} = I - \bar{P}_n$. We have used the fact that $\bar{P}_n$ maps H onto G_n and $\bar{\Psi}_n = \Psi_n$ on G_n. To verify (17.44) it suffices to note that

$$(\bar{P}^{(n)}z, \psi_i) = (z - \tilde{P}_n^{-1}P_nz, P_n\psi_i) = (P_nz - P_nz, \psi_i) = 0 \qquad (i = 1, \ldots, n),$$

and so $\bar{\Psi}_n\bar{P}^{(n)} = 0$.

In matrix notation, the perturbed system (17.35) is

$$A_n\tilde{\xi}_n = B_n\tilde{\xi}_n + \Gamma_n\tilde{\xi}_n + \bar{\Psi}_nf + \gamma^{(n)}. \tag{17.45}$$

Here

$$B_n = \begin{pmatrix} (T\phi_1, \psi_1) & (T\phi_2, \psi_1) & \ldots & (T\phi_n, \psi_1) \\ . & . & . & . \\ (T\phi_1, \psi_n) & (T\phi_2, \psi_n) & \ldots & (T\phi_n, \psi_n) \end{pmatrix};$$

it follows from the definition of the operators $\bar{\Psi}_n$ and B_n that

$$\bar{\Psi}_nTz_n = B_n\zeta^{(n)} \qquad \left(z_n = \sum_{j=1}^{n} \zeta_j\phi_j \in G_n \,;\, \zeta^{(n)} = (\zeta_1, \ldots, \zeta_n) \right). \tag{17.46}$$

Using (17.24) and (17.46), we write equation (17.45) in the form

$$\Psi_n\tilde{x}_n = \bar{\Psi}_nT\tilde{x}_n + \Gamma_nA_n^{-1}\Psi_n\tilde{x}_n + \bar{\Psi}_nf + \gamma^{(n)}.$$

Applying the operator Ψ_n^{-1} to both sides and using the relation (17.44), we finally get equation (17.42). The solution of the perturbed system (17.35) and equation (17.42) satisfy the relation (17.36).

By Lemma 17.4, the perturbations S_n and g_n given by formulas (17.43) are

$$\|S_n\| \leqq \|\Psi_n^{-1}\| \cdot \|\Gamma_n\| \cdot \|A_n^{-1}\Psi_n\| \leqq \frac{\alpha^n}{\sqrt{\lambda_n}\sqrt{\mu_n}} \|\Gamma_n\|,$$

$$\|g_n\| \leqq \|\Psi_n^{-1}\| \cdot \|\gamma^{(n)}\| \leqq \frac{\alpha_n}{\sqrt{\mu_n}} \|\gamma^{(n)}\|,$$

where $\overline{\lim_{n\to\infty}} \alpha_n < \infty$ by condition (17.37). It is easy to construct specific perturbations Γ_n and $\gamma^{(n)}$ such that the corresponding S_n and g_n satisfy the equalities

$$\|S_n\| = \|\Psi_n^{-1}\| \cdot \|\Gamma_n\| \cdot \|A_n^{-1}\Psi_n\|, \quad \|g_n\| = \|\Psi_n^{-1}\| \cdot \|\gamma^{(n)}\|.$$

By Lemma 17.4,

$$\|S_n\| \geq \frac{1}{\sqrt{\lambda_n}\sqrt{\mu_n}} \|\Gamma_n\|, \quad \|g_n\| \geq \frac{1}{\sqrt{\mu_n}} \|\gamma^{(n)}\|.$$

All assertions of our theorem now follow directly from Theorem 17.3, applied to equations (17.1), (17.41) and (17.42). Note only that in proving inequality (17.38) one should use the fact that the norms $\|\bar{P}_n\|$ ($n = 1$, $2, \ldots$) are uniformly bounded (see the proof of Theorem 16.1), so that

$$\|\bar{P}^n f\| \leq c\|f\| \qquad (f \in H \; ; \; n \geq n_0) . \quad \blacksquare$$

Exercise 17.14. Prove the inequality

$$\|\tilde{\xi}^{(n)} - \xi^{(n)}\| \leq \frac{1}{\sqrt{\lambda_n}} \|\tilde{x}_n - x_n\|,$$

where $\xi^{(n)}$ and $\tilde{\xi}^{(n)}$ are solutions of the systems (17.34) and (17.35), respectively, while x_n and $\tilde{x}_n$ are the corresponding approximations (17.32) and (17.36).

17.9. *Convergent quadrature processes.* We shall now prove an auxiliary result, to be used later (subsection 17.10) in studying the convergence of mechanical quadrature methods for integral equations.

Consider the quadrature formula

$$\int_a^b z(s)\, ds = \sum_{j=1}^n \alpha_{jn} z(s_{jn}) + R_n(z) \tag{17.47}$$

with remainder term $R_n(z)$, positive coefficients α_{jn}, and interpolation points in $[a, b]$:

$$a \leq s_{1n} < s_{2n} < \ldots < s_{nn} \leq b.$$

Our subject here is a special index of the convergence of the quadrature process (17.47).

Denote

$$\beta_{jn} = \frac{b - a}{\displaystyle\sum_{j=1}^n \alpha_{jn}} \alpha_{jn} .$$

It is clear that $\beta_{jn} > 0$ and $\displaystyle\sum_{j=1}^n \beta_{jn} = b - a$. Divide the interval $[a, b]$

into n disjoint subintervals

$$J_{1n} = [a, a + \beta_{1n}),$$

$$J_{2n} = [a + \beta_{1n}, a + \beta_{1n} + \beta_{2n}), \ldots, J_{nn} = [b - \beta_{nn}, b].$$

Consider the subset

$$D_{jn} = \left(J_{jn} \setminus \bigcup_{j=1}^{n} \{s_{in}\} \right) \cup \{s_{jn}\} \qquad (j = 1, \ldots, n),$$

where $\{s_{jn}\}$ denotes the singleton whose element is s_{jn}. In other words, we reconstruct the sets J_{jn}, transferring the point s_{1n} to J_{1n}, the point s_{2n} to J_{2n}, etc. It is clear that

$$s_{jn} \in D_{jn}, \ \bigcup_{j=1}^{n} D_{jn} = [a, b], \ D_{in} \cap D_{jn} = \emptyset \text{ for } i \neq j, \qquad (17.48)$$

$$\text{mes } D_{jn} = \text{mes } J_{jn} = \beta_{jn} \qquad (j = 1, \ldots, n).$$

Lemma 17.5. Assume that the quadrature process (17.47) converges, i.e., $R_n(z) \to 0$ as $n \to \infty$, for any function $z(s)$ continuous on $[a, b]$. Then the diameters of the sets D_{jn} converge to zero:

$$\max_{1 \leq j \leq n} \ \sup_{s', s'' \in D_{jn}} |s' - s''| \to 0 \qquad as \qquad n \to \infty. \qquad (17.49)$$

Proof. Denote

$$\gamma_{jn} = \alpha_{jn} - \beta_{jn}. \qquad (17.50)$$

Apply the quadrature formula (17.47) to the function $z_1(s) \equiv 1$:

$$b - a = \sum_{j=1}^{n} \alpha_{jn} + R_n(z_1).$$

Since $b - a = \sum_{j=1}^{n} \beta_{jn}$ and all the numbers γ_{jn} ($j = 1, \ldots, n$) have the same sign,

$$\sum_{j=1}^{n} |\gamma_{jn}| = |R_n(z_1)| \to 0 \qquad as \qquad n \to \infty. \qquad (17.51)$$

There there are two ways in which (17.49) may be false. Either the lengths of the intervals J_{jn} do not converge to zero as $n \to \infty$, or some of the points s_{jn} lie at a great distance from the corresponding J_{jn}. We shall prove that in either case the quadrature process (17.47) diverges.

The length of the interval J_{jn} is β_{jn}, and in the first case there is a subsequence $\beta_{j_k n_k}$ such that

$$\beta_{j_k n_k} \geq \eta \qquad (k = 1, 2, \ldots), \tag{17.52}$$

where η is a positive constant. Extracting another subsequence (if necessary), we may assume that the corresponding sequence of interpolation points converges:

$$s_{j_k n_k} \to s_0 \in [a, b] \qquad \text{as} \qquad k \to \infty .$$

Let $s_0 \in (a, b)$; by making the number η smaller (if necessary), we may assume that (a, b) also contains a neighborhood $(s_0 - \eta/2, s_0 + \eta/2)$ of s_0. Let $z(s)$ be the continuous function on $[a, b]$ defined as 1 for $|s - s_0| \leq \eta/4$, 0 for $|s - s_0| \geq \eta/2$, and linear for $\eta/4 \leq |s - s_0| \leq \eta/2$; then

$$\int_a^b z(s)\,ds = \tfrac{3}{4}\eta . \tag{17.53}$$

On the other hand, for sufficiently large k the point $s_{j_k n_k}$ is in the interval $|s - s_0| \leq \eta/4$ in which $z(s) \equiv 1$, and so

$$\sum_{j=1}^{n_k} \alpha_{jn_k} z(s_{jn_k}) \geq \alpha_{j_k n_k} ;$$

in view of conditions (17.51) and (17.52),

$$\sum_{j=1}^{n_k} \alpha_{jn_k} z(s_{jn_k}) \geq \beta_{j_k n_k} - \frac{\eta}{8} \geq \frac{7}{8}\eta$$

for sufficiently large k. Comparing this with (17.53), we conclude that the quadrature process (17.47) diverges for this function $z(s)$. We have assumed that $s_0 \in (a, b)$. If $s_0 = a$ or $s_0 = b$, a function $z(s)$ for which the quadrature process (17.47) diverges is constructed in a similar way. This is left to the reader.

Now consider the second possibility: some of the points s_{jn} lie at a great distance from the corresponding J_{jn}. More precisely, there exists a subsequence of points $s_{j_k n_k}$ whose distances from the corresponding $J_{j_k n_k}$ exceed some positive number η. Again extracting a subsequence if necessary, we may assume that each point $s_{j_k n_k}$ lies on the same side of the corresponding $J_{j_k n_k}$, say on its right, and that the right endpoints of the intervals $J_{j_k n_k}$ converge to a limit $s_1 \in [a, b]$ as $k \to \infty$. Now the right endpoints of the intervals $J_{j_k n_k}$ are the numbers $a + \sum\limits_{j=1}^{j_k} \beta_{j n_k}$, and therefore

$$\sum_{j=1}^{j_k} \beta_{j n_k} \to s - a \qquad \text{as} \qquad k \to \infty,$$

and, in view of (17.51), also

$$\sum_{j=1}^{j_k} \alpha_{j n_k} \to s_1 - a \qquad \text{as} \qquad k \to \infty. \tag{17.54}$$

Clearly, $s_1 < b$ (otherwise the points $s_{j_k n_k}$ would lie outside $[a, b]$ for large k). Let η be so small that $s_1 + \eta \leq b$. Let $z(s)$ be the continuous function defined as 1 on $a \leq s \leq s_1$, 0 for $s_1 + \eta/2 \leq s \leq b$, and linear for $s_1 \leq s \leq s_1 + \eta/2$; then

$$\int_a^b z(s)\, ds = (s_1 - a) + \frac{\eta}{4}. \tag{17.55}$$

On the other hand, for sufficiently large k,

$$\sum_{j=1}^{n_k} \alpha_{j n_k} z(s_{j n_k}) \leq \sum_{j=1}^{j_k} \alpha_{j n_k}. \tag{17.56}$$

In fact, since the distance of the point $s_{j_k n_k}$ from $J_{j_k n_k}$ is $\geq \eta$ and the right endpoints of $J_{j_k n_k}$ converge to s_1, it follows that $s_{j_k n_k} > s_1 + \eta/2$ for large k, and *a fortiori* $s_{j n_k} > s_1 + \eta/2$ for $j > j_k$. In other words, for $j > j_k$, the point $s_{j n_k}$ is in the region in which $z(s) \equiv 0$.

It follows from (17.54), (17.56) and (17.55) that the quadrature process (17.47) diverges for this function $z(s)$. ∎

Exercise 17.15. Show that (17.49) is a necessary and sufficient condition for convergence of the quadrature process (17.47), provided $\sum\limits_{j=1}^{n} \alpha_{jn} \to b - a$ as $n \to \infty$.

17.10. *The method of mechanical quadratures for integral equations.** Consider the linear integral equation

$$x(t) = \int_a^b K(t, s)\, x(s)\, ds + f(t). \tag{17.57}$$

---

* A bibliography on the method of mechanical quadratures may be found in Kantorovich and Krylov [1]. We shall reproduce a convergence proof due to Vainikko [11] for nonlinear integral equations.

It is assumed that the kernel $K(t, s)$ is continuous in the square $a \leqq t$, $s \leqq b$ and the free term $f(t)$ is continuous on $[a, b]$.

Basing our discussion on the quadrature formula (17.47), instead of a solution $x_0(t)$ of equation (17.57) we shall try to find its values $x_0(s_{jn})$ at the interpolation points s_{jn} $(j = 1, \ldots, n)$. The approximations $\xi_{jn} \approx x_0(s_{jn})$ will be determined from the system of equations

$$\xi_{in} = \sum_{j=1}^{n} \alpha_{jn} K(s_{in}, s_{jn}) \xi_{jn} + f(s_{in}) \qquad (i = 1, \ldots, n). \qquad (17.58)$$

This system is derived from (17.57) by replacing the integral of $K(t, s) x(s)$ by its approximation (17.47), dropping the remainder term, and then successively replacing the variable t by $s_{1n}, \ldots, s_{nn}$.

Theorem 17.5. Assume that equation (17.57) has a unique continuous solution $x_0(t)$, and that the quadrature process (17.47) converges.

Then, for sufficiently large n, the system (17.58) has a unique solution $(\xi_{1n}, \ldots, \xi_{nn})$, and

$$\max_{1 \leqq j \leqq n} |\xi_{jn} - x_0(s_{jn})| \to 0 \qquad as \qquad n \to \infty. \qquad (17.59)$$

The rate of convergence is described by the inequalities

$$c_1 r_n \leqq \max_{1 \leqq j \leqq n} |\xi_{jn} - x_0(s_{jn})| \leqq c_2 r_n, \qquad (17.60)$$

where $c_1, c_2 = \text{const} > 0$,

$$r_n = \max_{1 \leqq i \leqq n} |R_n(z_{in})|, \qquad z_{in}(s) = K(s_{in}, s) x_0(s),$$

$R_n(z)$ being the remainder term in (17.47).

Proof. Let us regard equation (17.57) as an operator equation (17.1) in the Banach space E of bounded measureable functions $x(t)$ on $[a, b]$, with norm
$$\|x\| = \sup_{a \leqq t \leqq b} |x(t)|.$$

Since the kernel $K(t, s)$ is continuous, the operator

$$Tx = \int_a^b K(t, s) x(s) \, ds$$

is compact, as an operator mapping E into the space C of functions continuous on $[a, b]$. *A fortiori,* T is compact as an operator in E. The operator $I - T$ is invertible in E. In fact, it follows from $x_1 \in E$, $x_1 = Tx_1$ that $x_1 \in C$, since T maps E into C. But since equation (17.57) is uniquely solvable, it follows that $x_1 = 0$, as required.

Let $\chi_{jn}(s)$ denote the characteristic function of the set D_{jn} constructed in subsection 17.9:

$$\chi_{jn}(s) = \begin{cases} 1 & \text{if} & s \in D_{jn}, \\ 0 & \text{if} & s \notin D_{jn} \end{cases} \quad (j = 1, \ldots, n).$$

Let E_n be the linear span of the functions χ_{jn} $(j = 1, \ldots, n)$. It is obvious that E_n is a closed subspace of E. Since the functions χ_{jn} $(j = 1, \ldots, n)$ are linearly independent, the system (17.58) is equivalent to equation (17.2) with

$$f_n = \sum_{i=1}^{n} f(s_{in}) \chi_{in}, \tag{17.61}$$

$$T_n z_n = \sum_{i=1}^{n} \left\{ \sum_{j=1}^{n} \alpha_{jn} K(s_{in}, s_{jn}) \zeta_j \right\} \chi_{in} \quad \left(z_n = \sum_{j=1}^{n} \zeta_j \chi_{jn} \in E_n \right). \tag{17.62}$$

By "equivalence" we mean here that a vector $(\xi_{1n}, \ldots, \xi_{nn})$ is a solution of the system (17.58) if and only if $x_n = \sum_{j=1}^{n} \xi_{jn} \chi_{jn}$ is a solution of equation (17.2).

In order to apply Theorem 17.1, we introduce a projection P_n mapping E onto E_n:

$$P_n x = \sum_{j=1}^{n} x(s_{jn} (\chi_{jn} \quad (x \in E).$$

It follows from (17.48) that $P_n E = E_n$ and $P_n z_n = z_n$ for $z_n \in E_n$, i.e., P_n is indeed a projection. Moreover, P_n is bounded, $\|P_n\| = 1$. Further, by Lemma 17.5, any continuous (therefore also uniformly continuous) function $z(s)$ on $[a, b]$ satisfies the condition

$$\max_{a \le s \le b} \left| z(s) - \sum_{j=1}^{n} z(s_{jn}) \chi_{jn}(s) \right| = \max_{1 \le j \le n} \sup_{s \in D_{jn}} |z(s) - z(s_{jn})| \to 0$$

as $n \to \infty$. In other words, the projections P_n, considered as operators mapping C into E, converge strongly to an operator embedding C in E. Recalling that T is compact as an operator mapping E into C, we see from Lemma 15.5 that

$$\|U_n\| \to 0 \quad \text{as} \quad n \to \infty \quad (U_n = T - P_n T).$$

Now, $f_n = P_n f$ (see (17.61)), $\|f - P_n F\| \to 0$ as $n \to \infty$. To apply Theorem 17.1, we must still prove that

$$\|S_n\| \to 0 \quad \text{as} \quad n \to \infty \quad (S_n = T_n - P_n T).$$

For $z_n = \sum_{j=1}^{n} \zeta_j \chi_{jn} \in E_n$, we have

$$P_n T z_n = \sum_{i=1}^{n} \left\{ \int_a^b K(s_{in}, s) z_n(s)\, ds \right\} \chi_{in} =$$

$$= \sum_{i=1}^{n} \left\{ \sum_{j=1}^{n} \int_{D_{jn}} K(s_{in}, s)\, ds\, \zeta_j \right\} \chi_{in}.$$

On the other hand, using (17.48) and (17.50) we can write (17.62) as

$$T_n z_n = \sum_{i=1}^{n} \left\{ \sum_{j=1}^{n} \int_{D_{jn}} K(s_{in}, s_{jn})\, ds\, \zeta_j \right\} \chi_{in} +$$

$$+ \sum_{i=1}^{n} \left\{ \sum_{j=1}^{n} \gamma_{jn} K(s_{in}, s_{jn}) \zeta_j \right\} \chi_{in}.$$

Thus,

$$S_n z_n = \sum_{i=1}^{n} \left\{ \sum_{j=1}^{n} \int_{D_{jn}} [K(s_{in}, s_{jn}) - K(s_{in}, s)]\, ds\, \zeta_j \right\} \chi_{in} +$$

$$+ \sum_{i=1}^{n} \left\{ \sum_{j=1}^{n} \gamma_{jn} K(s_{in}, s_{jn}) \zeta_j \right\} \chi_{in}.$$

Hence,

$$\|S_n z_n\| \leq \left[\max_{1 \leq i \leq n} \sum_{j=1}^{n} \int_{D_{jn}} |K(s_{in}, s_{jn}) - K(s_{in}, s)|\, ds + \right.$$

$$\left. + M \sum_{j=1}^{n} |\gamma_{jn}| \right] \|z_n\|,$$

where $M = \max |K(t, s)|$. Since the function $K(t, s)$ is uniformly continuous, formulas (17.49) and (17.51) imply that $\|S_n\| \to 0$ as $n \to \infty$. Theorem 17.1 is thus applicable. For $n \geqq n_0$, equation (17.2) has a unique solution $x_n = \sum_{j=1}^{n} \xi_{jn}\chi_{jn}$; as we have already stated, this is true if and only if the system (17.58) is uniquely solvable. The fact that $\|x - x_0\| \to 0$ (see Theorem 17.1) and the inequality

$$\max_{1 \leqq j \leqq n} |\xi_{jn} - x_0(s_{jn})| = \|x_n - P_n x_0\| \leqq \|x_n - x_0\| \qquad (17.63)$$

imply (17.59). The estimate (17.60) follows from (17.9) and (17.63). We need only note that in this case $g_n = 0$ and $\|(P_n T - T_n P_n)x_0\| = r_n$, where r_n is the number defined in the statement of the theorem. ∎

There are various ways of constructing analytic approximations $\bar{x}_n(t)$ to $x_0(t)$, based on the solution $(\xi_{1n}, \ldots, \xi_{nn})$ of the system (17.58). The most natural definition is

$$\bar{x}_n(t) = \sum_{j=1}^{n} \alpha_{jn}K(t, s_{jn}) \xi_{jn} + f(t).$$

No additional computations are required to construct this function.

Exercise 17.16. Under the assumptions of Theorem 17.5, show that $\|\bar{x}_n - x_0\| \to 0$ as $n \to \infty$, and prove the estimate

$$\max_{a \leqq t \leqq b} |\bar{x}_n(t) - x_0(t)| \leqq c \max_{a \leqq t \leqq b} |R_n(z_t)|, \quad z_t(s) = K(t, s) x_0(s).$$

Exercise 17.17. Let $K(t, s)$ and $f(t)$ have m continuous derivatives with respect to t. Under the assumptions of Theorem 17.5, show that $\max_{a \leqq t \leqq b} |\bar{x}_n^{(i)}(t) - x_0^{(i)}(t)| \to 0$ as $n \to \infty$ ($i = 0, 1, \ldots, m$), and establish convergence estimates.

§ 18. Projection methods in the eigenvalue problem

18.1. *The eigenvalue problem.* In a complex Banach space E (compare subsection 18.9), consider the equation

$$x = \mu T x. \qquad (18.1)$$

Throughout this section the operator T will be assumed linear and compact. We wish to determine the eigenvalues, eigenelements and generalized eigenelements of equation (18.1). We recall the basic concepts.

The eigenvalues of the equation (18.1) are those values of the numerical (complex) parameter μ for which the equation has nonzero solutions; the latter are called *eigenelements*. The set of eigenelements of equation (18.1) is at most countable and has no finite accumulation points.

It should be emphasized that the concept used above was that of the eigenvalues of an operator T; these are related to the eigenvalues of equation (18.1) by the equality $\lambda\mu = 1$.

Exercise 18.1. Let $\overline{\Omega}$ be a closed bounded set in the complex plane not containing any eigenvalues of equation (18.1). Show that the norm $\|(I - T)^{-1}\|$ is bounded for any constant $\mu \in \overline{\Omega}$.

Exercise 18.2. Assume that equation (18.1) has no eigenvalues in the disk $|\mu - \mu_1| \leq \delta$. Show that $\sup_{|\mu - \mu_1| \leq \delta} \|(I - \mu T)^{-1}\|$ is attained at some point of the circle $|\mu - \mu_1| = \delta$.

Let μ_0 be an eigenvalue of equation (18.1). The eigenelements associated with μ_0 form a subspace

$$X_0^{(1)} = \{x_0 \in E : (I - \mu_0 T) x_0 = 0\} .$$

Consider the ascending sequence of subspaces

$$X_0^{(i)} = \{x_0 \in E : (I - \mu_0 T)^i x_0 = 0\} \qquad (i = 1, 2, \ldots).$$

For compact operators, all the subspaces $X_0^{(i)}$ $(i = 1, 2, \ldots)$ are finite-dimensional, and only finitely many of them are different; in other words, there is a natural number l such that $X_0^{(l-1)} \neq X_0^{(l)}$ and $X_0^{(i)} = X_0^{(l)}$ for $i \geq l$. The number l is known as the *rank* of the eigenvalue μ_0 and the number dim $X_0^{(l)}$ its *multiplicity*. If $l = 1$ and dim $X_0^{(1)} = 1$, the eigenvalue μ_0 is said to be *simple*; otherwise, it is *multiple*. The elements $x_0 \in X_0^{(l)}$ $(x_0 \neq 0)$ are known as *generalized eigenelements** of equation (18.1). Finally, $X_0^{(1)}$ is known as an *eigensubspace* of equation (18.1), and $X_0^{(l)}$ a *generalized eigensubspace*.

There exists a bounded projection Q_0 such that Q_0 and $Q^{(0)} = I - Q_0$ project E onto $X_0^{(l)}$ and the closed subspace

$$\tilde{X}_0 = (I - \mu_0 T)^l E ,$$

---

* [The Russian term is *root elements*.]

respectively. In other words, the space E splits into a direct sum

$$E = X_0^{(l)} + \tilde{X}_0 .$$

Each of the subspaces $X_0^{(l)}$ and $\tilde{X}_0$ is T-invariant: $TX_0^{(l)} \subset X_0^{(l)}$, $T\tilde{X}_0 \subset \tilde{X}_0$. The operator $I - \mu_0 T$ is continuously invertible in $\tilde{X}_0$, since $x_0 = \mu_0 T x_0$, $x_0 \neq 0$ implies that $x_0 \in X_0^{(l)}$, and by the same token $x_0 \notin \tilde{X}_0$.

Exercise 18.3. Prove that $Q_0 T = T Q_0$, $Q^{(0)} T = T Q^{(0)}$.

Eigenvalue problems are most naturally treated by replacing equation (18.1) by "neighboring" (in a certain sense) equations in finite-dimensional spaces; the original problem is thus reduced to an eigenvalue problem for matrices. For practical methods of computing the eigenvalues, eigenvectors and generalized eigenvectors of matrices, see, e.g., Berezin and Zhidkov [1], or, in a more exhaustive exposition, Faddeev and Faddeeva [1].

In the following subsection we shall justify the transition from equation (18.1) to equations in finite- or infinite-dimensional subspaces of E.

18.2 *Convergence of the perturbed Galerkin method.** Let $\{E_n\}$ be a sequence of closed subspaces of E. Let us replace equation (18.1) by a sequence of "approximate" equations

$$x_n = \mu T_n x_n , \tag{18.2}$$

where T_n are compact linear operators in $E_n (n = 1, 2, \dots)$. To characterize the closeness of equations (18.1) and (18.2), we define (possibly unbounded) projections P_n of the space onto the corresponding E_n:

$$P_n z \in E_n \ (z \in D(P_n)) , \ P_n z_n = z_n \ (z_n \in E_n \subset D(P_n)) .$$

It is assumed that $TE \subset D(P_n) (n = 1, 2, \dots)$ and $P_n T$ is bounded in E (therefore also in E_n). All the basic facts will be proved under the assumption that**

---

* The results of this subsection were established (in somewhat different situations) by Pol'skii [1, 2] and Troitskaya [1]. See also, Kantorovich and Akilov [1], from which, among other things, we have borrowed the idea of the proof of Theorem 18.1.

** Recall that if the projections P_n are bounded, a sufficient condition for (18.3) to hold is that $\xi_n \to 0$, $\|P_n\|\eta_n \to 0$ as $n \to \infty$, where ξ_n and η_n are the numbers figuring in Kantorovich's conditions I and II (see p. 249).

$$\|S_n\| \to 0, \quad \|\bar{U}_n\| \to 0 \quad \text{as} \quad n \to \infty$$

$$(S_n = T_n - P_nT, U_n = T - P_nT). \tag{18.3}$$

The reader should note that the operators S_n are defined in E_n, and U_n in E.

Theorem 18.1. Assume that (18.3) *holds. Then, for every eigenvalue* μ_0 *of equation* (18.1) *there is a sequence* μ_n *of eigenvalues of equations* (18.2) *such that* $\mu_n \to \mu_0$ *as* $n \to \infty$.

Conversely, every limit point of any sequence μ_n *of eigenvalues of equations* (18.2) *is an eigenvalue of equation* (18.1).

Proof. Let μ_0 be an eigenvalue of equation (18.1), and $\delta > 0$ a number such that the disk $|\mu - \mu_0| \leq \delta$ contains no eigenvalues of equation (18.1) other than μ_0. Let ε be arbitrary, $0 < \varepsilon \leq \delta$. The norm $\|(I - \mu T)^{-1}\|$ is bounded on the circle $|\mu - \mu_0| = \varepsilon$, independently of the constant μ:

$$\|(I - \mu T)^{-1}\| \leq \kappa \qquad (\kappa = \text{const}; |\mu - \mu_0| = \varepsilon).$$

We see from (18.3) and Lemma 17.1 that the operators $I - \mu T_n$ are invertible in E_n for $|\mu - \mu_0| = \varepsilon$ and sufficiently large n, and the norms of their inverses are uniformly bounded in μ and n:

$$\|(I - \mu T_n)^{-1}\| \leq \kappa' \qquad (\kappa' = \text{const}; |\mu - \mu_0| = \varepsilon; \, n \geq n_0). \tag{18.4}$$

Suppose that there is a subsequence $n = n_i$ $(i = 1, 2, \dots)$ for which equations (18.2) have no eigenvalues in the disk $|\mu - \mu_0| \leq \varepsilon$. For these values of n, inequality (18.4) extends to the entire disk $|\mu - \mu_0| < \varepsilon$ (see Exercise 18.2). In particular,

$$\|(I - \mu_0 T_{n_i})^{-1}\| \leq \kappa'.$$

If n_i is chosen sufficiently large, it follows from Lemma 17.2 that the operator $I - \mu_0 T$ is invertible in E, i.e., μ_0 is not an eigenvalue of equation (18.1), contradicting the assumption. Thus, from some $n = n(\varepsilon)$ on equation (18.2) has at least one eigenvalue μ_n in the disk $|\mu - \mu_0| \leq \varepsilon$. Since ε is arbitrary, this is equivalent to the first assertion of our theorem.

As for the second assertion of the theorem, let R be an arbitrarily large number and $\varepsilon > 0$ arbitrarily small. The disk $|\mu| \leq R$ contains finitely many eigenvalues of equation (18.1), say $\mu_0^{(1)}, \dots, \mu_0^{(m)}$. Let $\Omega_{R, \varepsilon}$ denote the closed set in the complex plane obtained by cutting ε-neighborhoods of the points $\mu_0^{(1)}, \ldots, \mu_0^{(m)}$ from the disk $|\mu| \leq R$. The operator $(I - \mu T)^{-1}$ is uniformly bounded with respect to μ in the set $\Omega_{R,\varepsilon}$:

$$\|(I - \mu T)^{-1}\| \leq \kappa \qquad (\kappa = \text{const}; \ \mu \in \Omega_{R,\varepsilon}).$$

It follows from Lemma 17.1 that for sufficiently large n (say for $n \geq n_1$) and any $\mu \in \Omega_{R,\varepsilon}$ the operators $I - \mu T_n$ are invertible in E_n. Thus, for $n \geq n_1$ those eigenvalues of equation (18.2) whose absolute values do not exceed R must lie in ε-neighborhoods of the points $\mu_0^{(1)}, \ldots, \mu_0^{(m)}$. Since R and ε are arbitrary, this is equivalent to the second assertion of the theorem. ∎

A multiple eigenvalue μ_0 of equation (18.1) is usually associated with several different eigenvalues $\mu_n^{(1)}, \ldots, \mu_n^{(k_n)}$ of equation (18.2), $\mu_n^{(i_n)} \to \mu_0$ as $n \to \infty$ $(1 \leq i_n \leq k_n)$. It is easily proved, using Theorem 18.3 below, that for sufficiently large n the multiplicity of μ_0 is the sum of multiplicities of $\mu_n^{(1)}, \ldots, \mu_n^{(k_n)}$.

Exercise 18.4. Let μ_0 be an eigenvalue of equation (18.1), and let

$$\|(I - \mu T)^{-1}\| \leq \kappa \qquad (|\mu - \mu_0| = \varepsilon),$$

$$[|\mu_0|(1 + |\mu_0|\kappa\|P_n T\|) + (|\mu_0| + \varepsilon)\kappa]\|U_n\| + (2|\mu_0| + \varepsilon)\|S_n\| < 1.$$

Show that equation (18.2) has an eigenvalue in the disk $|\mu - \mu_0| \leq \varepsilon$.

Exercise 18.5. Let μ_n be an eigenvalue of equation (18.2), and let

$$\|(I - \mu T_n)^{-1}\| \leq \kappa_n \qquad (|\mu - \mu_n| = \varepsilon),$$

$$(2|\mu_n| + \varepsilon)\tau_n\|U_n\| + [|\mu_n|\tau_n + (|\mu_n| + \varepsilon)\kappa_n]\|S_n\| < 1,$$

where $\tau_n = 1 + (|\mu_n| + \varepsilon)\kappa_n\|P_n T\|$. Show that equation (18.1) has an eigenvalue in the disk $|\mu - \mu_n| \leq \varepsilon$.

Theorem 18.2. Assume that condition (18.3) is satisfied.

Then every sequence x_n of normalized eigenelements of equations (18.2) associated with eigenvalues $\mu_n \to \mu_0$ contains a convergent subsequence; the limit of any convergent subsequence x_{n_k} is an eigenelement of equation (18.1) associated with the eigenvalue μ_0.

Proof. Since the operator T is compact, the sequence Tx_n contains a convergent subsequence:

$$Tx_{n_k} \to y_0 \in E \quad \text{as} \quad k \to \infty.$$

In view of (18.3), moreover,

$$T_{n_k} x_{n_k} = T x_{n_k} - U_{n_k} x_{n_k} + S_{n_k} x_{n_k} \to y_0 \quad \text{as} \quad k \to \infty,$$

and since $\mu_n \to \mu_0$ and $x_n = \mu_n T_n x_n$,

$$x_{n_k} \to \mu_0 y_0 \quad \text{as} \quad k \to \infty,$$

i.e., the subsequence x_{n_k} is also convergent.

Now let x_{n_k} be any convergent subsequence,

$$x_{n_k} \to x_0 \quad \text{as} \quad k \to \infty.$$

Then $T x_{n_k} \to y_0 = T x_0$, and, repeating the above arguments, we get

$$x_{n_k} \to \mu_0 y_0 = \mu_0 T x_0 \quad \text{as} \quad k \to \infty.$$

Thus $x_{n_k} \to x_0$ and $x_{n_k} \to \mu_0 T x_0$, so that $x_0 = \mu_0 T x_0$. ∎

It is not always true that all eigenelements of equation (18.1) are limits of sequences of eigenelements of equations (18.2). A simple counter-example is

$$E = E_n = R_2, \quad T = \begin{pmatrix} 1 & 0 \\ 0 & 1 \end{pmatrix}, \quad T_n = \begin{pmatrix} 1 & 1/n \\ 0 & 1 \end{pmatrix} \quad (n = 1, 2, \ldots).$$

Theorem 18.3. *Assume that condition (18.3) is satisfied.*

Then all generalized eigenelements of equation (18.1) associated with an eigenvalue μ_0 are limits of linear combinations of generalized eigenelements of equations (18.2) associated with eigenvalues that converge to μ_0 as $n \to \infty$; moreover, any limit of such a sequence is a generalized eigenelement of equation (18.1).

We shall not need Theorem 18.3 in the sequel. The proof may be reconstructed by slightly modifying the reasoning used by Pol'skii [2] in proving an analogous statement for the Galerkin method (see also Troitskaya [1]).

Exercise 18.6. State and prove analogs of Theorems 18.1 to 18.3 if (18.2) is replaced by the equation

$$x_n + S'_n x_n = \mu T_n x_n,$$

where S'_n is a bounded linear operator in E_n such that $\|S'_n\| \to 0$ as $n \to \infty$.

18.3. *Rate of convergence.* Let μ_0 and μ_n be eigenvalues of equations (18.1) and (18.2), respectively. We retain the previous notation:

$$P^{(n)} = I - P_n,$$

$$X_0^{(i)} = \{x_0 \in E: \ (I - \mu_0 T)^i x_0 = 0\},$$

$$F_0^{(i)} = \{f_0 \in E^*: \ [(I - \mu_0 T)^i]^* f_0 = 0\},$$

$$X_n^{(i)} = \{x_n \in E_n: \ (I - \mu_n T_n)^i x_n = 0\},$$

$$\rho(x, X_0^{(i)}) = \inf_{x_0 \in X_0^{(i)}} \|x - x_0\|.$$

Note that if $X_0^{(l)}$ is a generalized eigensubspace of equation (18.1), then $F_0^{(l)}$ is a generalized eigensubspace of the dual equation $f = \mu T^* f$, and moreover[†] dim $X_0^{(l)} = $ dim $F_0^{(l)}$. The operator T^* is defined and compact in the space E^* dual to E.

Theorem 18.4. Assume that condition (18.3) is satisfied. Let $\mu_n \to \mu_0$, where μ_0 and μ_n are eigenvalues of equations (18.1) and (18.2), respectively, μ_0 of rank l.

Then

$$|\mu_n - \mu_0| \leqq c(\|S_n\| + \varepsilon_n)^{1/l} \tag{18.5}$$

and, if the projections P_n are bounded,

$$|\mu_n - \mu_0| \leqq c(\|S_n\| + \varepsilon_n \varepsilon_n^*)^{1/l}, \tag{18.6}$$

where

$$\varepsilon_n = \sup_{x_0 \in X_0^{(l)}, \|x_0\| = 1} \|P^{(n)} x_0\|, \qquad \varepsilon_n^* = \sup_{f_0 \in F_0^{(l)}, \|f_0\| = 1} \|(P^{(n)})^* f_0\|. \tag{18.7}$$

Moreover, for every $x_n^{(k)} \in X_n^{(k)}$ ($\|x_n^{(k)}\| = 1, \ k \geqq 1$),

$$\rho(x_n^{(k)}, X_0^{(l)}) \leqq c(\|S_n\| + \varepsilon_n) \tag{18.8}$$

[†] There is a certain difference between the properties of the duals of Banach and Hilbert spaces. For Banach spaces, $(\mu T)^* = \mu T^*$, while for Hilbert spaces $(\mu T)^* = \bar{\mu} T^*$, where $\bar{\mu}$ is the complex conjugate of μ. Here the subspaces F_0 correspond to the eigenvalue μ_0 of the equation $f = \mu T^* f$. If E is a Hilbert space, F_0 corresponds to the eigenvalue $\bar{\mu}_0$.

and for $k \leq i \leq l$

$$\rho(x_n^{(k)}, X_0^{(i)}) \leq c(|\mu_n - \mu_0|^{i-k+1} + \|S_n\| + \varepsilon_n). \tag{18.9}$$

Exercise 18.7. Show that $\varepsilon_n \leq c \|U_n\|$ ($c = $ const).

Proof of Theorem 18.4. The proof will be given for the case $l = 1$, $k = 1$ only.

For $x_n \in E_n$,

$$(I - \mu_0 T)x_n = (I - \mu_n T_n)x_n + (\mu_n - \mu_0) T_n x_n - \mu_0(T - T_n)x_n.$$

Setting $x_n = x_n^{(1)} \in X_n^{(1)}$ ($\|x_n^{(1)}\| = 1$), we get

$$(I - \mu_0 T) x_n^{(1)} = \frac{\mu_n - \mu_0}{\mu_n} x_n^{(1)} - \mu_0(T - T_n) x_n^{(1)}. \tag{18.10}$$

Now apply a functional $f_0 \in F_0^{(1)}$ ($\|f_0\| = 1$) to both sides of (18.10). Since

$$f_0(I - \mu_0 T) = (I - \mu_0 T)^* f_0 = 0,$$

it follows that

$$\frac{\mu_n - \mu_0}{\mu_n} f_0(x_n^{(1)}) = \mu_0 f_0(T - T_n) x_n^{(1)}$$

and

$$|f_0(x_n^{(1)})| \cdot |\mu_n - \mu_0| = |\mu_0| \cdot |\mu_n| \cdot |f_0(T - T_n) x_n^{(1)}| \quad (n \geq n_0). \tag{18.11}$$

We have yet to choose the functional $f_0 \in F_0^{(1)}$ ($\|f_0\| = 1$). Assume that this has been done so that, for $n \geq n_0$,

$$|f_0(x_n^{(1)})| \geq c_0 > 0 \quad (c_0 = \text{const}). \tag{18.12}$$

This is possible, for otherwise there would be a subsequence $\{x_{n_j}^{(1)}\}$ such that

$$\sup_{f_0 F_0^{(1)}, \|f_0\|=1} |f_0(x_{n_j}^{(1)})| \to 0 \quad \text{as} \quad j \to \infty.$$

By Theorem 18.2, we may assume that this subsequence is convergent:

$$x_{n_j} \to x_0 \in X_0^{(1)} \quad \text{as} \quad j \to \infty \quad (\|x_0\| = 1).$$

Now the limit x_0 is such that $f_0(x_0) = 0$ for every $f_0 \in F_0$. Thus the free term of the equation $x = \mu_0 Tx + x_0$ is orthogonal to the eigensubspace F_0 of the homogeneous adjoint equation, and, by a theorem of Fredholm, this equation has a solution x'. Now

$$(I - \mu_0 T)x' = x_0 \neq 0, \ (I - \mu_0 T)^2 x' = (I - \mu_0 T)x_0 = 0.$$

This means that the rank of the eigenvalue μ_0 is greater than unity, a contradiction.

It follows from (18.11) and (18.12) that

$$|\mu_n - \mu_0| \leq c_1 \sup_{f_0 \in F_0^{(1)}, \|f_0\| = 1} |f_0(T - T_n)x_n^{(1)}| \qquad (c_1 = \text{const}). \quad (18.13)$$

We need a similar estimate for $\rho(x_n^{(1)}, X_0^{(1)})$. Let Q_0 be a projection such that Q_0 and $Q^{(0)} = I - Q_0$ project E onto the subspaces $X_0^{(1)}$ and $X_0 = (I - \mu_0 T)E$, respectively (see subsection 18.1). Apply $Q^{(0)}$ to both sides of (18.10):

$$(I - \mu_0 T)Q^{(0)}x_n^{(1)} = \frac{\mu_n - \mu_0}{\mu_n} Q^{(0)}x_n^{(1)} - \mu_0 Q^{(0)}(T - T_n)x_n^{(1)}$$

(recall that T commutes with Q_0 and $Q^{(0)}$). Since the operator $I - \mu_0 T$ is continuously invertible in $\tilde{X}_0$ and $\mu_n \to \mu_0 \neq 0$ as $n \to \infty$, we get the estimate

$$\|Q^{(0)}x_n^{(1)}\| \leq c_2 \left[|\mu_n - \mu_0| + \|(T - T_n)x_n^{(1)}\|\right] \qquad (c_2 = \text{const}).$$

Together with (18.13), this gives

$$\|Q^{(0)}x_n^{(1)}\| \leq c_3 \|(T - T_n)x_n^{(1)}\| \qquad (c_3 = \text{const}). \quad (18.14)$$

Obviously, $\rho(x_n^{(1)}, X_0^{(1)}) \leq \|Q^{(0)}x_n^{(1)}\|$.

Let us estimate $\|(T - T_n)x_n^{(1)}\|$:

$$(T - T_n)x_n^{(1)} = U_n x_n^{(1)} - S_n x_n^{(1)} = U_n Q_0 x_n^{(1)} + U_n Q^{(0)}x_n^{(1)} - S_n x_n^{(1)}$$

and, since $Q_0 x_n^{(1)} \in X_0^{(1)}$,

$$U_n Q_0 x_n^{(1)} = P^{(n)} T(Q_0 x_n^{(1)}) = \frac{1}{\mu_0} P^{(n)} (Q_0 x_n^{(1)}).$$

Consequently,

$$(T - T_n)x_n^{(1)} = \frac{1}{\mu_0} P^{(n)}(Q_0 x_n^{(1)}) + U_n(Q^{(0)}x_n^{(1)}) - S_n x_n^{(1)}. \quad (18.15)$$

Hence

$$\left\| (T - T_n) x_n^{(1)} \right\| \leq \frac{\| Q_0 \|}{| \mu_0 |} \varepsilon_n + \| U_n \| \cdot \left\| Q^{(0)} x_n^{(1)} \right\| + \| S_n \|,$$

where ε_n is the number defined in (18.7). Replace $\left\| Q^{(0)} x_n^{(1)} \right\|$ by (18.14) in the right-hand side of this inequality. Using the fact that $\| U_n \| \to 0$ as $n \to \infty$, we get, after some manipulation,

$$\left\| (T - T_n) x_n^{(1)} \right\| \leq c_4 (\varepsilon_n + \| S_n \|) \qquad (c_4 = \text{const}). \qquad (18.16)$$

Estimates (18.5) and (18.8) follow from (18.13), (18.14) and (18.16). As for the estimate (18.9), in the present case ($l = 1$, $k = 1$) it provides no more information than (18.8) and is a consequence thereof.

Let the projections P_n be bounded. Let us estimate $\left| f_0 (T - T_n) x_n^{(1)} \right|$. Clearly, $U_n = P^{(n)} U_n$, and by (18.15)

$$f_0 (T - T_n) x_n^{(1)} = f_0 P^{(n)} \left[\frac{1}{\mu_0} P^{(n)} (Q_0 x_n^{(1)}) + U_n (Q^{(0)} x_n^{(1)}) \right] - f_0 S_n x_n^{(1)}.$$

Since $\left\| f_0 P^{(n)} \right\| = \left\| (P^{(n)})^* f_0 \right\| \leq \varepsilon_n^*$ (see (18.17)), we have

$$\sup_{f_0 \in F_0^{(1)}, \| f_0 \| = 1} \left| f_0 (T - T_n) x_n^{(1)} \right| \leq$$

$$\leq \varepsilon_n^* \left(\frac{\| Q_0 \|}{| \mu_0 |} \varepsilon_n + \| U_n \| \cdot \left\| Q^{(0)} x_n^{(1)} \right\| \right) + \| S_n \|,$$

and, estimating $\left\| Q^{(0)} x_n^{(1)} \right\|$ via (18.14) and (18.16),

$$\sup_{f_0 \in F_0^{(1)}, \| f_0 \| = 1} \left| f_0 (T - T_n) x_n^{(1)} \right| \leq c_5 (\varepsilon_n^* \varepsilon_n + \varepsilon_n^* \| U_n \| \cdot \| S_n \| + \| S_n \|).$$

$$(18.17)$$

The estimate (18.6) follows from (18.5) for values of n such that $\varepsilon_n^* \geq 1$, and from (18.13) and (18.17) for values of n such that $\varepsilon_n^* \leq 1$.

This completes the proof for the case $l = 1$, $k = 1$.

The complete proof of Theorem 18.4 follows the same plan, and the only difficulties are technical. ∎

Exercise 18.8. Assume that Kantorovich's conditions I and II (see p. 249) are satisfied, and moreover $\xi_n \to 0$, $\| P_n \| \eta_n \to 0$ as $n \to \infty$. Let $\mu_n \to \mu_0$ as $n \to \infty$, where μ_0 and μ_n are eigenvalues of equations (18.1) and (18.2), respectively, μ_0 of rank l. Establish the estimates

$$| \mu_n - \mu_0 | \leq c (\xi_n + \| P_n \| \zeta_n)^{1/t},$$

$$|\mu_n - \mu_0| \le c(\xi_n + \|P_n\|^2 \zeta_n \zeta_n^*)^{1/l},$$

$$\rho(x_n^{(k)}, X_0^{(l)}) \le c(\xi_n + \|P_n\| \zeta_n) \quad (x_n^{(k)} \in X_n^{(k)}, \|x_n^{(k)}\| = 1),$$

where

$$\zeta_n = \sup_{x_0 \in X_0^{(l)}, \|x_0\|=1} \rho(x_0, E_n), \quad \zeta_n^* = \sup_{f_0 \in F_0^{(l)}, \|f_0\|=1} \rho(f_0, P_n^* E^*).$$

Show that $\zeta_n \le c\eta_n$.

Exercise 18.9. Let T_n be a sequence of compact operators in E, $\|T_n - T\| \to 0$ as $n \to \infty$. Let μ_0 be an eigenvalue of the equation $x = \mu Tx$, of rank l. Show that for sufficiently large n the equation $x = \mu T_n x$ has an eigenvalue μ_n, $|\mu_n - \mu_0| \le c\|T - T_n\|^{1/l}$.

Exercise 18.10. Generally speaking, the exponent $1/l$ in estimates (18.5) and (18.6) cannot be enlarged. Discuss the simple example

$$E = E_n = R_2, \quad T = \begin{pmatrix} 0 & 1 \\ 0 & 0 \end{pmatrix}, \quad T_n = \begin{pmatrix} 0 & 1 \\ 1/n & 0 \end{pmatrix}.$$

For simple eigenvalues, the rate of convergence of projection methods in the eigenvalue problem has been studied by Krasnosel'skii [2, 4, 8]; the general case has been studied by Vainikko [4, 5, 10, 13]. The last-mentioned paper includes a complete proof of Theorem 18.4. Estimates of another type have been obtained by Vainikko and Dement'eva [1]. The results of subsections 18.4 to 18.8 are due to Vainikko.

18.4. *The Bubnov–Galerkin method; eigenvalue estimates.* We shall discuss estimates relating to the error in the n-th approximation.

Let T be a compact selfadjoint linear operator in a Hilbert space H. Then the eigenvalues of the equation

$$x = \mu Tx \tag{18.18}$$

are real and of rank $l = 1$. Denote them by μ_k ($k = \pm 1, \pm 2, \ldots$), where the number of occurrences of each eigenvalue in this sequence is equal to its multiplicity:

$$\cdots \le \mu_{-3} \le \mu_{-2} \le \mu_{-1} < 0 < \mu_1 \le \mu_2 \le \mu_3 \le \cdots$$

The sequence may terminate in one or both directions. For example, if $(Tx, x) \ge 0$ for any $x \in H$, all the eigenvalues of equation (18.18) are positive.

Let x_k denote an eigenelement corresponding to μ_k ($k = \pm 1, \pm 2, \ldots$).

If $\mu \neq \mu_j$ $(j = \pm 1, \pm 2, \ldots)$, the operator $I - \mu T$ is invertible and

$$\|(I - \mu T)^{-1}\| = \max \left\{ 1, \sup_{j = \pm 1, \pm 2, \ldots} \frac{|\mu_j|}{|\mu_j - \mu|} \right\}. \qquad (18.19)$$

Let $\{H_n\}$ be an ultimately dense sequence of closed subspaces in H, P_n the corresponding orthogonal projections. The Bubnov–Galerkin method for the eigenvalue problem amounts to determining the approximate eigenvalues $\mu = \mu_{k,n}$ and eigenelements $x = x_{k,n}$ from the conditions

$$P_n(x - \mu T x) = 0 \qquad (x \in H_n).$$

This leads to the equation

$$x = \mu P_n T x, \qquad (18.20)$$

which may be considered both in H and in H_n. The operator $P_n T$ is selfadjoint and compact in H_n, and the eigenvalues $\mu_{k,n}$ of equation (18.20) are therefore real; list them in a sequence

$$\ldots \leqq \mu_{-2,n} \leqq \mu_{-1,n} < 0 < \mu_{1,n} \leqq \mu_{2,n} \leqq \ldots$$

This sequence terminates in both directions if H_n is finite-dimensional.

Exercise 18.11. Show that $|\mu_k| \leqq |\mu_{k,n}|$.

Exercise 18.12. Let $H_n \subset H_{n+1}$. Show that $|\mu_{k,n+1}| \leqq |\mu_{k,n}|$.

Exercise 18.13. Assume that H_n is the linear span of elements $\phi_1, \ldots, \phi_n$. Show that the point $\mu_{k,n}$ is a solution of the algebraic equation

$$\det (A_n - \mu B_n) = 0,$$

where A_n and B_n are matrices with elements $a_{ij} = (\phi_j, \phi_i)$ and $b_{ij} = (T\phi_j, \phi_i)$ $(i, j = 1, \ldots, n)$, respectively. Show that the coefficients ξ_j of the approximate eigenelement

$$x_{k,n} = \sum_{j=1}^{n} \xi_j \phi_j$$

must be a nontrivial equation of the system

$$(A_n - \mu_{k,n} B_n) \xi^{(n)} = 0 \qquad (\xi^{(n)} = (\xi_1, \ldots, \xi_n)).$$

Theorem 18.5. *The following error estimate holds*:

$$|\mu_{k,n} - \mu_k| \leqq \sqrt{2} \, |\mu_k| \cdot |\mu_{k,n}| \cdot \|P^{(n)} T\|, \qquad (18.21)$$

where

$$\|P^{(n)} T\| \to 0 \quad as \quad n \to \infty \qquad (P^{(n)} = I - P_n). \qquad (18.22)$$

Proof.

$$Tx_k = \frac{1}{\mu_k} x_k, \qquad P_n T P_n x_{k,n} = \frac{1}{\mu_{k,n}} x_{k,n}.$$

Like the operator T, the operator $P_n T P_n$ is compact and selfadjoint in H. By a theorem of Weil (see, e.g., Riesz and Nagy [1]), the difference between the k-th (in absolute values) eigenvalues of equal sign of two compact selfadjoint operators does not exceed the norm of the difference between the operators:

$$\left| \frac{1}{\mu_k} - \frac{1}{\mu_{k,n}} \right| \leq \| T - P_n T P_n \|. \qquad (18.23)$$

Let us estimate the norm $\| T - P_n T P_n \|$. Setting

$$z = (T - P_n T P_n) x$$

for any $x \in H$, we have

$$P_n z = P_n T P^{(n)} x, \qquad P^{(n)} z = P^{(n)} T x.$$

Hence, in view of the fact that $T P^{(n)}$ is the adjoint of $P^{(n)} T$, we get the estimate

$$\| P_n z \| \leq \| P^{(n)} T \| \cdot \| x \|, \qquad \| P^{(n)} z \| \leq \| P^{(n)} T \| \cdot \| x \|,$$

$$\| z \|^2 = \| P_n z \|^2 + \| P^{(n)} z \|^2 \leq 2 \| P^{(n)} T \|^2 \| x \|^2.$$

This inequality means that $\| T - P_n T P_n \| \leq \sqrt{2} \| P^{(n)} T \|$.

The estimate (18.21) now follows from (18.23).

The relation (18.22) is proved by an application of Lemma 15.4. ∎

Exercise 18.14. Let $(Tx, x) > 0$ for any $x \in H$. Show that the factor $\sqrt{2}$ in (18.21) may then be omitted:

$$0 \leq \mu_{k,n} - \mu_k \leq \mu_k \mu_{k,n} \| P^{(n)} T \|.$$

It is not known whether this improvement is valid in the general case.

Hint. Let H_T be the completion of H with respect to the inner product $(x, y)_T = (Tx, y)$. Show that T and $P_n T$ are compact and symmetric in H_T and map H_T into H. Use Weil's theorem, and show that $\| P^{(n)} T \|_T \leq \| P^{(n)} T \|$.

Exercise 18.15. Show that for every $\mu_{k,n}$ there exists μ_l (l may be different from k) such that

$$| \mu_{k,n} - \mu_l | \leq | \mu_l | \cdot | \mu_{k,n} | \cdot \| P^{(n)} T \|.$$

Hint. Consider the equation $x = \mu (P_n T P_n + P^{(n)} T P^{(n)}) x$.

18.5. *Estimates for eigenelements.* We continue our investigation of the Bubnov–Galerkin method (18.20) for equation (18.18). We shall assume that the multiplicity of the eigenvalue μ_k of equation (18.18) is r:

$$\mu_{k-1} < \mu_k = \mu_{k+1} = \ldots = \mu_{k+r-1} < \mu_{k+r}.$$

Let X_k and $X_{k,n}(X_k, X_{k,n} \subset H, \dim X_k = \dim X_{k,n} = r)$ denote the true and approximate eigensubspaces of equation (18.18) corresponding to μ_k, with bases $x_k, \ldots, x_{k+r-1}$ and $x_{k,n}, \ldots, x_{k+r-1,n}$, respectively; let Q_k and $Q_{k,n}$ be the corresponding orthogonal projections. We wish to find estimates for the errors

$$\|x_{i,n} - Q_k x_{i,n}\| = \rho(x_{i,n}, X_k) \qquad (i = k, \ldots, k + r - 1),$$

$$\|x_k - Q_{k,n} x_k\| = \rho(x_k, X_{k,n}) \qquad (x_k \in X_k).$$

Theorem 18.6. The following error estimates hold (x_k is any element of X_k; $i = k, \ldots, k + r - 1$):

$$\|x_{i,n} - Q_k x_{i,n}\| \leqq c_{i,n} \|P^{(n)} T x_{i,n}\|, \tag{18.24}$$

$$\|x_k - Q_{k,n} x_k\| \leqq c'_{i,n} \|P^{(n)} T x_k\|. \tag{18.25}$$

Here

$$c_{i,n} = \frac{|\mu_k| \cdot |\mu_{i,n}| (1 + \gamma_k |\mu_k| \cdot \|P^{(n)} T\|)}{|\mu_{i,n}| - \gamma_k |\mu_{i,n} - \mu_k|} \to |\mu_k| \quad as^* \ n \to \infty, \tag{18.26}$$

where

$$\gamma_k = \max \left\{ 1, \sup_{\substack{\mu_j \\ j \neq k, \ldots, k+r-1}} \frac{|\mu_j|}{|\mu_j - \mu_k|} \right\}, \tag{18.27}$$

$$c'_{i,n} = \frac{|\mu_k| \cdot |\mu_{i,n}| (1 + \gamma_{i,n} |\mu_k| \cdot \|P^{(n)} T\|)}{|\mu_k| - \gamma_{i,n} |\mu_{i,n} - \mu_k|} \to |\mu_k| \quad as \ n \to \infty, \tag{18.28}$$

---

* We are considering only values of n so large that $|\mu_{i,n}| - \gamma_k |\mu_{i,n} - \mu_k| > 0$ and $|\mu_k| - \gamma_{i,n} |\mu_{i,n} - \mu_k| > 0$.

The quantities $|\mu_{i,n} - \mu_k|$ may be replaced by the estimate (18.21) or by the sharper estimates (18.35), (18.37).

where

$$\gamma_{i,n} = \max\left\{1, \sup_{\substack{\mu_{j,n} \\ j \neq k, \ldots, k+r-1}} \frac{|\mu_{j,n}|}{|\mu_{j,n} - \mu_{i,n}|}\right\} \to \gamma_k \quad as \quad n \to \infty. \quad (18.29)$$

Proof. We first prove (18.24). We have

$$(I - \mu_k T)x_{i,n} = (I - \mu_{i,n}P_n T)x_{i,n} + (\mu_{i,n} - \mu_k)\, P_n T x_{i,n} - \mu_k P^{(n)} T x_{i,n}$$

or, since $x_{i,n} = \mu_{i,n} P_n T x_{i,n}$,

$$(I - \mu_k T)x_{i,n} = \frac{\mu_{i,n} - \mu_k}{\mu_{i,n}} x_{i,n} - \mu_k P^{(n)} T x_{i,n}. \quad (18.30)$$

The operators Q_k and T, therefore also $Q^{(k)} = I - Q_k$ and T, commute, and so $Q^{(k)}T = Q^{(k)}Q^{(k)}T = Q^{(k)}TQ^{(k)}$. Apply the operator $Q^{(k)}$ to both sides of (18.30):

$$(I - \mu_k Q^{(k)}T)\, Q^{(k)}x_{i,n} = \frac{\mu_{i,n} - \mu_k}{\mu_{i,n}} Q^{(k)}x_{i,n} - \mu_k Q^{(k)}P^{(n)} T x_{i,n}. \quad (18.31)$$

The operator $Q^{(k)}T$ is selfadjoint in H; the eigenvalues of the equation $x = \mu Q^{(k)}Tx$ are the numbers $\mu_j\, (j = \pm 1, \pm 2, \ldots; j \neq k, \ldots, k + r - 1)$, and so the operator $I - \mu_k Q^{(k)}T$ is invertible and

$$\left\|(I - \mu_k Q^{(k)}T)^{-1}\right\| = \gamma_k,$$

where γ_k is the constant defined by (18.27). Further,

$$(I - \mu_k Q^{(k)}T)^{-1} = I + \mu_k (I - \mu_k Q^{(k)}T)^{-1} Q^{(k)}T,$$

and since

$$TQ^{(k)} = Q^{(k)}T, \quad \left\|TP^{(n)}Tx\right\| = \left\|TP^{(n)}P^{(n)}Tx\right\| \leqq \left\|P^{(n)}T\right\| \cdot \left\|P^{(n)}Tx\right\|,$$

it follows that

$$\left\|(I - \mu_k Q^{(k)}T)^{-1} Q^{(k)}P^{(n)}Tx_{i,n}\right\| \leqq$$

$$\leqq \left\|P^{(n)}Tx_{i,n}\right\| + |\mu_k| \cdot \gamma_k \left\|P^{(n)}T\right\| \cdot \left\|P^{(n)}Tx_{i,n}\right\|.$$

It therefore follows from (18.31) that

$$\|Q^{(k)}x_{i,n}\| \le \gamma_k \frac{|\mu_{i,n} - \mu_k|}{|\mu_{i,n}|} \|Q^{(k)}x_{i,n}\| +$$

$$+ |\mu_k|(1 + \gamma_k|\mu_k| \cdot \|P^{(n)}T\|) \|P^{(n)}Tx_{i,n}\|.$$

Collecting like terms, we get (18.24).

The proof of (18.25) is analogous:

$$(I - \mu_{i,n}P_nTP_n)x_k = (I - \mu_kT)x_k + (\mu_k - \mu_{i,n})Tx_k + \mu_{i,n}(T - P_nTP_n)x_k,$$

or, since $x_k = \mu_kTx_k$ and $T - P_nTP_n = P^{(n)}T + P_nTP^{(n)}$,

$$(I - \mu_{i,n}P_nTP_n)x_k =$$

$$= \frac{\mu_k - \mu_{i,n}}{\mu_k} x_k + \mu_{i,n}P^{(n)}Tx_k + \mu_{i,n}P_nTP^{(n)}x_k. \qquad (18.32)$$

Apply the operator $Q^{(k,n)} = I - Q_{k,n}$ to both sides of (18.32). Since $Q^{(k,n)}$ and P_nTP_n commute, we get

$$(I - \mu_{i,n}Q^{(k,n)}P_nTP_n)Q^{(k,n)}x_k =$$

$$= \frac{\mu_k - \mu_{i,n}}{\mu_k} Q^{(k,n)}x_k + \mu_{i,n}Q^{(k,n)}P^{(n)}Tx_k + \mu_{i,n}Q^{(k,n)}P_nTP^{(n)}x_k. \qquad (18.33)$$

As before, we have

$$\|(I - \mu_{i,n}Q^{(k,n)}P_nTP_n)^{-1}\| = \gamma_{i,n},$$

where $\gamma_{i,n}$ is the constant defined by (18.29). Further,

$$(I - \mu_{i,n}Q^{(k,n)}P_nTP_n)^{-1} = I + \mu_{i,n}(I - \mu_{i,n}Q^{(k,n)}P_nTP_n)^{-1}Q^{(k,n)}P_nTP_n.$$

Multiplying both sides of this equality on the right by $Q^{(k,n)}P^{(n)}T$ and using the fact that $P_nTP_nQ^{(k,n)}P^{(n)}T = Q^{(k,n)}P_nTP_nP^{(n)}T = 0$, we see that

$$(I - \mu_{i,n}Q^{(k,n)}P_nTP_n)^{-1}Q^{(k,n)}P^{(n)}T = Q^{(k,n)}P^{(n)}T.$$

Consequently, it follows from (18.33) that

$$\|Q^{(k,n)}x_k\| \le \gamma_{i,n} \frac{|\mu_k - \mu_{i,n}|}{|\mu_k|} \|Q^{(k,n)}x_k\| + |\mu_{i,n}| \cdot \|P^{(n)}Tx_k\| +$$

$$+ \gamma_{i,n}|\mu_{i,n}| \cdot \|TP^{(n)}x_k\|.$$

Collecting like terms, and noting that

$$\left\| TP^{(n)}x_k \right\| = \left| \mu_k \right| \cdot \left\| TP^{(n)}Tx_k \right\| \leq \left| \mu_k \right| \cdot \left\| P^{(n)}T \right\| \cdot \left\| P^{(n)}Tx_k \right\| ,$$

we get the estimate (18.25). ∎

Exercise 18.16. Prove that the estimates (18.24) and (18.25) are *asymptotically exact*, i.e., the ratio of the error to its estimate converges to unity as $n \to \infty$:

$$\frac{\left\| x_{i,n} - Q_k x_{i,n} \right\|}{c_{i,n} \left\| P^{(n)}Tx_{i,n} \right\|} \to 1 , \quad \frac{\left\| x_k - Q_{k,n}x_k \right\|}{c'_{i,n} \left\| P^{(n)}Tx_k \right\|} \to 1 .$$

18.6. *Improved eigenvalue estimates.* In this subsection we shall conclude our study of the Bubnov–Galerkin method (18.20) for equation (18.18). We retain the notation and assumptions of subsection 18.5. We shall establish sharper estimates than in subsection 18.4 for $\left| \mu_{k,n} - \mu_k \right|$. These new estimates presuppose the availability of error estimates for the approximate eigenelements (e.g., those following from Theorem 18.6).

Theorem 18.7. *Assume that for some* i $(k \leq i \leq k + r - 1)$

$$\left\| x_{i,n} - Q_k x_{i,n} \right\| \leq \delta_{i,n} \left\| x_{i,n} \right\| \quad (\delta_{i,n} < 1); \tag{18.34}$$

then, for this i,

$$\left| \mu_{i,n} - \mu_k \right| \leq \frac{\mu_k^2 \left| \mu_{i,n} \right| \cdot \left\| P^{(n)}Tx_{i,n} \right\| \cdot \left\| P^{(n)}T(Q_k x_{i,n}) \right\|}{[1 - \delta_{i,n}^2]^{1/2} \left\| x_{i,n} \right\| \cdot \left\| Q_k x_{i,n} \right\|} . \tag{18.35}$$

If

$$\left\| x_k - Q_{k,n}x_k \right\| \leq \delta_n \left\| x_k \right\| \quad (\delta_n < 1) \tag{18.36}$$

for any $x_k \in X_k$, *then for all* $i = k, \dots, k + r - 1$

$$\left| \mu_{i,n} - \mu_k \right| \leq \frac{\mu_k^2 \left| \mu_{i,n} \right| \cdot \left\| P^{(n)}Tx_{i,n} \right\| \cdot \left\| P^{(n)}Tx_k^{(i)} \right\|}{[1 - \delta_n^2]^{1/2} \left\| x_{i,n} \right\| \cdot \left\| x_k^{(i)} \right\|} , \tag{18.37}$$

where $x_k^{(i)}$ *is an element of* X_k *such that* $Q_{k,n}x_k^{(i)} = x_{i,n}$ *(by (18.36), such an element exists).*

Proof. We first prove (18.35). To compute the inner product of $Q_k x_{i,n}$ and both sides of (18.30), we treat each term separately:

$$\left((I - \mu_k T) x_{i,n}, Q_k x_{i,n} \right) = (x_{i,n}, (I - \mu_k T) Q_k x_{i,n}) = (x_{i,n}, 0) = 0 ,$$

$$(x_{i,n}, Q_k x_{i,n}) = \left\| Q_k x_{i,n} \right\|^2 ,$$

$$(P^{(n)}Tx_{i,n}, Q_k x_{i,n}) = (P^{(n)}Tx_{i,n}, \mu_k TQ_k x_{i,n}) = \mu_k(P^{(n)}Tx_{i,n}, P^{(n)}TQ_k x_{i,n}).$$

The result is

$$\mu_{i,n} - \mu_k = \mu_k^2 \mu_{i,n} \frac{(P^{(n)}Tx_{i,n}, P^{(n)}TQ_k x_{i,n})}{\|Q_k x_{i,n}\|^2}.$$

To derive the estimate (18.35), we need only note that, by (18.34),

$$\|Q_k x_{i,n}\| = [\|x_{i,n}\|^2 - \|Q^{(k)}x_{i,n}\|^2]^{1/2} \geqq [1 - \delta_{i,n}^2]^{1/2} \|x_{i,n}\|.$$

The proof of (18.37) is analogous. Set $x_k = x_k^{(i)}$ in (18.32) and then compute the inner product of both sides with $x_{i,n} = Q_{k,n} x_k^{(i)}$; simple manipulations then yield

$$\mu_{i,n} - \mu_k = \mu_k^2 \mu_{i,n} \frac{(P^{(n)}Tx_k^{(i)}, P^{(n)}Tx_{i,n})}{\|x_{i,n}\|^2}.$$

The estimate (18.37) now follows, in view of the fact that, by (18.36),

$$\|x_{i,n}\| = \|Q_{k,n} x_k^{(i)}\| \geqq [1 - \delta_n^2]^{1/2} \|x_k^{(i)}\|. \quad \blacksquare$$

Exercise 18.17. Let $\delta_{i,n} \to 0$, $\delta_n \to 0$ as $n \to \infty$. Show that the estimates (18.35) and (18.37) are then asymptotically exact (see Exercise 18.16).

Exercise 18.18. Establish the relations

$$\mu_{i,n} - \mu_k = \mu_{i,n}(1 + \varepsilon_{i,n}) \frac{\|P^{(n)}(Q_k x_{i,n})\|^2}{\|Q_k x_{i,n}\|^2},$$

$$\|x_{i,n} - Q_k x_{i,n}\| = (1 + \varepsilon'_{i,n})\|P^{(n)}(Q_k x_{i,n})\|,$$

$$\|x_k - Q_{k,n} x_k\| = (1 + \varepsilon''_{k,n})\|P^{(n)}x_k\|,$$

where $\varepsilon_{i,n} \to 0$, $\varepsilon'_{i,n} \to 0$, $\varepsilon''_{k,n} \to 0$ as $n \to \infty$.

Exercise 18.19. Show that if μ_k is a simple eigenvalue then

$$\mu_{k,n} - \mu_k = \mu_{k,n}(1 + \varepsilon_{k,n})\|P^{(n)}x_k\|^2 \quad (\|x_k\| = 1; \quad \varepsilon_{k,n} \to 0 \quad \text{as} \quad n \to \infty),$$

$$\|x_{k,n} - x_k\| = (1 + \varepsilon'_{k,n})\|P^{(n)}x_k\| \quad (\|x_k\| = \|x_{k,n}\| = 1; \quad \varepsilon'_{k,n} \to 0 \quad \text{as} \quad n \to \infty).$$

Exercise 18.20. Establish the following *a posteriori* error estimates $(i = k, \ldots, k + r - 1)$:

$$|\mu_{i,n} - \mu_k| \leqq \frac{\mu_k^2 |\mu_{i,n}| \alpha_{i,n}^2}{1 - c'_{i,n}\|P^{(n)}T\|},$$

$$\|x_{i,n} - Q_k x_{i,n}\| \leqq c_{i,n}\alpha_{i,n}\|x_{i,n}\|,$$

where

$$\alpha_{i,n} = \frac{\|x_{i,n} - \mu_{i,n}Tx_{i,n}\|}{|\mu_{i,n}| \cdot \|x_{i,n}\|},$$

and $c_{i,n}$ and $c'_{i,n}$ are defined by (18.26) and (18.28). Prove that these estimates are asymptotically exact.

18.7. *The Galerkin–Petrov method.* Let T be a compact (not necessarily selfadjoint) linear operator in a Hilbert space H. We wish to solve the eigenvalue problem for the equation

$$x = \mu Tx \qquad (18.38)$$

by the Galerkin–Petrov method.

Let $\{G_n\}$ and $\{H_n\}$ be two sequences of closed subspaces of H such that

$$\varlimsup_{n \to \infty} \Theta(G_n, H_n) < 1, \qquad (18.39)$$

and Q_n and P_n the orthogonal projections onto G_n and H_n, respectively, $Q^{(n)} = I - Q_n$, $P^{(n)} = I - P_n$. In the Galerkin–Petrov method, the approximate eigenvalues $\mu = \mu_n$ and eigenelements x_n of equation (18.38) are determined from the equation

$$P_n(x_n - \mu Tx_n) = 0 \qquad (x_n \in G_n). \qquad (18.40)$$

By (18.39), there exist a natural number n_0 and constants q, $0 \le q < 1$, such that $\Theta(Q_n, H_n) \le q$ for $n \ge n_0$. For $n \ge n_0$, condition (18.40) is equivalent to the equation (see subsection 16.2)

$$x_n = \mu \bar{P}Tx_n, \qquad (18.41)$$

where $\bar{P}_n = \tilde{P}_n^{-1}P_n$ is a bounded projection mapping H onto G_n, $\tilde{P}_n$ is the restriction of P_n to G_n. Let the sequence $\{G_n\}$ be ultimately dense in H. It then follows from (18.39) (see subsection 16.2) that $\|\bar{P}^{(n)}T\| \to 0$ as $n \to \infty$ ($\bar{P}^{(n)} = I - \bar{P}_n$). Thus the operators T and $T_n = \bar{P}_nT$ satisfy the conditions (18.3):

$$\|U_n\| \to 0 \quad \text{as} \quad n \to \infty \quad (U_n = T - \bar{P}_nT), \qquad S_n = 0.$$

We have thus proved that Theorems 18.1 through 18.4 are applicable to equation (18.41), that is, to the Galerkin–Petrov method (18.40). In the estimates (18.6), (18.8) and (18.9) we must set

$$\varepsilon_n = \sup_{x_0 \in X_0^{(l)}, \|x_0\| = 1} \left\| \bar{P}^{(n)} x_0 \right\|, \qquad \varepsilon_n^* = \sup_{f_0 \in F_0^{(l)}, \|f_0\| = 1} \left\| (\bar{P}^{(n)})^* f_0 \right\|.$$

Let us put these quantities in a more convenient form. It follows from the definition of the operator $\bar{P}^{(n)}$ that $\bar{P}^{(n)} Q_n = 0$ and $P_n \bar{P}^{(n)} = 0$, i.e., $(\bar{P}^{(n)})^* P_n = 0$. Thus

$$\bar{P}^{(n)} = \bar{P}^{(n)} Q^{(n)}, \qquad (\bar{P}^{(n)})^* = (\bar{P}^{(n)})^* P^{(n)}.$$

But since $\left\| \bar{P}^{(n)} \right\| = \left\| (\bar{P}^{(n)})^* \right\| \leq 1 + \dfrac{1}{\sqrt{1 - q^2}} = c_0$ (see (16.6)),

$$\varepsilon_n \leq c_0 \sup_{x_0 \in X_0^{(l)}, \|x_0\| = 1} \left\| Q^{(n)} x_0 \right\|, \qquad \varepsilon_n^* \leq c_0 \sup_{f_0 \in F_0^{(l)}, \|f_0\| = 1} \left\| P^{(n)} f_0 \right\|.$$

Summarizing, we have

Theorem 18.8. *Let* $\{G_n\}$ *and* $\{H_n\}$ *be sequences of closed subspaces satisfying condition* (18.39), *with* $\{G_n\}$ *ultimately dense in* H. *Then the Galerkin–Petrov method* (18.40) *is convergent in the sense indicated in Theorems* 18.1 *through* 18.3.

Let μ_0 *be an exact eigenvalue of equation* (18.38), *of rank* l, *and* μ_n *approximate eigenvalues determined by the Galerkin–Petrov method* (18.40), *such that* $\mu_n \to \mu_0$ *as* $n \to \infty$. *Then*

$$|\mu_n - \mu_0| \leq c \sup_{\substack{x_0 \in X_0^{(l)}, \|x_0\| = 1 \\ f_0 \in F_0^{(l)}, \|f_0\| = 1}} \left(\left\| Q^{(n)} x_0 \right\| \cdot \left\| P^{(n)} f_0 \right\| \right)^{1/l},$$

$$\rho(x_n^{(k)}, X_0^{(l)}) \leq c \sup_{x_0 \in X_0^{(l)}, \|x_0\| = 1} \left\| Q^{(n)} x_0 \right\|,$$

$$\rho(x_n^{(k)}, X_0^{(i)}) \leq c \left(|\mu_n - \mu_0|^{i-k+1} + \sup_{x_0 \in X_0^{(l)}, \|x_0\| = 1} \left\| Q^{(n)} x_0 \right\| \right)$$

$$(1 \leq k \leq i \leq l),$$

where $x_n^{(k)}$ *is any normalized element of* $X_n^{(k)}$,

$$X_0^{(i)} = \{x_0 \in H : (I - \mu_0 T)^i x_0 = 0\} \qquad (i = 1, \dots, l),$$

$$F_0^{(l)} = \{f_0 \in H : (I - \bar{\mu}_0 T^*)^l f_0 = 0\},$$

$$X_n^{(k)} = \{x_n \in G_n : (I - \mu_n \bar{P}_n T)^k x_n = 0\} \qquad (k \geq 1).$$

Exercise 18.21. Let $\{\phi_k\}$ and $\{\psi_k\}$ be two complete coordinate sequences in a separable Hilbert space H, and G_n and H_n the linear spans of the elements $\phi_1, \ldots, \phi_n$ and $\psi_1, \ldots, \psi_n$, respectively. Prove the validity of the following plan for numerical implementation of the Galerkin–Petrov method.

1) Determine the approximate eigenvalues as solutions of the algebraic equation

$$\det (A_n - \mu B_n) = 0,$$

where A_n and B_n are the matrices with elements

$$a_{ij} = (\phi_j, \psi_i), \quad b_{ij} = (T\phi_j, \psi_i) \qquad (i, j = 1, \ldots, n),$$

respectively.

2) Determine the eigenelements $x_n = \sum_{j=1}^{n} \xi_j \phi_j \in X_n^{(1)}$ from the condition

$$(A_n - \mu_n B_n) \xi^{(n)} = 0 \qquad (\xi^{(n)} = (\xi_1, \ldots, \xi_n)).$$

3) Determine the elements $x_n = \sum_{j=1}^{n} \xi_j \phi_j \in X_n^{(k)}$ for $k \geqq 1$ from the condition

$$(I_n - \mu_n A_n^{-1} B_n)^k \xi^{(n)} = 0,$$

where I_n is the unit matrix.

Exercise 18.22. Let the assumptions of Exercise 18.21 hold, and also condition (18.39). Assume that the matrices A_n and B_n have been computed with errors $\Gamma_n = (\gamma_{ij})$ and $\Delta_n = (\delta_{ij})$, respectively, and the approximations μ_n and x_n thus replaced by perturbed approximations $\bar{\mu}_n$ and $\tilde{x}_n$. Prove that, if the coordinate sequences $\{\phi_k\}$ and $\{\psi_k\}$ are strongly minimal, then $\bar{\mu}_n$ and $\tilde{x}_n$ are stable with respect to the perturbations Γ_n and Δ_n in the following sense: if $\|\Gamma_n\|_{R_n} \to 0$, $\|\Delta_n\|_{R_n} \to 0$ as $n \to \infty$, then the convergence Theorems 18.1 through 18.3 remain valid for $\bar{\mu}_n$ and $\tilde{x}_n$.

18.8. *Case of unbounded operators.* By various transformations of variables or choice of new function spaces, the above results carry over to the eigenvalue problem for equations with unbounded operators. We shall consider the Galerkin–Petrov method.

Consider the eigenvalue problem for the equation

$$Lu = \mu Mu, \qquad (18.42)$$

where L and M are linear operators, generally unbounded, mapping a Banach space E into a Hilbert space H. We shall assume that the domains of L and M are dense in E, $D(L) \subset D(M)$, the operator L is continuously invertible $(D(L^{-1}) = H)$, the operator ML^{-1} is compact in H.

Also given are two sequences of subspaces $\{E_n\}$ and $\{H_n\}$,

$$E_n \subset D(L) \subset E, \qquad H_n \subset H \qquad (n = 1, 2, \ldots),$$

such that the subspaces LE_n and H_n $(n = 1, 2, \ldots)$ are closed in H, the sequence of subspaces $\{LE_n\}$ is ultimately dense in H, and

$$\varlimsup_{n \to \infty} \Theta(LE_n, H_n) < 1. \tag{18.43}$$

The approximate eigenvalues $\mu = \mu_n$ and eigenelements u_n of equation (18.42) will be determined from the equation

$$P_n(Lu_n - \mu M u_n) = 0 \qquad (u_n \in E_n), \tag{18.44}$$

where P_n is the orthogonal projection onto H_n.

With the substitution $x = Lu$, equation (18.42) becomes

$$x = \mu M L^{-1} x, \tag{18.45}$$

and equation (18.44)

$$P_n(x_n - \mu M L^{-1} x_n) = 0 \qquad (x_n \in G_n), \tag{18.46}$$

where $G_n = LE_n$. Conditions (18.46) are precisely the Galerkin–Petrov method applied to equation (18.45) with the compact operator ML^{-1}. By Theorem 18.8 the Galerkin–Petrov method converges, and the convergence estimates of that Theorem are valid (with the sole difference that $T = ML^{-1}$ in the definition of the subspaces $X_0^{(i)}$, $F_0^{(l)}$ and $X_n^{(k)}$). Recall that in these estimates $Q^{(n)} = I - Q_n$, $P^{(n)} = I - P_n$, where Q_n and P_n are the orthogonal projections onto the subspaces $G_n = LE_n$ and H_n.

Exercise 18.23. Let the operator $L^{-1}M$ be compact in E. Show that $F_0^{(l)} \subset D(L^*)$.

Exercise 18.24. Show that the operator $L^{-1}M$ is compact in E if and only if $D(L^*) \subset D(M^*)$ and the operator $M^*(L^*)^{-1}$ is compact in E^*.

If the operator L is complicated in structure, direct estimation of the norms T may be difficult. It is then desirable to isolate a simple principal part A, i.e., to express L as $L = A + K$, where A is continuously invertible and KA^{-1} compact in H. The operator A should be so chosen that norms of the type $\|R^{(n)}x\|$ are easily estimated, where $R^{(n)} = I - R_n$, R_n being the orthogonal projection onto the subspace AE_n.

Exercise 18.25. Prove the following statements.

1) The subspace LE_n is closed in H if and only if the subspace AE_n is closed in H.

2) The sequence $\{LE_n\}$ is ultimately dense in H if and only if the sequence $\{AE_n\}$ is ultimately dense in H.

3) Condition (18.43) is equivalent to the condition

$$\overline{\lim_{n \to \infty}} \; \Theta(AE_n, H_n) < 1.$$

4) For any $x \in E$,

$$c_1 \|R^{(n)} A L^{-1} x\| \leq \|Q^{(n)} x\| \leq c_2 \|R^{(n)} A L^{-1} x\| \qquad (c_1, c_2 = \text{const} > 0).$$

It is clear from parts 1–3 of Exercise 18.25 that isolation of a simple principal part of L facilitates not only estimation of the norms $\|Q^{(n)} x_0\|$ ($x_0 \in X_0^{(l)}$), but also verification of the conditions that guarantee convergence of the Galerkin–Petrov method (18.44).

Exercise 18.26. Let us solve the eigenvalue problem for the differential equation

$$Lu \equiv u^{(m)} + \sum_{j=0}^{m-1} p_j(t) u^{(j)} = \mu \, q(t) \, u$$

with boundary conditions (16.38) by the Galerkin–Petrov method (method of moments). As the first coordinate sequence, choose the polynomials $\phi_j(t)$ (see subsections 15.7 and 16.8), and as the second—the functions t^j ($j = 0, 1, 2, \ldots$). Let $p_j(t) \in L_2^{(r_j)}$, $q(t) \in L_2^{(r_0)}$, where $r_j \geq j$ ($j = 0, \ldots, m-1$). Assume that the problem has a simple eigenvalue μ_0. Prove that

$$|\mu_n - \mu_0| = o(n^{-(m+r+r')}), \qquad \rho(x_n, X_0^{(1)}) = o(n^{-r}),$$

where $x_n = Lu_n$ ($\|x_n\|_{L_2} = 1$), $\rho(x, y)$ is the metric in $L_2 [a, b]$,

$$r = \min_{0 \leq j \leq m-1} r_j, \qquad r' = \min_{0 \leq j \leq m-1} (r_j - j).$$

Find estimates for the rate of convergence, when the eigenvalue μ_0 is multiple.

18.9. *Operators in real spaces.* Hitherto our discussion has dealt with operators in complex spaces. However, all the theorems may be reformulated for compact linear operators T in a real Banach space E. We now indicate some special features of this case.

It is natural to consider equation (18.1) for real μ only. If μ_0 is a *simple* real eigenvalue, all the theorems proved in this section remain valid for real spaces. The situation is more complicated for *multiple* eigenvalues μ_0. For example, if the multiplicity of μ_0 is even, one can construct operators T_ε arbitrarily close to T which have no real eigenvalues close to μ_0 (so that one cannot expect to find real eigenvalues μ close to μ_0, for which equation (18.2) has nontrivial solutions).

For a complete solution to the problem of eigenvalues and eigenelements as described above, one must first extend the operator T to the complex extension of the space E (see Kantorovich and Akilov [1]). If T is a selfadjoint operator in a Hilbert space H and the operators T_n in equations (18.2) are selfadjoint in the corresponding subspaces H_n, it is of course unnecessary to consider the complex extension of H.

§ 19. Projection methods for solution of nonlinear equations*

19.1. *Statement of the problem.* In this section we shall study projection methods applied to the equation

$$x = Tx. \tag{19.1}$$

where T is a nonlinear operator in a Banach space E, defined and continuous over some nonempty open set $\Omega \subset E$.

Let $\{E_n\}$ be a sequence of closed subspaces of E; $\Omega_n = \Omega \cap E_n$.

Let T_n be a continuous operator mapping Ω_n into E_n, which is, in some sense (to be defined below), close to T. Then a solution of the equation

$$x_n = T_n x_n \tag{19.2}$$

may be considered an approximate solution of equation (19.1).

Let P_n be a sequence of linear projections, each mapping E onto the corresponding E_n. The operators P_n may also be unbounded; if this is so we assume that $T(\Omega) \subset D(P_n)$ and that the operators $P_n T$ are continuous on Ω.

As done earlier in this chapter, we shall characterize the degree to which equation (19.2) approximates (19.1) by the "smallness" of the operators

$$S_n = T_n - P_n T, \qquad U_n = T - P_n T$$

(the first of these is defined in E_n, the second in E).

The equation

$$x_n = P_n T x_n$$

---

* Projection methods have been applied to the solution of nonlinear equations by several authors, among the first of whom were Lur'e and Chekmarev, and Panov. Krasnosel'skii has justified application of the Galerkin method to equations with compact nonlinear operators [2, 4, 8]. More exhaustive results for equations with compact operators have been established by Vainikko [11], who has also investigated equations with continuous (not necessarily compact) operators.

is known as the *Galerkin equation*. A natural name for equation (19.2) is the *perturbed Galerkin equation*, and the approximate method employing this equation is known as the *perturbed Galerkin method*.

We shall present two convergence proofs for the perturbed Galerkin method. The first proof is based on the contracting mapping principle applied to a certain equation equivalent to equation (19.2). The second uses the concept of the rotation of a compact vector field.

19.2. *Solvability of nonlinear equations.* Our first convergence proof for the perturbed Galerkin method is based on the following lemma.

Lemma 19.1. Let A be an operator in a Banach space F, which is Fréchet-differentiable for $\|x - x_*\| \leq \delta_*$, where x_* is a fixed point of F and $\delta_* > 0$. Assume that the linear operator $A'(x_*)$ is continuously invertible in F, and for some δ_0 and q $(0 < \delta_0 \leq \delta_*; 0 \leq q < 1)$

$$\sup_{\|x - x_*\| \leq \delta_0} \left\| [A'(x_*)]^{-1} [A'(x) - A'(x_*)] \right\| \leq q,$$ (19.3)

$$\alpha \equiv \left\| [A'(x_*)]^{-1} A x_* \right\| \leq \delta_0 (1 - q).$$ (19.4)

Then the equation $Ax = 0$ has a unique solution in the ball $\|x - x_*\| \leq \delta_0$, and the solution satisfies the estimate

$$\frac{\alpha}{1 + q} \leq \|x_0 - x_*\| \leq \frac{\alpha}{1 - q}.$$ (19.5)

Proof. If the operator T is differentiable at each point of the interval $[x_1, x_1 + h]$, then

$$\|T(x_1 + h) - Tx_1\| \leq \|h\| \sup_{0 < \theta < 1} \|T'(x_1 + \theta h)\|.$$

Let $T = V(A - W)$, where V and W are arbitrary continuous linear operators in F. Then the above inequality becomes

$$\|V[A(x_1 + h) - Ax_1 - Wh]\| \leq \|h\| \sup_{0 < \theta < 1} \|V[A'(x_1 + \theta h) - W]\|.$$ (19.6)

It is easily seen that the equation $Ax = 0$ is equivalent to the equation $x = Bx$, where

$$Bx = x_* - [A'(x_*)]^{-1} \{Ax_* + [Ax - Ax_* - A'(x_*)(x - x_*)]\}.$$

We claim that B is a contraction in the ball $\|x - x_*\| \leq \delta_0$. Indeed, it follows from (19.3), (19.4) and (19.6) that, for $\|x - x_*\| \leq \delta_0$,

$$\|Bx - x_*\| \leq \delta_0 (1 - q) +$$

$$+ \|x - x_*\| \sup_{0 < \theta < 1} \left\| [A'(x_*)]^{-1} [A'(x_* + \theta (x - x_*)) - A'(x_*)] \right\| \leq$$

$$\leq \delta_0 (1 - q) + \delta_0 q = \delta_0, \qquad (19.7)$$

i.e., B maps the ball $\|x - x_*\| \leq \delta_0$ into itself. Let x_1 and x_2 be two points in the ball $\|x - x_*\| \leq \delta_0$. Then

$$Bx_1 - Bx_2 = [A'(x_*)]^{-1} [Ax_2 - Ax_1 - A'(x_*)(x_2 - x_1)],$$

and it follows from (19.3) and (19.6) that

$$\|Bx_1 - Bx_2\| \leq$$

$$\leq \|x_2 - x_1\| \sup_{0 < \theta < 1} \left\| [A'(x_*)]^{-1} [A'(x_1 + \theta (x_2 - x_1)) - A'(x_*)] \right\| \leq$$

$$\leq q \|x_1 - x_2\|,$$

i.e., B contracts the ball $\|x - x_*\| \leq \delta_0$.

By the contracting mapping principle, B has a unique fixed point x_0 in the ball $\|x - x_*\| \leq \delta_0$. x_0 is obviously the unique solution of the equation $Ax = 0$ in this ball. The definition of the operator B implies the following inequalities for the norms $\|x_0 - x_*\| = \|Bx_0 - x_*\|$ (compare (19.7)):

$$\|x_0 - x_*\| \leq \alpha + \|x_0 - x_*\| q, \qquad \|x_0 - x_*\| \geq \alpha - \|x_0 - x_*\| q.$$

These inequalities are equivalent to the estimates (19.5). ∎

Exercise 19.1. Show that the estimates (19.5) are the best possible.

19.3. *Convergence of the perturbed Galerkin method.* We return to equations (19.1) and (19.2). Denote $P^{(n)} = I - P_n$, and note that if $x \in \Omega$ and $P^{(n)}x \to 0$ as $n \to \infty$, then $P_n x \in \Omega_n$ for sufficiently large n.

Theorem 19.1. *Let the operators T and $P_n T$ be Fréchet-differentiable**

---

* We also assume that $(P_n T)'(x) = P_n T'(x)$. If the projection P_n is bounded, the differentiability of $P_n T$ follows from that of T and this equality holds automatically.

in Ω, and T_n Fréchet-differentiable in Ω_n. Assume that equation (19.1) has a solution $x_0 \in \Omega$ and the linear operator $I - T'(x_0)$ is continuously invertible in E. Let

$$\|P^{(n)}x_0\| \to 0,\tag{19.8}$$

$$\|P_nTP_nx_0 - Tx_0\| \to 0, \quad \|P_nT'(P_nx_0) - T'(x_0)\| \to 0,\tag{19.9}$$

$$\|S_nP_nx_0\| \to 0, \quad \|S'_n(P_nx_0)\| \to 0 \quad (S_n = T_n - P_nT)\tag{19.10}$$

as $n \to \infty$. Finally, assume that for any $\varepsilon > 0$ there exist n_ε and $\delta_\varepsilon > 0$ such that

$$\|T'_n(x) - T'_n(P_nx_0)\| \leq \varepsilon \quad (n \geq n_\varepsilon;\ \|x - P_nx_0\| \leq \delta_\varepsilon, x \in \Omega_n).\tag{19.11}$$

Then there exist n_0 and $\delta_0 > 0$ such that when $n \geq n_0$ equation (19.2) has a unique solution x_n in the ball $\|x - x_0\| \leq \delta_0$. Moreover,

$$\|x_n - x_0\| \leq \|P^{(n)}x_0\| + \|x_n - P_nx_0\| \to 0 \quad \text{as} \quad n \to \infty,\tag{19.12}$$

and $\|x_n - P_nx_0\|$ satisfies the following two-sided estimate ($c_1, c_2 > 0$):

$$c_1\|P_nTx_0 - T_nP_nx_0\| \leq \|x_n - P_nx_0\| \leq c_2\|P_nTx_0 - T_nP_nx_0\|.\tag{19.13}$$

Proof. We shall use Lemma 19.1 with $F = E_n$, $A = I - T_n$, $x_* = P_nx_0$.

The open set Ω contains an entire neighborhood of x_0. Since $P_nx_0 \to x_0$ as $n \to \infty$, it follows that for sufficiently large n (say $n \geq n_*$) P_nx_0 is contained in Ω together with some ball $\|x - P_nx_0\| \leq \delta_*$ of E, and in Ω_n with the same ball of E_n. Thus, the operator $I - T_n$ is differentiable in the ball $\|x - P_nx_0\| \leq \delta_*$ for $n \geq n_*$.

It follows from the fact that $I - T'(x_0)$ is invertible and from (19.9) that for sufficiently large n the operators $I - P_nT'(P_nx_0)$ are invertible in E, and the norms of their inverses are uniformly bounded. It follows that the operators $I - P_nT'(P_nx_0)$ are invertible in E_n, and the norms of their inverses are uniformly bounded. Hence we conclude (using (19.10)) that for sufficiently large n the operators $I - T'_n(P_nx_0)$ are also invertible in E_n, and the norms of their inverses are uniformly bounded:

$$\|[I - T'_n(P_nx_0)]^{-1}\| \leq \kappa \quad (n \geq n_*).\tag{19.14}$$

It follows from (19.9) and (19.10) that the norms of the operators themselves are uniformly bounded:

$$\left\| I - T'_n(P_n x_0) \right\| \leqq \kappa' \qquad (n \geqq n_*).$$

Thus, the numbers

$$\alpha_n = \left\| [I - T'_n(P_n x_0)]^{-1} (I - T_n) P_n x_0 \right\|$$

satisfy the two-sided estimate

$$\frac{1}{\kappa'} \left\| P_n T x_0 - T_n P_n x_0 \right\| \leqq \alpha_n \leqq \kappa \left\| P_n T x_0 - T_n P_n x_0 \right\| \qquad (n \geqq n_*). \quad (19.15)$$

Since $T x_0 = x_0$ and

$$P_n T x_0 - T_n P_n x_0 = (P_n T x_0 - T x_0) + (T x_0 - P_n T P_n x_0) - S_n P_n x_0,$$

it follows from (19.8), (19.9) and (19.10) that

$$\left\| P_n T x_0 - T_n P_n x_0 \right\| \to 0 \quad \text{as} \quad n \to \infty. \quad (19.16)$$

Fixing some q $(0 < q < 1)$, set $\varepsilon_0 = q/\kappa$, and find numbers n_0 and δ_0 $(n_0 \geqq n_* \; ; \; 0 < \delta_0 \leqq \delta_*)$ such that, when $n \geqq n_0$ and $\left\| x - P_n x_0 \right\| \leqq \delta_0$, inequality (19.11) holds with $\varepsilon = \varepsilon_0$. Then condition (19.3) of Lemma 19.1 holds; by increasing n_0 we may assume (see (19.15) and (19.16)) that the second condition (19.4) of Lemma 19.1 also holds. It now follows from the lemma that for $n \geqq n_0$ equation (19.2) has a unique solution x_n in the ball $\left\| x - P_n x_0 \right\| \leqq \delta_0$.

In view of (19.15), the estimate (19.5) may be rewritten as (19.13), with $c_1 = 1/[\kappa' (1 + q)]$, $c_2 = \kappa/(1 - q)$. Inequality (19.12) is obvious, and the fact that its right-hand side converges to zero has already been proved (see (19.8) and (19.16)). Thus $x_n \to x_0$ as $n \to \infty$. By making δ_0 smaller, we may assume that x_n is the only solution of equation (19.2) in the ball $\left\| x - x_0 \right\| \leqq \delta_0$ of E. ∎

It is clear from the proof that condition (19.11) need hold only for some sufficiently small ε_0, not necessarily for all $\varepsilon > 0$. Neither is it difficult to see that the second relation in each of (19.9) and (19.10) may be replaced by the assumption that the norms in question be sufficiently small:

$$\left\| P_n T'(P_n x_0) - T'(x_0) \right\| \leqq \varepsilon_1, \qquad \left\| S'_n(P_n x_0) \right\| \leqq \varepsilon_2 \quad \text{for} \quad n \geqq n_*;$$

the numbers ε_1 and ε_2 must be such that $\kappa_0(\varepsilon_1 + \varepsilon_2) < 1$, where $\kappa_0 = \|[I - T'(x_0)]^{-1}\|$. Finally, the domains in which the operators are differentiable may be reduced. It suffices that T be differentiable at the point x_0, P_nT and S_n at $P_n x_0$, and T_n in the ball $\|x - P_n x_0\| \leqq \delta_0$.

Exercise 19.2. Under the assumptions of Theorem 19.1, show that the solution x_0 of equation (19.1) is isolated (x_0 has a neighborhood containing no other solutions).

A few of the simplest cases, in which the assumptions of Theorem 19.1 are satisfied, are indicated in Exercises 19.3 to 19.5. The reader will find it easier to solve these problems after rereading subsection 15.5. It is also useful to recall that the Fréchet-derivative of a compact operator is a compact linear operator.

Exercise 19.3. Let the operator T be compact on Ω (as an operator in E) and continuously differentiable at the point x_0 ($x_0 = Tx_0$). Let the projections P_n ($n = 1, 2, \ldots$) be bounded in E and $P_n \to I$ strongly. Show that conditions (19.8) and (19.9) are then satisfied.

Assume, moreover, that for any $\varepsilon > 0$ there exist n_ε and $\delta_\varepsilon > 0$ such that

$$\|S_n x\| \leqq \varepsilon, \quad \|S_n'(x)\| \leqq \varepsilon \quad (n \geqq n_\varepsilon \;;\; \|x - P_n x_0\| \leqq \delta_\varepsilon, x \in \Omega_n). \tag{19.17}$$

Show that conditions (19.10) and (19.11) are also satisfied.

Exercise 19.4. Let E' be a Banach space which is continuously embeddable in E. Let T be compact on Ω and continuously differentiable at x_0 ($x_0 = Tx_0$) as an operator mapping E into E'. Let the projections P_n ($n = 1, 2, \ldots$) be bounded as operators mapping E' into E, and $P_n \to P$ strongly, where P is the operator embedding E' into E. Show that conditions (19.8) and (19.9) hold. Show that if condition (19.17) also holds, then (19.10) and (19.11) also hold.

Exercise 19.5. Let T be continuously differentiable at the point x_0 ($x_0 = Tx_0$) as an operator in E. Suppose that for $\|x - x_0\| \leqq \delta_0$ and any $z \in E$

$$\rho(T'(x) z, E_n) \leqq \eta_n \|z\|.$$

Let the projections P_n be bounded in E, and

$$\|P_n\| \eta_n \to 0, \|P_n\| \rho(x_0, E_n) \to 0 \quad \text{as} \quad n \to \infty.$$

Finally, assume that condition (19.17) holds. Show that conditions (19.8) to (19.11) of Theorem 19.1 are satisfied.

19.4. A posteriori *error estimate.* Suppose that we have found an exact or approximate solution $\tilde{x}_n \in E_n$ of equation (19.2). Our problem is to determine whether equation (19.1) has a solution x_0, and to estimate the error $\|\tilde{x}_n - x_0\|$ (see §14).

Theorem 19.2. *Let the operators T, P_nT and T_n be Fréchet-differentiable* *in some neighborhood* of the point $\tilde{x}_n \in \Omega_n$, and $I - T'_n(\tilde{x}_n)$ continuously invertible in E_n,*

$$\left\| [I - T'_n(\tilde{x}_n)]^{-1} \right\| = \kappa_n . \tag{19.18}$$

Let**

$$\gamma_n \equiv (1 + \kappa_n \| P_n T'(\tilde{x}_n) \|) \| U'_n(\tilde{x}_n) \| + \kappa_n \| S'_n(\tilde{x}_n) \| < 1 , \tag{19.19}$$

and, for some δ_n and q_n ($\delta_n > 0$; $0 \leq q_n < 1$),

$$\sup_{\|x - \tilde{x}_n\| \leq \delta_n} \| T'(x) - T'(\tilde{x}_n) \| \leq \frac{q_n}{\kappa'_n} , \tag{19.20}$$

$$\| \tilde{x}_n - T\tilde{x}_n \| \leq \frac{\delta_n(1 - q_n)}{\kappa'_n} , \tag{19.21}$$

where

$$\kappa'_n = \frac{1 + \kappa_n \| P_n T'(\tilde{x}_n) \|}{1 - \gamma_n} . \tag{19.22}$$

Then equation (19.1) has a unique solution x_0 in the ball $\|x - \tilde{x}_n\| \leq \delta_n$, and we have the error estimate

$$\frac{\alpha_n}{1 + q_n} \leq \| \tilde{x}_n - x_0 \| \leq \frac{\alpha_n}{1 - q_n} , \tag{19.23}$$

where

$$\alpha_n \equiv \left\| [I - T'(\tilde{x}_n)]^{-1} (\tilde{x}_n - T\tilde{x}_n) \right\| \leq \kappa'_n \| \tilde{x}_n - T\tilde{x}_n \| . \tag{19.24}$$

Proof. Using Lemma 17.2, we conclude from (19.18) and (19.19) that the operator $I - T'(\tilde{x}_n)$ is invertible in E, and

$$\left\| [I - T'(\tilde{x}_n)]^{-1} \right\| \leq \kappa'_n , \tag{19.25}$$

where x'_n is the number defined by (19.22). It now remains to apply Lemma 19.1 with $F = E$, $A = I - T$, $x_* = \tilde{x}_n$. Conditions (19.20) and (19.21)

---

* It suffices that $P_n T$ and T_n be differentiable at the point x_n alone, $P_n T$ as an operator in E. We assume that $(P_n T)'(\tilde{x}_n) = P_n T'(\tilde{x}_n)$.

** Recall that $S_n = T_n - P_n T$, $U_n = T - P_n T$.

imply conditions (19.3) and (19.4) of the lemma in an obvious way, and estimate (19.5) of the lemma is equivalent to (19.23); the estimate (19.24) for α_n follows from (19.25). ∎

Exercise 19.6. Let $\tilde{x}_n \to x_0$ as $n \to \infty$; assume that the operator $I - T'(x_0)$ is continuously invertible in E and the assumptions of any one of Exercises 19.3 to 19.5 hold. Show that the required numbers q_n may be so chosen that $q_n \to 0$ as $n \to \infty$. Estimate (19.23) then becomes asymptotically exact.

19.5. *Rotation of a compact vector field.* Let Ω be a nonempty open bounded set in a real Banach space F, and Γ its boundary. Let A be an operator in F which is compact on the set $\overline{\Omega} = \Omega \cup \Gamma$ and has no fixed points on Γ. The rotation $\gamma(I - A ; \Gamma)$ of the compact vector field $(I - A) x$ on the boundary Γ is a certain integer which is a topological invariant of the operator A on the set Ω (see §14). To understand the contents of this section, a reader confining his attention to projection methods need only know the following facts; on a first reading, they may be regarded as axioms.

(1) If $\gamma(I - A; \Gamma) \neq 0$, then the equation $x = Ax$ has at least one solution in Ω.

(2) Let B be another operator in F, compact on $\overline{\Omega}$, such that

$$\sup_{x\in\Gamma} \|Bx\| < \inf_{x\in\Gamma} \|x - Ax\| . \tag{19.26}$$

Then $\gamma(I - A ; \Gamma) = \gamma(I - A - B ; \Gamma)$.

By way of explanation, we note that $\inf_{x\in\Gamma} \|x - Ax\| > 0$. Indeed, otherwise there exists a sequence $x_n \in \Gamma$ such that $x_n - Ax_n \to 0$ as $n \to \infty$. Since A is compact, we may assume that the sequence Ax_n converges to some limit z. But then this is also the limit of x_n, and it follows that $z \in \Gamma$, $z - Az = 0$, i.e., A has a fixed point on Γ, a contradiction.

Proposition (2) is a corollary of a more general statement according to which homotopic vector fields have the same rotation.

(3) Let $A\overline{\Omega} \subset \tilde{F}$, where $\tilde{F}$ is a closed subspace of F. Then $\gamma(I - A ; \Gamma) = \gamma(I - \tilde{A} ; \tilde{\Gamma})$, where $\tilde{A}$ is the restriction of A to $\tilde{F}$ and $\tilde{\Gamma}$ the boundary of the set $\Omega \cap \tilde{F}$ in $\tilde{F}$. (It may happen that $\Omega \cap \tilde{F} = \emptyset$. If so, $\gamma(I - A ; \Gamma) = 0$ and the assertion remains valid if we set $\gamma(I - \tilde{A}; \emptyset) = 0$ by definition).

(4) Let $x_0 \in \Omega$ be the only solution of the equation $x = Ax$ in a

ball $\|x - x_0\| \leqq \delta_0$ entirely contained in Ω. Let Γ_δ denote the sphere $\|x - x_0\| = \delta$. Then $\gamma(I - A; \Gamma_\delta) = \gamma(I - A; \Gamma_{\delta_0})$ for any $\delta, 0 < \delta \leqq \delta_0$. This common value of $\gamma(I - A ; \Gamma_\delta)$ is known as the *index* of the isolated solution x_0.

(5) Let $x_0 \in \Omega$ be a solution of the equation $x = Ax$, and assume that the operator A is Fréchet-differentiable at x_0 and the linear operator $I - A'(x_0)$ invertible. Then x_0 is an isolated solution of nonzero index; moreover, $|\gamma(I - A ; \Gamma_\delta)| = 1$.

(6) Let the domain Ω be convex and assume that A maps the closure $\bar{\Omega}$ of Ω into itself. It then follows from the Schauder principle that the equation $x = Ax$ has at least one solution in $\bar{\Omega}$. If A has no fixed points on Γ, then $\gamma(I - A ; \Gamma) = 1$.

We recall one more theorem for real Hilbert spaces: If $(Ax, x) < (x, x)$ for $\|x\| = r$, then $\gamma(I - A ; \Gamma_r) = 1$, where $\Gamma_r = \{x : \|x\| = r\}$.

We have mentioned only the simplest statements on the nonvanishing of the rotation (index). Extensions of this type of theorem (see §14) increase the significance of the theorems (19.3 and 19.4) proved below.

Exercise 19.7. Let $x_0 \in \Omega$ be an isolated solution of the equation $x = Ax$, of nonzero index, unique in the ball $\|x - x_0\| \leqq \delta_0$. Let B be a compact operator such that

$$\sup_{\|x - x_0\| = \delta} \|Bx\| < \inf_{\|x - x_0\| = \delta} \|x - Ax\| . \tag{19.27}$$

Show that the equation $x = Ax + Bx$ has at least one solution in the ball $\|x - x_0\| < \delta$. (The solution need not be unique.)

19.6. *Second convergence proof.* We resume our study of equations (19.1) and (19.2), assuming that E is a real space. We shall also assume that the operators T and $P_n T$ are defined not only over the open bounded set $\Omega \subset E$, but also over its closure $\bar{\Omega} = \Omega \cup \Gamma$, where Γ is the boundary of Ω. The operator T_n will be defined over $\bar{\Omega}_n = \Omega_n \cup \Gamma_n$, where $\Omega_n = \Omega \cap E_n$ and Γ_n is the boundary of Ω_n in E_n.

Exercise 19.8. Show that $\Gamma_n \subset \Gamma \cap E_n$. Construct an example for which this is a proper inclusion.

Theorem 19.3. *Let T and $P_n T$ be compact on $\bar{\Omega}$ as operators in E, T_n compact on $\bar{\Omega}_n$ as an operator in E_n, and*

$$\sup_{x \in \bar{\Omega}} \|U_n x\| \to 0 \quad as \quad n \to \infty . \quad (U_n = T - P_n T), \tag{19.28}$$

$$\sup_{x \in \bar{\Omega}_n} \|S_n x\| \to 0 \quad as \quad n \to \infty \qquad (S_n = T_n - P_n T). \qquad (19.29)$$

Assume that equation (19.1) has no solution on the boundary Γ *of* Ω, *and*

$$\gamma(I - T ; \Gamma) \neq 0. \qquad (19.30)$$

Then, for sufficiently large n, the set X_n *of solutions of equation (19.2) in* Ω_n *is not empty and*

$$\sup_{x_n \in X_n} \rho(x_n, X_0) \to 0 \quad as \quad n \to \infty, \qquad (19.31)$$

where X_0 *is the set of solutions of equation (19.1) in* Ω *(that* X_0 *is not empty follows from (19.30)).*

Proof. Denote $\alpha = \inf_{x \in \Gamma} \|x - Tx\|$. Since T has no fixed points on Γ, it follows that $\alpha > 0$ (for a proof see subsection 19.5). For sufficiently large n (say $n \geq n_0$), it follows from (19.28) that $\sup_{x \in \bar{\Omega}} \|U_n x\| \leq \alpha/2$, and from Proposition (2) we conclude that

$$\gamma(I - T ; \Gamma) = \gamma(I - P_n T ; \Gamma).$$

By Proposition (3),

$$\gamma(I - P_n T ; \Gamma) = \gamma(I - P_n T ; \Gamma_n)$$

(in the left-hand side of this equality $P_n T$ is an operator in E, in the right-hand side—in E_n). Since $\Gamma_n \subset \Gamma$,

$$\inf_{x \in \Gamma_n} \|x - P_n Tx\| \geq \inf_{x \in \Gamma} \|x - P_n Tx\| \geq \inf_{x \in \Gamma} \|x - Tx\| - \sup_{x \in \Gamma} \|U_n x\| \geq$$

$$\geq \alpha - \alpha/2 = \alpha/2 \qquad (n \geq n_0).$$

Increasing n_0 (if necessary), we assume that $\sup_{x \in \Omega_n} \|S_n x\| < \alpha/2$ for $n \geq n_0$ (see (19.29)). Again applying Proposition (2), we get

$$\gamma(I - P_n T ; \Gamma_n) = \gamma(I - T_n ; \Gamma_n).$$

Thus, for $n \geq n_0$,

$$\gamma(I - T_n ; \Gamma_n) = \gamma(I - T ; \Gamma) \neq 0.$$

It now follows from Proposition (1) that, for $n \geq n_0$, equation (19.2) has at least one solution $x_n \in \Omega_n$, i.e., the set X_n of solutions in Ω_n is not empty.

We now prove (19.31). Given $\varepsilon > 0$, describe a ball of radius $\leqq \varepsilon$, entirely contained in the set Ω, about each point $x_0 \in X_0$. Denote the resulting open covering of X_0 by X_0^ε. It is clear that the closed set $\overline{\Omega}\backslash X_0^\varepsilon$ contains no fixed points of the operator T, and therefore

$$\alpha_\varepsilon \equiv \inf_{x \in \overline{\Omega}\backslash X_0^\varepsilon} \|x - Tx\| > 0 .$$

If $x \in E_n \cap (\overline{\Omega}\backslash X_0^\varepsilon)$, then

$$\|x - T_n x\| \geqq \alpha_\varepsilon - \|U_n x\| - \|S_n x\| > 0 ,$$

if n is sufficiently large (say $n \geqq n_\varepsilon$). But since $x_n - T_n x_n = 0$ for $x_n \in X_n$, it follows that when $n \geqq n_\varepsilon$ the set $X_n \subset \overline{\Omega}$ is disjoint from $\overline{\Omega}\backslash X_0^\varepsilon$, i.e., $X_n \subset X_0^\varepsilon$. Since ε is arbitrary, this implies (19.31). ∎

The following theorem is an immediate corollary of Theorem 19.3.

Theorem 19.4. Let T and $P_n T$ be compact operators in the ball $\|x - x_0\| \leqq \delta$, and T_n compact on the intersection of this ball with E_n. Let x_0 be an isolated solution of equation (19.1), of nonzero index, unique in the above ball. Assume that conditions (19.28) and (19.29) hold, where $\overline{\Omega}$ and $\overline{\Omega}_n$ denote the above ball and its intersection with E_n.

Then there exists n_0 such that, for $n \geqq n_0$, equation (19.2) has at least one solution x_n in the ball $\|x - x_0\| < \delta$, and any sequence of these solutions x_n converges in norm to x_0 as $n \to \infty$.

Exercise 19.9. Let T be an operator that maps $\overline{\Omega}$ into a compact subset of E; let the projections P_n be bounded in E and $P_n \to I$ strongly. Show that condition (19.28) is satisfied.

Exercise 19.10. Let E' be a Banach space, continuously embeddable in E (i.e., $E' \subset E$ and $\|x\|_E \leqq c\|x\|_{E'}$ for $x \in E'$). Assume that the operator T maps $\overline{\Omega}$ into a compact subset of E', the projections P_n are bounded as operators mapping E' into E, and $P_n \to P$ strongly, where P is the operator embedding E' into E. Show that condition (19.28) holds.

Exercise 19.11. Let the projections P_n be bounded in E, and

$$\|P_n\| \sup_{x \in \Omega} \rho(Tx, E_n) \to 0 \qquad \text{as} \qquad n \to \infty .$$

Show that condition (19.28) holds.

Exercise 19.12. Under the assumptions of Theorem 19.4, let T be Fréchet-differentiable at the point x_0 and the operator $I - T'(x_0)$ invertible. Assume, moreover, that the operator T and the projections P_n satisfy the conditions of Exercise 19.9 or 19.10. Prove the estimate

$$c_1 \|P^{(n)} x_0 - S_n x_n\| \leqq \|x_n - x_0\| \leqq c_2 \|P^{(n)} x_0 - S_n x_n\| ,$$

where $c_1, c_2 = \text{const} > 0$, $P^{(n)} = I - P_n$.

One of the best-known methods for proving existence theorems is to apply various fixed point principles. The basic fixed point principles follow from the fact that the rotation of a suitable vector field does not vanish (see Krasnosel'skii [8]). The import of Theorem 19.3 is that the conditions for applicability of these fixed point principles also guarantee that the solutions may actually be approximated by projection methods. In particular, projection methods generally apply if the existence of a solution follows from Schauder's principle or from the one-sided estimate $(Ax, x) < (x, x)$ for equations in Hilbert spaces.

19.7. *The method of mechanical quadratures for nonlinear integral equations.* Consider the integral equation

$$x(t) = \int_a^b K(t, s, x(s)) \, ds + f(t) \tag{19.32}$$

with continuous kernel $K(t, s, x)$ and continuous free term $f(t)$. Define approximate values $\xi_{jn} \approx x_0(s_{jn})$ $(j = 1, \ldots, n)$ of the required solution $x_0(t)$ by the system of equations

$$\xi_{in} = \sum_{j=1}^n \alpha_{jn} K(s_{in}, s_{jn}, \xi_{jn}) + f(s_{in}) \qquad (i = 1, \ldots, n). \tag{19.33}$$

This system is based (see subsection 17.10) on the quadrature formula

$$\int_a^b z(s) \, ds = \sum_{j=1}^n \alpha_{jn} z(s_{jn}) + R_n(z). \tag{19.34}$$

Assume that $\alpha_{jn} > 0$ $(j = 1, \ldots, n)$ and

$$a \leqq s_{1n} < s_{2n} < \ldots < s_{nn} \leqq b.$$

Theorem 19.5. Assume that the quadrature process (19.34) *is convergent, i.e.,* $R_n(z) \to 0$ *as* $n \to \infty$ *for any function* $z(t)$ *continuous on* $[a, b]$. *Let equation* (19.32) *have a solution* $x_0(t)$; *let the kernel* $K(t, s, x)$ *be jointly continuous for*

$$a \leqq t, s \leqq b, \; |x - x_0(s)| \leqq \delta \qquad (\delta = \text{const} > 0), \tag{19.35}$$

and the free term $f(t)$ *continuous on* $[a, b]$.

Then the following two assertions are valid:

a) If the solution $x_0(t)$ *is isolated and has nonzero index (in the space C),* *then, for sufficiently large n, the system* (19.33) *is solvable, and**

$$\max_{1 \le j \le n} |\xi_{jn} - x_0(s_{jn})| \to 0 \quad as \quad n \to \infty. \tag{19.36}$$

b) *If the kernel* $K(t, s, x)$ *has a partial derivative with respect to* x *which is jointly continuous in the domain* (19.35), *and the homogeneous linear integral equation*

$$h(t) = \int_a^b K_0(t, s) h(s) \, ds \quad \left(K_0(t, s) = \frac{\partial K(t, s, x_0(s))}{\partial x} \right)$$

has no nontrivial solutions, then, for sufficiently large n, *equation* (19.33) *has a unique solution,*** *the convergence relation* (19.36) *holds, and we have the two-sided estimate*

$$c_1 r_n \le \max_{1 \le j \le n} |\xi_{jn} - x_0(s_{jn})| \le c_2 r_n, \tag{19.37}$$

where $c_1, c_2 = const > 0$,

$$r_n = \max_{1 \le i \le n} |R_n(z_{in})|, \quad z_{in}(s) = K(s_{in}, s, x_0(s)),$$

$R_n(z)$ *is the remainder term of the quadrature formula* (19.34).

Proof. Construct the spaces E and E_n and the projections P_n just as for the proof of the analogous theorem for linear integral equations in subsection 17.10. Regard equation (19.32) as an operator equation of type (19.1). The fact that the kernel $K(t, s, x)$ is continuous in the domain (19.35) implies that the operator

$$Tx = \int_a^b K(t, s, x(s)) \, ds + f(t)$$

is compact in the ball $\bar{\Omega}$ ($\|x - x_0\| \le \delta$) as an operator mapping E into C, therefore also as an operator in E. It was proved in subsection 17.10 that, as operators mapping C into E, the projections P_n converge strongly to an operator embedding C into E. It follows (see Exercise 19.10) that condition (19.28) holds.

---

* If $x_0(t)$ is not the unique solution of equation (19.32), the convergence relation (19.36) is valid only for solutions of the system (19.33) which are "close" to $x_0(t)$.

** I.e., there exist n_0 and δ_0 such that, for $n \ge n_0$, there is a unique solution of the system (19.33) such that

$$\max_{1 \le j \le n} |\xi_{jn} - x_0(s_{jn})| \le \delta_0.$$

The system of equations (19.33) is equivalent to equation (19.2), where the operator T_n is defined for elements $z_n = \sum_{j=1}^{n} \zeta_j \chi_{jn}$ in the set $\bar{\Omega}_n = \bar{\Omega} \cap E_n$ by

$$T_n z_n = \sum_{i=1}^{n} \left\{ \sum_{j=1}^{n} \alpha_{jn} K(s_{in}, s_{jn}, \zeta_j) + f(s_{in}) \right\} \chi_{in}.$$

As in subsection 17.10, we write the difference $T_n z_n - P_n T z_n = S_n z_n$ in the form

$$S_n z_n = \sum_{i=1}^{n} \left\{ \sum_{j=1}^{n} \int_{D_{jn}} [K(s_{in}, s_{jn}, \zeta_j) - K(s_{in}, s, \zeta_j)] \, ds \right\} \chi_{in} +$$

$$+ \sum_{i=1}^{n} \left\{ \sum_{j=1}^{n} \gamma_{jn} K(s_{in}, s_{jn}, \zeta_j) \right\} \chi_{in}.$$

Hence, using (17.48), (17.49) and (17.51) and the fact that $K(t, s, x)$ is uniformly continuous and bounded in the closed domain (19.35), we get (19.29).

By proposition (3) of subsection 19.5, the index of the solution $x_0(t)$ of equation (19.32) is the same in the spaces E and C; by assumption, it is not zero. Assertion a) now follows from Theorem 19.4. It suffices to note that the solution x_n of equation (19.2) and the solution $(\xi_{1n}, \ldots, \xi_{nn})$ of the system (19.33) satisfy the obvious equality $x_n = \sum_{j=1}^{n} \xi_{jn} \chi_{jn}$, and

$$\max_{1 \leq j \leq n} |\xi_{jn} - x_0(s_{jn})| = \|x_n - P_n x_0\| \leq \|x_n - x_0\|. \qquad (19.38)$$

We now prove assertion b). Since the partial derivative $\partial K(t, s, x)/x\partial$ exists and is continuous in the domain (19.35), T is continuously differentiable in Ω as an operator mapping E into C, a fortiori as an operator in E. Since the operators P_n converge strongly to the operator embedding C into E, it follows that conditions (19.8) and (19.9) of Theorem 19.1 are satisfied. Since the operator

$$T'(x_0) h = \int_a^b K_0(t, s) h(s) \, ds$$

is compact in E, and by assumption the equation $h = T'(x_0) h$ has no nontrivial solutions, the operator $I - T'(x_0)$ is continuously invertible.

The operators T_n and S_n are continuously differentiable in Ω_n, and moreover, for any $z_n = \sum\limits_{j=1}^{n} \zeta_j \chi_{jn} \in \Omega_n$ and $y_n = \sum\limits_{j=1}^{n} \eta_j \chi_{jn} \in E_n$,

$$T'_n(z_n)\, y_n = \sum_{i=1}^{n} \left\{ \sum_{j=1}^{n} \alpha_{jn} \frac{\partial K(s_{in},\, s_{jn},\, \zeta_j)}{\partial x} \eta_j \right\} \chi_{in},$$

$$S'_n(z_n)\, y_n = \sum_{i=1}^{n} \left\{ \sum_{j=1}^{n} \int_{D_{jn}} \frac{\partial [K(s_{in},\, s_{jn},\, \zeta_j) - K(s_{in},\, s,\, \zeta_j)]}{\partial x} ds\, \eta_j \right\} \chi_{in} +$$

$$+ \sum_{i=1}^{n} \left\{ \sum_{j=1}^{n} \gamma_{jn} \frac{\partial K(s_{in},\, s_{jn},\, \zeta_j)}{\partial x} \eta_j \right\} \chi_{in}.$$

In view of the fact that $\partial K(t, s, x)/\partial x$ is bounded and uniformly continuous in the closed domain (19.35), it is not difficult to see that conditions (19.10) and (19.11) of Theorem 19.1 are also satisfied.

Assertion b) now follows from Theorem 19.1. In particular, the estimate (19.37) follows from (19.13) and (19.38); we need only note that $\| P_n T x_0 - T_n P_n x_0 \| = r_n$. ∎

The solution $(\xi_{1n}, \ldots, \xi_{nn})$ of the system (19.33) may be used to construct an analytic approximation $x_n(t)$ to the solution $x_0(t)$ of equation (19.32), by setting

$$\tilde{x}_n(t) = \sum_{j=1}^{n} \alpha_{jn} K(t,\, s_{jn},\, \xi_{jn}) + f(t).$$

Exercise 19.13. Under the assumptions of Theorem 19.5, show that

$$\max_{a \leq t \leq b} |\tilde{x}_n(t) - x_0(t)| \leq c \max_{a \leq t \leq b} |R_n(z_t)|, \quad z_t(s) = K(t, s, x_0(s)).$$

19.8. *The method of finite differences.* As our next example of the application of the general theorems, we consider a problem of Bernstein concerning the solutions of the equation

$$u'' = f(t, u), \tag{19.39}$$

satisfying the boundary conditions

$$u(0) = u(1) = 0. \tag{19.40}$$

The problem (19.39)–(19.40) is equivalent to the integral equation

$$u(t) = \int_0^1 G(t, s) f(s, u(s)) \, ds, \tag{19.41}$$

where

$$G(t, s) = \begin{cases} -t(1 - s) & \text{if } t \leqq s, \\ -s(1 - t) & \text{if } t \geqq s. \end{cases} \tag{19.42}$$

is the Green's function of the operator u'' for boundary conditions (19.40).

Let $t_{in} = i/n$ $(i = 0, 1, \ldots, n)$, $h_n = 1/n$. Replace the boundary-value problem (19.39)–(19.40) by the finite-difference problem

$$\frac{u_{i-1,n} - 2u_{in} + u_{i+1,n}}{h_n^2} = f(t_{in}, u_{in}) \qquad (i = 1, \ldots, n - 1), \tag{19.43}$$

$$u_{0n} = u_{nn} = 0, \tag{19.44}$$

where u_{in} are approximate values of the required solution at the points t_{in} $(i = 0, 1, \ldots, n)$. We claim that the method of finite differences (19.43) is equivalent to the method of mechanical quadratures for equation (19.41), using the quadrature formula

$$\int_0^1 z(s) \, ds = \frac{1}{n} \sum_{j=1}^{n-1} z\left(\frac{j}{n}\right) + R_n(z). \tag{19.45}$$

Our proof will be based on the following simple result.

Lemma 19.2. *The system of linear equations*

$$\frac{u_{i-1,n} - 2u_{in} + u_{i+1,n}}{h_n^2} = \xi_{in} \qquad (i = 1, \ldots, n - 1) \, ; \qquad u_{0n} = u_{nn} = 0$$

has a unique solution

$$u_{in} = \frac{1}{n} \sum_{j=1}^{n-1} G(t_{in}, t_{jn}) \xi_{jn} \qquad (i = 0, 1, \ldots, n)$$

for any $\xi_{1n} \ldots, \xi_{n-1, n}$.

The proof is left to the reader.

By virtue of Lemma 19.2, the system of equations (19.43)–(19.44) is equivalent to the system

$$u_{in} = \sum_{j=1}^{n-1} G(t_{in}, t_{jn}) f(t_{jn}, u_{jn}) \quad (i = 1, \ldots, n - 1). \tag{19.46}$$

But the system (19.46) is the final result of the method of mechanical quadratures for equation (19.41), using formula (19.45). The quadrature process (19.45) is convergent, and Theorem 19.5 implies

Theorem 19.6. *Assume that the boundary-value problem* (19.39)–(19.40) *has an isolated solution $u_*(t)$ with nonzero index (as a solution of the integral equation* (19.41)). *Let the function $f(t, u)$ be continuous in the domain*

$$0 \leqq t \leqq 1, \quad |\mu - u_*(t)| \leqq \delta \qquad (\delta = \mathrm{const} > 0). \qquad (19.47)$$

Then there exist n_0 and δ_0 such that, for $n \geqq n_0$, the system (19.43)–(19.44) *has at least one solution satisfying the inequalities*

$$|u_{in} - u_*(t_{in})| \leqq \delta_0 \qquad (i = 1, \ldots, n - 1). \qquad (19.48)$$

Moreover,

$$\max_{0 \leqq i \leqq n} |u_{in} - u_*(t_{in})| \to 0 \quad as \quad n \to \infty.$$

If the function $f(t, u)$ has a continuous partial derivative with respect to u in the domain (19.47), *and the linearized problem*

$$u'' = q(t)\, u \qquad (q(t) = f(t, u_*(t)))$$

with boundary conditions (19.40) *has no nontrivial solutions, then the solution of the system* (19.43) *satisfying conditions* (19.48) *is unique. The rate of convergence is described by the relation*

$$\max_{0 \leqq i \leqq n} |u_{in} - u_*(t_{in})| = c_n r_n,$$

where

$$0 < c' \leqq c_n \leqq c'' < \infty \qquad (n = n_0, n_0 + 1, \ldots),$$

$$r_n = \max_{1 \leqq i \leqq n-1} |R_n(z_{in})|, \qquad z_{in}(s) = G(t_{in}, s)\, f(s, u_*(s)),$$

$R_n(z)$ *is the remainder term of the quadrature formula* (19.45).

Exercise 19.14. Let the function $f(t, u_*(t))$ be twice continuously differentiable. Show that

$$\max_{0 \leqq i \leqq n} |u_{in} - u_*(t_{in})| \leqq c/n^2 \qquad (c = \mathrm{const}).$$

Exercise 19.15. Under the assumptions of Theorem 19.6, show that

$$\max_{1 \leq i \leq n-1} \left| \frac{u_{i+1,n} - u_{i-1,n}}{2h_n} - u'_*(t_{in}) \right| \to 0,$$

$$\max_{1 \leq i \leq n-1} \left| \frac{u_{i-1,n} - 2u_{in} + u_{i+1,n}}{h_n^2} - u''_*(t_{in}) \right| \to 0.$$

Hint. The substitution $u'' - x$ transforms the boundary-value problem (19.39)–(19.40) into an integral equation

$$x(t) = f\left(t, \int_0^1 G(t, s) x(s) \, ds \right).$$

Using Lemma 19.2, replace the problem (19.43)–(19.44) by the system

$$\xi_{in} = f\left(t_{in}, \frac{1}{n} \sum_{j=1}^{n-1} G(t_{in}, t_{jn}) \xi_{jn} \right) \qquad (i = 1, \ldots, n - 1).$$

Vainikko [16] has proved an analog of Theorem 19.6 for the equation $u'' = f(t, u, u')$, with boundary conditions

$$\int_0^1 u(t) \, d\mu_{i0}(t) + \int_0^1 u'(t) \, d\mu_{i1}(t) = \gamma_i \qquad (i = 1, 2),$$

where $\mu_{i0}(t)$ and $\mu_{i1}(t)$ are functions of bounded variation.

19.9. *The Galerkin method.* Let us return to equation (19.1), replacing (19.2) by the Galerkin equation

$$x_n = P_n T x_n, \tag{19.49}$$

where P_n is a projection of E onto a closed subspace $E_n \subset E$. This equation is a special case of equation (19.2), obtained by setting $T_n = P_n T$.

The main assertions of the following theorem follow directly from Theorems 19.1 and 19.4.

Theorem 19.7. *Let T be compact on Ω (as an operator in E), and assume that equation (19.1) has an isolated solution $x_0 \in \Omega$ with nonzero index. Let the projections P_n be bounded in E and $P_n \to I$ strongly as $n \to \infty$.*

Then there exist n_0 and δ_0 such that, for $n \geq n_0$, equation (19.49) has at least one solution x_n in the ball $\|x - x_0\| \leq \delta_0$, and these solutions x_n all converge in norm to x_0 as $n \to \infty$.

If T is Fréchet-differentiable at x_0 and the homogeneous equation $h = T'(x_0) h$ has no nontrivial solutions,* then

---

* Recall that this already implies that x_0 is an isolated solution and has nonzero index.

$$c_1 \| P^{(n)} x_0 \| \leqq \| x_n - x_0 \| \leqq c_2 \| P^{(n)} x_0 \| . \qquad (19.50)$$

If T is continuously Fréchet-differentiable at x_0, then, for sufficiently large n, the solution x_n of equation (19.49) is unique in a ball $\| x - x_0 \| \leqq \delta$, where δ is sufficiently small.

Only the estimate (19.50) requires proof. Since the linear operator $T'(x_0)$ is compact and $P_n \to I$ strongly, it follows from Lemma 15.4 that

$$\| T'(x_0) - P_n T'(x_0) \| \to 0 \quad \text{as} \quad n \to \infty$$

Thus, together with $I - T'(x_0)$ the operators $I - P_n T'(x_0)$ are also invertible in E (for sufficiently large n); their norms and those of their inverses are uniformly bounded:

$$\| I - P_n T'(x_0) \| \leqq 1/c' , \quad \| [I - P_n T'(x_0)]^{-1} \| \leqq c'' \qquad (n \geqq n_0) .$$

Let x_0 and x_n be solutions of equations (19.1) and (19.49), respectively, $x_n \to x_0$ as $n \to \infty$. Since

$$[I - P_n T'(x_0)] (x_0 - x_n) =$$

$$= \{ P^{(n)} x_0 - P_n [T x_n - T x_0 - T'(x_0) (x_n - x_0)] \} ,$$

it follows that

$$c' (\| P^{(n)} x_0 \| - \| P_n \| \tau_n) \leqq \| x_n - x_0 \| \leqq c'' (\| P^{(n)} x_0 \| + \| P_n \| \tau_n),$$

where $\tau_n = \| T x_n - T x_0 - T'(x_0) (x_n - x_0) \|$. All that remains now to prove (19.50) is to note that the norms $\| P_n \|$ $(n = 1, 2, \ldots)$ are uniformly bounded, and

$$\frac{\tau_n}{\| x_n - x_0 \|} \to 0 \quad \text{as} \quad n \to \infty$$

since T is differentiable at x_0.

Exercise 19.16. Let $\| T'(x_0) P^{(n)} \| \to 0$ as $n \to \infty$. Show that the estimate (19.50) may be sharpened:

$$(1 - \varepsilon_n) \| P^{(n)} x_0 \| \leqq \| x_n - x_0 \| \leqq (1 + \varepsilon_n) \| P^{(n)} x_0 \| \qquad (\varepsilon_n \to 0 \quad \text{as} \quad n \to \infty).$$

Exercise 19.17. Let E' be a Banach space, continuously embedded in E. Show that Theorem 19.7 remains valid if T is compact and differentiable as an operator mapping E into E', the projections P_n are bounded as operators mapping E' into E and they converge strongly to the operator embedding E' into E.

Exercise 19.18. Extend the results of subsection 15.7 on the convergence of the collocation method to the nonlinear equation

$$u^{(m)} = f(t, u, u', \ldots, u^{(m-1)})$$

with boundary conditions (15.37).

Now compare equation (19.49) with the perturbed equation

$$x_n = P_n T x_n + S_n x_n, \tag{19.51}$$

where S_n is a (generally) nonlinear operator in E_n, continuous on $\Omega_n = \Omega \cap E_n$. We wish to study the distance between the solutions of (19.49) and (19.51).

Theorem 19.8. Assume that equation (19.1) has a solution $x_0 \in \Omega$; T is continuously Fréchet-differentiable at x_0 (as an operator in E), and the equation $h = T'(x_0) h$ has no nontrivial solutions. Let the projections P_n be bounded in E and $P_n \to I$ strongly as $n \to \infty$. Let S_n $(n = 1, 2, \ldots)$ be Fréchet-differentiable in the ball $\|x - x_n\| \leqq \delta$ of the space E, where x_n is a solution of equation (19.49). Let $x_n \to x_0$ as $n \to \infty$.

Then there exist positive constants δ_0 $(\delta_0 \leqq \delta)$, σ, σ' and a natural number n_0, all independent of n, such that if $n \geqq n_0$ and

$$\|S_n x_n\| \leqq \sigma, \quad \sup_{x \in E_n, \|x - x_n\| \leqq \delta_0} \|S_n'(x)\| \leqq \sigma', \tag{19.52}$$

then equation (19.51) has a unique solution $\tilde{x}_n$ in the ball $\|x - x_n\| \leqq \delta_0$, and

$$c_1 \|S_n x_n\| \leqq \|\tilde{x}_n - x_n\| \leqq c_2 \|S_n x_n\|. \tag{19.53}$$

Proof. We apply Lemma 19.1 with $F = E_n$, $A = I - P_n T - S_n$, $x_* = x_n$.

Since T is continuously differentiable at x_0 and $x_n \to x_0$ as $n \to \infty$,

$$\|P_n T'(x_n) - T'(x_0)\| \leqq \|P_n\| \cdot \|T'(x_n) - T'(x_0)\| + \|P^{(n)} T'(x_0)\| \to 0,$$

and the invertibility of $I - T'(x_0)$ implies the invertibility of $I - P_n T'(x_n)$ in E (therefore also in E_n) for sufficiently large n; the norms of the inverses are uniformly bounded. Choose a number $\sigma' > 0$ so small that, for sufficiently large n, $\|S_n'(x_n)\| \leqq \sigma'$ implies that operators $I - P_n T'(x_n) - S_n'(x_n)$ are invertible and their inverses are uniformly bounded:

$$\|[I - P_n T'(x_n) - S_n'(x_n)]^{-1}\| \leqq \kappa.$$

It is clear that the norms of the operators themselves are also uniformly bounded:

$$\left\| I - P_n T'(x_n) - S'_n(x_n) \right\| \leqq \kappa'.$$

Obviously,

$$\left\| P_n T'(x) - P_n T'(x_n) \right\| \leqq \left\| P_n \right\| \left(\left\| T'(x) - T'(x_0) \right\| + \left\| T'(x_n) - T'(x_0) \right\| \right).$$

The right-hand side of this inequality may be made arbitrarily small by making $\left\| x - x_n \right\|$ small and n large (the norm $\left\| x_n - x_0 \right\|$ is then also small). Fix q arbitrarily $(0 < q < 1)$, and choose δ_0 and n_0 so that, whenever $\left\| x - x_n \right\| \leqq \delta_0, n \geqq n_0$,

$$\left\| P_n T'(x) - P_n T'(x_n) \right\| \leqq \frac{q}{2\kappa}.$$

By making the number σ' smaller (if necessary), we may assume that $\sigma' \leqq q/4\kappa$. Then condition (19.3) of Lemma 19.1 holds: for $n \geqq n_0$ and $\left\| x - x_n \right\| \leqq \delta_0$,

$$\left\| \left[I - P_n T'(x_n) - S'_n(x_n) \right]^{-1} \left\{ \left[I - P_n T'(x) - S'_n(x) \right] - \left[I - P_n T'(x_n) - \right. \right. \right.$$

$$\left. \left. \left. - S'_n(x_n) \right] \right\} \right\| \leqq \kappa \left[\left\| P_n T'(x) - P_n T'(x_n) \right\| + \left\| S'_n(x) \right\| + \left\| S'_n(x_n) \right\| \right] \leqq$$

$$\leqq \kappa \left(\frac{q}{2\kappa} + \frac{q}{4\kappa} + \frac{q}{4\kappa} \right) = q.$$

If we set $\sigma = \left[\delta_0 (1 - q) \right]/\kappa$ in (19.52), condition (19.4) of Lemma 19.1 is also satisfied. All the assertions of our theorem follow directly from this lemma. In particular, the estimate (19.53) follows from (19.5), with $c_1 = 1/\left[\kappa'(1 + q) \right], c_2 = \kappa/(1 - q)$. ∎

Exercise 19.19. Under the assumptions of Theorem 19.7, let the operators S_n be compact on Ω_n (but not necessarily differentiable). Show that for any $\delta, 0 < \delta \leqq \delta_0$, there exist n_δ and σ such that if $n \geqq n_\delta$ and

$$\sup_{x \in E_n, \left\| x - x_0 \right\| \leqq \delta_0} \left\| S_n x \right\| \leqq \sigma,$$

equation (19.51) has at least one solution $\tilde{x}_n$ in the ball $\left\| x - x_0 \right\| \leqq \delta$, and no solutions in the spherical layer $\delta \leqq \left\| x - x_0 \right\| \leqq \delta_0$.

Exercise 19.20. State and prove an analog of Theorem 19.8 for perturbations of equation (19.2).

19.10. *The Galerkin–Petrov method. Stability.* We return to equation (19.1). Assuming that $E = H$ is a Hilbert space, we shall solve this equation by the Galerkin–Petrov method. Let $\{G_n\}$ and $\{H_n\}$ be sequences of closed subspaces of H. Approximate solutions x_n of equation (19.1) will be determined from the equations

$$P_n(x_n - Tx_n) = 0 \qquad (x_n \in \Omega_n = \Omega \cap G_n),\qquad (19.54)$$

where P_n is the orthogonal projection onto H_n.

If the aperture satisfies the inequality

$$\theta_n \equiv \Theta(G_n, H_n) < 1,$$

then (see subsection 16.2) equation (19.54) is equivalent to the Galerkin equation

$$x_n = \bar{P}_n Tx_n,\qquad (19.55)$$

where $\bar{P}_n = \tilde{P}_n^{-1} P_n$ is a continuous projection of H onto G_n, with $\tilde{P}_n$ the restriction of P_n to G_n. In this situation

$$\left\| \bar{P}^{(n)}x \right\| \leqq \frac{1}{\sqrt{1 - \theta_n^2}} \left\| Q^{(n)}x \right\| \qquad (x \in H),\qquad (19.56)$$

where $\bar{P}^{(n)} = I - \bar{P}_n$, $Q^{(n)} = I - Q_n$, with Q_n the orthogonal projection onto G_n. It is clear that $\bar{P}_n \to I$ strongly, provided the sequence of subspaces $\{G_n\}$ is ultimately dense in H and

$$\overline{\lim_{n \to \infty}}\ \theta_n < 1 \qquad (\theta_n = \Theta(G_n, H_n)).\qquad (19.57)$$

Applying Theorem 19.7 to the Galerkin method (19.55), we get

Theorem 19.9. Let T be compact on Ω (as an operator in the Hilbert space H), and assume that equation (19.1) has an isolated solution x_0 of nonzero index. Let the sequence of subspaces $\{G_n\}$ be ultimately dense in H and assume that (19.57) holds.

Then there exist n_0 and δ_0 such that, when $n \geqq n_0$, equation (19.54) has at least one solution x_n in the ball $\|x - x_0\| \leqq \delta_0$, and all these solutions converge in norm to x_0 as $n \to \infty$.

If the operator T is Fréchet-differentiable at x_0 and the homogeneous equation $h = T'(x_0)h$ has no nontrivial solutions, then

$$\left\| Q^{(n)}x_0 \right\| \leqq \left\| x_n - x_0 \right\| \leqq c \left\| Q^{(n)}x_0 \right\|,$$

where $Q^{(n)} = I - Q_n$, with Q_n the orthogonal projection onto G_n.

If T is also continuously Fréchet-differentiable at x_0, then, for sufficiently large n, the solution x_n of equation (19.54) is unique in any ball $\|x - x_0\| \leq \delta$ of sufficiently small radius δ.

Assume that the space H is separable. Let G_n and H_n be the linear spans of elements $\phi_1, \ldots, \phi_n$ and $\psi_1, \ldots, \psi_n$, respectively, where $\{\phi_k\}$ and $\{\psi_k\}$ are two complete coordinate sequences in H. Then condition (19.54) means that

$$(x_n - Tx_n, \psi_i) = 0 \qquad \left(i = 1, \ldots, n \, ; \, x_n = \sum_{j=1}^{n} \xi_j \phi_j \in \Omega \right);$$

this is equivalent to a system of scalar linear equations

$$\sum_{j=1}^{n} (\phi_j, \psi_i) \, \xi_j = f_{in}(\xi_1, \ldots, \xi_n) \qquad (i = 1, \ldots, n). \qquad (19.58)$$

Here $f_{in}(\zeta_1, \ldots, \zeta_n)$ are continuous functions of n variables, such that, for any $z_n = \sum_{j=1}^{n} \zeta_j \phi_j \in \Omega$,

$$f_{in}(\zeta_1, \ldots, \zeta_n) = (Tz_n, \psi_i) \qquad (i = 1, \ldots, n).$$

Owing to errors involved in setting up the equation system (19.58), one must actually solve a certain perturbed system

$$\sum_{j=1}^{n} [(\phi_j, \psi_i) + \gamma_{ij}] \tilde{\xi}_j = f_{in}(\tilde{\xi}_1, \ldots, \tilde{\xi}_n) + \delta_{in}(\tilde{\xi}_1, \ldots, \tilde{\xi}_n) \qquad (19.59)$$

$$(i = 1, \ldots, n),$$

so that, instead of the "exact" approximation x_n, the result is a perturbed approximation

$$\tilde{x}_n = \sum_{j=1}^{n} \tilde{\xi}_j \phi_j.$$

Let us investigate the dependence of the errors $\|\tilde{x}_n - x_n\|$ on the perturbations γ_{ij} and δ_{in} $(i, j = 1, \ldots, n)$. This problem was considered for linear equations in subsection 17.8. In operator notation, the system (19.59) is

$$(A_n + \Gamma_n) \tilde{\xi}^{(n)} = F_n \tilde{\xi}^{(n)} + \Delta_n \tilde{\xi}^{(n)} \qquad (19.60)$$

in the space R_n, where A_n and Γ_n are matrices with elements (ϕ_j, ψ_i) and γ_{ij} $(i, j = 1, \ldots, n)$, respectively, and F_n and Δ_n are nonlinear operators mapping $(\zeta_1, \ldots, \zeta_n)$ onto the vectors

$$(f_{1n}(\zeta_1, \ldots, \zeta_n), \ldots, f_{nn}(\zeta_1, \ldots, \zeta_n))$$

and

$$(\delta_{1n}(\zeta_1, \ldots, \zeta_n), \ldots, \delta_{nn}(\zeta_1, \ldots, \zeta_n)),$$

respectively.

It is easy to see (compare subsection 17.8) that the system (19.60) is equivalent to the perturbed Galerkin equation

$$\tilde{x}_n = \bar{P}_n T \tilde{x}_n + S_n \tilde{x}_n, \tag{19.61}$$

where

$$S_n = \Psi_n^{-1} (\Delta_n - \Gamma_n) A_n^{-1} \Psi_n$$

is an operator in G_n; the operator Ψ_n maps G_n into R_n:

$$\Psi_n x = ((x, \psi_1), \ldots, (x, \psi_n)) \qquad (x \in G_n).$$

The operator S_n is Fréchet-differentiable if the functions $\delta_{in}(\zeta_1, \ldots, \zeta_n)$ $(i = 1, \ldots, n)$ are differentiable, and

$$S_n'(z_n) = \Psi_n^{-1} \Delta_n' (A_n^{-1} \Psi_n z_n) A_n^{-1} \Psi_n - \Psi_n^{-1} \Gamma_n A_n^{-1} \Psi_n.$$

Applying Theorem 19.8 to equations (19.55) and (19.61), and using (17.24), (17.25) and (17.26), we get the following

Theorem 19.10. Assume that equation (19.1) has a solution $x_0 \in \Omega$, T is continuously Fréchet-differentiable at x_0 (as an operator in the Hilbert space H), and the homogeneous equation $h = T'(x_0) h$ has no nontrivial solutions. Assume that the linear spans G_n and H_n of the first n elements of two complete coordinate sequences $\{\phi_k\}$ and $\{\psi_k\}$ satisfy condition (19.57). Let λ_n and μ_n be the smallest eigenvalues of the Gram matrices $(\phi_j, \phi_i)_{i,j=1}^n$ and $(\psi_j, \psi_i)_{i,j=1}^n$, respectively, and $\xi^{(n)}$ a solution of the system (19.58) (which exists and is locally unique for sufficiently large n, by virtue of Theorem 19.9). Finally, let the operator Δ_n be Fréchet-differentiable in the ball $\|\zeta^{(n)} - \xi^{(n)}\| \leq \delta / \sqrt{\lambda_n}$ $(n = 1, 2, \ldots)$, where δ is a fixed (arbitrarily small) positive number.

Then there exist positive constants δ_0 ($\delta_0 \leqq \delta$), α, β and β', all independent of n, and a natural number n_0, such that, whenever $n \geqq n_0$ and

$$\alpha_n \equiv \frac{\|\Gamma_n\|}{\sqrt{\lambda_n}\sqrt{\mu_n}} \leqq \alpha,$$

$$\beta_n \equiv \frac{1}{\sqrt{\mu_n}} \sup_{\|\zeta^{(n)} - \xi^{(n)}\| \leqq \delta_0/\sqrt{\lambda_n}} \|\Delta_n \zeta^{(n)}\| \leqq \beta,$$

$$\beta'_n = \frac{1}{\sqrt{\lambda_n}\sqrt{\mu_n}} \sup_{\|\zeta^{(n)} - \xi^{(n)}\| \leqq \delta_0/\sqrt{\lambda_n}} \|\Delta'_n(\zeta_n)\| \leqq \beta',$$

then the system of equations (19.59) *has a unique solution $\tilde{\xi}^{(n)}$ in the ball $\|\zeta^{(n)} - \xi^{(n)}\| \leqq \delta_0/\sqrt{\lambda_n}$. The approximations*

$$x_n = \sum_{j=1}^{n} \xi_j \phi_j, \qquad \tilde{x}_n = \sum_{j=1}^{n} \tilde{\xi}_j \phi_j$$

satisfy the estimate

$$\|\tilde{x}_n - x_n\| \leqq c(\alpha_n + \beta_n) \qquad (c = \text{const}).$$

Exercise 19.21. Formulate an analog of Theorem 19.10 for the case in which the only available information on the solution x_0 is that it has nonzero index.

19.11. *Projection methods assuming one-sided estimates.* Let E be a real separable Banach space, E^* the dual space to E. Let F be a continuous operator mapping E into E^*, bounded in the ball $\|x\| \leqq r$, such that

$$(Fx, x) > 0 \qquad (\|x\| = r) \tag{19.62}$$

(where (x^*, x) denotes the value of the functional $x^* \in E^*$ for $x \in E$). Consider the equation

$$Fx = 0. \tag{19.63}$$

The most natural device for approximating solutions of equation (19.63) is the following projection method.

Let $\{E_n\}$ be a given sequence of finite-dimensional subspaces of E, ultimately dense in E. Let P_n ($n = 1, 2, \ldots$) be continuous linear projections of E onto E_n. Replace equation (19.63) by the approximate equation

$$P_n^* F x_n = 0 \qquad (x_n \in E_n), \tag{19.64}$$

where P_n^* is the adjoint of P_n (P_n^* is obviously also a projection).
It follows from the one-sided estimate (19.62) that

$$(P_n^* F x, x) = (F x, P_n x) = (F x, x) > 0 \qquad (\|x\| = r; \; x \in E_n).$$

Thus equation (19.64) always has a solution $x_n \in E_n$ such that $\|x_n\| \leqq r$.
We now study the convergence of the approximations x_n. As we have
mentioned, convergence proofs may be regarded as a justification (in a
"comforting" sense!) for the use of equations in finite-dimensional
subspaces. We shall prove only weak convergence of the approximations
x_n. This is of course less "comforting" than strong convergence.

We claim that the sequence $F x_n$ (where x_n is a solution of equation
(19.64)) converges weakly to zero. In fact, since $\{E_n\}$ is ultimately dense
in E, any $y \in E$ may be expressed in the form $y = u_n + v_n$, where $u_n \in E_n$
and $\|v_n\| \to 0$ as $n \to \infty$. But then, by (19.64),

$$(F x_n, y) = (F x_n, u_n + v_n) = (P_n^* F x_n, u_n) + (F x_n, v_n) = (F x_n, v_n)$$

and

$$\left| (F x_n, y) \right| \leqq \|F x_n\| \cdot \|v_n\| \to 0 \quad \text{as} \quad n \to \infty \qquad (y \in E).$$

Theorem 19.11 (*Browder–Minty*). *Assume that the space E is reflexive,
and the operator F satisfies the condition*

$$(F x - F y, x - y) \geqq 0 \qquad (x, y \in E). \tag{19.65}$$

Then any weak limit of solutions x_n of equations (19.64) *is a solution of
equation* (19.63). *In particular, if x_0 is the unique solution of equation* (19.63),
the sequence x_n converges weakly to x_0.

*Recall that any bounded set in a reflexive Banach space is weakly
compact. Thus the sequence x_n indeed contains a weakly convergent
subsequence.*

Proof of Theorem 19.11. For any $x \in E$ and $n = 1, 2, \ldots$,

$$(F x, x - x_n) = (F x - F x_n, x - x_n) + (F x_n, x) - (F x_n, x_n) \geqq (F x_n, x); \tag{19.66}$$

here we have used inequality (19.65) and the fact that the solution x_n
of equation (19.64) satisfies the equality

$$(F x_n, x_n) = (F x_n, P_n x_n) = (P_n^* F x_n, x_n) = 0.$$

Let x_{n_k} be a subsequence converging weakly to x_0. Since Fx_n converges weakly to zero, it follows from (19.66) that

$$(Fx, x - x_0) \geqq 0.$$

Setting $x = x_0 + \varepsilon h$, we see that $(F(x_0 + \varepsilon h), h) \geqq 0$ for any $\varepsilon > 0$ and $h \in E$; since F is continuous, this implies that

$$(Fx_0, h) \geqq 0 \qquad (h \in E).$$

Since $h \in E$ is arbitrary, it follows that $Fx_0 = 0$. ∎

> *Exercise* 19.22. Let the space E be reflexive and the operator F weakly continuous. Show that every weak limit point of a sequence of solutions x_n of equation (19.64) is a solution of equation (19.63) (Kachurovskii).
>
> *Exercise* 19.23. Let P_n and Q_n be projections of E onto E_n. Show that equation (19.64) and the equation
>
> $$Q_n^* F x_n = 0 \qquad (x_n \in E_n)$$
>
> have the same solutions.

19.12. *Minimum problems for functionals.* Let E be a real separable Banach space, and Φ a continuous functional on E which increases to infinity ($\Phi(x) \to \infty$ when $\|x\| \to \infty$) and has a minimum on E:

$$m = \min_{x \in E} \Phi(x).$$

Let $\{E_n\}$ be a sequence of finite-dimensional subspaces, ultimately dense in E,

$$m_n = \min_{x \in E_n} \Phi(x),$$

and x_n a point in E_n at which the minimum m_n is attained. It is natural to ask whether the sequence x_n converges in some sense to the minimum point of Φ in E. Note that the fact that $\{E_n\}$ is ultimately dense in E implies that

$$\lim_{n \to \infty} \Phi(x_n) = \lim_{n \to \infty} m_n = m.$$

It is easy to prove that x_n converges to the minimum point of Φ on E if the values of Φ increase rapidly as it "leaves" the minimum, i.e.,

$$|\Phi(x) - m| \geqq \alpha [\rho(x, X_0)], \qquad (19.67)$$

where X_0 is the set of all minimum points of Φ on E and $\alpha(\rho)$ is a monotone increasing function which is positive for $\rho > 0$. Indeed, it follows from (19.67) that

$$\rho(x_n, X_0) \leq \alpha^{-1} \left[|\, \Phi(x_n) - m \,| \right],$$

and therefore

$$\lim_{n \to \infty} \rho(x_n, X_0) = 0.$$

In particular, if X_0 consists of a single point x_0 condition (19.67) becomes

$$|\Phi(x) - \Phi(x_0)| \geq \alpha(\|x - x_0\|);$$
(19.68)

if this condition holds, the sequence x_n converges strongly to x_0.

As an example, consider a quadratic functional defined in a Hilbert space H:

$$\Phi(x) = \tfrac{1}{2}(Ax, x) - (f, x),$$
(19.69)

where A is a positive definite selfadjoint bounded operator ($(Ax, x) \geq \gamma^2 \|x\|^2$, $\gamma > 0$), and $f \in H$. Let x_0 be a solution of the equation $Ax = f$. Then

$$|\Phi(x) - \Phi(x_0)| = \tfrac{1}{2} \|A^{1/2}(x - x_0)\|^2 \geq \frac{\gamma^2}{2} \|x - x_0\|^2.$$

It follows that x_0 is the unique minimum point of the functional Φ, and also that (19.68) holds with $\alpha(\rho) = (\gamma^2/2)\rho^2$. Thus, to approximate the minimum point of the functional (19.69) it suffices to find its minimum in some finite-dimensional subspace (compare subsection 16.3).

A second example is the functional

$$\Phi(x) = \|x - Ax\|,$$
(19.70)

where A is a compact operator in E, and

$$\lim_{\|x\| \to \infty} \|x - Ax\| = \infty.$$

Let the equation

$$x = Ax$$
(19.71)

have a set of solutions X_0. X_0 clearly coincides with the set of minimum points of the functional (19.70). The functional (19.70) always satisfies inequality (19.67). One can thus approximate the minimum of (19.70) (or the solutions of equations (19.71)) by determining its minimum points in some finite-dimensional subspace.

Inequality (19.67) is quite restrictive, and holds comparatively rarely. On the assumption that the space E is reflexive and the functional Φ weakly lower semicontinuous, it can be proved that the approximations x_n "converge" weakly to the set of minimum points of Φ; more precisely, the limit x_0 of any weakly convergent subsequence x_{n_k} is a minimum point of Φ. Indeed, let x_0 be a weak limit of the subsequences x_{n_k} and $\Phi(x_0) > m + \varepsilon$. Choose n_0 so that, for $n \geqq n_0$, $m_n < m + \varepsilon$. But then if $n_k > n_0$ we have $\Phi(x_{n_k}) < m + \varepsilon$. Thus, since Φ is lower semicontinuous, $\Phi(x_0) < m + \varepsilon$, which is impossible.

The technique presented in this subsection for approximation of minimum points is quite well known. It is often used in solving various linear and nonlinear equations.

Exercise 19.24. Let the functional Φ be a Fréchet-differentiable:

$$\Phi(x + h) - \Phi(x) = (Fx, h) + \omega(x;h),$$

where F is an operator mapping E into E^*, and

$$\frac{|\omega(x;h)|}{\|h\|} \to 0 \quad \text{as} \quad \|h\| \to 0.$$

Show that
  a) the minimum points of Φ on E are solutions of the equation $Fx = 0$;
  b) the minimum points of Φ on E_n are solutions of the equation

$$P_n^* Fx = 0 \qquad (x \in E_n),$$

where P_n is any continuous linear projection onto E_n ($P_n E = E_n$).

19.13. *The factor method for nonlinear equations.* Many of the theorems proved in this chapter carry over to the factor method, using the concept of compact approximations of an operator (Vainikko [17–19]). We state one such result here.

Let E be a Banach space, $E^{(n)}$ ($n = 1, 2, \ldots$) a closed subspace of E, p_n ($n = 1, 2, \ldots$) the canonical mapping of E onto the factor space $E/E^{(n)}$. We shall say that a sequence of compact operators T_n defined in $E/E^{(n)}$ is a sequence of *compact approximations* to a compact operator T defined in E if a) $\|p_n Tx - T_n p_n x\| \to 0$ as $n \to \infty$, for every $x \in E$; b) for any $\xi_n \in E/E^{(n)}$, $\|\xi_n\| \leqq c = \text{const}$ ($n = 1, 2, \ldots$), there exist $y_n \in T_n \xi_n$ such that the sequence $\{y_n\}$ is compact in E.

Theorem 19.12. *Let* x_0 *be an isolated solution of the equation* $x = Tx$, *of nonzero index. Suppose that if* $\|\xi_n - p_n x\| \to 0$ $(\xi_n \in E/E^{(n)}, x \in E)$, *then* $\|T_n\xi_n - T_n p_n x\| \to 0$ *as* $n \to \infty$, *and assume that the operators* T_n *are compact approximations to* T. *Finally, let* $\|p_n x\| \to \|x\|$ *as* $n \to \infty$, *for any* $x \in E$.

Then there exist n_0 *and* $\delta_0 > 0$ *such that, for* $n \geqq n_0$, *the equation* $\xi_n = T_n\xi_n$ *has at least one solution* ξ_n^* *in the ball* $\|\xi_n - p_n x_0\| \leqq \delta_0$ *and* $\|\xi_n^* - p_n x_0\| \to 0$ *as* $n \to \infty$.

Small solutions of operator equations

§ 20. Approximation of implicit functions*

20.1. *Fundamental implicit function theorem.* Let E, F and Λ be Banach spaces, $f(\lambda, x)$ an operator defined for $\|\lambda - \lambda_0\|_\Lambda \leqq a$, $\|x - x_0\|_E \leqq b$ with values in F. Consider the equation

$$f(\lambda, x) = 0 . \tag{20.1}$$

Assuming that

$$f(\lambda_0, x_0) = 0 , \tag{20.2}$$

we wish to find a solution $x_*(\lambda)$ of equation (20.1) which is close to x_0 when λ is close to λ_0.

Simple examples show that equation (20.1) sometimes has a unique solution $x_*(\lambda)$ close to x_0; equation (20.1) is then said to define a single-valued implicit function. In other cases $x_*(\lambda)$ is multiple-valued, and the solutions are said to branch. Finally, equation (20.1) may have one or more solutions for certain values of λ close to λ_0 and no solutions for other values, and so on. In Chapter 1 we presented sufficient conditions for the existence of a single-valued implicit function. This section is devoted to a more detailed investigation, including methods for approximation of implicit functions.

---

* The first implicit function theorems for equations in Banach spaces are apparently due to Lamson [1], followed by Hildebrandt and Graves [1]. Various implicit function theorems have been proposed by Lyusternik and Sobolev [1], Kantorovich and Akilov [1], Dunford and Schwartz [1], Hille and Phillips [1] and other authors. Implicit function theorems for equations with analytic operators have been proved by Michael and Clifford [1], Gel'man [2, 3] and Tamme [1].

Throughout the sequel, we shall assume that the derivative $f_x''(\lambda_0, x_0)$ exists. The properties of the linear operator $f_x'(\lambda_0, x_0)$ have a decisive effect on the function $x_*(\lambda)$. In this section we shall consider the case in which the linear operator

$$\Gamma = [f_x'(\lambda_0, x_0)]^{-1} \tag{20.3}$$

mapping F into E, exists and is continuous. This case is called *non-degenerate*.

As usual, $S(x_0, r)$ denotes the ball $\|x - x_0\|_E \leqq r$; the notation $S(\lambda_0, r)$, etc., has an analogous meaning.

The principal technique for proving the existence of a single-valued implicit function is to replace equation (20.1) by an equivalent equation

$$x = T(\lambda, x), \tag{20.4}$$

where

$$T(\lambda, x) = x - \Gamma f(\lambda, x), \tag{20.5}$$

and to determine conditions under which the operator $T(\lambda, x)$ (for fixed λ) satisfies the assumptions of the contracting mapping principle. To prove the existence of an implicit function, therefore, it suffices to establish the inequalities

$$\|x_1 - x_2 - \Gamma f(\lambda, x_1) + \Gamma f(\lambda, x_2)\|_E \leqq m(\rho, r)\|x_1 - x_2\|_E$$

$$(\lambda \in S(\lambda_0, \rho) ; \; x_1, x_2 \in S(x_0, r)),$$

$$\|x - x_0 - \Gamma f(\lambda, x)\|_E \leqq n(\rho, r) \qquad (\lambda \in S(\lambda_0, \rho) ; \; x \in S(x_0, r))$$

and to show that, for some $a_0 \in (0, a]$, $b_0 \in (0, b]$,

$$m(a_0, b_0) < 1, \qquad n(a_0, b_0) \leqq b_0.$$

The above inequalities are easily checked (the details are left to the reader) when the assumptions of the following proposition hold.

Theorem 20.1. Let

$$\|f(\lambda_0, x_1) - f(\lambda_0, x_2) - f_x'(\lambda_0, x_0)(x_1 - x_2)\|_F \leqq p(r)\|x_1 - x_2\|_E, \tag{20.6}$$

$$\|[f(\lambda, x_1) - f(\lambda, x_2)] - [f(\lambda_0, x_1) - f(\lambda_0, x_2)]\|_F \leqq p_1(\rho, r)\|x_1 - x_2\|_E$$

$$(\lambda \in S(\lambda_0, \rho) ; \; x_1, x_2 \in S(x_0, r)), \tag{20.7}$$

and

$$c_0 = \lim_{\rho, r \to 0} \|\Gamma\| \, [p(r) + p_1(\rho, r)] < 1. \tag{20.8}$$

Let $f(\lambda, x_0)$ be continuous at λ_0.

Then there exist $a_0 \in (0, a]$ and $b_0 \in (0, b]$ such that when $\lambda \in S(\lambda_0, a_0)$ equation (20.1) has a unique solution $x_*(\lambda)$ in the ball $S(x_0, b_0)$, and moreover

$$\lim_{\lambda \to \lambda_0} \|x_*(\lambda) - x_0\|_E = 0. \tag{20.9}$$

The last assertion of this theorem follows from a simple observation: for any sufficiently small positive ε, the operator (20.5) satisfies the assumptions of the contracting mapping principle in the ball $S(x_0, \varepsilon)$, for all λ sufficiently close to λ_0.

The assumptions of Theorem 20.1 are of course satisfied if the operator $f(\lambda, x)$ is continuously differentiable with respect to x. However, in many important function spaces (L_2, L_p, Orlicz spaces, etc.) this assumption is extremely restrictive. For example, nonlinear integral operators defined in such spaces are usually differentiable only at isolated points (see Krasnosel'skii, Zabreiko, Pustyl'nik and Sobolevskii [1]).

Theorem 20.1 implies

Theorem 20.2. Under the assumptions of Theorem 20.1, let $f(\lambda, x)$ be continuous with respect to λ.

Then the implicit function $x_*(\lambda)$ whose existence is given by Theorem 20.1 is continuous in the ball $S(\lambda_0, a_0)$.

Exercise 20.1. Let the function $K(t, s, u)$ ($0 \le t, s \le 1$; $-\infty < u < \infty$) be measurable with respect to t and s and differentiable with respect to u,

$$|K'_u(t, s, u)| \le K_0(t, s) \, [a(s) + b \, |u|\,]$$

and

$$a(s) \in L_2, \quad \int_0^1 |K_0(t, s)| \, dt \in L_2, \quad \int_0^1 K(t, s, 0) \, ds \in L_2.$$

Show that for small λ the equation

$$x(t) = \lambda \int_0^1 K[t, s, x(s)] \, ds$$

has a unique small solution $x_\lambda(t) \in L_2$.

Exercise 20.2. Construct a function $K(t, s, u)$ satisfying the assumptions of Exercise 20.1 such that the operator

$$Ax(t) = \int_0^1 K[t, s, x(s)]\, ds$$

is neither compact nor differentiable over L_2.

Exercise 20.3. Show that Theorem 20.1 implies that the trivial solution of the system of ordinary differential equations

$$\frac{dx}{dt} = Bx + \omega(t, x),$$

where B is a matrix whose spectrum is in the left half-plane, $\omega(t, 0) = 0$, and

$$\|\omega(t, x_1) - \omega(t, x_2)\| \leq q(r)\, \|x_1 - x_2\| \qquad (\|x_1\|, \|x_2\| \leq r),$$

where $q(+0) = 0$, is asymptotically stable in the Lyapunov sense.

Many generalizations and modifications of the above implicit function theorems are known. In particular, Zabreiko and Krasnosel'skii [3] consider the case in which equation (20.1) has the form $x = A(\lambda, x)$, where A satisfies various "improving" conditions.

20.2. *Multilinear operators and Taylor's formula.** An operator $D(x_1, \ldots, x_k)$ in k variables $x_1, \ldots, x_k \in E$ is said to be *multilinear* or *k-linear* if it is linear in each variable separately. Examples are the operators

$$D(x_1, \ldots, x_k) = \int_0^1 \ldots \int_0^1 K(t, s_1, \ldots, s_k) x_1(s_1) \ldots x_k(s_k)\, ds_1 \ldots ds_k$$
$$(20.10)$$

and

$$D(x_1, \ldots, x_k) = \int_0^1 K(t, s) x_1(s) \ldots x_k(s)\, ds \qquad (20.11)$$

in the space of functions continuous on $[0, 1]$, provided the kernels $K(t, s_1, \ldots, s_k)$ and $K(t, s)$ are continuous. The number

$$\|D\| = \sup_{\|x_1\|, \ldots, \|x_k\| \leq 1} \|D(x_1, \ldots, x_k)\| \qquad (20.12)$$

is known as the *norm* of the multilinear operator D.

---

* For a detailed exposition, see Lyusternik and Sobolev [1], Hille and Phillips [1]. The basic results in this field are due to Mazur and Orlicz [1] and Banach. See also Gavurin [2], Alexiewicz and Orlicz [1], Taylor [1, 2].

Exercise 20.4. Compute or estimate the norms of the operators (20.10) and (20.11) in the space of continuous functions.

An operator $C = C_k$ is said to be a *homogeneous form of order* k, or a k-*form*, if

$$Cx = D(x, \ldots, x), \qquad (20.13)$$

where $D(x_1, \ldots, x_k)$ is a multilinear operator. It is clear that different k-linear operators D may give rise to the same homogeneous form of order k. It is sometimes convenient to regard a constant operator as a 0-form.

A multilinear operator is said to be *symmetric* if it is invariant under any permutation of its arguments. The operator (20.11) is symmetric; so is (20.10), provided the kernel $K(t, s_1, \ldots, s_k)$ is a symmetric function of $s_1, \ldots, s_k$. It is easy to see that every homogeneous form C of order k is defined by some symmetric k-linear operator D; this operator D is unique, and is expressed in terms of the form C, e.g., by the formula

$$D(x_1, \ldots, x_k) = \frac{1}{2^k k!} \sum_{\varepsilon_i = \pm 1} \varepsilon_1 \ldots \varepsilon_k C(\varepsilon_1 x_1 + \ldots + \varepsilon_k x_k). \qquad (20.14)$$

Henceforth we shall denote the operator (20.14) by $\tilde{C}$, while $\tilde{C} x_1^{\alpha_1} x_2^{\alpha_2} \ldots x_l^{\alpha_l}$ will denote the value of the symmetric operator (20.14) when α_1 arguments assume the same value x_1, another α_2 arguments the same value x_2, and so on; clearly, $\alpha_1 + \alpha_2 + \ldots + \alpha_k = k$. This notation greatly simplifies the statement of the so-called *addition theorem*; this term is generally applied to the formula

$$C(x_1 + \ldots + x_l) = \sum_{\alpha_1 + \ldots + \alpha_l = k} \frac{k!}{\alpha_1! \ldots \alpha_l!} \tilde{C} x_1^{\alpha_1} \ldots x_l^{\alpha_l}. \qquad (20.15)$$

With every k-form C we associate two norms: the usual norm

$$\|C\| = \sup_{\|x\| = 1} \|Cx\| \qquad (20.16)$$

and the $*$-norm $\|C\|_*$ defined by (20.12) where D is the operator (20.14) (i.e., $\|C\|_* = \|\tilde{C}\|$). It follows from (20.14) that for any homogeneous form of order k

$$\|C\| \leq \|C\|_* \leq \frac{k^k}{k!} \|C\|. \qquad (20.17)$$

A *polynomial* is an operator $B_0 + B_1 + \ldots + B_m$, where each B_i is a homogeneous form of order i; the highest order of nonzero terms in a polynomial is called its *degree*.

Exercise 20.5. Show that the superposition of two polynomials of degree m and n is a polynomial of degree at most mn.

Now let A be an arbitrary operator mapping some neighborhood of a point $x_0 \in E$ into the space F. Suppose that in this neighborhood

$$Ax = Ax_0 + B_1(x - x_0) + \ldots + B_m(x - x_0) + \omega(x - x_0), \quad (20.18)$$

where the B_k are homogeneous forms of order k and

$$\lim_{\|x - x_0\| \to 0} \frac{\|\omega(x - x_0)\|}{\|x - x_0\|^m} = 0.$$

The operator A is then said to have derivatives of order up to m at the point x_0 (in the sense of Taylor's formula); these derivatives are defined by

$$A^{(k)}(x_0) = k! B_k;$$

they are operators with domain in E and range in F.

The representation (20.18) is called *Taylor's formula*, and we write

$$Ax = Ax_0 + A'(x_0)(x - x_0) + \ldots + \frac{1}{m!} A^{(m)}(x_0)(x - x_0) + \omega(x - x_0).$$
$$(20.19)$$

Exercise 20.6. Show that the operator B_1 in (20.18) is precisely the Fréchet-derivative of an operator.

Exercise 20.7. Show that if A is a compact operator then so are $A'(x_0), \ldots, A^{(m)}(x_0)$ (Melamed and Perov [1]).

Exercise 20.8. Construct a nonlinear operator which has a compact derivative at each point of the space but is not itself compact.

Exercise 20.9. Let $A = FG$, where F and G have derivatives of order up to m at the points Gx_0 and x_0, respectively. Show that A has derivatives of order up to m at x_0, and for $k = 1, \ldots, m$

$$A^{(k)}(x_0) h = \sum_{\alpha_1 + 2\alpha_2 + \ldots + k\alpha_k = k} \frac{k!}{(1!)^{\alpha_1} \alpha_1! \ldots (k!)^{\alpha_k} \alpha_k!} \times$$

$$\times \tilde{F}^{(\alpha_1 + \ldots + \alpha_k)}(Gx_0) [G'(x_0) h]^{\alpha_1} \ldots [G^{(k)}(x_0) h]^{\alpha_k}.$$

An operator A is said to be *analytic at a point* x_0 if, in some neighborhood of this point, it may be represented by a uniformly convergent (in norm) series

$$Ax = Ax_0 + B_1(x - x_0) + \ldots + B_n(x - x_0) + \ldots, \qquad (20.20)$$

where B_n is a homogenous form of order n. When this is so,

$$\overline{\lim_{n \to \infty}} \sqrt[n]{\|B_n\|} = a < \infty. \qquad (20.21)$$

The series (20.20) converges uniformly in any ball $S(x_0, r)$, $ra < 1$. Simple estimates show that if A is analytic at x_0 it is analytic at any interior point of the ball $S(x_0, 1/a)$.

Exercise 20.10. Let E and F be complex spaces. Show that if A is differentiable in some domain in E it is analytic at all points of this domain.

We now consider operators $f(\lambda, x)$ depending on two variables $\lambda \in \Lambda$, $x \in E$ (see subsection 20.1).

The concept of a multilinear operator has a natural extension to the case of an operator some of whose arguments belong to one space Λ, the others to another space E. An operator $U(\lambda_1, \ldots, \lambda_k; x_1, \ldots, x_m)$, which is linear in each variable is said to be (k, m)-*linear*. The operator

$$V(\lambda, x) = U(\lambda, \ldots, \lambda; x, \ldots, x) \qquad (20.22)$$

is called a *homogeneous* (k, m)-*form*, and the sum of a finite number of operators of this type is a *polynomial in two variables*.

If λ is fixed, every (k, m)-form $V(\lambda; x)$ is an m-form in x. Let $\tilde{V}(\lambda; x_1, \ldots, x_m)$ denote an m-linear operator, symmetric in $x_1, \ldots, x_m$, such that $V(\lambda; x) = \tilde{V}(\lambda; x, \ldots, x)$ (see formula (20.14)). For fixed $x_1, \ldots, x_m$ the operator $\tilde{V}(\lambda; x_1, \ldots, x_m)$ is a k-form in λ.

A representation of the operator $f(\lambda; x)$ in the form

$$f(\lambda, x) = f(\lambda_0, x_0) + \sum_{k+m=1}^{n} V_{k,m}(\lambda - \lambda_0; x - x_0) + \omega(\lambda - \lambda_0; x - x_0) \qquad (20.23)$$

where $V_{k,m}$ are homogeneous (k, m)-forms and

$$\lim_{\|\lambda - \lambda_0\| + \|x - x_0\| \to 0} \frac{\|\omega(\lambda - \lambda_0; x - x_0)\|}{\|\lambda - \lambda_0\|^n + \|x - x_0\|^n} = 0, \qquad (20.24)$$

is called a *Taylor expansion*. Introducing the notation

$$f^{(k+m)}_{\lambda^k, x^m}(\lambda_0, x_0) = k!\, m!\, V_{k,m}, \qquad (20.25)$$

we get the usual version of Taylor's formula:

$$f(\lambda, x) = f(\lambda_0, x_0) +$$

$$+ \sum_{k=1}^{n} \frac{1}{k!} \left[\sum_{i=0}^{k} \frac{k!}{i!\,(k-i)!} f^{(k)}_{\lambda^i,\,x^{k-i}}(\lambda_0, x_0)\,(\lambda - \lambda_0\,;x - x_0) \right] +$$

$$+ \omega(\lambda - \lambda_0\,;x - x_0). \qquad (20.26)$$

The operators (20.25) are known as the *mixed derivatives of the operator* f. If (20.23) is valid, $f(\lambda, x)$ is said to have *derivatives of order up to n*.

Exercise 20.11. Assume that the operators $f(\lambda, x)$ and $x(\lambda)$ have derivatives of order up to n at the points $\{\lambda_0, x_0\}$ ($x_0 = x(\lambda_0)$) and λ_0, respectively. Show that the operator $y(\lambda) = f[\lambda, x(\lambda)]$ has derivatives of order up to n at λ_0, and for $k = 1, \ldots, n$

$$y^{(k)}(\lambda_0)\,v = \sum_{i+\alpha_1+2\alpha_2+\ldots+k\alpha_k=k} \frac{k!}{i!\,(1!)^{\alpha_1}\alpha_1!\ldots(k!)^{\alpha_k}\alpha_k!} \times$$

$$\times f^{(i+\alpha_1+\ldots+\alpha_k)}_{\lambda^i,\,x^{\alpha_1+\ldots+\alpha_k}}(\lambda_0, x_0)\,v^i\,[x'(\lambda_0)\,v]^{\alpha_1} \ldots [x^{(k)}(\lambda_0)\,v]^{\alpha_k}.$$

The concept of analyticity carries over to operators $f(\lambda, x)$ without difficulty: the operator $f(\lambda, x)$ is analytic at a point $\{\lambda_0, x_0\}$ if, in some neighborhood of this point, it is representable by a uniformly convergent series

$$f(\lambda, x) = f(\lambda_0, x_0) + \sum_{n=1}^{\infty} \left[\sum_{k=0}^{n} V_{k,\,n-k}(\lambda - \lambda_0\,;x - x_0) \right], \qquad (20.27)$$

where $V_{k,\,n-k}$ is a homogeneous $(k, n-k)$-form.

20.3. Differentiation of an implicit function.

Theorem 20.3. *Under the assumptions of Theorem 20.1, let the operator* f *have derivatives of order up to m at the point* $\{\lambda_0, x_0\}$.

Then the implicit function $x_*(\lambda)$ *whose existence follows from Theorem 20.1 has derivatives of order up to m at the point* λ_0. *These derivatives* $x_*^{(k)}(\lambda_0)$ ($k = 1, \ldots, m$) *may be defined by the recurrence relation*

$$x_*^{(k)}(\lambda_0)\,v = -\frac{1}{k!}\,\Gamma\,\frac{d^k}{d\varepsilon^k} f \left[\lambda_0 + \varepsilon v\,;x_0 + \varepsilon x_*'(\lambda_0)\,v + \ldots \right.$$

$$\left. \ldots + \frac{\varepsilon^{k-1}}{(k-1)!}\,x_*^{(k-1)}(\lambda_0)\,v \right]. \qquad (20.28)$$

The theorem is proved by induction. It is trivial for $k = 0$. Assuming that it holds for all $k < k_0 \leqq m$, we shall show that it also holds for $k = k_0$.

Consider the auxiliary equation

$$g(\lambda, u) = 0 \, ,$$

where

$$g(\lambda, u) = f\left[\lambda, x_0 + x'_*(\lambda_0)(\lambda - \lambda_0) + \dots \right.$$

$$\left. \dots + \frac{1}{(k_0 - 1)!} x^{(k-1)}_*(\lambda_0)(\lambda - \lambda_0) + u \right].$$

An obvious solution of this equation is

$$u_*(\lambda) = x_*(\lambda) - x_0 - x'_*(\lambda_0)(\lambda - \lambda_0) - \dots$$

$$\dots - \frac{1}{(k_0 - 1)!} x^{(k_0-1)}_*(\lambda_0)(\lambda - \lambda_0) \, ,$$

and, by the induction hypothesis, this solution satisfies the condition

$$\lim_{\lambda \to \lambda_0} \frac{\|u_*(\lambda)\|}{\|\lambda - \lambda_0\|^{k_0 - 1}} = 0 \, . \tag{20.29}$$

It is easily seen that the operator $g(\lambda, u)$ has derivatives of order up to $k_0 \leqq n$ at the point $\{\lambda_0, 0\}$, i.e.,

$$g(\lambda, u) = g_0(\lambda, u) + \omega(\lambda - \lambda_0, u) \, , \tag{20.30}$$

where $g_0(\lambda, u)$ is a polynomial and

$$\omega(\lambda - \lambda_0, u) = o\left[\|\lambda - \lambda_0\|^{k_0} + \|u\|^{k_0}\right] \, .$$

Since a polynomial satisfies a Lipschitz condition on any ball, the relations (20.29) and (20.30) imply the estimate

$$\|g(\lambda, 0)\| = o\left(\|\lambda - \lambda_0\|^{k_0 - 1}\right) \, .$$

This means that the derivatives $g'_\lambda(\lambda_0, 0), \dots, g^{(k_0-1)}_{\lambda^{k_0-1}}(\lambda_0, 0)$ vanish.

We now claim that the function $u(\lambda)$ has a derivative

$$u_*^{(k_0)}(\lambda_0) = -\frac{1}{k_0!}\,\Gamma g_{\lambda^{k_0}}^{(k_0)}(\lambda_0, 0)\,. \tag{20.31}$$

Obviously,

$$g(\lambda, u) - g(\lambda_0, 0) = \frac{1}{k_0!}\,g_{\lambda^{k_0}}^{(k_0)}(\lambda_0, 0)\,(\lambda - \lambda_0) + f_x'(\lambda_0, x_0)\,u +$$

$$+ \tilde\omega(\lambda - \lambda_0, u)\,,$$

where $\tilde\omega(\lambda, u)$ is an operator such that

$$\lim_{\|v\| + \|u\| \to 0} \frac{\|\tilde\omega(v, u)\|}{\|v\|^{k_0} + \|u\|} = 0\,.$$

Let ε be any number in $(0, 1)$, and $a_{k_0} \in (0, a]$, $b_{k_0} \in (0, b]$ numbers such that

$$\|\Gamma\,\bar\Psi \cdot \|\tilde\omega(v, u)\| \leq \varepsilon\,[\|v\|^{k_0} + \|u\|] \qquad (\|v\| \leq a_{k_0}, \|u\| \leq b_{k_0})$$

and

$$\|u(\lambda)\| \leq b_{k_0} \qquad (\|\lambda - \lambda_0\| \leq a_{k_0})\,.$$

Then

$$\left\| u_*(\lambda) + \frac{1}{k_0!}\,\Gamma g_{\lambda^{k_0}}^{(k_0)}(\lambda_0, 0)\,(\lambda - \lambda_0) \right\| \leq \|\Gamma\| \cdot \|\tilde\omega\,[\lambda - \lambda_0\,;\, u_*(\lambda)]\| \leq$$

$$\leq \varepsilon\,[\|\lambda - \lambda_0\|^{k_0} + \|u_*(\lambda)\|] \leq \varepsilon \left\| u_*(\lambda) + \frac{1}{k_0!}\,\Gamma g_{\lambda^{k_0}}^{(k_0)}(\lambda_0, 0)\,(\lambda - \lambda_0) \right\| +$$

$$+ \varepsilon \left[1 + \frac{1}{k_0!}\,\left\| \Gamma g_{\lambda^{k_0}}^{(k_0)}(\lambda_0, 0) \right\| \right] \|\lambda - \lambda_0\|^{k_0}\,,$$

and hence the estimate

$$\left\| u_*(\lambda) + \frac{1}{k_0!}\,\Gamma g_{\lambda^{k_0}}^{(k_0)}(\lambda_0, 0)\,(\lambda - \lambda_0) \right\| \leq$$

$$\leq \frac{\varepsilon}{1 - \varepsilon} \left[1 + \frac{1}{k_0!}\,\left\| \Gamma g_{\lambda^{k_0}}^{(k_0)}(\lambda_0, 0) \right\| \right] \|\lambda - \lambda_0\|^{k_0}\,.$$

Since ε is arbitrary,

$$\lim_{\lambda \to \lambda_0} \left\| \lambda - \lambda_0 \right\|^{-k_0} \left\| u_*(\lambda) + \frac{1}{k_0!} \Gamma g_{\lambda k_0}^{(k_0)}(\lambda_0, 0)(\lambda - \lambda_0) \right\| = 0 .$$

This equality is equivalent to (20.31).

It remains to remark that $x_*^{(k_0)}(\lambda_0) = u_*^{(k_0)}(\lambda_0)$ and, apart from the notation, equalities (20.28) and (20.31) coincide. ∎

Now let Λ be the real axis. Theorem 20.3 then means that the implicit function $x_*(\lambda)$ has a representation

$$x_*(\lambda) = x_0 + x_1(\lambda - \lambda_0) + \ldots + x_m(\lambda - \lambda_0)^m + o\left(\left| \lambda - \lambda_0 \right|^m\right), \quad (20.32)$$

where x_k are certain elements of E. Let $\mathscr{L}$ denote the curve

$$x = x_0(\lambda) = x_0 + x_1(\lambda - \lambda_0) + \ldots + x_m(\lambda - \lambda_0)^m \quad (20.33)$$

in the space $\Lambda + E$, and $\mathscr{L}_\varepsilon$ the set of points $\{\lambda, x\} \in \Lambda + E$ such that $\|x - x_0(\lambda)\| \leqq \varepsilon \|\lambda - \lambda_0\|^m$ (see Fig. 20.1); we shall call this set a *horn** with aperture ε. The geometrical meaning of Theorem 20.3 is that, for any $\varepsilon > 0$ and λ sufficiently close to λ_0, the curve $x = x(\lambda)$ lies inside the horn $\mathscr{L}_\varepsilon$.

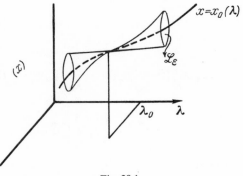

Fig. 20.1

---

* See Il'f and Petrov [1]; the role of the concept "hoof" will be described by the authors elsewhere.

Exercise 20.12. Show that for sufficiently small ε the operator

$$\tilde{T}(\lambda, x) = \{\lambda; T(\lambda, x)\},$$

defined in $\Lambda + E$, where $T(\lambda, x)$ is the operator (20.4), maps the intersection of the horn $\mathscr{L}_\varepsilon$ ($\varepsilon > 0$) with the strip $|\lambda| \leqq \delta$ into itself.

20.4. Method of undetermined coefficients.

Formula (29.28) is convenient for various theoretical constructions. For the actual construction of the Taylor expansion of implicit functions one usually employs the method of undetermined coefficients.

Assume that the conditions of Theorem 20.3 are satisfied. Then the implicit function $x_*(\lambda)$ may be expressed as

$$x_*(\lambda) = x_0 + x_1(\lambda - \lambda_0) + \ldots + x_m(\lambda - \lambda_0) + o\left(\|\lambda - \lambda_0\|^m\right) \quad (20.34)$$

and to compute the derivatives (20.28) we must determine the homogeneous operators $x_1, \ldots, x_m$. We shall use the representation (20.23) of the operator $f(\lambda, x)$:

$$f(\lambda, x) = f(\lambda_0, x) + \sum_{k+l=1}^{m} V_{k,l}(\lambda - \lambda_0, x - x_0) + \omega(\lambda, x). \quad (20.35)$$

Set

$$x_0(\lambda) = x_0 + x_1(\lambda - \lambda_0) + \ldots + x_m(\lambda - \lambda_0), \quad (20.36)$$

$$f_0(\lambda, x) = \sum_{k+l=1}^{m} V_{k,l}(\lambda - \lambda_0, x - x_0). \quad (20.37)$$

In proving Theorem 20.3, we in fact showed that all derivatives of order up to m of the polynomial $f_0\left[\lambda, x_0(\lambda)\right]$ vanish at the point λ_0. Let

$$f_0\left[\lambda, x_0(\lambda)\right] = y_1(\lambda - \lambda_0) + \ldots + y_m(\lambda - \lambda_0) + o\left(\|\lambda - \lambda_0\|^m\right).$$

It follows from (20.35) and (20.36) that for $l = 1, \ldots, m$

$$y_l v = \sum_{k+\alpha_1+2\alpha_2+\ldots+l\alpha_l=l} \frac{(\alpha_1 + \ldots + \alpha_l)!}{\alpha_1! \ldots \alpha_l!} \tilde{V}_{k, \alpha_1+\ldots+\alpha_l} v^k (x_1 v)^{\alpha_1} \ldots (x_l v)^{\alpha_l},$$

and, since

$$V_{0,1} = f_x'(\lambda_0, x_0), \qquad \Gamma = \left[f_x'(\lambda_0, x_0)\right]^{-1},$$

it follows that

$$\Gamma y_l v = x_l v + \sum_{k + \alpha_1 + \ldots + (l-1)\alpha_{l-1} = l} \frac{(\alpha_1 + \ldots + \alpha_{l-1})}{\alpha_1! \ldots \alpha_{l-1}!} \times$$

$$\times \Gamma \tilde{V}_{k, \alpha_1 + \ldots + \alpha_{l-1}} v^k (x_1 v)^{\alpha_1} \ldots (x_{l-1} v)^{\alpha_{l-1}}.$$

By equating the right-hand sides to zero, we obtain recurrence relations for the operators $x_1, \ldots, x_m$. Note that these recurrence relations have the form

$$x_l = \Delta_l(x_1, \ldots, x_{l-1}), \tag{20.38}$$

where

$$\Delta_l(x_1, \ldots, x_{l-1}) v = - \sum_{k + \alpha_1 + \ldots + (l-1)\alpha_{l-1} = l} \frac{(\alpha_1 + \ldots + \alpha_{l-1})}{\alpha_1! \ldots \alpha_{l-1}!} \times$$

$$\times \Gamma \tilde{V}_{k, \alpha_1 + \ldots + \alpha_{l-1}} v(x_1 v)^{\alpha_1} \ldots (x_{l-1} v)^{\alpha_{l-1}}. \tag{20.39}$$

Obviously,

$$x_1 v = - \Gamma \tilde{V}_{1,0} v,$$

$$x_2 v = - \Gamma \tilde{V}_{2,0} v^2 - \Gamma \tilde{V}_{1,1} v \cdot x_1(v) - \Gamma \tilde{V}_{0,2} [x_1(v)]^2.$$

Exercise 20.13. Write down formulas for x_3 and x_4.

Exercise 20.14. Let $f(\lambda, x) = f(x) - \lambda$. In this important special case, determination of the implicit function is precisely local inversion of the operator $f(x)$. Derive formulas (Lagrange's formulas) expressing the derivatives of the inverse to $f(x)$ in terms of the derivatives of $f(x)$.

Exercise 20.15. Let $f(\lambda, x) = x - \lambda Ax$, where A has derivatives of order up to m at $x_0 = 0$. Derive formulas for the derivatives of the implicit function $x_*(\lambda)$ at $\lambda_0 = 0$.

20.5. *Successive approximations.* To approximate an implicit function we need only approximate the solution of equation (20.1). All the methods described in previous chapters for approximate solution of nonlinear operator equations therefore apply to the approximation of implicit functions.

The proof of Theorem 20.1 was based on the fact that, for sufficiently small a_0, b_0 x_0 (the choice of a_0 depends on b_0) and for $\lambda \in S(\lambda_0, a_0)$ the operator (20.4) satisfies the assumptions of the contracting mapping principle in the ball $S(x_0, b_0)$, with contraction factor

$$q = \|\Gamma\| [p(b_0) + p_1(a_0, b_0)] < 1 . \tag{20.40}$$

This implies

Theorem 20.4. *Under the assumptions of Theorem* 20.1, *the successive approximations*

$$x_{n+1}(\lambda) = x_n(\lambda) - \Gamma f[\lambda, x_n(\lambda)] \qquad (\lambda \in S(\lambda_0, a_0) ; \ n = 0, 1, 2, \ldots) \tag{20.41}$$

are defined and converge to the implicit function $x_*(\lambda)$, *provided* $x_0(\lambda) \in S(x_0, b_0)$. *When this is so,*

$$\|x_n(\lambda) - x_*(\lambda)\| \leqq q^n \|x_0(\lambda) - x_*(\lambda)\| \tag{20.42}$$

and

$$\|x_n(\lambda) - x_*(\lambda)\| \leqq \frac{q^n}{1 - q} \|x_1(\lambda) - x_0(\lambda)\| \leqq \frac{q^n}{1 - q} \|\Gamma\| \cdot \|f[\lambda, x_0(\lambda)]\| .$$

$$\tag{20.43}$$

Now consider the iterative processes

$$x_{n+1}(\lambda) = x_n(\lambda) - [f_x'(\lambda, x_0)]^{-1} f[\lambda, x_n(\lambda)] \qquad (n = 0, 1, 2, \ldots) \tag{20.44}$$

and

$$x_{n+1}(\lambda) = x_n(\lambda) - \{f_x'[\lambda, x_n(\lambda)]\}^{-1} f[\lambda, x_n(\lambda)] \quad (n = 0, 1, 2, \ldots) . \tag{20.45}$$

These formulas correspond to approximation of implicit functions by the modified and ordinary Newton–Kantorovich methods, respectively (§ 11 and § 12). In studying the processes (20.44) and (20.45), we find it convenient to investigate convergence questions independently of the general theorems proved in Chapter 3.

Note, first and foremost, that the assumptions of Theorem 20.1 are not sufficient for these processes to be feasible.

Assume that the derivatives $f_x'(\lambda, x_0)$ $(\lambda \in S(\lambda_0, a))$ exist, and that for $\lambda \in S(\lambda_0, \rho)$, $x_1, x_2 \in S(x_0, r)$

$$\|f(\lambda, x_1) - f(\lambda, x_2) - f_x'(\lambda_0, x_0)(x_1 - x_2)\| \leqq p(\rho, r) \|x_1 - x_2\| , \tag{20.46}$$

where

$$c_0 = \lim_{\rho, r \to 0} \|\Gamma\| p(\rho, r) < 1 , \tag{20.47}$$

and assume, moreover, that $f(\lambda, x_0)$ and $f_x'(\lambda, x_0)$ are continuous at λ_0. Then, for sufficiently small a_0, b_0 and ε (the choice of a_0 depends on b_0 and ε) and for $\lambda \in S(\lambda_0, a_0)$ the operator

$$T(\lambda. x) = x - [f_x'(\lambda, x_0)]^{-1} f(\lambda, x) \qquad (20.48)$$

satisfies the assumptions of the contracting mapping principle, with constant

$$q = \frac{\|\Gamma\|}{1 - \varepsilon \|\Gamma\|} p(a_0, b_0) < 1. \qquad (20.49)$$

In fact, assume that

$$\varepsilon \|\Gamma\| < 1, \quad \frac{\|\Gamma\|}{1 - \varepsilon \|\Gamma\|} p(a_0, b_0) < 1,$$

and that for $\lambda \in S(\lambda_0, a_0)$

$$\|f_x'(\lambda, x_0) - f_x'(\lambda_0, x_0)\| < \varepsilon, \quad \|f(\lambda, x_0)\| \leq \frac{(1 - q) b_0 (1 - \varepsilon \|\Gamma\|)}{\|\Gamma\|}.$$

Then for $\lambda \in S(\lambda_0, a_0)$ and $x_1, x_2 \in S(x_0, b_0)$

$$\|T(\lambda, x_1) - T(\lambda, x_2)\| = \|x_1 - x_2 - [f_x'(\lambda, x_0)]^{-1} [f(\lambda, x_1) -$$

$$- f(\lambda, x_2)]\| \leq \|[f_x'(\lambda, x_0)]^{-1}\| \cdot \|f(\lambda, x_1) - f(\lambda, x_2) -$$

$$- f_x'(\lambda, x_0)(x_1 - x_2)\| \leq \frac{\|\Gamma\|}{1 - \varepsilon \|\Gamma\|} p(a_0, b_0) \|x_1 - x_2\| = q \|x_1 - x_2\|$$

and

$$\|T(\lambda, x) - x_0\| \leq qb_0 + \frac{\|\Gamma\|}{1 - \varepsilon \|\Gamma\|} \|f(\lambda, x_0)\| \leq b_0.$$

These arguments immediately imply

Theorem 20.5. *Assume that* (20.46) *holds, and let* $f(\lambda, x_0)$, $f_x'(\lambda, x_0)$ *be continuous at the point* λ_0.

Then for $\lambda \in S(\lambda_0, a_0)$ *the successive approximations* (20.44) *are defined and converge to the implicit function* $x_*(\lambda)$, *provided* $x_0(\lambda) \in S(x_0, b_0)$. *When this is so,*

$$\|x_n(\lambda) - x_*(\lambda)\| \leq q^n \|x_0(\lambda) - x_*(\lambda)\| \qquad (20.50)$$

and

$$\|x_n(\lambda) - x_*(\lambda)\| \leq \frac{q^n}{1 - q}\|x_1(\lambda) - x_0(\lambda)\| \leq$$

$$\leq \frac{q^n}{1 - q} \cdot \frac{\|\Gamma\|}{1 - \varepsilon\|\Gamma\|}\|f[\lambda, x_0(\lambda)]\|. \quad (20.51)$$

We cite one more theorem on the iterative process (20.45).

Theorem 20.6. *Let the derivative* $f_x'(\lambda, x)$ *exist in a neighborhood of the point* $\{\lambda_0, x_0\}$, *satisfying a Hölder condition*

$$\|f_x'(\lambda, x_1) - f_x'(\lambda, x_2)\| \leq c\|x_1 - x_2\|^\delta, \quad (20.52)$$

where $0 < \delta \leq 1$. *Let* $f(\lambda, x_0)$ *and* $f_x'(\lambda, x_0)$ *be continuous in* λ *at the point* λ_0.

Then there exist a_0 *and* b_0 *such that for* $\lambda \in S(\lambda_0, a_0)$ *the successive approximations* (20.45) *are defined and converge to the implicit function* $x_*(\lambda)$, *provided* $x_0(\lambda) \in S(x_0, b_0)$. *The rate of convergence is described by the inequality*

$$\|x_n(\lambda) - x_*(\lambda)\| \leq \frac{1}{k}[k\|x_0(\lambda) - x_*(\lambda)\|]^{(1+\delta)^n}, \quad (20.53)$$

where k *is a constant which can be taken arbitrarily close to* $[\|\Gamma\|c(1 + \delta)^{-1}]^{1/\delta}$.

We shall prove only the estimate (20.53). First, assume that a_0 and b_0 have been chosen so that, for $\lambda \in S(\lambda_0, a_0)$, $x \in S(x_0, b_0)$, the operators $f_x'(\lambda, x)$ have continuous inverses with norms not exceeding $(1 + \varepsilon)\|\Gamma\|$, where ε is arbitrarily small. Inequality (20.52) then implies the estimate

$$\|x_n(\lambda) - x_*(\lambda)\| =$$

$$= \|x_{n-1}(\lambda) - x_*(\lambda) - \{f_x'[\lambda, x_{n-1}(\lambda)]\}^{-1}\{f[\lambda, x_{n-1}(\lambda)] -$$

$$- f[\lambda, x_*(\lambda)]\}\| \leq (1 + \varepsilon)\|\Gamma\|\left\|\int_0^1 \{f_x'[\lambda, x_{n-1}(\lambda)] -$$

$$- f_x'[\lambda, (1 - \theta)x_*(\lambda) + \theta x_{n-1}(\lambda)]\}\, d\theta\right\|\ \|x_{n-1}(\lambda) - x_*(\lambda)\| \leq$$

$$\leqq (1 + \varepsilon) \|\Gamma\| c \int_0^1 (1 - \theta)^\delta \, d\theta \, \|x_{n-1}(\lambda) - x_*(\lambda)\|^{1+\delta} =$$

$$= \frac{(1 + \varepsilon) \|\Gamma\| c}{1 + \delta} \|x_{n-1}(\lambda) - x_*(\lambda)\|^{1+\delta},$$

which we rewrite as

$$k\|x_n(\lambda) - x_*(\lambda)\| \leqq [k\|x_{n-1}(\lambda) - x_*(\lambda)\|]^{1+\delta}, \qquad (20.54)$$

where

$$k = \left[\frac{(1 + \varepsilon) \|\Gamma\| c}{1 + \delta}\right]^{1/\delta}.$$

The estimate (20.53) now follows from (20.54).

To complete Theorem 20.6, we remark that for $x_0(\lambda) \equiv x_0$ and sufficiently small a_0 the product $k\|x_*(\lambda) - x_0\|$ may be assumed arbitrarily small. The estimate (20.53) therefore implies that the iterations (20.45) are "superconvergent."

Exercise 20.16. Discuss the following iterative process for approximation of implicit functions:

$$x_{n+1}(\lambda) = x_n(\lambda) - \Gamma [I + \Delta_n + \ldots + \Delta_n^{m(n)}] f[\lambda, x_n(\lambda)],$$

where Δ_n is the linear operator defined by

$$\Delta_n = \{f_x'[\lambda, x_n(\lambda)] - f_x'(\lambda_0, x_0)\} \Gamma,$$

and $m(n)$ an increasing sequence.

Exercise 20.17. Let $E = F$ be a Hilbert space. Discuss the following iterative processes for approximation of implicit functions:

$$a) \quad x_{n+1}(\lambda) = x_n(\lambda) - \frac{\|f[\lambda, x_n(\lambda)]\|^2}{\|[f_x'(\lambda_0, x_0)]^* f[\lambda, x_n(\lambda)]\|^2} \, f[\lambda, x_n(\lambda)],$$

$$b) \quad x_{n+1}(\lambda) = x_n(\lambda) - \frac{\|f[\lambda, x_n(\lambda)]\|^2}{\|[f_x'(\lambda, x_0)]^* f[\lambda, x_n(\lambda)]\|^2} \, f[\lambda, x_n(\lambda)],$$

$$c) \quad x_{n+1}(\lambda) = x_n(\lambda) - \frac{\|f[\lambda, x_n(\lambda)]\|^2}{\|\{f_x'[\lambda, x_n(\lambda)]\}^* f[\lambda, x_n(\lambda)]\|^2} \, f[\lambda, x_n(\lambda)].$$

We conclude this subsection with a few remarks.

One method for approximation of implicit functions is first to set up differential equations, whose solutions are the required functions.

This method (with λ a scalar) was discussed briefly in subsection 13.4. When equation (20.1) has the specific form

$$x = A(\lambda, x),$$

where $A(\lambda, x)$ is a compact operator, projection methods (see §19) or the modification of the Newton–Kantorovich method described in subsection 12.5 are often convenient for approximation of the implicit function $x(\lambda)$. We emphasize, in particular, that the methods considered in subsection 12.5 require inversion of operators in finite-dimensional subspaces (i.e., matrix inversion).

20.6. *Asymptotic approximations to implicit functions.** Let $\omega(\lambda)$ be a nonnegative functional, continuous at 0 and nonzero at every nonzero point. Examples are the functionals ($\alpha > 0$), $\|\lambda\|^{\alpha} \, | \ln \|\lambda\| \, |$ ($\alpha > 0$, which are all functions of $\|\lambda\|$, and the rather more complicated functions $\|P\lambda\|^{\alpha} + \|\lambda - P\lambda\|^{\alpha_2}$ ($\alpha_1, \alpha_2 > 0$, P a projection in Λ), and so on.

A function $y(\lambda)$ is said to be an *asymptotic approximation of order ω* to the function $x(\lambda)$ at λ_0 if

$$x(\lambda) = y(\lambda) + o\left[\omega(\lambda - \lambda_0)\right],$$

i.e.,

$$\lim_{\|\lambda - \lambda_0\| \to 0} \frac{\|x(\lambda) - y(\lambda)\|}{\omega(\lambda - \lambda_0)} = 0.$$

It is clear that $x(\lambda)$ is also an asymptotic approximation of order ω to $y(\lambda)$ at λ_0. If

$$x(\lambda) = y(\lambda) + O\left[\omega(\lambda - \lambda_0)\right],$$

i.e.,

$$\overline{\lim_{\|\lambda - \lambda_0\| \to 0}} \frac{\|x(\lambda) - y(\lambda)\|}{\omega(\lambda - \lambda_0)} < \infty,$$

we shall call $y(\lambda)$ an *asymptotic approximation of preorder ω* to $x(\lambda)$ at λ_0.

If the function $x(\lambda)$ has a Taylor expansion (20.34) in a neighborhood of λ_0, the function (20.33) is an asymptotic approximation of order

* Zabreiko and Krasnosel'skii [5, 6].

$\omega(\lambda) = \|\lambda\|^m$ to $x(\lambda)$, or, as one often says in such cases, of order m. Thus the method of subsection 20.4 may be regarded as the construction of asymptotic approximations of order m at λ_0 to the implicit function $x(\lambda)$, provided the latter has derivatives of order up to m at λ_0 (e.g., if the conditions of Theorem 20.3 hold). It is clear that if $x(\lambda)$ has derivatives of order up to m at λ_0, then any asymptotic approximation to $x(\lambda)$ of order m has derivatives of order up to m at λ_0, and these derivatives coincide with those of $x(\lambda)$.

In this subsection we intend to show that iterative processes of the type studied in subsection 20.5 lead easily to various asymptotic approximations to implicit functions.

Exercise 20.18. Under the assumptions of Theorem 20.1, show that a function $x(\lambda)$ is an asymptotic approximation of order (preorder) ω at a point λ_0 to an implicit function $x_*(\lambda)$ if and only if the zero function is an asymptotic approximation of order (preorder) ω at λ_0 to the function $f[\lambda, x(\lambda)]$.

Consider the auxiliary equation

$$x = \Pi(\lambda, x), \tag{20.55}$$

where $\Pi(\lambda, x)$ $(\lambda \in S(\lambda_0, a), x \in S(x_0, b))$ is an operator with values in E, $\Pi(\lambda_0, x_0) = x_0$. Let us assume that equation (20.55) has a unique solution $x_*(\lambda) \in S(x_0, b)$ when $\lambda \in S(\lambda_0, a)$ and, moreover, that when $\lambda \in S(\lambda_0, a_0)$, $x \in S(x_0, b_0)$

$$\|\Pi(\lambda, x) - \Pi[\lambda, x_*(\lambda)]\| \leqq q\|x - x_*(\lambda)\|, \tag{20.56}$$

where $q < 1$, and

$$\|\Pi(\lambda, x) - x_0\| \leqq b_0 ; \tag{20.57}$$

where a_0 and b_0 are numbers from the intervals $(0, a]$ and $(0, b]$, respectively. It is then obvious that the successive approximations

$$x_{n+1}(\lambda) = \Pi[\lambda, x_n(\lambda)] \qquad (\lambda \in S(\lambda_0, a_0) ; \; n = 0, 1, 2, \dots)$$

converge to $x_*(\lambda)$ if $x_0(\lambda) \in S(x_0, b_0)$, and

$$\|x_n(\lambda) - x_*(\lambda)\| \leqq q^n \|x_0(\lambda) - x_*(\lambda)\| .$$

We are interested in the behavior of the functions

$$\delta_n(\lambda) = \|x_n(\lambda) - x_*(\lambda)\|, \quad \mu_n(\lambda) = \|x_n(\lambda) - x_0\|, \quad \mu_*(\lambda) = \|x_*(\lambda) - x_0\|$$

as $\lambda \to \lambda_0$.

Lemma 20.1. *Let* $\Pi(\lambda, x)$ *satisfy the condition*

$$\|\Pi(\lambda, x) - x_0\| \leqq \pi(\lambda - \lambda_0, \|x - x_0\|)$$

$$(\lambda \in S(\lambda_0, a) \; ; \; x \in S(x_0, b)), \qquad (20.58)$$

where $\pi(\lambda, t)$ *is a nondecreasing function of t. Let the initial approximation* $x_0(\lambda) \in S(x_0, b)$ *satisfy the condition*

$$\mu_0(\lambda) = \|x_0(\lambda) - x_0\| \leqq x_0 \omega(\lambda - \lambda_0) \qquad (\lambda \in S(\lambda_0, a)). \quad (20.59)$$

Finally, let

$$\pi[v, c\omega(v)] \leqq k(c) \, \omega(v) \qquad (\|v\| \leqq a, \, 0 < c < \infty). \qquad (20.60)$$

Then

$$\mu_n(\lambda) \leqq c_n \omega(\lambda - \lambda_0) \qquad (n = 1, 2, \ldots), \quad \mu_*(\lambda) \leqq c_* \omega(\lambda - \lambda_0). \quad (20.61)$$

Proof. The first inequality of (20.61) is proved by induction: if $u_i(\lambda) \leqq c_i \omega(\lambda - \lambda_0)$, then, by (20.58),

$$\mu_{i+1}(\lambda) =$$

$$= \|\Pi[\lambda \; ; \; x_i(\lambda)] - x_0\| \leqq \pi[\lambda - \lambda_0, \mu_i(\lambda)] \leqq \pi[\lambda - \lambda_0, c_i \omega(\lambda - \lambda_0)]$$

and (20.60) implies the estimate $\mu_{i+1}(\lambda) \leqq c_{i+1} \omega(\lambda - \lambda_0)$.

Obviously,

$$\mu_*(\lambda) = \|\Pi | \lambda, x_*(\lambda)] - x_0\| \leqq$$

$$\leqq \|\Pi[\lambda, x_*(\lambda)] - \Pi[\lambda, x_0(\lambda)]\| + \|\Pi[\lambda, x_0(\lambda)] - x_0\|$$

and by (20.56) and (20.58)

$$\mu_*(\lambda) \leqq q\|x_*(\lambda) - x_0(\lambda)\| + \pi[\lambda - \lambda_0, \|x_0(\lambda) - x_0\|] \leqq$$

$$\leqq q\mu_*(\lambda) + q\mu_0(\lambda) + \pi[\lambda - \lambda_0, \mu_0(\lambda)],$$

and hence, by (20.59),

$$\mu_*(\lambda) \leqq \frac{qc_0}{1 - q} \omega(\lambda - \lambda_0) + \frac{1}{1 - q} \pi[\lambda - \lambda_0, c_0 \omega(\lambda - \lambda_0)].$$

This inequality, together with (20.60), implies the second estimate of (20.61). ∎

We now proceed to investigate the implicit function $x_*(\lambda)$ defined by equation (20.1), under the assumption (20.2). For simplicity, we shall assume that the operator $f(\lambda, x)$ is differentiable with respect to x in a neighborhood of $\{\lambda_0, x_0\}$.

Theorem 20.7. Assume that $f'_x(\lambda, x)$ satisfies the condition

$$\|f'_x(\lambda, x) - f'_x(\lambda_0, x_0)\| \leqq \phi(\lambda - \lambda_0, \|x - x_0\|) \qquad (20.62)$$

$$(\lambda \in S(\lambda_0, a) \; ; \; x \in S(x_0, b)),$$

where $\phi(v, t)$ is a nondecreasing function of t, continuous at $\{0, 0\}$,

$$\phi\left[v, c\omega(v)\right] \leqq k_1(c)\,\omega_0(v) \qquad (\|v\| \leqq a \; ; \; 0 < c < \infty). \qquad (20.63)$$

Let

$$\|f(\lambda, x_0)\| \leqq c\omega(\lambda - \lambda_0) \qquad (\lambda \in S(\lambda_0, a)). \qquad (20.64)$$

Finally, let the function $x_0(\lambda)$ satisfy condition (20.59) (e.g., $x_0(\lambda) \equiv x_0$).

Then the function $x_n(\lambda)$ defined by the recurrence relation (20.41) is an asymptotic approximation at λ_0 to the implicit function $x_*(\lambda)$ of preorder $\omega_0^n\omega$:

$$\varlimsup_{\|\lambda - \lambda_0\| \to 0} \frac{\|x_n(\lambda) - x_*(\lambda)\|}{\left[\omega_0(\lambda - \lambda_0)\right]^n \omega(\lambda - \lambda_0)} < \infty. \qquad (20.65)$$

Proof. Setting

$$\Pi(\lambda, x) = x - \Gamma f(\lambda, x)$$

we shall prove that this operator satisfies the assumptions of Lemma 20.1, with

$$\pi(v, t) = q_0 t + c_0 \omega(v), \qquad (20.66)$$

where q_0, c_0 are constants. It follows from (20.62) and (20.64) that $f(\lambda, x)$ satisfies the assumptions of Theorem 20.1. Thus, for $\lambda \in S(\lambda_0, a_0)$ the operator $\Pi(\lambda; x)$ is a contraction, with constant q_0, in the ball $S(x_0, b_0)$ (a_0 and b_0 are sufficiently small, a_0 depends on b_0). This implies inequalities (20.56) and (20.57). Condition (20.58) follows from the chain of inequalities

$$\|\Pi(\lambda, x) - x_0\| \leqq \|\Pi(\lambda, x) - \Pi(\lambda, x_0)\| + \|\Pi(\lambda, x_0) - x_0\| \leqq$$

$$\leqq q_0\|x - x_0\| + c_0\omega(\lambda - \lambda_0) = \pi(\lambda - \lambda_0, \|x - x_0\|).$$

Inequality (20.59) is one of the assumptions of the theorem, and inequality (20.60) is obvious for the function (20.66). Lemma 20.1 now implies the estimate (20.61).

To prove (20.65), we shall use induction to establish the estimate

$$\delta_n(\lambda) = \|x_n(\lambda) - x_*(\lambda)\| \leqq d_n[\omega_0(\lambda - \lambda_0)]^n \omega(\lambda - \lambda_0) \quad (20.67)$$

$$(n = 0, 1, 2, \ldots).$$

For $n = 0$, this follows from (20.59) and the second inequality of (20.61):

$$\delta_0(\lambda) = \|x_0(\lambda) - x_*(\lambda)\| \leqq \|x_0(\lambda) - x_0\| + \mu_*(\lambda) \leqq (c_0 + c_*) \omega(\lambda - \lambda_0).$$

Assume inequality (20.67) to hold for $n = l$. The identity

$$x_{l+1}(\lambda) - x_*(\lambda) = \Gamma \{ f_x'(\lambda_0, x_0) [x_l(\lambda) - x_*(\lambda)] - f[\lambda, x_l(\lambda)] +$$

$$+ f[\lambda, x_*(\lambda)] \}$$

implies that*

$$\delta_{l+1}(\lambda) \leqq \|\Gamma\| \sup_{0 \leqq \theta \leqq 0} \|f_x'(\lambda_0, x_0) - f_x'[\lambda, (1 - \theta) x_l(\lambda) + \theta x_*(\lambda)]\| \delta_l(\lambda),$$

and, by (20.62) and (20.61),

$$\delta_{l+1}(\lambda) \leqq \|\Gamma\| \phi [\lambda - \lambda_0, \mu_l(\lambda) + \mu_*(\lambda)] \delta_l(\lambda) \leqq$$

$$\leqq \|\Gamma\| \phi [\lambda - \lambda_0, (c_l + c_*) \omega(\lambda)] \delta_l(\lambda).$$

It now follows from the induction hypothesis and from (20.63) that

$$\delta_{l+1}(\lambda) \leqq \|\Gamma\| k_1 (c_l + c_*) \omega_0(\lambda - \lambda_0) d_l[\omega_0(\lambda - \lambda_0)]^l \omega(\lambda - \lambda_0) =$$

$$= d_{l+1} [\omega_0(\lambda - \lambda_0)]^{l+1} \omega(\lambda - \lambda_0). \quad \blacksquare$$

Exercise 20.19. Show that Theorem 20.7 remains valid if condition (20.62) is replaced by the inequalities

$$\|f(\lambda_0, x_1) - f(\lambda_0, x_2) - f_x'(\lambda_0, x_0)(x_1 - x_2)\| \leqq p(r) \|x_1 - x_2\| \quad (x_1, x_2 \in S(x_0, r)),$$

---

* Here we use the following elementary fact. If a continuous nonlinear operator A is differentiable at all interior points of a segment $(1 - \theta) x_0 + \theta x_1$ $(0 < \theta < 1)$, there exists $\theta^* \in (0, 1)$ such that

$$\|Ax_1 - Ax_0\| \leqq \|A'[(1 - \theta^*) x_0 + \theta^* x_1]\| \cdot \|x_1 - x_0\|.$$

$$\left\| [f(\lambda, x_1) - f(\lambda, x_2)] - [f(\lambda_0, x_1) - f(\lambda_0, x_2)] \right\| \leq p_1(\lambda - \lambda_0, r) \left\| x_1 - x_2 \right\|$$

$$(\lambda \in S(\lambda_0, \rho) \; ; \; x_1, x_2 \in S(x_0, r)),$$

$$p(t) + p_1(v, t) \leq \phi(v, t).$$

Finally, let condition (20.62) have the form

$$\left\| f_x'(\lambda, x) - f_x'(\lambda_0, x_0) \right\| \leq M(\left\| \lambda - \lambda_0 \right\| + \left\| x - x_0 \right\|),$$

and

$$\left\| f(\lambda, x_0) \right\| \leq M_1 \left\| \lambda - \lambda_0 \right\|, \qquad \left\| x_0(\lambda) - x_0 \right\| \leq M_2 \left\| \lambda - \lambda_0 \right\|.$$

We can then set

$$\phi(v, t) = M \left\| v \right\| + Mt, \qquad \omega(v) = \omega_0(v) = \left\| v \right\|,$$

and it will follow from Theorem 20.7 that any fixed approximation $x_n(\lambda)$ defined by (20.41) is an asymptotic approximation of preorder $n + 1$ (and thus, of course, an asymptotic approximation of any order $n + 1 - \varepsilon$) to the implicit function $x_*(\lambda)$. It should be emphasized that this construction of asymptotic approximations of arbitrary order requires neither the implicit function $x_*(\lambda)$ nor the operator $f(\lambda, x)$ to have derivatives of high orders. If the function $x_*(\lambda)$ has derivatives of order up to n at the point λ_0, the same holds for all asymptotic approximations of order n. The procedure of constructing approximations (20.41) and differentiating them may be used to compute the derivatives of the implicit function at the point λ_0. These remarks are also valid for other constructions considered in this subsection for asymptotic approximations to implicit functions.

Exercise 20.20. Estimate the preorder of the asymptotic approximations (20.41) to the implicit function $x_*(\lambda)$, if

$$\left\| f_x'(\lambda, x) - f_x'(\lambda_0, x_0) \right\| \leq M(\left\| \lambda - \lambda_0 \right\|^{\alpha_1} + \left\| x - x_0 \right\|^{\alpha_2}),$$

$$\left\| f(\lambda, x_0) \right\| \leq M_1 \left\| \lambda - \lambda_0 \right\|^{\alpha_3}, \quad \left\| x_0(\lambda) - x_0 \right\| \leq M_2 \left\| \lambda - \lambda_0 \right\|^{\alpha_4},$$

where $0 < \alpha_1, \alpha_2, \alpha_3, \alpha_4 \leq 1$.

The proof of the following theorem is analogous to that of Theorem 20.7.

Theorem 20.8. *Let* $f_x'(\lambda, x)$ *satisfy the condition*

$$\|f_x'(\lambda, x) - f_x'(\lambda, x_0)\| \leq \phi(\lambda - \lambda_0, \|x - x_0\|) \qquad (20.68)$$

$$(\lambda \in S(\lambda_0, a) \; ; \; x \in S(x_0, b)),$$

where $\phi(v, t)$ *is a nondecreasing function of* t, *continuous at* $\{0, 0\}$ *and satisfying condition* (20.63). *Assume that condition* (20.64) *holds, and let the function* $x_0(\lambda)$ *satisfy condition* (20.59). *Finally, let* $f_x'(\lambda, x_0)$ *be continuous at* λ_0.

Then the function $x_n(\lambda)$ *defined by the recurrence relations* (20.44) *is an asymptotic approximation at* λ_0 *to the implicit function* $x_*(\lambda)$, *of preorder* $\omega_0^n \omega$.

Exercise 20.21. Show that Theorem 20.8 remains valid if condition (20.68) is replaced by the inequality

$$\|f(\lambda, x_1) - f(\lambda, x_2) - f_x'(\lambda, x_0)(x_1 - x_2)\| \leq \phi(\lambda - \lambda_0, r)\|x_1 - x_2\|$$

$$(\lambda \in S(\lambda_0, \rho) \; ; \quad x \in S(x_0, r)).$$

Theorem 20.9. *Under the assumptions of Theorem* 20.6, *assume that condition* (20.64) *holds, and let the initial approximation* $x_0(\lambda)$ *satisfy condition* (20.59).

Then the function $x_n(\lambda)$ *defined by the recurrence relations* (20.45) *is an asymptotic approximation at* λ_0 *to the implicit function* $x_*(\lambda)$, *of preorder* $\omega^{(1+\delta)^n}$:

$$\varlimsup_{\lambda \to \lambda_0} \frac{\|x_n(\lambda) - x_*(\lambda)\|}{[\omega(\lambda - \lambda_0)]^{(1+\delta)^n}} < \infty. \qquad (20.69)$$

Proof. By Theorem 20.6, the approximations $x_n(\lambda)$ $(n = 1, 2, \ldots;$ $\lambda \in S(\lambda_0, a_0)$, a_0 sufficiently small) are defined and satisfy inequalities (20.53). To prove the theorem, therefore, it will suffice to show that

$$\|x_0(\lambda) - x_*(\lambda)\| \leq d_0 \omega(\lambda - \lambda_0).$$

By (20.59), this, in turn, will follow from the simple inequality

$$\|x_*(\lambda) - x_0\| \leq c_* \omega(\lambda - \lambda_0).$$

This is easily proved by the same reasoning as in the proof of Lemma 20.1 (the details are left to the reader). ∎

The estimate (20.69) indicates that the method (20.45) yields asymptotic approximations to the implicit function of "super-rapidly" increasing orders. For example, if $\omega(v) = M\|v\|$, the function $x_n(\lambda)$ is an asymptotic approximation of order $(1 + \delta)^n$.

Exercise 20.22. Estimate the order of the asymptotic approximation $x_n(\lambda)$ to the implicit function $x_*(\lambda)$ defined by (20.45), when the Hölder condition (20.52) is replaced by the condition

$$\|f'_x(\lambda, x_1) - f'_x(\lambda, x_2)\| \leqq \psi(\lambda - \lambda_0 \,;\, r \,;\, \|x_1 - x_2\|) \qquad (\lambda \in S(\lambda_0, \rho) \,;\, x_1, x_2 \in S(x_0, r)).$$

20.7. *Formal series and formal implicit functions.** Let Λ and E be real or complex Banach spaces. An expression of the form

$$x(v) = \sum_{k=1}^{\infty} A_k v, \tag{20.70}$$

where A_k is a homogeneous k-form (a homogeneous form of order k mapping Λ into E) is known as a *formal power series* "mapping" Λ into E. Note that the definition includes no stipulations about convergence of the formal power series.

The set of formal series mapping Λ into E forms a linear space under the natural (term-by-term) definitions of multiplication by a scalar and addition; the zero of this space is the formal power series of zero k-forms.

The *order* $\sigma(x)$ of a formal power series is the minimal order of the nonzero forms appearing in the series. The order $\sigma(0)$ of the zero series is by definition ∞. The following relations are obvious:

$$\sigma(\alpha x) = \sigma(x) \quad (\alpha \neq 0), \quad \sigma(x_1 + x_2) \geqq \min\left\{\sigma(x_1), \sigma(x_2)\right\}.$$

Polynomials (see subsection 20.2) may be regarded as a special class of formal power series; the degree of a polynomial should not be confused with its order (as a series).

It is sometimes convenient to consider the sum of infinitely many formal series $x_l(v)$ $(l = 1, 2, \ldots)$. A sum of this type is defined when $\sigma_l = \sigma(x_l) \to \infty$; when this condition holds, for each fixed k there are only finitely many series

$$x_l(v) = \sum_{k=1}^{\infty} A_k^l(v) \qquad (l = 1, 2, \ldots)$$

---

* For the case of finite dimensional spaces see Bochner and Martin [1].

in which the operators A_k^l are not zero; denote their sum by B_k. The sum $x_1(v) + \ldots + x_l(v) + \ldots$ is defined as the formal series

$$x_1(v) + \ldots + x_l(v) + \ldots = \sum_{k=1}^{\infty} B_k v.$$

If v is a scalar parameter, the formal power series has the form

$$x(v) = a_1 v + a_2 v^2 + \ldots + a_k v^k + \ldots,$$

where $a_1, a_2, \ldots, a_k, \ldots$ are arbitrary elements of E.

We shall need formal power series in two variables:

$$g(v, h) = \sum_{k+m=1}^{\infty} V_{k,m}(v, h), \qquad (20.71)$$

where $V_{k,m}$ are homogeneous (k, m)-forms. One works with these series in exactly the same way as with series in one variable. The *order* $\sigma(g)$ of a series of this type is the minimal number $k + m$ such that $V_{k,m}$ is not identically zero.

An important concept here is that of the superposition of two power series, or the substitution of one formal series in another. The series $g[v, x(v)]$ obtained by substituting the series (20.70) for h in (20.71) is defined as the infinite sum

$$g[v, x(v)] = \sum_{k+m=1}^{\infty} V_{k,m}[v, x(v)] \qquad (20.72)$$

of formal power series $V_{k,m}[v, x(v)]$. In analogy with (20.15), each of the latter is defined by

$$v_{k,m}(v) = V_{k,m}[v, x(v)] = \sum_{j=1}^{\infty} v_{k,m,j}(v), \qquad (20.73)$$

where

$$v_{k,m,j}(v) = \sum_{\alpha_1 + 2\alpha_2 + \ldots + j\alpha_j = j} \frac{(\alpha_1 + \ldots + \alpha_j)!}{\alpha_1! \ldots \alpha_j!} \tilde{V}_{k,m} v^k (A_1 v)^{\alpha_1} \ldots (A_j v)^{\alpha_j}.$$

$$(20.74)$$

The sum (20.72) is meaningful, since $\sigma(v_{k,m}) \geqq k + m \to \infty$. The cumbersome formulas (20.74) need not deter the reader—the coefficients of the formal series $g[v, x(v)]$ are obtained via the usual algebraic rules for substitution of series.

We now return to our study of implicit functions. In this subsection we shall look for a solution of type (20.70) to the equation

$$g(v, h) = 0, \tag{20.75}$$

where $g(v, h)$ is the formal series (20.71). Naturally, this solution is defined to be a formal series $h_*(v)$ which, when substituted in the left-hand side of (20.75), gives the zero series.

If the series (20.71) converges for small v, h, the linear operator $V_{0,1}$ coincides with the derivative of the sum with respect to h at the point $\{0, 0\}$, and we shall often use the notation

$$V_{0,1} = g_h'(0, 0).$$

Theorem 20.10. *Let the operator* $g_h'(0, 0)$ *have a continuous inverse* Γ. *Then equation* (20.75) *has a unique solution* $h_*(v)$ *in the class of formal power series.*

For the proof, substitute the series (20.70) with undetermined k-forms A_k in the series (20.71), and equate the coefficients to zero. This gives equations of the form

$$V_{0,1}A_k = U(A_1, \ldots, A_{k-1}) \qquad (k = 1, 2, \ldots),$$

from which the unique operators $A_1, A_2, \ldots$ are determined successively.

20.8. *Analyticity of an implicit function.* In this subsection we shall study the implicit function $x_*(\lambda)$ defined by equation (20.1) assuming the condition (20.2), when the operator $f(\lambda, x)$ is analytic at the point $\{\lambda_0, x_0\}$. If the operator (20.3) is also defined, it is easy to see that all assumptions of Theorem 20.3 are satisfied, and the implicit function $x_*(\lambda)$ therefore exists in a neighborhood $S(\lambda_0, a_0)$ of λ_0, and is differentiable there. It turns out that $x_*(\lambda)$ is analytic at λ_0 (see, e.g., Hille and Phillips [1]). In complex spaces Λ, E analyticity is a simple consequence of differentiability. In real spaces, one can, for example, first construct the analytic continuation of $f(\lambda, x)$ to the complex extensions $\Lambda + i\Lambda$ and $E + iE$ of the spaces Λ and E, and then employ the previous argument. For the sake of convenience, we formulate these assertions as a theorem.

Theorem 20.11. *Let the operator* $f(\lambda, x)$ *be analytic at the point* $\{\lambda_0, x_0\}$, *and assume that the operator* (20.3) *exists.*

Then the implicit function $x_*(\lambda)$ *exists, is unique, and is analytic at* λ_0.

Exercise 20.23. Show that any k-form A_k mapping a real space Λ into a real space E may be extended by the formula

$$A_k(x_1 + ix_2) = \sum_{l=0}^{k} \frac{k!}{l!(k-l)!} i^{k-l} \tilde{A}_k x_1^l x_2^{k-l}$$

to a k-form mapping $\Lambda + i\Lambda$ into $E + iE$. Estimate the factor by which the norm of A_k is increased by this procedure (Taylor [2], Alexiewicz and Orlicz [1]).

Under the assumptions of Theorem 20.11, the implicit function has derivatives of all orders. To determine them, or, equivalently, to determine the k-forms A_k in the representation

$$x_*(\lambda) = x_0 + \sum_{k=1}^{\infty} A_k(\lambda - \lambda_0) \qquad (20.76)$$

one can use the method of undetermined coefficients, described in subsection 20.4.

If the series

$$h(v) = \sum_{k=1}^{\infty} A_k v \qquad (20.77)$$

is regarded as a formal power series, it is a solution of equation (20.75), where $g(\lambda, h)$ is the expansion of the operator $f(\lambda_0 + v, x_0 + h)$, regarded as a formal power series in the two variables v, h.

It thus follows from Theorem 20.10 that the forms A_k are uniquely determined when the method of undetermined coefficients is used. This reasoning implies the following obvious, though important

Theorem 20.12. *Under the assumptions of Theorem* 20.11, *let the formal series* (20.77) *be a solution of equation* (20.75), *where* $g(v, h)$ *is the expansion of the operator* $f(\lambda_0 + v, x_0 + h)$ *regarded as a formal series.*

Then the series (20.77) *converges for small v, and* (20.76) *defines a unique implicit function.*

Theorems 20.11 and 20.12 are insufficient for many applications. It is important to have estimates for domains in the space Λ in which the implicit function $x_*(\lambda)$ is defined and analytic. We shall look for a ball $S(\lambda_0, r_*)$ in a domain of this type. The basic procedure for estimating the radius r_* of this ball is based on the widely used method of majorant series.*

---

* Our presentation will employ several arguments due to Gel'man.

Let

$$\eta = \sum_{k=1}^{\infty} \beta_k \xi^k \qquad (20.78)$$

be a scalar series with nonnegative coefficients β_k (ξ being a scalar variable), which is a majorant of the expansion (20.76) of the implicit function, in the sense that

$$\|A_k\| \leqq \beta_k \qquad (k = 1, 2, \ldots). \qquad (20.79)$$

If r_0 is the radius of convergence of the series (20.78), it follows from (20.79) that the series (20.76) converges in norm for $\|\lambda - \lambda_0\| < r_0$. Of course, we do not yet know whether the implicit function $x_*(\lambda)$ is defined at all interior points of the ball $S(\lambda_0, r_0)$.

Exercise 20.24. Construct a scalar function $f(\xi, \eta)$ ($f(0, 0) = 0$, $f'_\eta(0, 0) = 1$), analytic in the bicylinder $|\xi| < 1, |\eta| < 1$, which has no analytic continuation beyond the cylinder, while the implicit function $\eta_*(\xi)$ defined by the equation $f(\xi, \eta) = 0$ has an expansion which is convergent throughout the complex ξ-plane.

In determining the radius r_*, one must bear in mind that if, as is usually the case, the operator $f(\lambda, x)$ is defined for $\lambda \in S(\lambda_0, a)$, $x \in S(x_0, b)$, then necessarily $r_* \leqq a$.

Finally, the radius r_* must be chosen in such a way that for $\lambda \in S(\lambda_0, r_*)$ the sum of the series (20.76) belongs to the ball $S(x_0, b)$. To estimate this sum it suffices to notice that

$$\left\| \sum_{k=1}^{\infty} A_k(\lambda - \lambda_0) \right\| \leqq \sum_{k=1}^{\infty} \beta_k \|\lambda - \lambda_0\|^k = \eta(\|\lambda - \lambda_0\|),$$

where

$$\eta(\xi) = \sum_{k=1}^{\infty} \beta_k \xi^k \qquad (0 \leqq \xi < r_0).$$

$\eta(\xi)$ is an increasing function. Assume that it is defined on the interval $[0, r_{\eta, b}]$, where it is bounded above by b. Then the sum of the series (20.76) defines an implicit function $x_*(\lambda)$ in tha ball $S(\lambda_0, r_*)$ if

$$r_* \leqq a, \quad r_* \leqq r_{\eta, b}, \quad r_* < r_0. \qquad (20.80)$$

We shall now describe a general construction for majorant series of type (20.78).

Set

$$\Pi(\lambda, x) = x - \Gamma(\lambda; x) f(\lambda, x), \tag{20.81}$$

where

$$\Gamma(\lambda; x) = \sum_{i+j=1}^{\infty} \Gamma_{i, j}(\lambda - \lambda_0, x - x_0), \qquad \Gamma_{0, 1} = \left[f_x'(\lambda_0, x_0) \right]^{-1}$$

(for example, $\Gamma(\lambda; x)$ may be defined as one of the operators $\left[f_x'(\lambda_0, x_0) \right]^{-1}$, $\left[f_x'(\lambda, x_0) \right]^{-1}$, $\left[f_x'(\lambda, x) \right]^{-1}$). Then

$$\Pi(\lambda; x) = \sum_{i+j=1}^{\infty} \Pi_{i, j}(\lambda - \lambda_0, x - x_0), \tag{20.82}$$

and $\Pi_{0, 1} = 0$.

Suppose that we have estimates

$$\left\| \Pi_{i, j} \right\|_* \leqq t_{i, j}, \tag{20.83}$$

we may assume that $t_{0, 1} = 0$. Define a scalar function $t(\xi, \eta)$ by

$$t(\xi, \eta) = \sum_{i+j=1}^{\infty} t_{i, j} \xi^i \eta^j \tag{20.84}$$

and consider the equation

$$\eta = t(\xi, \eta). \tag{20.85}$$

We shall say that this equation is a majorant of the equation

$$x = \Pi(\lambda; x).$$

In the subsequent reasoning the domain of convergence of the series (20.84) will be immaterial; moreover, one can construct a majorant equation in the class of equations in formal power series.

It follows from Theorem 20.10 (or Theorem 20.11) that there exists a unique implicit function

$$\eta_*(\xi) = \sum_{i=1}^{\infty} \beta_i \xi^i, \tag{20.86}$$

defined by equation (20.85) (with the condition $\eta_*(0) = 0$). One can show that the series (20.86) is a majorant for the implicit function; the proof follows from the formulas $A_1 = 0$, $\beta_1 = 0$,

$$A_{k+1}v = \sum_{i+\alpha_1+2\alpha_2+\ldots+k\alpha_k=k+1} \frac{(\alpha_1 + \ldots + \alpha_k)!}{\alpha_1!\ldots\alpha_k!} \times$$

$$\times \tilde{\Pi}_{i,\,\alpha_1+\ldots+\alpha_k}v^i(A_1v)^{\alpha_1}\ldots(A_kv)^{\alpha_k}, \quad (20.87)$$

$$\beta_{k+1} = \sum_{i+\alpha_1+2\alpha_2+\ldots+k\alpha_k=k+1} \frac{(\alpha_1 + \ldots + \alpha_k)!}{\alpha_1!\ldots\alpha_k!} t_{i,\,\alpha_1+\ldots+\alpha_k}\beta_1^{\alpha_1}\ldots\beta_k^{\alpha_k}.$$
$$(20.88)$$

For example, let

$$\|\Pi_{ij}\|_* \leqq Mu^iv^j.$$

Then

$$t(\xi, \eta) = \sum_{i,\,j=0}^{\infty} Mu^iv^j\xi^i\eta^j - M - Mv\eta =$$

$$= \frac{M}{(1 - u\xi)(1 - v\eta)} - M - Mv\eta.$$

The majorant equation

$$\eta = \frac{M}{(1 - u\xi)(1 - v\eta)} - M - Mv\eta$$

has two solutions. One of them—the solution $\eta_*(\xi)$—vanishes when $\xi = 0$:

$$\eta_*(\xi) = \frac{1 - (1 - u\xi)^{-1/2}\left[1 - u(1 + 2Mv)^2\,\xi\right]^{1/2}}{2v(1 + Mv)}.$$

It is clear that the radius of convergence r_0 of the series expansion (20.86) of this function is conditioned by its singular points, and so

$$r_0 = \frac{1}{u(1 + 2Mv)^2}.$$

Since $\eta_*(\xi)$ is an increasing function on $[0, r_0]$, it follows that $r_{\eta,\,b} = r_0$ if

$$\frac{1}{2v(1 + Mv)} \leqq b.$$

But if the converse inequality is true, then $r_{\eta,\,b} < r_0$ and $r_{\eta,\,b}$ is defined by the equation $\eta_*(r) = b$. Solving this equation, we get

$$r_{\eta,\,b} = \frac{b - b^2 v - M b^2 v^2}{u\,(1 - bv)\,(M + b + Mbv)}\,.$$

Exercise 20.25. Estimate r_* for the following form of inequality (20.83):

$$\|\Pi_{i,\,j}\|_* \leqq M \frac{u^i v^j}{i!\,j!}\,.$$

§ 21. Finite systems of equations

21.1. *Review of ring theory.* In this section we shall utilize several general theorems from the theory of commutative rings (see, e.g., van der Waerden [1], Grave [2], Zariski and Samuel [1], Walker [1], Hodge and Pedoe [1]).

Let K be a commutative ring with identity 1. Throughout the sequel, we shall assume that the elements $a \in K$ are all of *infinite* (*additive*) *order*, i.e., for any nonzero a in K and any integer m,

$$m \cdot a = a + a + \dots a \neq 0\,.$$

The set K_0 of invertible elements $a \in K$ forms a multiplicative group. It is clear that $ab \in K_0$ implies that $a,\ b \in K_0$.

Two elements a and b in K are said to be *associates* if $a = \varepsilon b$, where $\varepsilon \in K_0$. If $a = bc$, b is a *divisor* of a, or b divides a. The elements of K_0 divide any $a \in K$; divisors from K_0 are said to be *trivial* divisors.

A ring K is called an *integral domain* if $ab = 0$ implies that either $a = 0$ or $b = 0$. An element a of an integral domain is said to be *irreducible* if it is not invertible and has no nontrivial divisors.

An integral domain is said to be a *unique factorization* (or *Gaussian*) *domain* if: 1) every element a of K can be expressed as a product

$$a = \varepsilon a_1 \dots a_r\,, \tag{21.1}$$

where $\varepsilon \in K_0$; $a_1,\ \dots,\ a_r$ are irreducible elements; 2) if

$$a = \varepsilon' a_1' \dots a_{r'}'$$

is another representation of this type, then $r = r'$ and the factors $a_1',\ \dots,$

a'_r (possibly rearranged) are associates of $a_1, \ldots, a_r$, respectively. It is not hard to see that an integral domain K in which each element a has a representation (21.1) *is an integral domain if and only if K has the following property: if $p \in K$ is an irreducible divisor of ab, then p divides either a or b*. If K is a unique factorization domain, the greatest common divisor (g.c.d.) $d = (a_1, \ldots, a_r)$ may be defined (up to associates) for any elements $a_1, \ldots, a_r \in K$.

Let K be an integral domain. $K[x]$ denotes the ring of polynomials

$$f(x) = a_0 x^m + a_1 x^{m-1} + \ldots + a_m \tag{21.2}$$

with coefficients $a_0, a_1, \ldots, a_m$ in K. If $f(x)$ is a polynomial (21.2) and $a_0 \neq 0$, the number m is called the *degree* of $f(x)$. It is clear that the degree of the product of two polynomials of Kx is the sum of their degrees. It follows immediately that $K[x]$ is also an integral domain. We denote the ring of polynomials in n variables $x_1, \ldots, x_n$ over K by $K[x_1, \ldots, x_n]$.

The following lemma, which is easily proved by induction, will play an important role in the sequel.

Lemma 21.1. *Let $f(x)$ and $g(x)$ be two polynomials in $K[x]$, of degrees m and n, respectively:*

$$f(x) = a_0 x^m + a_1 x^{m-1} + \ldots + a_m, \quad g(x) = b_0 x^n + b_1 x^{n-1} + \ldots + b_n;$$

$$\tag{21.3}$$

let $k = \max \{m - n + 1, 0\}$.

Then $K[x]$ contains polynomials $q(x)$ and $r(x)$ such that

$$b_0^k f(x) = q(x) g(x) + r(x); \tag{21.4}$$

where either $r(x)$ is of degree $< n$ or $r(x) \equiv 0$. The polynomials $q(x)$ and $r(x)$ are unique.

Now let K be a unique factorization domain and $f(x)$ the polynomial (21.2). The g.c.d. $\kappa(f) = (a_0, \ldots, a_m)$ of the coefficients $a_0, \ldots, a_m$ is called the *content* of the polynomial f. If $\kappa(f) \in K_0$, the polynomial $f(x)$ is said to be *primitive*. Every polynomial $f(x)$ in $K[x]$ can be expressed as a product

$$f(x) = \kappa(f) \cdot \tilde{f}(x), \tag{21.5}$$

where $\tilde{f}$ is primitive; $\tilde{f}$ is known as the reduction of f [it is unique up to a factor in K_0]. We have the following equalities:

$$\kappa(fg) = \varepsilon_1 \kappa(f) \kappa(g), \tag{21.6}$$

$$(\widetilde{fg}) = \varepsilon_2 \tilde{f}\tilde{g}, \tag{21.7}$$

where ε_1, $\varepsilon_2 \in K_0$. Both these equalities follow from *Gauss's Lemma*, which states that *the product fg of primitive polynomials f and g is a primitive polynomial.*

Exercise 21.1. Prove Gauss's Lemma.

Lemma 21.2. *Let K be a unique factorization domain.*
Then the polynomial ring $K[x]$ is also a unique factorization domain.
Proof. We first show (by induction) that every polynomial f over K may be expressed as

$$f = \varepsilon f_1 \ldots f_r, \tag{21.8}$$

where $\varepsilon \in K_0$ and $f_1, \ldots, f_r$ are irreducible polynomials (i.e., irreducible elements of the ring $K[x]$).

This is true for polynomials of degree zero (i.e., elements of K), for K is a unique factorization domain. Assume that every polynomial of degree smaller than m has a representation (21.8); we claim that the same holds for any polynomial $f(x)$ of degree m.

Let $\kappa(f) = c_1 \ldots, c_l$ be the factorization of the content of f into irreducible elements. If $\tilde{f}$ is irreducible, then

$$f(x) = c_1 \ldots c_l \tilde{f}(x)$$

is a representation (21.8) of f. If $\tilde{f}$ is reducible, say $\tilde{f} = g_1 g_2$, then both polynomials g_1 and g_2 are primitive and their degrees are positive, therefore smaller than m. By the induction hypothesis, g_1 and g_2 may be factorized into irreducible elements. But this gives an irreducible factorization of the polynomial $\tilde{f} : \tilde{f}(x) = f_1(x) \ldots f_r(x)$, and so

$$f(x) = c_1 \ldots c_l f_1(x) \ldots f_r(x)$$

is a representation (21.8) of the polynomial f.

Now let $p(x)$ be an irreducible polynomial dividing a product $f(x) g(x)$ of polynomials $f(x)$ and $g(x)$ ($p(x) \lambda(x) = f(x) g(x)$), and suppose that $p(x)$ does not divide $f(x)$. Let $\mathfrak{N}$ denote the set of all nonzero polynomials $d(x)$ of the form

$$d(x) = a(x) f(x) + b(x) p(x) \qquad (a(x), b(x) \in K[x]). \tag{21.9}$$

Let $d_0(x)$ be a polynomial of minimal degree in $\mathfrak{N}$. If the degree of $d_0(x)$

is positive, then, by Lemma 21.1,

$$\delta p(x) = q(x) d_0(x) + r(x), \tag{21.10}$$

where $\delta \in K$, $q(x)$, $r(x) \in K[x]$ and either the degree of $r(x)$ is smaller than that of $d_0(x)$ or $r(x) \equiv 0$. But it follows from (21.9) and (21.10) that $r(x) \in \mathfrak{N}$ if $r(x) \not\equiv 0$. Thus $r(x) \equiv 0$ and $\delta p(x) \equiv q(x) d_0(x)$. By (21.7), this equality implies that $d_0(x)$ is a divisor of $p(x)$. Similarly one proves that $d_0(x)$ divides $f(x)$. Since $p(x)$ is irreducible and not a divisior of $f(x)$, it follows that $d_0(x)$ is of degree zero, i.e., $d_0(x) = d_0 \in K$.

Let

$$d_0 = a_0(x) f(x) = b_0(x) p(x).$$

Multiplying this equality by $g(x)$, we get

$$d_0 g(x) = [a_0(x) \lambda(x) + b_0(x) g(x)] p(x),$$

whence it follows, by (21.7), that $p(x)$ is a divisor of $g(x)$. ∎

Exercise 21.2. Prove that if K is a unique factorization domain, so is $K[x_1, \ldots, x_n]$.

Since the ring of polynomials $K[x]$ over a unique factorization domain K is itself a unique factorization domain, any polynomials $f_1(x), \ldots, f_r(x)$ have a well-defined (up to a factor in K_0) g.c.d. $d = (f_1, \ldots, f_r)$. Obviously,

$$(f_1, \ldots, f_r) = \varepsilon (\kappa(f_1), \ldots, \kappa(f_r))(\tilde{f}_1, \ldots, \tilde{f}_r) \quad (\varepsilon \in K_0). \tag{21.11}$$

An immediate corollary of Lemma 21.2 is

Lemma 21.3. *Let $f(x)$ and $g(x)$ be two polynomials in $K[x]$, of degrees m and n, respectively.*

Then the degree of the g.c.d. $d(x)$ of $f(x)$ and $g(x)$ is positive if and only if there exist polynomials $f_1(x)$ and $g_1(x)$ of degrees smaller than m and n, respectively, such that

$$g_1(x) f(x) = f_1(x) g(x). \tag{21.12}$$

Exercise 21.3. Show that the g.c.d. $d(x)$ of two polynomials $f(x)$ and $g(x)$, whose degrees are m and n, respectively, has degree $\geq k$ if and only if there exist polynomials $f_1(x)$ and $g_1(x)$ of degrees $m - k$ and $n - k$, respectively, satisfying (21.12).

Let $f(x)$ and $g(x)$ be the polynomials (21.3). The determinant

$$
R = \begin{vmatrix}
a_0 & a_1 & \dots & & a_m & 0 & \dots & 0 \\
0 & a_0 & a_1 \dots & & & a_m & \dots & 0 \\
 & & & \cdot & \cdot & \cdot & & \\
0 & 0 & 0 & & a_0 & a_1 & \dots & a_m \\
b_0 & b_1 & \dots & b_n & 0 & \dots & & 0 \\
0 & b_0 & b_1 & \dots & b_n & 0 & \dots & 0 \\
 & & & \cdot & \cdot & \cdot & & \\
0 & 0 & 0 \dots & & b_0 & b_1 & \dots & b_n
\end{vmatrix}
\begin{matrix} \left.\rule{0pt}{30pt}\right\} n \text{ rows} \\ \\ \left.\rule{0pt}{30pt}\right\} m \text{ rows} \end{matrix}
\qquad (21.13)
$$

is called the *resultant* (in Sylvester's form) of the polynomials $f(x)$ and $g(x)$.

Theorem 21.1. *Let* $f(x)$ *and* $g(x)$ *be two polynomials in* $K[x]$ *of degrees m and n, respectively, and* $d(x)$ *their g.c.d.*
Then the degree of $d(x)$ *is positive if and only if* $R = 0$.
Proof. Let

$$
f_1(x) = -u_0 x^{m-1} - u_1 x^{m-2} - \dots - u_{m-1},
$$

$$
g_1(x) = v_0 x^{n-1} + v_1 x^{n-2} + \dots + v_{n-1}.
$$

Then the equality (21.12) is clearly equivalent to the system of equations

$$
\left.
\begin{aligned}
a_0 v_0 && + \, b_0 u_0 && = 0, \\
a_1 v_0 + a_0 v_1 && + \, b_1 u_0 + b_0 u_1 && = 0, \\
a_2 v_0 + a_1 v_1 + a_0 v_2 && + \, b_2 u_0 + b_1 u_1 + b_0 u_2 && = 0, \\
\cdot \quad \cdot \quad \cdot \quad \cdot \quad \cdot && \cdot \quad \cdot \quad \cdot \quad \cdot \quad \cdot && \\
a_m v_{n-1} && + \, b_n u_{m-1} && = 0.
\end{aligned}
\right\}
\qquad (21.14)
$$

Now the assertion of the theorem holds if and only if this system of linear equations in $v_0, v_1, \dots, v_{n-1}, u_0, u_1, \dots, u_{m-1}$ has a nontrivial solution. This is true if and only if its determinant—which is precisely R—vanishes. ∎

Exercise 21.4. Let $f(x)$ and $g(x)$ be two polynomials. Show that there exist polynomials $a(x)$ and $b(x)$ such that

$$
Rd(x) = a(x)f(x) + b(x)g(x).
$$

Exercise 21.5. Find necessary and sufficient conditions on the coefficients of two polynomials $f(x)$ and $g(x)$ for their g.c.d. to have degree k.

Let $m_1, \ldots, m_r$ be positive integers. A system of polynomials $R_1, \ldots,$ R_s in all the coefficients of polynomials $f_1(x), \ldots, f_r(x)$ of degrees $m_1, \ldots,$ m_r, respectively,

$$f_1(x) = a_0^{(1)} x^{m_1} + a_1^{(1)} x^{m_1 - 1} + \ldots + a_{m_1}^{(1)},$$

$$\cdot \ \cdot \ \cdot \ \cdot \ \cdot \ \cdot \ \cdot \ \cdot \ \cdot \ \cdot \ \cdot \ \cdot \ \cdot \ \cdot \ \cdot$$

$$f_r(x) = a_0^{(r)} x^{m_r} + a_0^{(r)} x^{m_r - 1} + \ldots + a_{m_r}^{(r)},$$

is called an $(m_1, \ldots, m_r)$-*system of resultants*, if the equalities

$$R_1 = 0, \ldots, R_s = 0$$

constitute a necessary and sufficient condition for the polynomials $f_1(x), \ldots, f_r(x)$ to have a g.c.d. of positive degree.

Theorem 21.2. For any set of integers $m_1, \ldots, m_r,$ *there exists an* $(m_1, \ldots, m_r)$-*system of resultants* $R_1, \ldots, R_s.$

Our proof of this theorem is due to Kronecker. Let $K^* = K[w_1, \ldots,$ $w_{r-1}]$ be the ring of polynomials in $r - 1$ variables $w_1, \ldots, w_{r-1}$ over K, and let $f_1(x), \ldots, f_r(x)$ be polynomials of degrees $m_1, \ldots, m_r$. Set

$$f(x) = w_1 f_1(x) + \ldots + w_{r-1} f_{r-1}(x), \quad g(x) = f_r(x); \quad (21.15)$$

$f(x)$ and $g(x)$ are polynomials over K^*. It is easy to see that the g.c.d. $d^*(x)$ of $f(x)$ and $g(x)$ is a polynomial over K, and it coincides (up to a factor in K_0) with the g.c.d. of $f_1(x), \ldots, f_r(x)$. This implies the following assertion: a necessary and sufficient condition for the degree of the g.c.d. of $f_1(x), \ldots, f_r(x)$ to be positive is that the resultant R^* of the polynomials (21.15) vanish in K^*.

The resultant R^* is clearly a homogeneous polynomial in $w_1, \ldots, w_{r-1}$ (over K), of degree l, say:

$$R^* = \sum_{\alpha_1 + \ldots + \alpha_{r-1} = l} R_{\alpha_1, \ldots, \alpha_{r-1}} w_1^{\alpha_1} \ldots w_{r-1}^{\alpha_{r-1}}.$$

It vanishes in K^* if and only if $R_{\alpha_1, \ldots, \alpha_{r-1}} = 0$ for all $\alpha_1, \ldots, \alpha_{r-1}$. Thus the polynomials $R_{\alpha_1, \ldots, \alpha_{r-1}}$ form an $(m_1, \ldots, m_{r-1})$-system of resultants. ∎

How can one calculate the g.c.d. of polynomials $f_1(x), f_2(x), \ldots, f_r(x)$? By (21.11), it suffices to find the g.c.d. of the polynomials $\tilde{f}_1(x), \tilde{f}_2(x), \ldots,$ $\tilde{f}_r(x)$. To do this, we use Lemma 21.1; we may confine ourselves to the case of two polynomials.

Let $f_1(x)$ and $f_2(x)$ be two given polynomials. Using Lemma 21.1, we can construct polynomials $g_0(x), g_1(x), \ldots, g_{k+1}(x)$ such that $g_0(x) =$

$f_1(x), g_1(x) = f_2(x)$ and

$$\left.\begin{array}{l} \delta_1 g_0(x) = q_1(x) g_1(x) + g_2(x), \\ \cdots \cdots \cdots \cdots \cdots \cdots \cdots \\ \delta_k g_{k-1}(x) = q_k(x) g_k(x) + g_{k+1}(x) ; \end{array}\right\} \quad (21.16)$$

where the degrees of the polynomials $g_1(x), \ldots, g_k(x)$ form a strictly decreasing sequence, and the polynomial $g_{k+1}(x)$ either vanishes or is of degree 0: $g_{k+1}(x) \equiv g_{k+1} \in K$.

Theorem 21.3. The g.c.d. $d(x)$ *of the reductions* $\tilde{f}_1(x)$ *and* $\tilde{f}_2(x)$ *of the polynomials* $f_1(x)$ *and* $f_2(x)$ *has positive degree if and only if* $g_{k+1} = 0$. *If* $g_{k+1} = 0$, *then* $d(x) = \tilde{g}_k(x)$.

Proof. Let $d(x)$ be the g.c.d. of $\tilde{f}_1(x)$ and $\tilde{f}_2(x)$. Then it follows from (21.16) that $d(x)$ divides all the polynomials $g_0(x), g_1(x), \ldots, g_{k+1}(x)$. If the degree of $d(x)$ is positive, then $g_{k+1}(x) = 0$, since $d(x)$ is a divisor of $g_{k+1}(x) = g_{k+1}$. Since the polynomial $d(x)$ is primitive, it must divide $\tilde{g}_k(x)$ (by (21.7)).

Now let $g_{k+1} = 0$. Then $\tilde{g}_k(x)$ divides $\delta_k g_{k-1}(x)$ and, by (21.7), also divides $\tilde{g}_{k-1}(x)$. But then it follows from (21.16) that $\tilde{g}_k(x)$ divides $\delta_{k-1} g_{k-2}(x)$, whence, again by (21.7), it follows that $\tilde{g}_k(x)$ divides $\tilde{g}_{k-2}(x)$. Continuing in this way, we see, after finitely many steps, that $\tilde{g}_k(x)$ is a common divisor of $\tilde{g}_0(x)$ and $\tilde{g}_1(x)$. Since every divisor of $\tilde{g}_0(x)$ and $\tilde{g}_1(x)$ is also a divisor of $\tilde{g}_k(x)$, it follows that $\tilde{g}_k(x)$ is the g.c.d.∎

Let $f(x)$ be the polynomial (21.2). Its [(formal)] *derivative* $f'(x)$ is defined as the polynomial

$$f'(x) = ma_0 x^{m-1} + (m-1) a_1 x^{m-2} + \ldots + a_{m-1}. \quad (21.17)$$

It is easy to see that

$$[f(x) + g(x)]' = f'(x) + g'(x), \qquad [f(x) g(x)]' = f'(x) g(x) + f(x) g'(x).$$

$$(21.18)$$

Let

$$f(x) = \kappa(f) f_1(x) \ldots f_l(x), \quad (21.19)$$

where $f_1(x), \ldots, f_l(x)$ are irreducible polynomials. Some of the polynomials $f_1(x), \ldots, f_l(x)$ may be identical, and so the factorization (21.19) is conveniently written as

$$f(x) = \kappa(f) f_1^{\gamma_1}(x) \ldots f_r^{\gamma_r}(x), \quad (21.20)$$

where $\gamma_1, \ldots, \gamma_r$ are integers and $f_1(x), \ldots, f_r(x)$ are distinct irreducible polynomials. If $\gamma_1 = \ldots = \gamma_r = 1$, we say that the polynomial $f(x)$ *has no multiple factors.*

It follows from (21.20) and (21.18) that

$$f'(x) = \kappa(f) \sum_{\rho=1}^{r} \gamma_\rho f_1^{\gamma_1}(x) \ldots f_\rho^{\gamma_\rho - 1}(x) \ldots f_r^{\gamma_r}(x) f_\rho'(x). \qquad (21.21)$$

It follows from (21.20) and (21.21) that the g.c.d. of the polynomials $f(x)$ and $f'(x)$ is defined by

$$(f(x), f'(x)) = \kappa(f) f_1^{\gamma_1 - 1}(x) \ldots f^{\gamma_r - 1}(x). \qquad (21.22)$$

This implies that the g.c.d. has positive degree if and only if at least one of the numbers $\gamma_1, \ldots, \gamma_r$ is greater than 1, i.e., the polynomial (21.20) has multiple factors.

The resultant of the polynomials $f(x)$ and $f'(x)$ is called the *discriminant* D of $f(x)$. Theorems 21.1 and formula (21.22) imply

Theorem 21.4. *A polynomial $f(x)$ has no multiple factors if and only if $D \neq 0$.*

21.2. *Rings of power series.* Besides the rings of real and complex numbers, we shall make systematic use of various rings of power series and polynomial rings over rings of power series.

In subsection 20.7 we considered formal power series in Banach spaces. In this section and hereafter we shall use more general classes of formal power series in several scalar variables.

A *formal power series in n variables* $z = \{\zeta_1, \ldots, \zeta_n\}$ is an expression

$$f(z) = \sum_{k_1, \ldots, k_n = 0}^{\infty} a_{k_1, \ldots, k_n} \zeta_1^{k_1} \ldots \zeta_n^{k_n} \qquad (21.23)$$

where $a_{k_1, \ldots, k_n}$ are numerical coefficients. In contrast to subsection 20.7, the coefficient $a_{0,\ldots,0}$ need not vanish. It is often convenient to write the series (21.23) in the form

$$f = \sum_{k=0}^{\infty} f_k, \qquad (21.24)$$

where f_k is a homogeneous polynomial of degree k:

$$f_k = \sum_{k_1 + \ldots + k_n = k} a_{k_1, \ldots, k_n} \zeta_1^{k_1} \ldots \zeta_n^{k_n}. \qquad (21.25)$$

The smallest k for which f_k is not identically zero is called the *order* of the formal power series f, denoted by $\sigma(f)$.

Later on, the following fact will be essential: the series (21.23) converges in some neighborhood of zero if and only if its coefficients satisfy inequalities

$$\left| a_{k_1, \ldots, k_n} \right| \leqq M c_1^{k_1} \ldots c_n^{k_n} ; \tag{21.26}$$

where $M, c_1, \ldots, c_n$ are constants.

Exercise 21.6. Prove that if (21.26) holds, then the series (21.23) converges in the multi-cylinder $|\zeta_1| < 1/c_1, \ldots, |\zeta_n| < 1/c_n$. Is the converse true?

The set of formal power series forms a ring $K\{z\} = K\{\zeta_1, \ldots, \zeta_n\}$ with respect to the natural operations of addition and multiplication. Note the relations

$$\sigma(f + g) \leqq \min \{\sigma(f), \sigma(g)\} ; \quad \sigma(fg) = \sigma(f) + \sigma(g).$$

As usual, f^{-1} denotes a formal series such that $ff^{-1} = f^{-1}f = 1$, where 1 is regarded as a formal power series. We leave it to the reader to prove that the series (21.23) has an inverse f^{-1} if and only if $a_{0,\ldots,0} \neq 0$; the series f^{-1} may be found by the method of undetermined coefficients, or simply by long division.

Those series (21.23) that converge in a neighborhood of zero form a subring $K\{z\}_a$ of the ring $K\{z\}$. It is not difficult to see that a series f in $K\{z\}_a$ is invertible in $K\{z\}$ if and only if it is invertible in $K\{z\}_a$.

The rings $K\{z\}$ and $K\{z\}_a$ are integral domains. They are also unique factorization domains, but this statement is by no means trivial.

Exercise 21.7. Prove that the rings $K\{\zeta\}$ and $K\{\zeta\}_a$ of series in one variable are unique factorization domains.

The fact that the coefficients $a_{k_1, \ldots, k_n}$ of the series (21.23) are numbers was irrelevant for the definition of a formal power series. The fundamental properties just discussed are also valid when the coefficients $a_{k_1, \ldots, k_n}$ are elements of an arbitrary integral domain K. This more general approach will be used when the coefficients are elements of a certain fixed ring of formal power series.

It is sometimes convenient to interpret the series (21.23) as a formal power series in one of the variables ζ_i, with coefficients in the ring of formal series in the remaining variables. To fix ideas, consider the variable

$\zeta_n = w$; then the series (21.23) becomes

$$f = \sum_{k=0}^{\infty} a_k w^k,$$ (21.27)

where

$$a_k = \sum_{k_1,\ldots,k_{n-1}=0}^{\infty} a_{k_1,\ldots,k_{n-1}} \zeta_1^{k_1} \cdots \zeta_{n-1}^{k_{n-1}}.$$ (21.28)

We shall consider power series of the type

$$f(z, w) = \sum_{k=0}^{\infty} f_k(z) w^k,$$ (21.29)

whose coefficients $f_k(z) = \sum_{m=0}^{\infty} f_{k,m}(z)$ are power series with complex coefficients in n variables $z = \{\varsigma_1, \ldots, \varsigma_n\}$. The series (21.29) is said to be *regular* if the series

$$f(0, w) = \sum_{k=0}^{\infty} f_k(0) w^k$$ (21.30)

is not zero. The smallest k for which $f_k(0) = f_{k,0} \neq 0$ is called the *order of regularity* of the series (21.29). If (21.29) is the expansion of an analytic function $f(z, w)$, it is regular if and only if the function $f(0, w)$ is not identically zero. Its order of regularity coincides with the order of the first nonzero derivative $f_{w^k}^{(k)}(0, 0)$.

Consider the power series

$$f(z, w) = \sum_{k=0}^{\infty} f_k(z) w^k, \qquad g(z, w) = \sum_{k=0}^{\infty} g_k(z) w^k,$$ (21.31)

where

$$f_k(z) = \sum_{m=0}^{\infty} f_{k,m}(z), \qquad g_k(z) = \sum_{m=0}^{\infty} g_{k,m}(z).$$ (21.32)

Can one divide the series $g(z, w)$ by the series $f(z, w)$ (with remainder)? It is immediate that this problem is nontrivial only when $f^{-1}(z, w)$ does not exist (i.e., $f_{0,0} = 0$).

*Lemma 21.4.** Let the series (21.29) be regular, and s its order of regularity.*

* See Bochner and Martin [1].

Then there exists a polynomial

$$r(z, w) = \sum_{k=0}^{s-1} r_k(z) w^k, \tag{21.33}$$

whose coefficients are formal power series in z, such that

$$g(z, w) = q(z, w) f(z, w) + r(z, w), \tag{21.34}$$

where $q(z, w)$ is some formal power series. The polynomial $r(z, w)$ and the series $q(z, w)$ are unique.

If the series (21.31) *converge in some neighborhood of zero, the coefficients $r_k(z)$ $(k = 0, 1, \ldots, s - 1)$ and the series $q(z, w)$ also converge in some neighborhood of zero.*

Proof. The first assertion of the lemma is proved by direct construction of the series $r_k(z)$ and

$$q(z, w) = \sum_{k=0}^{\infty} q_k(z) w^k, \tag{21.35}$$

where

$$q_k(z) = \sum_{m=0}^{\infty} q_{k,m}(z). \tag{21.36}$$

To determine the series $q_k(z)$, we have no need of the values of the series $r_k(z)$ $(k = 0, 1, \ldots, s - 1)$. It follows from (21.34) that the series $q_k(z)$ must satisfy the infinite equation system

$$g_n(z) = \sum_{v=0}^{n} q_v(z) f_{n-v}(z) \qquad (n = s, s + 1, s + 2, \ldots). \tag{21.37}$$

It is easy to see that this system is equivalent to a system

$$g_{n,m}(z) = \sum_{v=0}^{n} \sum_{\mu=0}^{m} q_{v,\mu}(z) f_{n-v, m-\mu}(z)$$

$$(m = 0, 1, 2, \ldots; \; n = s, s + 1, s + 2, \ldots),$$

or, via the chain of equalities

$$\sum_{v=0}^{n} \sum_{\mu=0}^{m} q_{v,\mu}(z) f_{n-v,m-\mu}(z) =$$

$$= \sum_{v=0}^{n} \sum_{\mu=0}^{m-1} q_{v,\mu}(z) f_{n-v,m-\mu}(z) + \sum_{v=0}^{n} q_{v,m}(z) f_{n-v,0}(z) =$$

$$= \sum_{v=0}^{n} \sum_{\mu=0}^{m-1} q_{v,\mu}(z) f_{n-v,m-\mu}(z) + \sum_{v=0}^{n-s-1} q_{v,m}(z) f_{n-v,0}(z) +$$

$$+ q_{n-s,m}(z) f_{s,0},$$

to a system

$$q_{n-s,m}(z) =$$

$$= \frac{1}{f_{s,0}} \left[g_{n,m}(z) - \sum_{v=0}^{n} \sum_{\mu=0}^{m-1} q_{v,\mu}(z) f_{n-v,m-\mu}(z) - \sum_{v=0}^{n-s-1} q_{v,m}(z) f_{n-v,0} \right]$$

$$(n = s, s+1, s+2, \ldots; m = 0, 1, 2, \ldots). \qquad (21.38)$$

Setting $n = s$, $m = 0$ in these equalities, we get

$$q_{0,0} = \frac{1}{f_{s,0}} g_{s,0}. \qquad (21.39)$$

Now replacing $n - s$ by n, we get

$$q_{n,m}(z) =$$

$$= \frac{1}{f_{s,0}} \left[g_{n+s,m}(z) - \sum_{v=0}^{n+s} \sum_{\mu=0}^{m-1} q_{v,\mu}(z) f_{n+s-v,m-\mu}(z) - \right.$$

$$\left. - \sum_{v=0}^{n-1} q_{v,m}(z) f_{n+s-v,0} \right] \qquad (21.40)$$

$$(n = 0, 1, 2, \ldots; m = 0, 1, 2, \ldots).$$

Equation (21.40) may be regarded as a recursive definition of the polynomials $q_{n,m}(z)$. To verify this, we introduce the *index H* of the polynomial $q_{\nu,\mu}(z)$, and note that the index of the polynomial $q_{n,m}(z)$ is strictly greater than the indices of all polynomials $q_{\nu,\mu}(z)$ appearing in the right-hand side of (21.40). It follows that the series $q(z, w)$ is uniquely determined.

The proof of the first assertion is completed by the remark that the polynomial $r(z, w)$ is now determined by (21.34).

Now consider the second assertion of the lemma.

Let

$$| f_{n,m}(z)|, \; | g_{n,m}(z)| \leqq Ma^m b^n | z|^m \qquad (n, m = 0, 1, 2, \ldots), \qquad (21.41)$$

and assume that the numbers L, ξ and η satisfy the inequalities

$$\xi > 1, \; \eta > 1, \; \frac{Mb^s}{|f_{s,0}|} \left[1 + \frac{L\eta^{s+1}}{(\eta - 1)(\xi - 1)} + \frac{L}{\eta - 1} \right] \leqq L. \qquad (21.42)$$

We shall prove by induction that

$$| q_{n,m}(z)| \leqq L(\xi a)^m (\eta b)^n | z|^m \qquad (n, m = 0, 1, 2, \ldots). \qquad (21.43)$$

This will complete the proof of the lemma.

Inequality (21.43) is obviously true for $n, m = 0$. Assume that it has been proved for all n, m such that $\kappa(n, m) < n_0 + (s + 1) m_0$. It then follows from (21.40) that

$$| q_{n_0, m_0}(z)| \leqq$$

$$\leqq \frac{1}{|f_{s,0}|} \left[Ma^{m_0} b^{n_0 + s} + LM \sum_{\nu=0}^{n_0+s} \sum_{\mu=0}^{m_0-1} (\xi a)^\mu (\eta b)^\nu a^{m_0 - \mu} b^{n_0 + s - \nu} + \right.$$

$$\left. + LM \sum_{\nu=0}^{n_0-1} (\xi a)^{m_0} (\eta b)^\nu b^{n_0 + s - \nu} \right] | z|^{m_0} \leqq$$

$$\leqq \frac{M}{|f_{s,0}|} \left[a^{m_0} b^{n_0 + s} + La^{m_0} b^{n_0 + s} \frac{\xi^{m_0} \eta^{n_0 + s + 1}}{(\xi - 1)(\eta - 1)} + \right.$$

$$\left. + La^{m_0} b^{n_0 + s} \frac{\xi^{m_0} \eta^{n_0}}{\xi - 1} \right] | z|^{m_0} \leqq \frac{Mb^s}{|f_{s,0}|} \left[1 + L \frac{\eta^{s+1}}{(\xi - 1)(\eta - 1)} + \right.$$

$$\left. + \frac{L}{\eta - 1} \right] (\xi a)^{m_0} (\eta b)^{n_0} | z|^{m_0} \leqq L(\xi a)^{m_0} (\eta b)^{n_0} | z|^{m_0}. \; ∎$$

The following corollary of Lemma 21.4, usually known as the Weier-strass Preparation Theorem, is important.

Theorem 21.5. *Let the series* (21.29) *be regular, s its order of regularity. Then there is a unique polynomial*

$$*f(z, w) = w^s + a_1(z) w^{s-1} + \ldots + a_s(z),$$ (21.44)

whose coefficients are series in z such that

$$a_1(0) = a_2(0) = \ldots = a_s(0) = 0$$ (21.45)

and

$$f(z, w) = \varepsilon(z, w) *f(z, w),$$ (21.46)

where $\varepsilon(z, w)$ *is an invertible formal power series.*

If the series (21.29) *converges in some neighborhood of zero, the coefficients of the polynomial* (21.46) *and the series* $\varepsilon(z, w)$ *also converge in some neighborhood of zero.*

To prove this theorem, set

$$g(z, w) = w^s.$$ (21.47)

in Lemma 21.4. Then there exist a polynomial $r(z, w)$ of degree $s - 1$ and a series $q(z, w)$ such that

$$w^s = q(z, w) f(z, w) + r(z, w)$$

or

$$q(z, w) f(z, w) = *f(z, w),$$ (21.48)

where

$$*f(z, w) = w^s - r(z, w).$$

It remains to note that the series $q(z, w)$ is invertible, and to set $\varepsilon(z, w) = q^{-1}(z, w)$. ∎

We direct the reader's attention to the fact, important for applications, that the numerical coefficients of the series $a_n(z) = \sum_{m=0}^{\infty} a_{n,m}(z) \, (n = 1, \ldots, s)$ may actually be computed. One can use the method of undetermined coefficients, aided by the equalities

$$a_{n,m}(z) = \sum_{v=0}^{n} \sum_{\mu=0}^{m} q_{v,\mu}(z) f_{n-v,m-\mu}(z) \tag{21.49}$$

$$(n = 1, \ldots, s; \; m = 0, 1, 2, \ldots),$$

where the polynomials $q_{n,m}(z)$ are defined by the recurrence relations

$$q_{0,0}(z) = 1, \tag{21.50}$$

$$q_{n,0}(z) = - \sum_{v=0}^{n-1} q_{v,0}(z) f_{n+s-v,0} \qquad (n = 1, 2, \ldots), \tag{21.51}$$

$$q_{n,m}(z) = - \sum_{v=0}^{n+s} \sum_{\mu=0}^{m-1} q_{v,\mu}(z) f_{n+s-v,m-\mu}(z) - \sum_{v=0}^{n-1} q_{v,m}(z) f_{n+s-v,0}$$

$$(n = 0, 1, 2, \ldots ; \; m = 1, 2, \ldots). \tag{21.52}$$

These formulas follow from (21.40), in the special case $g(z, w) = w^s$. We consider series $f(x) = f(x_1, \ldots, x_n)$ in the ring $K \{x_1, \ldots, x_n\}$, i.e.,

$$f(x) = \sum_{k_1, \ldots, k_n = 0}^{\infty} a_{k_1, \ldots, k_n} x_1^{k_1} \ldots x_n^{k_n} \tag{21.53}$$

with coefficients in an integral domain K. The series (21.53) is said to be *regular in x_n* if the series

$$f(0, \ldots, 0, x_n) = \sum_{k=0}^{\infty} a_{0, \ldots, 0, k} x_n^k, \tag{21.54}$$

obtained from (20.53) by setting $x_1 = \ldots = x_{n-1} = 0$, is not zero. The *order of regularity of the series* (21.53) *in x_n* is defined as the order of regularity of the series (21.54). Not every series (21.53) is regular in one of its variables (an example is the series consisting of the single term $x_1 x_2 .. x_n$).

Now let $x = (x_1, \ldots, x_n)$ and $x' = (x'_1, \ldots, x'_n)$, and consider the linear transformation

$$x = qx' \tag{21.55}$$

or, more explicitly,

$$\left.\begin{aligned} x_1 &= q_{11}x_1' + \ldots + q_{1n}x_n' \\ &\cdot \quad \cdot \quad \cdot \quad \cdot \quad \cdot \quad \cdot \quad \cdot \quad \cdot \quad \cdot \\ x_n &= q_{n1}x_1' + \ldots + q_{nn}x_n' \end{aligned}\right\} \tag{21.56}$$

This change of variables transforms each power series $f(x)$ into a power series $Qf(x') = f(qx')$ in the new variables x'. If the matrix q is non-singular, this transformation is an isomorphism of the rings $K\{x\}$ and $K\{x'\}$ (and, of course, of the rings $K\{x\}_a$ and $K\{x'\}_a$).

Lemma 21.5. *Given any finite sequence of power series* $f_1(x), f_2(x), \ldots, f_r(x)$ *of orders* $s_1, s_2, \ldots, s_r$, *respectively, one can construct a nonsingular matrix* q *such that all power series* $Qf_1(x'), \ldots, Qf_r(x')$ *are regular in* x_n' *with orders of regularity* $s_1, s_2, \ldots, s_r$, *respectively.*

Proof. Let $g(x)$ denote the product of the series $f_1(x), \ldots, f_r(x)$. The order of the series $g(x)$ is $s = s_1 + \ldots + s_r$. To prove the lemma, we shall construct a matrix q such that the series $Qg(x')$ is regular in x_n' with order s.

Let

$$g(x) = g_s(x) + \sum_{k=s+1}^{\infty} g_k(x),$$

where $g_k(x)$ is a homogeneous polynomial of degree k. Obviously,

$$g(qx') = g_s(qx') + \sum_{k=s+1}^{\infty} g_k(qx').$$

It will thus suffice to choose the matrix q in such a way that the coefficient π of $(x_n')^s$ in the polynomial $g_s(qx')$ is not zero. It is not hard to see that

$$\pi = g_s(q_{1n}, \ldots, q_{nn}).$$

Since $g_s(x) \not\equiv 0$, we can choose a nonsingular matrix q such that $\pi \neq 0$. $\blacksquare$

Exercise 21.8. Prove that the ring of formal power series in n variables and the ring of convergent power series in n variables are unique factorization domains.

21.3. *The Newton diagram.** Let K be a ring, and $K^*\{\lambda\}$ the set of all

---

* The method of Newton diagrams plays an important role in many fields of algebra and analysis (see, e.g., Walker [1], Chebotarev [1, 2], Fuks and Shabat [1], Trenogin [3]).

formal power series of the form

$$x(\lambda) = \sum_{k=0}^{\infty} c_k \lambda^{\varepsilon_k}, \tag{21.57}$$

where $c_k \in K$, $0 = \varepsilon_0 < \varepsilon_1 < \varepsilon_2 < \ldots$. $K^*\{\lambda\}$ is a ring under the natural algebraic operations. $K^*\{\lambda\}$ is an integral domain if K is, and, similarly, a unique factorization domain if K is. In particular, if K is a field, then $K^*\{\lambda\}$ is a unique factorization domain.

We shall need certain important subrings of $K^*\{\lambda\}$. If r is a positive integer, $K^r\{\lambda\}$ will denote the subring of series

$$x(\lambda) = \sum_{k=0}^{\infty} c_k \lambda^{k/r}. \tag{21.58}$$

All these subrings are isomorphic; in particular, this is true of the rings $K\{\lambda\} = K^1\{\lambda\}$ of ordinary formal power series. The union $K^\infty\{\lambda\}$ of all the rings $K^r\{\lambda\}$ is also a subring of $K^*\{\lambda\}$.

If the series (21.57), belongs to $K^\infty\{\lambda\}$, there exists a smallest r such that (21.57) belongs to $K^r\{\lambda\}$. We call this number r the *radix* of the series (21.57) (this term is borrowed from Bobrov [1], who uses it in a different connection). If the series (21.57) has radix $r > 1$ and ε is a primitive r-th root of unity, then it is easily seen that the series

$$\left.\begin{aligned} x(\lambda) &= \sum_{k=0}^{\infty} c_k \lambda^{k/r}, \\[2mm] x_1(\lambda) &= \sum_{k=0}^{\infty} \varepsilon^k c_k \lambda^{k/r}, \\[2mm] &\cdot \quad \cdot \quad \cdot \quad \cdot \quad \cdot \quad \cdot \quad \cdot \\[2mm] x_{r-1}(\lambda) &= \sum_{k=0}^{\infty} \varepsilon^{(r-1)k} c_k \lambda^{k/r} \end{aligned}\right\} \tag{21.59}$$

are all different. We shall call the series (21.59) *conjugates.*

Throughout the rest of this subsection K will be the field of complex numbers, and so all our rings will be unique factorization domains.

Consider the equation

$$f(\lambda, x) = 0, \tag{21.60}$$

where $f(\lambda, x)$ is a formal power series

$$f(\lambda, x) = \sum_{m, n = 0}^{\infty} a_{m, n} \lambda^m x^n \tag{21.61}$$

over the complex field. We shall assume that

$$f(0, 0) = 0,$$

i.e., $a_{0, 0} = 0$. A solution of equation (21.60) is a formal series (21.57) in which $c_0 = 0$ and which, when substituted [for x] in the series (21.61), yields the zero series in λ. It is obvious ($c_0 = 0$!) that this substitution is possible.

The assumption that $c_0 = 0$ in the above definition of a solution corresponds to the fact that we wish to find a solution of (21.60) that vanishes for $\lambda = 0$.

Exercise 21.9. Show that equation (21.60) has a unique solution $x_*(\lambda)$ in $K^*\{\lambda\}$, if $a_{0, 1} \neq 0$. Show that $x_*(\lambda) \in K^1\{\lambda\}$.

Another notation for the series (21.61) is

$$f(\lambda, x) = \sum_{n = 0}^{\infty} a_n(\lambda) x^n, \tag{21.62}$$

where

$$a_n(\lambda) = \sum_{m = 0}^{\infty} a_{n, m} \lambda^m.$$

Assume that the series (21.62) is not identically zero. We can write each nonzero coefficient $a_n(\lambda)$ of (21.62) as

$$a_n(\lambda) = \lambda^{\rho_n} \sum_{m = 0}^{\infty} b_{n, m} \lambda^m \qquad (b_{n, m} = a_{n, m + \rho_n}),$$

where ρ_n are nonnegative integers ($\rho_0 > 0$!) and $b_{n, 0} \neq 0$. The minimum ρ_n is known as the *height* of the series (21.62), denoted by $h = h(f)$. Denote the minimum n for which $\rho_n = h$ by $s = s(f)$; $s(f)$ is called the *degree* of the series (21.62).

Without loss of generality, we may assume that the height $h(f)$ of the series (21.62) is zero (though this will not be needed in the sequel). In fact, otherwise, we can cancel out λ^h from equation (21.60); the resulting equation has the same solutions in $K^*\{\lambda\}$ as equation (21.60). If $h(f) = 0$, the series (21.62) is regular, and its degree is precisely its order of regularity.

The aim of this subsection is to describe Newton's procedure for constructing all solutions of equation (21.60) and to prove the following fundamental proposition.

Theorem 21.6. Assume that the degree of the series (21.62) is s.

Then equation (21.60) is solvable in $K^*\{\lambda\}$ if and only if $s > 0$. If $s > 0$, equation (21.60) has at most s solutions in $K^*\{\lambda\}$, and each of these is in $K^\infty\{\lambda\}$. The solutions may be divided into groups

$$\left.\begin{array}{c} x_{1,1}(\lambda), \ldots, x_{1,r_1}(\lambda), \\ \cdot \quad \cdot \quad \cdot \quad \cdot \quad \cdot \quad \cdot \quad \cdot \quad \cdot \\ x_{\tau,1}(\lambda), \ldots, x_{\tau,r_\tau}(\lambda) \end{array}\right\} \tag{21.63}$$

of conjugate solutions whose radixes $r_1, \ldots, r_\tau$ satisfy the inequality

$$r_1 + r_2 + \ldots + r_\tau \leqq s. \tag{21.64}$$

The multiplicity of a solution of equation (21.60) may be defined in the natural way, and each group in the table (21.63) repeated a number of times equal to the multiplicity of the solution that it contains. Inequality (21.64) then becomes an equality.

It will be convenient to study an equation

$$g(\lambda, x) = 0, \tag{21.65}$$

more general than (21.60), where $g(\lambda, x)$ is a formal power series

$$g(\lambda, x) = \sum_{n=0}^{\infty} g_n(\lambda) x^n \tag{21.66}$$

with coefficients in $K^r\{\lambda\}$, some of which may vanish. The nonzero coefficients $g_n(\lambda)$ of (21.66) may be expressed as

$$g_n(\lambda) = \lambda^{\rho_n} \sum_{m=0}^{\infty} b_{n,m} \lambda^{m/r}, \tag{21.67}$$

where $\rho_n = d_n r^{-1}$, d_n being nonnegative integers, $b_{n,0} \neq 0$. As before, we assume that

$$g(0,0) = 0,$$

i.e., $\rho_0 > 0$, and consider solutions of equation (21.65) in $K^*\{\lambda\}$ that "*vanish for* $\lambda = 0$." As in the previous special case ($r = 1$), the *height*

$h = h(g)$ of series (21.66) is the minimum number ρ_n, the *degree* $s = s(g)$ of the series (21.66) is the minimum n such that $\rho_n = h$.

Let $g(\lambda, x)$ be a series of type (21.66), of degree s. In the $\{n, \rho\}$-plane (Fig. 21.1), consider the set of points $\{n, \rho_n\}$. The *Newton polygon* $\mathscr{L}$ associated with the series (21.66) is defined as the lower convex envelope of the set W of points $\{n, \rho_n\}$ $(0 \leqq n \leqq s)$ (see Fig. 21.1, in which W is hatched). The Newton polygon consists of several *segments* $\mathscr{L}_1, \ldots \mathscr{L}_\tau$.

The *length* $l(\mathscr{L})$ of the Newton polygon is defined as the length of its projection onto the horizontal axis; the *lengths* $l(\mathscr{L}_i)$ of its segments are defined similarly. The length $l(\mathscr{L})$ of the Newton polygon is obviously an integer such that

$$0 \leqq l(\mathscr{L}) \leqq s.$$

The number $l_0(\mathscr{L}) = s - l(\mathscr{L})$ is called the *degeneracy order* of $\mathscr{L}$. We have

$$l(\mathscr{L}_1) + \ldots + l(\mathscr{L}_\tau) = l(\mathscr{L}). \tag{21.68}$$

Let

$$\rho + n\varepsilon^{(i)} = h^{(i)} \qquad (i = 1, \ldots, \tau) \tag{21.69}$$

be the equations of the straight lines through the segments $\mathscr{L}_1, \ldots, \mathscr{L}_\tau$ of the Newton polygon $\mathscr{L}$. The number $\varepsilon^{(i)}$ is called the *exponent* and the number $h^{(i)}$ the *height* associated with the segment $\mathscr{L}_i$. For each $i = 1, \ldots,$ τ, define the polynomial

$$\Phi_i(c) = \sum_{\rho_n + n\varepsilon^{(i)} = h^{(i)}} b_{n,0} c^n, \tag{21.70}$$

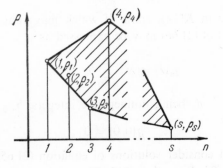

Fig. 21.1

which we call the *supporting polynomial of $\mathscr{L}$ associated with the segment* $\mathscr{L}_i$ (or with the exponent $\varepsilon^{(i)}$).

Lemma 21.6. *Assume that the length $l(\mathscr{L})$ of the polygon $\mathscr{L}$ is positive. Then the exponents $\varepsilon^{(i)}$ and heights $h^{(i)}$ are rational numbers, and*

$$h^{(i)} - h(f) \geqq \varepsilon^{(i)} . \tag{21.71}$$

If $\varepsilon^{(i)} = p/qr$, where p and q are relatively prime, then q divides $l(\mathscr{L}_i)$ and

$$\Phi_i(c) = c^a \Psi_i(c^q) , \tag{21.72}$$

where α is an integer and $\Psi_i(c)$ is a polynomial of degree $l(\mathscr{L}_i)/q$.

Proof. Let $\{\alpha, \rho_\alpha\}$ and $\{\beta, \rho_\beta\}$ be the endpoints of the segment $\mathscr{L}_i$, $\alpha < \beta$. Then

$$\rho_\alpha + \alpha\varepsilon^{(i)} = \rho_\beta + \beta\varepsilon^{(i)} = h^{(i)} ,$$

and so

$$\varepsilon^{(i)} = \frac{\rho_\alpha - \rho_\beta}{\beta - \alpha} = \frac{d_\alpha - d_\beta}{r(\beta - \alpha)} .$$

Consequently, $\varepsilon^{(i)}$ is a rational number. Let

$$\varepsilon^{(i)} = \frac{p}{rq} ,$$

where p and q are relatively prime. Then q is obviously a divisor of $\beta - \alpha = l(\mathscr{L}_i)$.

Inequality (21.71) is obvious.

If $\{n, \rho_n\}$ is a point on $\mathscr{L}_i$, then

$$\frac{p}{rq} = \varepsilon^{(i)} = \frac{\rho_\alpha - \rho_n}{n - \alpha} = \frac{d_\alpha - d_n}{r(n - \alpha)}$$

and so $p(n - \alpha) = q(d_\alpha - d_n)$. Hence q is a divisor of $n - \alpha$, i.e.,

$$n = \alpha + jq , \tag{21.73}$$

for some integer j.

Now consider the supporting polynomial of $\mathscr{L}_i$. Clearly,

$$\Phi_i(c) = \sum_{\rho_n + n\varepsilon^{(i)} = h^{(i)}} b_{n,0} c^n = c^\alpha \sum_{\rho_n + n\varepsilon^{(i)} = h^{(i)}} b_{n,0} c^{n-\alpha} .$$

By (21.73), all the numbers $n - \alpha$ are divisible by q. The last equality may therefore be expressed in the form (21.72), where the polynomial $\Psi_i(c)$ is defined by

$$\Psi_i(c) = \sum_{\rho_n + n\varepsilon^{(i)} = h^{(i)}} b_{n,0} c^{n-\alpha} = \sum_{j=0}^{(\beta-\alpha)q-1} \tilde{b}_j c^j,$$

where

$$\tilde{b}_j = \begin{cases} b_{\alpha+jq,0}, & \text{if } \rho_{\alpha+jq} + (\alpha + jq)\varepsilon^{(i)} = h^{(i)}, \\ 0, & \text{if } \rho_{\alpha+jq} + (\alpha + jq)\varepsilon^{(i)} > h^{(i)}. \end{cases}$$

The degree of this polynomial is obviously $(\beta - \alpha) q^{-1}$. ∎

We now proceed to investigate the solutions of equation (21.65).

*Lemma 21.7. If $l_0(\mathscr{L}) > 0$, equation (21.65) has a trivial solution; conversely, if equation (21.65) has a trivial solution, then $l_0(\mathscr{L}) > 0$. If equation (21.65) has a nontrivial solution, then $l(\mathscr{L}) > 0$. Each nontrivial solution $x(\lambda)$ of equation (21.65) in $K^*\{\lambda\}$ has the form**

$$x(\lambda) = c_1 \lambda^{\varepsilon_1} + o(\lambda^{\varepsilon_1}), \tag{21.74}$$

where ε_1 is one of the exponents $\varepsilon^{(1)}, \ldots, \varepsilon^{(\tau)}$ of the Newton polygon $\mathscr{L}$, and c_1 is a nonzero root of the corresponding supporting polynomial (21.70).*

Proof. The first assertion of the lemma is obvious.

Let $x(\lambda)$ be a nontrivial solution of equation (21.65). Then

$$x(\lambda) = c_1 \lambda^{\varepsilon_1} + o(\lambda^{\varepsilon_1}),$$

where $\varepsilon_1 > 0$, $c_1 \neq 0$. Since

$$[x(\lambda)]^n = c_1^n \lambda^{n\varepsilon_1} + o(\lambda^{n\varepsilon_1}),$$

we see, substituting $x(\lambda)$ in the series (21.66), that

$$g[\lambda, x(\lambda)] = \sum_{n=0}^{\infty} g_n(\lambda) [c_1^n \lambda^{n\varepsilon_1} + o(\lambda^{n\varepsilon_1})] = \sum_{n=0}^{\infty} [b_{n,0} c_1^n \lambda^{\rho_n + n\varepsilon_1} +$$

$$+ o(\lambda^{\rho_n + n\varepsilon_1})]. \tag{21.75}$$

---

* $o(\lambda^k)$ denotes a formal power series $\sum_{i=1}^{\infty} c_i \lambda^{\varepsilon_i}$ in which all ε_i are greater than k.

Let

$$h_1 = \min_n \{\rho_n + n\varepsilon_1\}. \tag{21.76}$$

Then it follows from (21.75) that

$$g[\lambda, x(\lambda)] = \Phi(c_1)\lambda^{h_1} + o(\lambda^{h_1}), \tag{21.77}$$

where

$$\Phi(c) = \sum_{\rho_n + n\varepsilon_1 = h_1} b_{n,0} c^n. \tag{21.78}$$

Since $x(\lambda)$ is a solution of equation (21.65), $g[\lambda, x(\lambda)] \equiv 0$. In particular, the number c_1 must be a root of the polynomial (21.78). This polynomial has nonzero roots only if it contains at least two nonzero terms, i.e., if the equation

$$\rho_n + n\varepsilon_1 = h_1 \tag{21.79}$$

has at least two solutions. By (21.76), the inequalities

$$\rho_n + n\varepsilon_1 \geqq h_1 \tag{21.80}$$

must hold for the remaining values of n.

Thus, ε_1 and h_1 are not arbitrary numbers: they must satisfy inequalities (21.80), and these inequalities must be equalities for at least two values of n. In combination with the definition of the Newton polygon, this implies that $l(\mathscr{L}) > 0$ and the straight line (21.79) must coincide with one of the lines (21.69). But then $\varepsilon_1 = \varepsilon^{(i)}$, and (21.78) is the supporting polynomial $\Phi_i(c)$ associated with ε_i. ∎

Lemma 21.7 enables one to determine whether equation (21.65) has a trivial solution, and to find the first terms of all its nontrivial solutions. To find the remaining terms of the expansions of the nontrivial solutions, a natural approach is to perform the substitution

$$x = (c_1 + x_1)\lambda^{\varepsilon_1} \tag{21.81}$$

in equation (21.65), and to look for x_1 as a solution in $K^*\{\lambda\}$ of the equation

$$g_1(\lambda, x_1) = 0, \tag{21.82}$$

where

$$g_1(\lambda, x_1) = g[\lambda, (c_1 + x_1)\lambda^{\varepsilon_1}]. \tag{21.83}$$

Lemma 21.8. *Let* $\varepsilon_1 = \varepsilon^{(i)}$ *(where i is fixed), c_1 an s_1-fold root of the supporting polynomial $\Phi_i(c)$ associated with $\varepsilon^{(i)}$, and h_1 the corresponding height.*

Then the formal power series (21.83) has height h_1 and degree s_1.

Proof. First note that the polynomial $\Phi_i(c)$ may be expressed as

$$\Phi_i(c) = (c - c_1)^{s_1} \tilde{\Phi}(c), \tag{21.84}$$

where $\tilde{\Phi}(c)$ is a polynomial such that $\tilde{\Phi}(c_1) \neq 0$.

Substitute (21.81) into the series (21.66). The result is

$$g\left[\lambda, (c_1 + x_1)\lambda^{\varepsilon_1}\right] = \sum_{n=0}^{\infty} g_n(\lambda)\, \lambda^{n\varepsilon_1}(c_1 + x_1)^n =$$

$$= \sum_{\rho_n + n\varepsilon_1 = h_1} g_n(\lambda)\, \lambda^{n\varepsilon_1}(c_1 + x_1)^n + \sum_{\rho_n + n\varepsilon_1 > h_1} g_n(\lambda)\, \lambda^{n\varepsilon_1}(c_1 + x_1)^n =$$

$$= \lambda^{h_1} \sum_{\rho_n + n\varepsilon_1 = h_1} b_{n,0}\,(c_1 + x_1)^n +$$

$$+ \sum_{\rho_n + n\varepsilon_1 = h_1} \left[g_n(\lambda) - b_{n,0}\lambda^{\rho_n}\right] \lambda^{n\varepsilon_1}(c_1 + x_1)^n +$$

$$+ \sum_{\rho_n + n\varepsilon_1 > h_1} g_n(\lambda)\, \lambda^{n\varepsilon_1}(c_1 + x_1)^n = \lambda^{h_1}\Phi_i(c_1 + x_1) + o(\lambda^{h_1}),$$

whence it follows that

$$\lambda^{-h_1} g_1(\lambda, x_1) = \sum_{n=0}^{\infty} \tilde{g}_n(\lambda)\, x_1^n,$$

where the coefficients $\tilde{g}_n(\lambda)$ are obviously in $K^{rq}\{\lambda\}$ and, by (21.84),

$$\tilde{g}_0(0) = \ldots = \tilde{g}_{s_1-1}(0) = 0, \quad \tilde{g}_{s_1}(0) = \tilde{\Phi}(c_1) \neq 0.$$

This implies the assertion of the lemma. ∎

To study equation (21.82) we again apply Lemma 21.7. If the degeneracy order $l_0(\mathcal{L}_1)$ of the Newton polygon $\mathcal{L}_1$ of the series (21.83) is positive, equation (21.82) has a trivial solution; by the same token, equation (21.65) will have a solution $x(\lambda) = c_1 \lambda^{\varepsilon_1}$. If the length $l(\mathcal{L}_1)$ of the Newton polygon is positive, equation (21.82) may also have nontrivial solutions $x_1(\lambda) = c_2 \lambda^{\bar{\varepsilon}_2} + o(\lambda^{\bar{\varepsilon}_2})$, where $\bar{\varepsilon}_2$ is one of the exponents of $\mathcal{L}_1$ and c_2

is a root of the corresponding supporting polynomial; equation (21.65) will then have solutions

$$x(\lambda) = c_1\lambda^{\varepsilon_1} + c_2\lambda^{\varepsilon_2} + o(\lambda^{\varepsilon_2}),$$

where $\varepsilon_2 = \varepsilon_1 + \bar{\varepsilon}_2$. This procedure may be continued indefinitely. The result is a set $\mathfrak{N}$ of formal power series

$$x(\lambda) = c_1\lambda^{\varepsilon_1} + c_2\lambda^{\varepsilon_2} + \ldots + c_n\lambda^{\varepsilon_n} + \ldots, \tag{21.85}$$

which contains all solutions of equation (21.65). Note that the exponents $\varepsilon_1, \ldots, \varepsilon_n, \ldots$ and the coefficients $c_1, \ldots, c_n, \ldots$ of the series (21.85) are determined successively, together with a sequence of formal power series

$$g(\lambda, x), g_1(\lambda, x), \ldots, g_n(\lambda, x), \ldots \tag{21.86}$$

(the sequence (21.68) is constructed separately for each series (21.85) in $\mathfrak{N}$). If the exponent $\varepsilon_1, \ldots, \varepsilon_k$ and the coefficients $c_1, \ldots, c_k$ have been determined, the series $g_k(\lambda, x_k)$ is defined by the equality

$$g_k(\lambda, x_k) = g_{k-1}\left[\lambda, (c_k + x_k)\lambda^{\bar{\varepsilon}_k}\right], \tag{21.87}$$

where $\bar{\varepsilon}_k = \varepsilon_k - \varepsilon_{k-1}$. If the degeneracy order $l_0(\mathscr{L}_k)$ of the Newton polygon $\mathscr{L}_k$ of the series (21.87) is positive, the series

$$x(\lambda) = c_1\lambda^{\varepsilon_1} + \ldots + c_k\lambda^{\varepsilon_k}$$

is in $\mathfrak{N}$. If the length $l(\mathscr{L}_k)$ of the Newton polygon is positive, then (for any value of $l(\mathscr{L}_k)$) $\mathfrak{N}$ will contain the series

$$x(\lambda) = c_1\lambda^{\varepsilon_1} + \ldots + c_k\lambda^{\varepsilon_k} + c_{k+1}\lambda^{\varepsilon_{k+1}} + o(\lambda^{\varepsilon_{k+1}});$$

here $\varepsilon_{k+1} = \bar{\varepsilon}_{k+1} - \varepsilon_k$, ε_{k+1} is one of the exponents of the Newton polygon $\mathscr{L}_k$ and c_{k+1} is a nonzero root of the corresponding supporting polynomial.

The above procedure for constructing a set $\mathfrak{N}$ of formal power series (21.85) is known as the *Newton procedure*. It turns out that $\mathfrak{N}$ contains at most s series, and each series (21.85) is a solution of equation (21.65).

The concept of *multiplicity* for a series (21.85) in the set $\mathfrak{N}$ may be introduced naturally.

Assume that a series (21.85) in $\mathfrak{N}$ has only finitely many terms, i.e.,

$$x(\lambda) = c_1\lambda^{\varepsilon_1} + \ldots + c_k\lambda^{\varepsilon_k};$$

the exponents $\varepsilon_1, \ldots \varepsilon_k$ and coefficients $c_1, \ldots, c_k$ of this series are determined successively, together with a finite sequence of formal power series (21.86)

$$g(\lambda, x), g_1(\lambda, x_1), \ldots, g_k(\lambda, x_k),$$

where the Newton polygon $\mathscr{L}_k$ of the last of these series has nonzero degeneracy $l_0(\mathscr{L}_k)$. This degeneracy is defined to be the *multiplicity* $\kappa(x)$ of the series (21.85).

Assume that a series (21.85) in $\mathfrak{N}$ has infinitely many terms. Its exponents $\varepsilon_1, \ldots, \varepsilon_k, \ldots$ and coefficients are determined successively together with an infinite sequence of series (21.86)

$$g(\lambda, x), \ g_1(\lambda, x_1), \ \ldots, \ g_k(\lambda, x_k), \ldots$$

The degrees $s, s_1, \ldots, s_k, \ldots$ are clearly positive, and

$$s \geqq s_1 \geqq s_2 \geqq \ldots \geqq s_k \geqq \ldots$$

The *multiplicity* $\kappa(x)$ of the series (21.85) is the smallest of these degrees.

The proof of the following simple proposition is left to the reader.

Lemma 21.9. *Let the degree $s(g)$ of the series* (21.66) *be positive.*

Then the sum of multiplicities of all series in the set $\mathfrak{N}$ is equal to s.

It follows from Lemma 21.9 that the Newton procedure, applied to equation (21.67), yields a finite number of series. If each series is counted a number of times equal to its multiplicity, the set $\mathfrak{N}$ will consist of exactly s elements.

Lemma 21.10. *Every formal power series* (21.85) *in $\mathfrak{N}$ is an element of the ring $K^\infty\{\lambda\}$ and is a solution of equation* (21.65).

Proof. The assertion is obvious if the series (21.85) has only finitely many terms.

Now let (21.85) be a series in $\mathfrak{N}$ with infinitely many terms, and κ its multiplicity. Let k_0 be a number such that, for $k \geqq k_0$, the degree of the series $g_k(\lambda, x_k)$ is κ. Then, for $k \geqq k_0$, the Newton polygon $\mathscr{L}_k$ consists of a single segment, and the associated supporting polynomial $\Phi_k(c)$ has the form

$$\Phi_{(k)}(c) = c^{\alpha_k}(c - c_k)^\kappa ; \tag{21.88}$$

where $c_k \neq 0$. Moreover, by Lemma 21.6 this polynomial has a representation (21.72):

$$\Phi_{(k)}(c) = c^{\alpha_k}\Psi(c^{q_k}) ; \tag{21.89}$$

the integers q_k are defined successively by the equalities

$$\varepsilon_k - \varepsilon_{k-1} = \frac{p_k}{rq_1 \ldots q_k}$$

and the condition that p_k and q_k are relatively prime. Comparing (21.88) and (21.89), we see that $q_k = 1$ for all $k \geq k_0$. But then $\bar{\varepsilon}_k$, thus also the exponents $\varepsilon_k = \bar{\varepsilon}_1 + \bar{\varepsilon}_2 + \ldots + \bar{\varepsilon}_k$, have the form

$$\varepsilon_k = t_k/r^*,$$

where $r^* = rq_1 \ldots q_{k_0}$, and t_k are integers. This means that the series (21.85) is an element of $K^\infty\{\lambda\}$.

We shall now show that the series (21.85) is a solution. It follows from (21.87) that

$$g\left[\lambda, x(\lambda)\right] = g_1\left[\lambda, x_1(\lambda)\right] = \ldots = g_k\left[\lambda, x_k(\lambda)\right], \tag{21.90}$$

where

$$x_k(\lambda) = \lambda^{-\varepsilon_k}\left[x(\lambda) - c_1\lambda^{\varepsilon_1} - \ldots - c_k\lambda^{\varepsilon_k}\right].$$

It follows from (21.90) that for any k

$$g\left[\lambda, x(\lambda)\right] = o(\lambda^{h_k}),$$

where h_k is the height of the series $g_k(\lambda, x_k)$. Thus, to prove that (21.85) is a solution of equation (21.65) it suffices to show that $h_k \to \infty$ as $k \to \infty$.

By Lemma 21.6,

$$h_1 - h(g) \geq \varepsilon_1, h_2 - h_1 \geq \varepsilon_2 - \varepsilon_1, \ldots, h_k - h_{k-1} \geq \varepsilon_k - \varepsilon_{k-1};$$

these inequalities imply that $h_k \geq \varepsilon_k$, and, since $\varepsilon_k \to \infty$ as $k \to \infty$, the same holds for h_k. ∎

Coming back to equation (21.60), we see from the above arguments that this equation has at most s solutions in $K^*\{\lambda\}$ (exactly s if each solution is counted in accordance with its multiplicity), and these solutions belong to $K^\infty\{\lambda\}$. To prove Theorem 21.6, it remains to show that these solutions may be divided into groups of conjugate solutions and to establish the inequality (21.64). This will be taken care of by the following obvious fact.

Lemma 21.11. Let the formal power series (21.58), *with radix* r, *be a solution of equation* (21.60).

Then all series conjugate to (21.58) *are also solutions of equation* (21.60).

Exercise 21.10. Show that a solution $x(\lambda)$ of equation (21.60) has multiplicity 1 if and only if the series $f'_x[\lambda, x(\lambda)]$ does not vanish.

Exercise 21.11. Show that the multiplicity of a solution $x(\lambda)$ of equation (21.60) is precisely the minimum order for which the derivatives

$$f'_x[\lambda, x(\lambda)], \ldots, f^{(k)}_{x^k}[\lambda, x(\lambda)], \ldots$$

do not vanish.

Exercise 21.12. Show that if the degree of the series (21.61) is s, equation (21.60) is equivalent to the algebraic equation

$$x^s + a_1(\lambda) x^{s-1} + \ldots + a_s(\lambda) = 0$$

of degree s with coefficients in $K\{\lambda\}$. Show that the multiplicity of each solution of equation (21.60) is precisely the multiplicity of the corresponding root of the algebraic equation.

Exercise 21.13. Show that a formal power series $x = x(\lambda)$ is a solution of equation (21.60) with multiplicity $\geq \kappa$ if and only if the series $[x - x(\lambda)]^\kappa$ divides the series (21.61).

Newton's procedure has one essential feature that plays a decisive role in its practical applications: to determine the first terms of the series-solutions, we need only finitely many coefficients of the series (21.61).

21.4. *Branching of solutions of scalar analytic equations.* In this subsection we consider the scalar equation

$$f(\lambda, x) = 0, \tag{21.91}$$

where $f(\lambda, x)$ is a function analytic in some neighborhood of the point $\{\lambda_0, x_0\}$, such that

$$f(\lambda_0, x_0) = 0. \tag{21.92}$$

We are interested in solutions $x = x(\lambda)$ which are close to x_0 when the parameter λ is close to λ_0. The case

$$f'_x(\lambda_0, x_0) \neq 0 \tag{21.93}$$

was studied in detail in §20. Here we shall consider the general case.

If condition (21.93) is not satisfied, equation (21.91) may have no solutions equal to x_0 when $\lambda = \lambda_0$. A simple example is the equation

$$\lambda + \lambda x = 0 \qquad (\lambda_0 = x_0 = 0).$$

If equation (21.91) has solutions equal to x_0 at $\lambda = \lambda_0$, they are, as a rule, not analytic at λ_0. A simple example is the equation $-\lambda + x^2 = 0$

$(\lambda_0 = x_0 = 0)$; the solution of this equation is $x(\lambda) = \sqrt{\lambda}$. It turns out that, under very general assumptions, the solutions of equation (21.91) are analytic functions of the variable $\mu = (\lambda - \lambda_0)^{1/r}$, where r is an integer.

Let $\mathscr{K}\{\lambda\}$ denote the set of single-valued and multiple-valued analytic functions $x(\lambda)$ ($x(0) = 0$), defined in some neighborhood of zero, such that zero is either a regular point or a branch point of finite order. The set $\mathscr{K}\{\lambda\}$ is a ring under the natural algebraic operations.

Let $x(\lambda)$ be a function of $\mathscr{K}\{\lambda\}$ which is r-valued for nonzero λ. Then the function $z(\mu) = x(\mu^r)$ is also r-valued in a neighborhood of zero, and its r different single-valued branches $z_1(\mu), \ldots, z^r(\mu)$ may be so chosen that

$$z_{j+1}(\mu) = z_j(e^{2\pi i/r}\mu) \qquad (j = 1, \ldots, r - 1). \tag{21.94}$$

Each of the analytic functions $z_1(\mu), \ldots, z_r(\mu)$ has a convergent expansion in power series. By (21.94), these series have the form

$$z_j(\mu) = \sum_{k=0}^{\infty} c_k e^{2\pi i k j/r} \mu^k \qquad (j = 1, \ldots, r).$$

Therefore, the single-valued functions

$$\sum_{k=0}^{\infty} c_k \lambda^{k/r}, \ldots, \sum_{k=0}^{\infty} c_k e^{2\pi i k (r-1)/r} \lambda^{k/r}$$

coincide with $x(\lambda)$. In other words, every r-valued function $x(\lambda)$ in $\mathscr{K}\{\lambda\}$ may be associated with r conjugate series (of radix r) in $K^r\{\lambda\}$.

Conversely, if a series

$$x(\lambda) = \sum_{k=0}^{\infty} c_k \lambda^{k/r} \tag{21.95}$$

of radix r in $K^{\infty}\{\lambda\}$ converges in a neighborhood of zero, then its sum is a function in $\mathscr{K}\{\lambda\}$, and all its conjugate series define the same function.

Let

$$f(\lambda, x) = \sum_{m,n=0}^{\infty} a_{m,n}(\lambda - \lambda_0)^m (x - x_0)^n \tag{21.96}$$

be the power-series expansion of the left-hand side of equation (21.91). The degree $s = s(f)$ of this series is called the *degree* of the function

(21.96) (at the point $\{\lambda_0, x_0\}$). If

$$f(\lambda, x) = (\lambda - \lambda_0)^\alpha g(\lambda, x), \tag{21.97}$$

where $g(\lambda, x)$ is a function analytic at $\{\lambda_0, x_0\}$ ($g(\lambda_0, x)$ is not identically zero), then the degree of the function (21.96) is precisely the order of the first nonzero derivative $g'_x(\lambda_0, x_0), \ldots, g^{(k)}_{x^k}(\lambda_0, x_0), \ldots$.

Theorem 21.7.* *Assume that the function $f(\lambda, x)$ satisfies condition (21.92) and has degree s at the point $\{\lambda_0, x_0\}$.*

Then equation (21.91) has a solution equal to x_0 at $\lambda = \lambda_0$ if and only if $s > 0$. If $s > 0$, equation (21.91) has at most s solutions equal to x_0 at $\lambda = \lambda_0$. Each of these solutions $x_1(\lambda), \ldots, x_\tau(\lambda)$ belongs to $\mathscr{H}\{\lambda - \lambda_0\}$:

$$x_j(\lambda) = x_0 + \sum_{k=1}^{\infty} c_{k,j}(\lambda - \lambda_0)^{k/r}j \qquad (j = 1, \ldots, \tau), \tag{21.98}$$

where $r_1, \ldots, r_\tau$ are positive integers such that

$$r_1 + \ldots + r_\tau \leqq s. \tag{21.99}$$

Before proceeding to the proof of this theorem, we note that (21.98) defines an at most r_j-valued function. Inequality (21.99) therefore means that the number of single-valued solutions of equation (21.91) is at most s. Had we counted the solutions in accordance with their multiplicities, inequality (21.99) would have been an equality.

We now prove the theorem. The first assertion is evident. To prove the second, replace equation (21.91) by the equation

$$g(\lambda, x) = 0, \tag{21.100}$$

where $g(\lambda, x)$ is the function defined by (21.97). Equations (21.91) and (21.100) have the same solutions equal to x_0 at $\lambda = \lambda_0$.

By Theorem 21.5, equation (21.100) is equivalent to the equation

$$*g(\lambda, x) = 0, \tag{21.101}$$

where $*g(\lambda, x)$ is a polynomial in $x - x_0$ of degree s:

$$*g(\lambda, x) = (x - x_0)^s + a_1(\lambda - \lambda_0)(x - x_0)^{s-1} + \ldots + a_s(\lambda - \lambda_0), \tag{21.102}$$

---

* See, e.g., Goursat [1], Markushevich [1], Fuks and Shabat [1].

whose coefficients $a_1(v), \ldots, a_s(v)$ are functions analytic at zero such that

$$a_1(0) = \ldots = a_s(0) = 0. \tag{21.103}$$

By Rouché's Theorem, there exist positive numbers ε and δ such that when $|\lambda - \lambda_0| < \delta$ equation (21.101) has s solutions $x_1(\lambda), \ldots, x_s(\lambda)$ (counting multiplicities) in the disk $|x - x_0| < \varepsilon$.

We first assume that all these solutions are simple for $0 < |\lambda - \lambda_0| < \delta$. By Theorem 20.11, the functions $x_1(\lambda), \ldots, x_s(\lambda)$ are analytic at each fixed point such that $0 < |\lambda - \lambda_0| < \delta$. Thus it is easily seen that the solutions $x_1(\lambda), \ldots, x_s(\lambda)$ fall into τ groups

$$x_{1,1}(\lambda), \ldots, x_{1,r_1}(\lambda),$$
$$\cdots \cdots \cdots$$
$$x_{\tau,1}(\lambda), \ldots, x_{\tau,r_\tau}(\lambda)$$

in such a way that when λ encircles λ_0 in the counterclockwise direction the solution $x_{j,1}(\lambda)$ transforms into the solution $x_{j,2}(\lambda)$, $x_{j,2}(\lambda)$ transforms into $x_{j,3}(\lambda), \ldots, x_{j,r_j}(\lambda)$ transforms into $x_{j,1}(\lambda)$ $(j = 1, 2, \ldots, \tau)$, where

$$r_1 + \ldots + r_\tau = s.$$

Consider the group of solutions $x_{j,1}(\lambda), \ldots, x_{j,r_j}(\lambda)$. Obviously, the functions $x_{j,1}(\lambda), \ldots, x_{j,r_j}(\lambda)$ are the branches of the multiple-valued function $x_j(\lambda)$ defined in the disk $|\lambda - \lambda_0| < \delta$ punctured at $\lambda = \lambda_0$. Since all branches of this function converge to x_0 as $\lambda \to \lambda_0$, we may consider $x_j(\lambda)$ to be analytic in the entire disk $|\lambda - \lambda_0| < \delta$ with a branch point $\lambda = \lambda_0$ of order $r_j - 1$. Consequently, $x_j(\lambda) \in \mathcal{H}\{\lambda - \lambda_0\}$.

This proves the assertion of the theorem when equation (21.101) has no multiple roots for small $\lambda - \lambda_0$.

Now assume that equation (21.101) has multiple roots for arbitrarily small $\lambda - \lambda_0$. Let $D(\lambda)$ be the discriminant of the polynomial (21.102). $D(\lambda)$ is clearly an analytic function of λ. If (21.101) has multiple roots for some λ, $D(\lambda)$ vanishes at this point. Since (21.102) has multiple roots for arbitrary small $\lambda - \lambda_0$, it follows that $D(\lambda)$ vanishes for a sequence of points λ_n which converges to λ_0. By the uniqueness theorem $D(\lambda)$ is identically zero.

Consider the polynomial (21.102) as a polynomial over the ring $K\{\lambda - \lambda_0\}_a$ of convergent power series in $\lambda - \lambda_0$. The discriminant D of this polynomial, considered as an element of $K\{\lambda - \lambda_0\}_a$, coincides with

$D(\lambda)$, and therefore vanishes. By Theorem 21.4 the polynomial (21.102) has multiple factors. Let

$$*g(\lambda, x) = [\pi_1(\lambda, x)]^{\gamma_1} \dots [\pi_l(\lambda, x)]^{\gamma_l}$$

be the irreducible factorization of (21.102) (in the ring of polynomials in $x - x_0$ over the ring $K\{\lambda - \lambda_0\}_a$). It is obvious that the equation

$$\pi(\lambda, x) = 0, \qquad (21.104)$$

where

$$\pi(\lambda, x) = \pi_1(\lambda, x) \dots \pi_l(\lambda, x),$$

has the same solutions as equation (21.101). However, in contrast to the latter, equation (21.104) has no multiple solutions for small $\lambda - \lambda_0$. Its solutions therefore have the required properties, and this completes the proof of Theorem 21.7. ∎

We shall find it convenient to use the concept of the multiplicity of a solution of equation (21.91). If $x(\lambda)$ is a solution of equation (21.91) equal to x_0 at $\lambda = \lambda_0$, then, for small $\lambda - \lambda_0$, the number $x(\lambda)$ is a root of the polynomial (21.102), of the same multiplicity for all sufficiently small $\lambda - \lambda_0$; this common multiplicity is defined to be the *multiplicity of the solution* $x = x(\lambda)$ of equation (21.91).

Exercise 21.14. Show that the multiplicity of a solution $x = x(\lambda)$ of equation (21.91) is precisely the multiplicity of the root $x(\lambda)$ of the polynomial (21.102), considered as a polynomial over $K\{\lambda - \lambda_0\}$.

Exercise 21.15. Show that the solution $x = x(\lambda)$ of equation (21.91) has multiplicity κ if and only if

$$f[\lambda, x(\lambda)] \equiv \dots \equiv f_x^{(\kappa-1)}[\lambda, x(\lambda)] \equiv 0$$

and $f_{x^\kappa}^{(\kappa)}[\lambda, x(\lambda)] \not\equiv 0$.

We now return to equation (21.91). Together with this equation, consider the equation

$$f(v, h) = 0, \qquad (21.105)$$

where the formal power series

$$f(v, h) = \sum_{m, n = 0}^{\infty} a_{m, n} v^m h^n \qquad (21.106)$$

is defined by the expansion of the function

$$f(\lambda, x) = \sum_{m, n = 0}^{\infty} a_{m, n} (\lambda - \lambda_0)^m (x - x_0)^n$$

in powers of $\lambda - \lambda_0$ and $x - x_0$. Each solution $x = x(\lambda)$ of equation (21.91) may be expressed as

$$x(\lambda) = x_0 + h(\lambda - \lambda_0), \tag{21.107}$$

where $h = h(v)$ is a power series

$$h(v) = \sum_{k = 1}^{\infty} c_k v^{k/r} \tag{21.108}$$

in $K^{\infty} \{v\}_a$ which is a solution of equation (21.105). It is natural to ask what formal solutions of equation (21.105) define solutions of equation (21.91). The answer is

Theorem 21.8.* *Under the assumptions of Theorem 21.7, let the formal power series* (21.108) *be a solution of equation* (21.105).

Then the series (21.108) *converges for small* v, *and formula* (21.107) *defines a solution of equation* (21.91) *which is equal to* x_0 *at* $\lambda - \lambda_0$.

Proof. It suffices to note that each solution (21.107) of multiplicity κ of equation (21.91) defines a formal solution (21.108) of equation (21.105), of the same multiplicity κ. The sum of multiplicities of all solutions of equation (21.91) is the degree s of the function $f(\lambda, x)$ at $\{\lambda_0, x_0\}$; the sum of multiplicities of all solutions of equation (21.105) is the degree of the series (21.106), i.e., the degree of the series (21.96), which is also s. ∎

21.5. *The method of elimination.* Let K_n denote complex n-space, and $K_n^* \{\lambda\}$ the set of all formal power series

$$x(\lambda) = \sum_{k = 0}^{\infty} c_k \lambda^{\varepsilon_k}, \tag{21.109}$$

where $c_k \in K_n$, $0 = \varepsilon_0 < \varepsilon_1 < \dots$. As usual, we shall say that the series (21.109) vanishes at $\lambda = 0$ if $c_0 = 0$.

---

* Pokornyi [1, 3], Melamed [3].

The set $K_n^*\{\lambda\}$ is a linear space under the natural algebraic operations. $K_n^r\{\lambda\}$ will denote the space of series

$$x(\lambda) = \sum_{k=0}^{\infty} c_k \lambda^{k/r}, \qquad (21.110)$$

where r is an integer, and $K_n^\infty\{\lambda\}$ is the union of all $K_n^r\{\lambda\}$ ($r = 1, 2, \ldots$). If $x(\lambda) \in K_n^\infty\{\lambda\}$, the minimum r for which $x(\lambda) \in K_n^r\{\lambda\}$ is called the *radix* of the series $x(\lambda)$. If the series (21.110) has radix r, then the series

$$\left.\begin{aligned}
x(\lambda) &= \sum_{k=0}^{\infty} c_k \lambda^{k/r}, \\[2mm]
x_1(\lambda) &= \sum_{k=0}^{\infty} c_k \varepsilon^k \lambda^{k/r}, \\[2mm]
&\cdot \ \cdot \ \cdot \ \cdot \ \cdot \ \cdot \ \cdot \\[2mm]
x_{r-1}(\lambda) &= \sum_{k=0}^{\infty} c_k \varepsilon^{(r-1)k} \lambda^{k/r},
\end{aligned}\right\} \qquad (21.111)$$

where ε is some primitive r-th root of unity, are all different. As in the scalar case, these series are known as *conjugates*.

Given a basis in K_n, each vector c of K_n is determined by an n-tuple of complex numbers. Thus each series $x(\lambda)$ in $K_n^*\{\lambda\}$ generates n component series $x_j(\lambda)$ in $K^*\{\lambda\}$, and $x(\lambda) \in K_n^\infty\{\lambda\}$ if and only if $x_1(\lambda), \ldots, x_n(\lambda) \in K^\infty\{\lambda\}$ and the radix of the series $x(\lambda)$ is the lowest common multiple of the radixes of the series $x_1(\lambda), \ldots, x_n(\lambda)$.

We shall study systems

$$f_\mu(\lambda; x_1, \ldots, x_n) = 0 \qquad (\mu = 1, \ldots, m), \qquad (21.112)$$

where $f_\mu(\lambda; x_1, \ldots, x_n)$ are formal power series with complex coefficients, in the variables $\lambda, x_1, \ldots, x_n$:

$$f_\mu(\lambda; x_1, \ldots, x_n) = \sum_{k_0, k_1, \ldots, k_n = 0}^{\infty} a_{k_0, k_1, \ldots, k_n}^\mu \lambda^{k_0} x_1^{k_1} \ldots x_n^{k_n}. \qquad (21.113)$$

We shall assume throughout this subsection that

$$f_\mu(0, 0, \ldots, 0) = 0, \qquad (21.114)$$

i.e., $a^\mu_{0,0,\ldots,0} = 0$. An abbreviated notation for the system (21.112) is

$$f(\lambda, x) = 0. \tag{21.115}$$

A solution $x = x(\lambda)$ of the system (21.112) is defined as a series $x(\lambda) = \{x_1(\lambda), \ldots, x_n(\lambda)\}$ in $K^\infty_n\{\lambda\}$, which vanishes at $\lambda = 0$, and, when substituted in the series (21.113), equates them to zero. Together with individual solutions of (21.112), it is convenient to consider *families* of solutions

$$x(\lambda) = X(\lambda; u_1, \ldots, u_l). \tag{21.116}$$

Here $u_1, \ldots, u_l$ are arbitrary formal power series in $K^\infty\{\lambda\}$ which vanish at $\lambda = 0$; they act as parameters. We call a family of solutions (21.116) *l-dimensional* if the operator $X(\lambda; u_1, \ldots, u_l)$ depends essentially on all the parameters $u_1, \ldots, u_l$. We are not assuming that the operator $X(\lambda; u_1, \ldots, u_l)$ is a formal power series in $u_1, \ldots, u_l$; all we require is that, for arbitrary formal power series $u_1, \ldots, u_l$ that vanish at $\lambda = 0$, formula (21.116) define a formal power series $x = x(\lambda)$ which solves the system (21.112).

Without loss of generality, we may assume that none of the series (21.113) vanishes. Let

$$f_\mu(\lambda; x_1, \ldots, x_n) = \lambda^{\alpha_\mu} \hat{f}_\mu(\lambda; x_1, \ldots, x_n), \tag{21.117}$$

where α_μ are nonnegative integers, while

$$\hat{f}_\mu(\lambda; x_1, \ldots, x_n) = \sum_{k_0, k_1, \ldots, k_n = 0}^{\infty} \hat{a}^\mu_{k_0, k_1, \ldots, k_n} \lambda^{k_0} x_1^{k_1} \ldots x_n^{k_n} \tag{21.118}$$

are formal power series such that the series

$$\hat{f}_\mu(0; x_1, \ldots, x_n) = \sum_{k_1, \ldots, k_n = 0}^{\infty} \hat{a}^\mu_{0, k_1, \ldots, k_n} x_1^{k_1} \ldots x_n^{k_n}, \tag{21.119}$$

derived from (21.118) by setting $\lambda = 0$, do not vanish. Let s^*_μ and s_μ ($\mu = 1, \ldots, m$) denote the orders of the series (21.118) and (21.119), respectively. Clearly, $s^*_\mu \leqq s_\mu$, and moreover $s^*_\mu = 0$ if and only if $s_\mu = 0$. The number $s = \min\{s_1, \ldots, s_m\}$ is known as the *rank* of the system (21.112).

By (21.117), the system (21.112) is equivalent to the system

$$\hat{f}_\mu(\lambda; x_1, \ldots, x_n) = 0 \qquad (\mu = 1, \ldots, m). \tag{21.120}$$

This system has no solutions in $K_n^\infty\{\lambda\}$ if any of the numbers s^* are zero. We shall therefore assume henceforth that $s_1^*, \ldots, s_m^*$, and thus also $s_1, \ldots, s_m$, are all positive. This means that the rank s of the system (21.112) is positive.

We may assume that each of the series (21.119) is regular in the variable x_n, with order of regularity s_μ; otherwise, we need only apply a linear transformation to the variables $x_1, \ldots, x_n$ (see Lemma 21.5). Then each of the series (21.118) is also regular in x_n, with order s_μ. Applying the Weierstrass Preparation Theorem (Theorem 21.5), we get

$$\hat{f}_\mu(\lambda; x_1, \ldots, x_n) = \varepsilon_\mu(\lambda; x_1, \ldots, x_n) \, {}^*\!f_\mu(\lambda; x_1, \ldots, x_n), \qquad (21.121)$$

where $\varepsilon_\mu(\lambda; x_1, \ldots, x_n)$ is a formal power series which is invertible ($\varepsilon_\mu(0; 0, \ldots, 0) \neq 0$) in the ring $K\{\lambda; x_1, \ldots, x_n\}$, while ${}^*\!f_\mu(\lambda; x_1, \ldots, x_n)$ is a formal power series which is a polynomial in x_n, of degree s_μ, with coefficients in the ring $K\{\lambda; x_1, \ldots, x_{n-1}\}$ and leading coefficient 1:

$$^*\!f_\mu(\lambda; x_1, \ldots, x_n) = x_n^{s_\mu} + a_1^\mu(\lambda; x_1, \ldots, x_{n-1}) x_n^{s_\mu - 1} + \cdots$$

$$\ldots + a_{s_\mu}^\mu(\lambda; x_1, \ldots, x_{n-1}). \qquad (21.122)$$

It follows from (21.121) and from the invertibility of the series $\varepsilon_\mu(\lambda; x_1, \ldots, x_n)$ that the system (21.120), and thus also the system (21.112), is equivalent to a system of algebraic equations

$$^*\!f_\mu(\lambda; x_1, \ldots, x_n) = 0 \qquad (\mu = 1, \ldots, m). \qquad (21.123)$$

Consider the resultants $R_1, \ldots, R_m$ of the polynomials (21.222) (see subsection 21.1). It is clear that each resultant R_i is a formal power series in $\lambda, x_1, \ldots, x_{n-1}$, in the ring $K\{\lambda; x_1, \ldots, x_{n-1}\}$:

$$R_i = f_i^{(1)}(\lambda; x_1, \ldots, x_{n-1}). \qquad (21.124)$$

We call the system

$$f_\mu^{(1)}(\lambda; x_1, \ldots, x_{n-1}) = 0 \qquad (\mu = 1, \ldots, m_1) \qquad (21.125)$$

the *first elimination system*.

Lemma 21.12. The system (21.112) *is solvable if and only if the system* (21.125) *is solvable. If* $x(\lambda) = \{x_1^0(\lambda), \ldots, x_n^0(\lambda)\}$ *is a solution of* (21.112), *then* $\bar{x}(\lambda) = \{x_1^0(\lambda), \ldots, x_{n-1}^0(\lambda)\}$ *is a solution of* (21.125). *If* $\bar{x}(\lambda) = \{x_1^0(\lambda), \ldots, x_{n-1}^0(\lambda)\}$ *is a solution of* (21.125), *there exists a finite (nonzero)*

number of power series $x_n^j(\lambda)$ *in* $K^\infty\{\lambda\}$ *such that* $x^j(\lambda) = \{x_1^0(\lambda), \ldots,$
$x_{n-1}^0(\lambda), x_n^j(\lambda)\}$ *is a solution of* (21.112).

Proof. The first two assertions of the lemma are direct consequences of the definition of a resultant system. We shall prove the last assertion. Let $\bar{x}(\lambda) = \{x_1^0(\lambda), \ldots, x_{n-1}^0(\lambda)\}$ be a solution of the system (21.125), and let

$$\Delta_\mu(\lambda, x_n) = {}^*f_\mu\left[\lambda; x_1^0(\lambda), \ldots, x_{n-1}^0(\lambda), x_n\right] \quad (\mu = 1, \ldots, m). \quad (21.126)$$

It is obvious that the n-tuple $\{x_1^0(\lambda), \ldots, x_{n-1}^0(\lambda), x_n^0(\lambda)\}$ is a solution of the system (21.123) if and only if the series $x_n^0(\lambda)$ is a solution of the scalar system

$$\Delta_\mu(\lambda, x_n) = 0 \quad (\mu = 1, \ldots, m). \quad (21.127)$$

The definition of the resultant system for the polynomials (21.126) implies that all their resultants vanish. Thus these polynomials have common divisors of positive degree; each of the latter is obviously a polynomial in x_n with leading coefficient 1. The system (21.127) is equivalent to the single equation

$$\Delta(\lambda, x_n) = 0, \quad (21.128)$$

where $\Delta(\lambda, x_n)$ is the g.c.d. of the polynomials (21.126). Applying Theorem 21.6 to this equation, we see that it has a finite (nonzero) number of solutions $x_n^j(\lambda)$ in $K^*\{\lambda\}$, each belonging to $K^\infty\{\lambda\}$. Thus the system (21.123), and therefore also the system (21.112), has a finite (nonzero) number of solutions

$$x(\lambda) = \{x_1^0(\lambda), \ldots, x_{n-1}^0(\lambda), x_n^j(\lambda)\}. \quad \blacksquare$$

This lemma plays an essential role in Melamed [5].

We have thus reduced the problem solving the system (21.112) to that of solving the system (21.125). This new system is similar in form to (21.112), but it involves one less variable.

Assume that the polynomials (21.122) have no common factors of positive degree; the system (21.123) is then said to be *nondegenerate*. By the definition of a resultant system, not all the series (21.124) vanish. We may therefore assume, without loss of generality, that *none* of the series (21.124) vanishes, and the system (21.125) may be studied in the same way as the system (21.112) above. The order of the system (21.125) may prove to be zero; in this case the system (21.125), and so also (21.122), is unsolvable. But if the order of the system (21.125) is positive, the system is equivalent to a system of algebraic equations in, say x_{n-1}:

$$*f_\mu^{(1)}(\lambda; x_1, \ldots, x_{n-1}) = 0 \qquad (\mu = 1, \ldots, m_1). \qquad (21.129)$$

In the nondegenerate case, one can reduce the study of this system to that of the second elimination system:

$$*f_\mu^{(2)}(\lambda; x_1, \ldots, x_{n-2}) = 0 \qquad (\mu = 1, \ldots, m_2). \qquad (21.130)$$

The procedure may be continued.

It is clear that this procedure does not yield *all* the solutions of the system (21.112).

In favorable cases, one either establishes that the original system (21.112) has no solutions (when the degree of one of the elimination systems is found to be zero), or derives a chain of systems of algebraic equations

$$\left. \begin{array}{ll} *f_\mu(\lambda; x_1, \ldots, x_{n-1}, x_n) = 0 & (\mu = 1, \ldots, m), \\ *f_\mu^{(1)}(\lambda; x_1, \ldots, x_{n-1}) = 0 & (\mu = 1, \ldots, m_1), \\ \cdot \ \cdot \ \cdot \ \cdot \ \cdot \ \cdot \ \cdot \ \cdot \ \cdot \ \cdot \ \cdot \ \cdot \ \cdot \ \cdot \ \cdot \\ *f_\mu^{(n-1)}(\lambda; x_1) = 0 & (\mu = 1, \ldots, m_{n-1}). \end{array} \right\} \qquad (21.131)$$

In the latter case, solution of the system (21.112) involves first finding solutions $x_1 = x_1(\lambda)$ of the last system in (21.131). These solutions are then substituted in the previous system, which is then solved. The procedure continues until the solutions of the first system in (21.131) are found—this gives the solutions of the original system (21.112). If the last system in (21.131) is unsolvable, so is the system (21.112).

A system (21.112) whose solutions may be determined by the above procedure is said to be *structurally stable*. It follows from the above arguments that *structurally stable systems always have finitely many solutions*.

Now consider the case in which the k-th system of algebraic equations

$$*f_\mu^{(k)}(\lambda; x_1, \ldots, x_{n-k}) = 0 \qquad (\mu = 1, \ldots, m_k) \qquad (21.132)$$

is degenerate, i.e., the left-hand sides have common divisors of positive degree. The system (21.132) then reduces to a single equation

$$d_k(\lambda; x_1, \ldots, x_{n-k}) = 0, \qquad (21.133)$$

where $d_k(\lambda; x_1, \ldots, x_{n-k})$ is the g.c.d. of the polynomials $*f_\mu^{(k)}(\lambda; x_1, \ldots, x_{n-k})$, and a system of algebraic equations

$$^0f_\mu^{(k)}(\lambda; x_1, \ldots, x_{n-k}) = 0 \qquad (\mu = 1, \ldots, m_k), \qquad (21.134)$$

where $^0f_\mu^{(k)}(\lambda; x_1, \ldots, x_{n-k})$ are polynomials defined by

$$*f_\mu^{(k)}(\lambda; x_1, \ldots, x_{n-k}) = d_k(\lambda; x_1, \ldots, x_{n-k}) \cdot {}^0f_\mu^k(\lambda; x_1, \ldots, x_{n-k}). \ (21.135)$$

It is not hard to show that the g.c.d. $d_k(\lambda; x_1, \ldots, x_{n-k})$ is a polynomial in x_{n-k} with leading coefficient 1. Thus equation (21.133) defines one or more $(n - k - 1)$-dimensional families of solutions, which are also solutions of the system (21.132). This gives a chain of systems

$$\left. \begin{array}{c} *f_\mu(\lambda; x_1, \ldots, x_n) = 0 \quad (\mu = 1, \ldots, m), \\[4pt] \cdots \cdots \cdots \cdots \cdots \cdots \cdots \cdots \cdots \cdots \cdots \cdots \\[4pt] *f_\mu^{(k-1)}(\lambda; x_1, \ldots, x_{n-k+1}) = 0 \quad (\mu = 1, \ldots, m_{k-1}), \\[4pt] d_k(\lambda; x_1, \ldots, x_{n-k}) = 0, \end{array} \right\} \qquad (21.136)$$

which defines one or more $(n - k - 1)$-dimensional families of solutions of the system (21.112).

Now consider the system (21.134). It may turn out that one of the polynomials $^0f_\mu^{(k)}(\lambda; x_1, \ldots, x_{n-k})$ has degree zero (if one of the polynomials $*f_i^{(k)}(\lambda; x_1, \ldots, x_{n-k})$ divides all the others). Then the system (21.134) is unsolvable, and the solutions of (21.112) are determined by (21.136) alone. If all equations (21.134) have positive degrees, the system (21.134) is solvable; to determine the solutions, we must set up an elimination system

$$*f_\mu^{(k+1)}(\lambda; x_1, \ldots, x_{n-k-1}) = 0 \qquad (\mu = 1, \ldots, m_{k+1}). \qquad (21.137)$$

Since the left-hand sides of equations (21.134) have no nontrivial common divisors, not all the series $f_\mu^{(k+1)}$ vanish. Thus, if the system (21.137) is of positive degree, it may also be reduced to a system of algebraic equations in x_{n-k-1}:

$$*f_\mu^{(k+1)}(\lambda; x_1, \ldots, x_{n-k-1}) = 0 \qquad (\mu = 1, \ldots, m_{k+1}) \qquad (21.138)$$

and a suitable elimination system is again constructed.

Note that if one of the elimination systems is degenerate, the original system (21.112) has infinitely many solutions. This leads to the following assertion (Melamed [4, 5]).

Theorem 21.9. The system (21.112) *is structurally stable if and only if it has finitely many solutions.*

One is naturally interested in sufficient conditions for the system (21.112) to be structurally stable. Conditions of this kind may be formulated in terms of the "isolatedness" of the trivial solution. Let us say that the trivial solution of the system (21.112) is *isolated* at $\lambda = 0$ if the system

$$\phi_\mu(\lambda; x_1, \ldots, x_n) = 0 \qquad (\mu = 1, \ldots, m), \tag{21.139}$$

where

$$\phi_\mu(\lambda; x_1, \ldots, x_n) = f_\mu(0; x_1, \ldots, x_n), \tag{21.140}$$

has no nontrivial solutions in $K_n^\infty \{\lambda\}$. Note that the parameter λ does not actually appear in the left-hand sides of the equations (21.139).

Theorem 21.10. Assume that the trivial solution of the system (21.112) *is isolated at $\lambda = 0$.*

Then the system is structurally stable and, in particular, has finitely many solutions in $K^\infty \{\lambda\}$.

This theorem is easily proved by analyzing the elimination method set forth above; we shall not go into details.

Note that the condition of Theorem 21.10 is only sufficient. For example, the system

$$(x_1 + x_2)(\lambda - x_1) = 0, \qquad \lambda(\lambda + x_2) = 0$$

is structurally stable: it has a unique solution in $K_2^\infty \{\lambda\}$, but its trivial solution is not isolated at $\lambda = 0$, since the system

$$-(x_1 + x_2) x_1 = 0, \qquad 0 = 0$$

has the solution $x_1 = \lambda$, $x_2 = -\lambda$.

Exercise 21.16. (Hilbert's Nullstellensatz). Let $g(\lambda; x_1, \ldots, x_n)$ be a formal power series which vanishes at all solutions of the system (21.112) in $K_n^\infty \{\lambda\}$. Show that there exist a number ρ and series $\psi_1, \ldots, \psi_m$ in $K\{\lambda; x_1, \ldots, x_n\}$ such that

$$g^\rho = \psi_1 f_1 + \ldots + \psi_m f_m.$$

Exercise 21.17. A system

$$\Phi_1(\lambda; x_1, \dots, x_l, x_{l+1}, \dots, x_{n-1}, x_n) = 0,$$
$$\Phi_2(\lambda; x_1, \dots, x_l, x_{l+1}, \dots, x_{n-1}) = 0,$$
$$\cdot \quad \cdot \quad \cdot \quad \cdot \quad \cdot \quad \cdot \quad \cdot \quad \cdot \quad \cdot \quad \cdot \quad \cdot \quad \cdot \quad \cdot \quad \cdot$$
$$\Phi_{n-l}(\lambda; x_1, \dots, x_l, x_{l+1}) = 0$$

is said to be a *regular system of order l* if each formal power series Φ_{n-j} is a polynomial in x_{j+1} of positive degree s_{n-j}, with coefficients in $K\{\lambda; x_1, \dots, x_j\}$, and the leading coefficient (with respect ot x_{j+1}) is 1; moreover, the series $\Phi_1, \dots, \Phi_{n-l}$ generate a prime ideal N in the ring $K\{\lambda; x_1, \dots, x_n\}$ (i.e., if $\Psi_1 \Psi_2 \in N$, then either $\Psi_1 \in N$ or $\Psi_2 \in N$). Show that a regular system of order zero has exactly $s = s_1 \dots s_n$ different solutions, each belonging to $K_n^s\{\lambda\}$; these solutions are mutually conjugate, and their radix is s. Show that the solutions of a regular system of nonzero order l form a finite number of l-dimensional families.

Exercise 21.18. Can any system (21.112) be split into a finite number of regular systems by a suitable linear transformation of the variables (Lefschetz)?

Exercise 21.19. Using Lemma 21.5, show that any structurally stable system (21.112) is equivalent to a system of algebraic equations over $K\{\lambda\}$ in all the variables $x_1, \dots, x_n$.

21.6. *Structurally stable case.* In the most frequently encountered systems of type (21.112) (see §22), one can determine only finitely many coefficients of the series (21.113), and so the method of elimination is not always practical. Even in cases in which the method is possible, substantial modifications are in order.

First and foremost, there are two basic obstacles to applications of the elimination method. The first is that, knowing only finitely many coefficients of the power series (21.113), one can determine only finitely many terms in the series expansions of the coefficients of the polynomials (21.123). This information cannot adequately indicate whether these polynomials have a nontrivial common divisor. Analogous difficulties arise at each step of the elimination procedure. Thus, one can expect a detailed analysis to be feasible only for structurally stable systems.

The second obstacle arises in determining whether the last system in (21.131) has a general solution. There is no algorithm (and there *can* be none!) which will do this using the finitely many known coefficients in the series expansions of the left-hand sides of the scalar equations in this system. True, this difficulty disappears if the basic system consists of two equations—there is then only one resultant.

It would thus seem that investigation and numerical solution of systems of type (21.112) require other methods. Here we shall describe a computation scheme for the elimination method which is often—though not always (even for structurally stable systems)—effective. We shall treat the scheme for systems of three equations in three unknowns; the extension to larger systems is obvious.

Given the equations

$$f_1(\lambda; x_1, x_2, x_3) = 0, \quad f_2(\lambda; x_1, x_2, x_3) = 0, \quad f_3(\lambda; x_1, x_2, x_3) = 0. \quad (21.141)$$

Suppose that, through a suitable linear transformation and the Weierstrass Preparation Theorem, we have been able to replace this system by an equivalent system whose left-hand sides are polynomials of degrees s_1, s_2, s_3 in x_3, with leading coefficients 1. This means that the system (21.141) is equivalent to a system

$$\left.\begin{array}{l}
{}^*f_1 \equiv x_3^{s_1} + a_{1,1}(\lambda; x_1, x_2) x_3^{s_1-1} + \ldots + a_{1,s_1}(\lambda; x_1, x_2) = 0, \\[4pt]
{}^*f_2 \equiv x_3^{s_2} + a_{2,1}(\lambda; x_1, x_2) x_3^{s_2-1} + \ldots + a_{2,s_2}(\lambda; x_1, x_2) = 0, \\[4pt]
{}^*f_3 \equiv x_3^{s_3} + a_{3,1}(\lambda; x_1, x_2) x_3^{s_3-1} + \ldots + a_{3,s_3}(\lambda; x_1, x_2) = 0.
\end{array}\right\} \quad (21.142)$$

The formal series $a_{i,j}(\lambda; x_1, x_2)$ may not be completely known; only a finite number of their coefficients will be used in the following construction. If all the coefficients of the series $f_i(\lambda; x_1, x_2, x_3)$ are known, one can determine any finite number of coefficients of the series $a_{i,j}(\lambda; x_1, x_2)$.

Let $f_{1,2}$ denote the resultant of the polynomials ${}^*f_1$ and ${}^*f_2, f_{1,3}$ that of ${}^*f_1$ and ${}^*f_3$. These resultants are formal power series in λ, x_1, x_2, and a finite number of their coefficients are known. Our first basic assumption will be that these series are regular in x_2, and that we have been able to prove this on the basis of the known coefficients (here one sometimes needs a suitable linear transformation of the variables x_1, x_2). Then, by the Weierstrass Preparation Theorem, the system

$$f_{1,2}(\lambda; x_1, x_2) = 0, \qquad f_{1,3}(\lambda; x_1, x_2) = 0 \qquad (21.143)$$

is equivalent to a system of algebraic equations

$$\left.\begin{array}{l}
x_2^{p_1} + b_{1,1}(\lambda; x_1) x_2^{p_1-1} + \ldots + b_{1,p_1}(\lambda; x_1) = 0, \\[4pt]
x_2^{p_2} + b_{2,1}(\lambda; x_1) x_2^{p_2-1} + \ldots + b_{2,p_2}(\lambda; x_1) = 0.
\end{array}\right\} \quad (21.144)$$

The first coefficients of the formal series $b_{i,j}(\lambda; x_1)$ are determined from the first coefficients of the series $a_{i,j}(\lambda; x_1, x_2)$.

Let $f_{1,2,3}(\lambda; x_1)$ denote the resultant of the left-hand sides of equations

(21.144). It is not difficult to determine finitely many coefficients of the formal series $f_{1,2,3}(\lambda; x_1)$.

Consider the scalar equation

$$f_{1,2,3}(\lambda; x_1) = 0. \tag{21.145}$$

It is natural to attack this equation with Newton polygons. Suppose that the known coefficients of the series $f_{1,2,3}$ suffice to determine the number of solutions of equation (21.145) and the first coefficients of their series expansions in integral or fractional powers of the parameter λ. The rest of the construction proceeds for each solution separately. Let

$$x_1 = x_1^0(\lambda) \tag{21.146}$$

be one of these solutions. By Theorem 21.1, the system

$$\left.\begin{array}{l} x_2^{p_1} + b_{1,1}\left[\lambda; x_1^0(\lambda)\right]x_2^{p_1-1} + \ldots + b_{1,p_1}\left[\lambda; x_1^0(\lambda)\right] = 0, \\ x_2^{p_2} + b_{2,1}\left[\lambda; x_1^0(\lambda)\right]x_2^{p_2-1} + \ldots + b_{2,p_2}\left[\lambda; x_1^0(\lambda)\right] = 0, \end{array}\right\} \tag{21.147}$$

derived from (21.144) when x_1 is replaced by the series (21.146), is solvable (note that the systems (21.144) and (21.143) are equivalent). Each equation in (21.144) may be solved by Newton polygons. Assume that the known coefficients of the series expansions of the left-hand sides of equations (21.147) determine the first terms in the series expansions of the solutions $x_2^{1,j}(\lambda)$ $(j = 1, \ldots, p_1)$ of the first equation in (21.147); similarly for the solutions $x_2^{2,j}$ $(j = 1, \ldots, p_2)$ of the second equation. Moreover, assume that the known coefficients show that all but one of the solutions $x_2^{1,j}(\lambda)$ are not solutions of the second equation (these constitute the second group of assumptions). Then the remaining solution $x_2^0(\lambda)$ is also a solution of the second equation (21.147). It is immaterial that the known coefficients may be insufficient to determine whether the solution $x_2^0(\lambda)$ is a simple solution of one or both equations (21.147).

Now substitute the series $x_1^0(\lambda)$ and $x_2^0(\lambda)$ in the left-hand sides of equations (21.142). The result is a system

$$\left.\begin{array}{l} x_3^{s_1} + a_{1,1}\left[\lambda; x_1^0(\lambda), x_2^0(\lambda)\right]x_3^{s_1-1} + \ldots + a_{1,s_1}\left[\lambda; x_1^0(\lambda), x_2^0(\lambda)\right] = 0, \\ x_3^{s_2} + a_{2,1}\left[\lambda; x_1^0(\lambda), x_2^0(\lambda)\right]x_3^{s_2-1} + \ldots + a_{2,s_2}\left[\lambda; x_1^0(\lambda), x_2^0(\lambda)\right] = 0, \\ x_3^{s_3} + a_{3,1}\left[\lambda; x_1^0(\lambda), x_2^0(\lambda)\right]x_3^{s_3-1} + \ldots + a_{3,s_3}\left[\lambda; x_1^0(\lambda), x_2^0(\lambda)\right] = 0. \end{array}\right\}$$

$$\tag{21.148}$$

By Theorem 21.1, the first two equations of this system have at least one common solution, which may be determined via Newton polygons, as described in the preceding paragraph. Assume that this common solution $x_3^0(\lambda)$ is unique and is a simple solution of the first equation (the next essential assumption is that the latter can be verified on the basis of the known coefficients). Similarly, we find a common solution of the first and third equation in (21.148). If this common solution $x_3^{00}(\lambda)$ is again unique, we are done. If (again, according to the known coefficients) $x_3^0(\lambda)$ coincides with $x_3^{00}(\lambda)$, the system (21.148) has the solution $x_3^0(\lambda)$, and the triple $\{x_1^0(\lambda), x_2^0(\lambda), x_3^0(\lambda)\}$ is a solution of the system (21.141). But if $x_3^0(\lambda)$ and $x_3^{00}(\lambda)$ are different, the system (21.141) has no solutions whose first component $x_1(\lambda)$ is $x_1^0(\lambda)$.

If the known coefficients of the series (21.148) prove insufficient for this procedure, one can try to determine additional coefficients in some way. Unfortunately, using finitely many coefficients, one cannot check whether there is some large number of coefficients which will yield a complete solution of the problem.

21.7. *Systems of analytic equations.* Consider the system

$$g_\mu(\lambda; x_1, \ldots, x_n) = 0 \qquad (\mu = 1, \ldots, m), \tag{21.149}$$

where

$$g_\mu(\lambda_0; x_1^0, \ldots, x_n^0) = 0 \tag{21.150}$$

and the functions $g(\lambda; x_1, \ldots, x_n)$ are analytic in a neighborhood of the point $\{\lambda_0, x_1^0, \ldots, x_n^0\}$. The system (21.149) may be rewritten as

$$f_\mu(v; h_1, \ldots, h_n) = 0 \qquad (\mu = 1, \ldots, m), \tag{21.151}$$

where

$$f_\mu(v; h_1, \ldots, h_n) = g_\mu(\lambda_0 + v; x_1^0 + h_1, \ldots, x_n^0 + h_n).$$

We wish to determine solutions $x_1(\lambda), \ldots, x_n(\lambda)$ of the system (21.149) which are close to $x_1^0, \ldots, x_n^0$ for λ close to λ_0, that is to say, small (for small v) solutions $h_1(v), \ldots, h_n(v)$ of the system (21.151).

If the left-hand sides of the system (21.151) possess power series expansions, they may be treated as a system (21.144) in formal power series. Simple examples show that the system (21.151) may have function-solutions which cannot be expandèd in either integral or fractional powers of v. There are also systems (21.144) whose left-hand sides are convergent series (i.e., analytic functions), and whose solutions are formal series convergent for $v = 0$ only. The method of elimination provides an immediate clarification of the situations in which these cases may occur.

All the constructions of subsection 21.4 carry over without modification to systems of equations in analytic functions. The only change is that the Weierstrass Preparation Theorem for normal power series is needed for analytic functions, and common factors in formal series must be replaced by common function-factors. Of course, the solution is then a system of functions (not series) satisfying the equation.

If the system (21.151) turns out to be structurally stable, each component $h_i(v)$ of a solution is determined by a scalar analytic equation. It may therefore be expanded in a series of powers of v, with some radix. If the equations of the elimination system possess a common factor at some stage of the elimination method (equivalently, if the system (21.151) has an infinite set of solutions), then, clearly, the set of solutions of (21.151) contains solutions which cannot be expanded in power series. This was apparently first noticed by Vainberg and Trenogin [1, 2]. Analogous reasoning leads to the following important assertion, first established, apparently, by Melamed.

Theorem 21.11. Regarding the system (21.151) as a system in formal power series, assume that it has finitely many formal solutions. Then each formal solution converges in a neighborhood of zero and its sum is a function-solution of the system.

Regarding the system (21.151) as a system in formal power series, assume that it has infinitely many solutions. Then some of these formal solutions converge only for $v = 0$.

It follows from this theorem that the procedure described in subsection 21.6 for solving structurally stable systems always yields true solutions of analytic equations.

A natural complement to Theorem 21.11 is the following assertion (Melamed [4, 5]).

Theorem 21.12. Assume that the trivial solution of the system (21.151) is isolated at $v = 0$. Then the system (21.151), regarded as a system in formal series (and also of analytic functions), possesses finitely many solutions.

The proof is left to the reader.

Suppose that the analytic system (21.151) has been solved by some method as a system of formal power series, and a solution $h^0(v) = \{h_1^0(v), \ldots, h_n^0(v)\}$ found. If these formal series converge in a neighborhood of zero, their sums are obviously solutions of the system. It is therefore important to have techniques for investigating individual formal solutions, without information as to how many other solutions the system has. The most frequently used technique for convergence proofs is that of majorant series. Theorems 21.11 and 21.12 imply the following statement.

Let $h_i^0(v) = h_i^{(r)}(v) + o(v^\varepsilon r)$, where

$$h_i^{(r)}(v) = \sum_{k=1}^{r} c_k^i v^{\varepsilon k}.$$

Consider the new system

$$f_\mu \left[v, h_1^{(r)}(v) + \xi_1 v^\varepsilon r, \ldots, h_n^{(r)}(v) + \xi_n v^\varepsilon r \right] = 0 \qquad (\mu = 1, \ldots, m).$$

If it has finitely many solutions, the formal solution $h^0(v)$ converges for small v. Note that the original system (21.151) may well have infinitely many formal solutions.

In the last three subsections, we have not discussed the relation between the number n of unknowns and the number m of equations in the systems. The most frequent case is $n = m$.

§ 22. Branching of solutions of operator equations

22.1. *Statement of the problem.* Let E, F, Λ be Banach spaces, $f = f(\lambda, x)$ an operator defined for $\lambda \in S(\lambda_0, a)$, $x \in S(x_0, b)$ with values in F (recall that $S(\lambda_0, a)$ denotes the ball $\|\lambda - \lambda_0\|_\Lambda \leqq a$ in Λ, and $S(x_0, b)$ the ball $\|x - x_0\|_E \leqq b$ in E, etc.).

In this section we shall continue our study of equation (20.1):

$$f(\lambda, x) = 0. \tag{22.1}$$

As in §20, we shall assume that

$$f(\lambda_0, x_0) = 0, \tag{22.2}$$

and look for solutions $x = x_*(\lambda)$ of equation (22.1) that are close to x_0 for λ close to λ_0.

The case in which the linear operator $f_x'(\lambda_0, x_0)$ exists and has a continuous inverse Γ was discussed in detail in § 20. We recall only that, under fairly broad assumptions on the smoothness of the operator f, equation (22.1) has a unique solution close to x_0 for small $\lambda - \lambda_0$; that is, equation (22.1) defines a unique implicit function equal to x_0 for $\lambda = \lambda_0$.

We shall now direct our attention to a considerably more complicated case: $f_x'(\lambda_0, x_0)$ is not continuously invertible. Even in the scalar case, equation (22.1) often exhibits a new phenomenon (see §21): Branching of solutions.

The main results known to date in the theory of branching of solutions of equation (22.1) are due to Lyapunov [1], Schmidt [1] and Nekrasov [1, 2]. Also worthy of mention are Lichtenstein [1], Iglisch [1, 2], Nazarov [1–4], Gel'man [1–6], Bartle [1], Cronin [1, 2], Pokornyi [1–4], Melamed [3–5], Trenogin [1–3], Akhmedov [1]. Several authors have studied special classes of equations. In this section we shall follow certain arguments of Zabreiko and Krasnosel'skii [4–6].

There are two essentially different techniques for investigating solutions of (22.1) close to x_0. The first technique includes methods whereby equation (22.1) is replaced by an equivalent finite-dimensional equation

$$\phi(v, u) = 0 \tag{22.3}$$

or, equivalently, by a system of m scalar equations in n scalar unknowns:

$$\left.\begin{array}{l} \phi_1(v; u_1, \ldots, u_n) = 0, \\ \cdot \quad \cdot \quad \cdot \quad \cdot \quad \cdot \quad \cdot \quad \cdot \quad \cdot \\ \phi_m(v; u_1, \ldots, u_n) = 0. \end{array}\right\} \tag{22.4}$$

Equation (22.3) or the system (22.4) is known as a *branching equation*. The solutions of equation (22.1) are simply expressed in terms of the solutions of the scalar system (22.4). The branching equation therefore summarizes all the properties of equation (22.1): the way in which the solutions depend on the parameters, the number of solutions, etc. The branching equation may be attacked by the theory of §21.

The second technique is based on direct construction of solutions and direct investigation of the operator $f(\lambda, x)$. Of course, this distinction between the two techniques is purely formal, and they are strongly interrelated.

22.2. *Decomposable operators.* We recall a few definitions.

A linear operator P defined in a Banach space E is called a *projection* if $P^2 = P$. Each projection generates a decomposition of the space E into a direct sum $E_0 \oplus E^0$, where $E_0 = PE$ is the space *onto which P projects E* and E^0 is the null subspace of P, *the direction of the projection P.* The operator $I - P$ is also a projection.

If E is expressed as a direct sum of subspaces E_0 and E^0, i.e., every element $x \in E$ has a unique representation in the form $x = u + v$ $(u \in E_0,$ $v \in E^0)$, the operator $P_0 x = u$ projects E onto E_0 in the direction of E^0, while $P^0 = I - P_0$ projects E onto E^0 in the direction of E_0.

Not every subspace of a Banach space is the range of a projection (see Dunford and Schwartz [1]). In a Hilbert space, every subspace is the range of a projection.

Exercise 22.1. Show that in L_∞ there is no projection whose range is the space C of continuous functions.

Exercise 22.2. Let L_M^* be an Orlicz space, defined by an N-function $M(u)$ which does not satisfy the Δ_2-condition (see Krasnosel'skii and Rutitskii [1]). Show that the space E_M (the closure of the set of bounded functions) is not the range of a projection.

Let E_0 be a finite-dimensional subspace. Let $e_1, \ldots, e_m$ be a basis of E_0 and define linear functionals $l_1, \ldots, l_m$ on E_0 by

$$l_1(\xi_1 e_1 + \ldots + \xi_m e_m) = \xi_i \qquad (i = 1, \ldots, m). \qquad (22.5)$$

Now extend the functionals $l_1, \ldots, l_m$ to the entire space E, preserving their norms (it suffices to preserve continuity), retaining the same notation for the extensions. Then the formula

$$P_0 x = \sum_{i=1}^{m} l_i(x) e_i \qquad (x \in E) \qquad (22.6)$$

defines a projection of E onto E_0. The operator $P^0 = I - P_0$ projects E onto the intersection E^0 of the null subspaces of the functionals $l_1, \ldots, l_m$.

Exercise 22.3. Let E_0 be finite-dimensional and $e_1, \ldots, e_m$ a basis of E_0. Let $\phi_1, \ldots, \phi_m$ be linear functionals such that

$$D = \begin{vmatrix} \phi_1(e_1) \ldots \phi_1(e_m) \\ \cdot \quad \cdot \quad \cdot \quad \cdot \quad \cdot \quad \cdot \quad \cdot \\ \phi_m(e_1) \ldots \phi_m(e_m) \end{vmatrix} \neq 0 . \tag{22.7}$$

Show that formula (22.6) defines a projection P_0 of E onto E_0, if each functional l_i is defined as a linear combination of the functionals $\phi_1, \ldots, \phi_m$.

A subspace $E^0 \subset E$ is said to have *defect m* (def $E^0 = m$) if it is the intersection of the null subspaces of m linearly independent functionals $\phi_1, \ldots, \phi_m$. By linear independence we mean that there exist elements $e_1, \ldots, e_m$ satisfying (22.7). Then, for any fixed $i = 1, \ldots, m$, the system

$$\alpha_{i1}\phi_1(e_1) + \ldots + \alpha_{im}\phi_m(e_1) = 0 ,$$

$$\cdot \quad \cdot \quad \cdot \quad \cdot \quad \cdot \quad \cdot \quad \cdot \quad \cdot \quad \cdot \quad \cdot \quad \cdot \quad \cdot$$

$$\alpha_{i1}\phi_1(e_{i-1}) + \ldots + \alpha_{im}\phi_m(e_{i-1}) = 0 ,$$

$$\alpha_{i1}\phi_1(e_i) + \ldots + \alpha_{im}\phi_m(e_i) = 1 ,$$

$$\alpha_{i1}\phi_1(e_{i+1}) + \ldots + \alpha_{im}\phi_m(e_{i+1}) = 0 ,$$

$$\cdot \quad \cdot \quad \cdot \quad \cdot \quad \cdot \quad \cdot \quad \cdot \quad \cdot \quad \cdot \quad \cdot \quad \cdot \quad \cdot$$

$$\alpha_{i1}\phi_1(e_m) + \ldots + \alpha_{im}\phi_m(e_m) = 0$$

is solvable. Now use solutions of this system to construct functionals

$$l_i(x) = \alpha_{i1}\phi_1(x) + \ldots + \alpha_{im}\phi_m(x) \qquad (x \in E) ;$$

these functionals satisfy (22.5). Thus formula (22.6) defines a projection of E onto the linear span E_0 of the points $e_1, \ldots, e_m$ in the direction of the subspace E^0. The operator $P^0 = I - P_0$ projects E onto E^0.

Let B be a linear operator mapping E into F. B is said to be *normally solvable* if its range $BE = F^0$ is closed. Let E_0 denote the *kernel* of the operator B, i.e., its null subspace. B is said to be *decomposable* if there exist projections mapping E onto E_0 and F onto F^0. If $\dim E_0$ and def F^0 are finite, the difference

$$\text{ind } B = \text{def } F^0 - \dim E_0 \tag{22.8}$$

is called the *index* (or *Noether index*) of the normally solvable operator B. Operators B for which the index (22.8) is defined are clearly decomposable and in a Hilbert space every normally solvable operator is decomposable.

An important example of a decomposable operator is any operator of the form $B = C + D$ where C is continuously invertible and D is compact.

Exercise 22.4. Show that if B is the sum of a continuously invertible operator and a compact operator then ind $B = 0$ (S. M. Nikol'skii).

22.3. *Matrix representations of decomposable operators.* Let P_1, $P^1 = I - P_1$, Q_1, $Q^1 = I - Q_1$ be projections defined in the spaces E and F, respectively, and B a linear operator mapping E into F. Obviously,

$$Bx = Q_1 B P_1 x + Q_1 B P^1 x + Q^1 B P_1 x + Q^1 B P^1 x. \qquad (22.9)$$

This decomposition of the operator B is usually written in *matrix* notation:

$$B = \begin{pmatrix} B_{11} & B_{12} \\ B_{21} & B_{22} \end{pmatrix}, \qquad (22.10)$$

where B_{11} is the restriction of the operator $Q_1 B P_1$ to $E_1 = P_1 E$; B_{12} is the restriction of $Q_1 B P^1$ to $E^1 = P^1 E$, etc. The operator B_{11} maps E_1 into $F_1 = Q_1 F$, B_{12} maps E^1 into F_1, B_{21} maps E_1 into F^1, and B_{22} maps E^1 into F^1. The representation (22.9) of an operator is said to be *nondegenerate* if B_{22} has a continuous inverse B_{22}^{-1} mapping F^1 into E^1. Finally, the representation (22.10) is said to be *regular* if it is nondegenerate and

$$B_{11} = B_{12} B_{22}^{-1} B_{21}. \qquad (21.11)$$

Theorem 22.1. An operator B has a regular representation if and only if it is decomposable.

If (22.10) is a regular representation of B, then its null subspace E_0 consists precisely of the elements $x \in E$ such that

$$P^1 x = - B_{22}^{-1} B_{21} P_1 x; \qquad (22.12)$$

the operator

$$P_0 x = P_1 x - B_{22}^{-1} B_{21} P_1 x \qquad (x \in E) \qquad (22.13)$$

projects E onto the subspace E_0; the range F^0 of B consists of the elements $f \in F$ such that

$$Q_1 f = B_{12} B_{22}^{-1} Q^1 f; \qquad (22.14)$$

the operator

$$Q^0 f = Q^1 f - B_{12} B_{22}^{-1} Q^1 f \qquad (f \in F) \qquad (22.15)$$

projects F onto F^0.

Proof. Let B be decomposable, P_0 a projection of E onto the null subspace E_0 of B, Q^0 a projection of F onto the range F^0 of B, $P^0 = I - P_0$, $Q_0 = I - Q^0$. Then the decomposition (22.9) is $Bx = Q^0 B P^0 x$, i.e.,

$$B = \begin{pmatrix} 0 & 0 \\ 0 & B_{22}^0 \end{pmatrix}, \qquad (22.16)$$

where B_{22}^0 is the restriction of B to $E^0 = (I - P_0)E$. The operator B_{22}^0 is a one-to-one mapping of E^0 onto $F^0 = BE$; by Banach's theorem, B_{22}^0 has a continuous inverse $(B_{22}^0)^{-1}$ on F^0; therefore, the representation (22.16) of B is nondegenerate. Equality (22.11) is obvious.

Let (22.10) be a regular representation of B, based on the four projections P_1, P^1, Q_1, Q^1. The zeros x of the operator B are determined by the system

$$B_{11} P_1 x + B_{12} P^1 x = 0, \quad B_{21} P_1 x + B_{22} P^1 x = 0. \qquad (22.17)$$

Solving the second equation, we get (22.12); by (22.11), the first equation of (22.17) then becomes an identity.

To prove that formula (22.13) defines a projection of E onto E_0, it suffices to show that $P_0 x \in E_0$ for $x \in E$, i.e.,

$$P^1 P_0 x \equiv -B_{22}^{-1} B_{21} P_1 P_0 x,$$

and $P_0 x = x$ for $x \in E_0$, i.e.,

$$P_0(P_1 x - B_{22}^{-1} B_{21} P_1 x) \equiv P_1 x - B_{22}^{-1} B_{21} P_1 x.$$

Both of these identities follow directly from the definition (22.13).

An element f is in the range of B if the system

$$B_{11} P_1 x + B_{12} P^1 x = Q_1 f, \quad B_{21} P_1 x + B_{22} P^1 x = Q^1 f$$

is solvable. The second of these equations determines $P^1 x$:

$$P^1 x = - B_{22}^{-1} B_{21} P_1 x + B_{22}^{-1} Q^1 f .$$

Substituting $P^1 x$ into the first equation and using (22.11), we get (22.14).

Finally, formula (22.15) defines a projection of F onto F^0, since $Q^0 Q^0 f \equiv Q^0 f$ ($f \in F$); the values of the operator Q^0 satisfy (22.14), and it follows from (22.11) that $Q^0 f = f$ for $f \in F^0$ (the details are left to the reader). ∎

Exercise 22.5. Let B be a normally solvable operator with finite index (therefore decomposable). Let E^1 be the complement of the null subspace E_0 of B, and P^1 a projection of E onto E^1. Let Q_1 be a projection of F onto the complement F_1 of the range F^0 of B. Show that the representation (22.16) of B defined by the decomposition (22.9) is regular.

22.4. *Branching equations.* We now return to equation (22.1), introducing the notation

$$B = f_x'(\lambda_0, x_0), \qquad h = x - x_0, \qquad v = \lambda - \lambda_0 .$$

Then equation (22.1) (with condition (22.2)) becomes

$$Bh = \Omega(v, h), \tag{22.18}$$

where $\Omega(0, 0) = 0$ and $\Omega_h'(0, 0) = 0$.

Let

$$B = \begin{pmatrix} B_{11} & B_{12} \\ B_{21} & B_{22} \end{pmatrix} \tag{22.19}$$

be a matrix representation of the operator B, generated by projections P_0, P^0 (of E) and Q_0, Q^0 (of F). Then equation (22.10) may be written as a system

$$\left. \begin{array}{l} B_{11} u + B_{12} v = Q_0 \Omega(v, u + v), \\ B_{21} u + B_{22} v = Q^0 \Omega(v, u + v), \end{array} \right\} \tag{22.20}$$

where

$$u = P_0 h \in E_0, \quad v = P^0 h \in E^0 \qquad (h \in E). \tag{22.21}$$

If the operator B_{22} is continuously invertible, and $\Omega(v, h)$ is sufficiently smooth, Theorem 20.1 implies that the second of equations (22.20) defines a unique implicit function

$$v = R(v, u), \tag{22.22}$$

defined for small v and u. Consequently, the problem of small solutions of the system (22.20) (or, equivalently, of equation (22.18)) reduces to solution of an equation in the space E_0; this equation is derived from the first of equations (22.20) by replacing v by its value (22.22):

$$\phi(v, u) = 0, \qquad (22.23)$$

where

$$\phi(v, u) = B_{11}u + B_{12}R(v, u) - Q_0\Omega\left[v, u + R(v, u)\right]. \qquad (22.24)$$

If we can find the solution (or solutions) $u(v)$ of equation (22.23), a solution $h(v)$ of equation (22.18) is defined by

$$h(v) = u(v) + R\left[v, u(v)\right]. \qquad (22.25)$$

The procedure reduces equation (22.18) in the space E to equation (22.23) in the subspace E_0. If E_0 and F_0 are finite-dimensional, equation (22.23) is a finite system of scalar equations.

One often considers cases in which B is decomposable and (22.19) is a regular representation of the operator B. Equation (22.23) is then called the *branching equation*. The reader should find it easy to show that for any regular representation (22.19) of B the derivative $\phi'_u(0, 0)$ of the operator (22.24) vanishes.

Investigation of the branching equation, its solution (exact or approximate), followed by application of formula (22.25) constitute a basic technique for studying small solutions of equation (22.18). At first sight, one might expect the derivation of the branching equation (for finite-dimensional E_0 and F_0) to furnish a complete solution of the problem, since all that remains is to deal with finite systems (applying, say, the methods of §21).* However, one must remember that there is no explicit expression for the branching equation, as there is none for the operator $R(v, u)$ of (22.22). Thus, one cannot always study the branching equation directly. It is in this connection that the importance of other methods, not employing equation (22.23), becomes evident; they are often successful when equation (22.23) is untractable.

---

* The first indication of the role of elimination theory in the theory of branching is apparently due to Lefschetz [1]. Applications of the method may be found in the work of many authors (Aizengendler, Vainberg, Melamed, Trenogin, etc.).

These arguments are also valid for other finite systems which provide theoretical treatment of the problem of small solutions.

Exercise 22.6. Let B be decomposable, F_0 a subspace of F complementary to the range F^0 of B. Let $\Pi(v)$ denote the surface consisting of all $h \in E$ for which the vector $Bh - \Omega(v, h)$ is parallel to F_0. Show that E has a direct-sum representation $E = E_0 \oplus E^0$, such that, for small v, every small $u \in E_0$ corresponds to a unique $v \in E^0$ such that $u + v \in \Pi(v)$.

Exercise 22.7. Under the assumptions of Exercise 22.6, parameterize the surface $\Pi(v)$ (for small v, in a neighborhood of the zero of E) by the projection u of its points $x \in \Pi(v)$ onto E_0. Then, for $x \in \Pi(v)$, the vectors $Bx - \Omega(v, x)$ may be regarded as vectors $\Phi(\gamma, u) \in F_0$. Construct a regular representation (22.19) of B such that the branching equation (22.23) coincides with the equation $\Phi(v, u) = 0$.

Apparently, equation (22.23) was first derived by Lyapunov, for nonlinear integral equations. Lyapunov in effect constructed regular representations (22.16) of an operator B. More general representations (22.19) of B are convenient in cases where the second equation of (22.20) may be simplified by suitable selection of the operator Q^0 (for example, if equation (22.18) has finite-dimensional nonlinearities).

We now proceed to the so-called *Schmidt branching equation* (Schmidt also constructed these for integral equation; as in the case of the Lyapunov branching equation, generalization to operator equations is immediate).

Consider equation (22.18). Let C be a linear operator mapping E into a subspace F_0 of F, such that $B + C$ is continuously invertible. Replace equation (22.18) by the equivalent system

$$\left.\begin{array}{r} (B + C)h = \Omega(v, h) + f, \\ Ch = f, \end{array}\right\} \tag{22.26}$$

where $f \in F_0$. If the operator $\Omega(v, h)$ is sufficiently smooth, the first of equations (22.26) can be solved in h (for small v and f). Substitute the solution

$$h = T(v, f) \tag{22.27}$$

in the second of equations (22.26); the result is the Schmidt branching equation

$$\psi(v, f) = 0, \tag{22.28}$$

where

$$\psi(v, f) = CT(v, f) - f. \tag{22.29}$$

To find the complete solution of equation (22.18), we must solve equation (22.29) and then substitute the solution (or solutions) $f = f(v)$ in the right-hand side of (22.27).

Let B be normally solvable and of finite index (22.8). As usual, let E_0 be the null subspace of the operator B, $F^0 = BE$, E^0 and F_0 the complements, P_0, P^0, Q_0, Q^0 the corresponding projections. A regular representation of B then has the form (22.16).

Let ind $B = 0$, i.e., E_0 and F_0 have equal finite dimension m. Then, the operator C may be constructed as follows. Choose bases $e_1, \ldots, e_m$ and $f_1, \ldots, f_m$ of E_0 and F_0, respectively; set

$$C_0(\xi_1 e_1 + \ldots + \xi_m e_m) = \xi_1 f_1 + \ldots + \xi_m f_m, \qquad (22.30)$$

and now extend C_0 to E by the formula

$$Cx = C_0 P_0 x \qquad (x \in E). \qquad (22.31)$$

The representation (22.19) of the operator $B + C$ has the form

$$B + C = \begin{pmatrix} C_0 & 0 \\ 0 & B_{22} \end{pmatrix}. \qquad (22.32)$$

Thus $B + C$ can be inverted if the inverses of B_{22} (in F^0) and C_0 (in F_0) are known.

Let us derive the Schmidt branching equation for the case in which C is defined by (22.30) and (22.31), using the notation (22.21). The first equation of the system (22.26) is equivalent to the system

$$\left. \begin{array}{l} C_0 u = Q_0 \Omega(v, u + v) + f, \\ B_{22} u = Q^0 \Omega(v, u + v), \end{array} \right\} \qquad (22.33)$$

and the second is the equation

$$C_0 u = f. \qquad (22.34)$$

The second equation of (22.23) coincides with the second equation of (22.20) (based, of course, on the representation (22.16) of B). The solution of the second equation of (22.33) is therefore precisely the function (22.22). Substituting this solution in the first equation of (22.33), we get

$$C_0 u + \phi(v, u) = f, \qquad (22.35)$$

where $\phi(v, u)$ is the left-hand side of the Lyapunov branching equation

(22.23). To construct the Schmidt branching equation, we must use equation (22.35), or, equivalently, the equation

$$u + C_0^{-1}\phi(v, u) = w \qquad (w = C_0^{-1}f \in E_0) \tag{22.36}$$

to determine $u = \chi(v, w)$, and then form the equality

$$\chi(v, w) = w \ ; \tag{22.37}$$

this is precisely the Schmidt branching equation (22.28).

For the sequel, we need the Lyapunov branching equation (22.23) as a system

$$\left.\begin{array}{r} u + C_0^{-1}\phi(v, u) = w, \\ u = w, \end{array}\right\} \tag{22.38}$$

and the Schmidt branching equation (22.37) as a system

$$\left.\begin{array}{r} \chi(v, w) = u, \\ u = w. \end{array}\right\} \tag{22.39}$$

Consider the space of pairs $\{u, v\}$ $(u, w \in E_0)$, and define operators

$$T_1\{u, w\} = \{w, u + C_0^{-1}\phi(v, u)\}, \quad T_2\{u, w\} = \{\chi(v, w), u\} . \tag{22.40}$$

It is clear that the solutions of the system (22.38) are the fixed points of the operator T_1, while those of the system (22.39) are the fixed points of T_2. But the operators T_1 and T_2 are mutual inverses (for small u, w), and they therefore have the same fixed points.

We have thus proved a simple, and somewhat surprising statement: *The Schmidt branching equation* (22.37) *and the Lyapunov branching equation* (22.23) *have the same solutions* (based on identical operators P_0, P^0, Q_0, Q^0).

Our proof also illustrates the procedure for transition from one form of the branching equation to the other.

We need not dwell on the case ind $B \neq 0$, since it reduces trivially to that just considered. Indeed, let ind $B = -k < 0$; then we must consider B and $\Omega(v, h)$ as operators with range in a space $\tilde{F}$ in which F is a subspace of defect k (i.e., $\tilde{F}$ is the direct sum of F and a k-dimensional subspace). The extended operator B has zero index. If ind $B = k > 0$, we consider a space $\tilde{E}$ in which E has defect k, and replace equation (22.18) by the equivalent equation

$$BRx = \Omega(v, Rx), \tag{22.41}$$

where R is any projection of $\tilde{E}$ onto E; the operator BR, which maps $\tilde{E}$ into F, has zero index. If ind $B = -k < 0$, equation (22.18) has small solutions which generally form a k-parameter family. If ind $B > 0$, then, in the general case, equation (22.18) has no solutions for nonzero v. The first direct analysis of the branching equation for ind $B \neq 0$ is apparently due to Trenogin.

Methods for investigating linear operators B of nonzero index, based on extending the spaces E and F by finite-dimensional complements, have been developed and applied in another situation by Krein and Krasnosel'skii [2].

There are other methods for deriving branching equations for equation (22.18) (the branching equations are finite-dimensional if the null subspace of B is finite-dimensional and ind B finite). The general idea of these branching equations for equation (22.18) may be described as follows.

Suppose that, besides the spaces E, F, Λ, we have two more spaces U and Z, and operators $S_0(v, u, z, x, f)$ and $S^0(v, u, z, x, f)$, defined for small $v \in \Lambda$, $u \in U$, $z \in Z$, $x \in E$, $f \in F$ (their ranges are unimportant), such that the equalities

$$S_0(v, u, z, x, f) = 0, \qquad S^0(v, u, z, x, f) = 0$$

imply that $f = 0$. Further, let $R(v, u, z)$ be an operator with values in E, defined for small $v \in \Lambda$, $u \in U$, $z \in Z$ such that the equation

$$S^0\{v, u, z, R(v, u, z), \quad BR(v, u, z) - \Omega[v, R(v, u, z)]\} = 0$$

has a unique small solution

$$z = T(v, u).$$

It is then obvious that the function

$$x = R\{v, u(v), \quad T[v, u(v)]\}$$

is a solution of equation (22.18), provided $u = u(v)$ is a solution of

$$\chi(v, u) = 0, \tag{22.42}$$

where

$$\chi(v, u) = S_0\{v, u, T(v, u), R[v, u, T(v, u)],$$

$$BR[v, u, T(v, u)] - \Omega[v, R[v, u, T(v, u)]]\}.$$

Now assume that the equalities $z = T(v, u)$ and

$$S_0(v, u, z, x, 0) = 0, \quad S^0(v, u, z, x, 0) = 0, \quad x = R(v, u, z)$$

imply that $u = L(v, x)$, where $L(v, x)$ ($L(0, 0) = 0$) is an operator with values in U, defined for small v and x. If $x = x(v)$ is a solution of equation (22.18) which is close to 0 for v close to 0, then, as is easily seen, $u = L[v, x(v)]$ is a solution of equation (22.42). By the same token, all the solutions of equation (22.18) may be constructed using the solutions of equation (22.42); the latter is the branching equation.

Exercise 22.8. Show that the branching equations (22.23) and (22.28) are special cases of equation (22.42).

22.5. *Asymptotic approximations of branching equations.* Only rarely can one construct the exact branching equation. Much importance therefore attaches to methods for constructing approximate branching equations, whose solutions provide sufficiently complete information on the solutions of the exact branching equation. This information should include, at the very least, a bound for the number of solutions, asymptotic approximations to the solutions, and so on.

Consider the two equations

$$\chi(v, u) = 0, \tag{22.43}$$

$$\tilde{\chi}(v, u) = 0, \tag{22.44}$$

where $v \in \Lambda$, $u \in E_0$; $\chi(v, u)$ and $\tilde{\chi}(v, u)$ are operators with values in F_0, defined for small v and u, such that

$$\chi(0, 0) = 0, \quad \tilde{\chi}(0, 0) = 0.$$

Assume that the operator $\tilde{\chi}(v, u)$ is an asymptotic approximation of the operator $\chi(v, u)$, of order (or preorder) $\omega(v, u)$ (see subsection 20.6):

$$\|\chi(v, u) - \tilde{\chi}(v, u)\| = o[\omega(v, u)] \quad (\|\chi(v, u) - \tilde{\chi}(v, u)\| = O[\omega(v, u)]).$$

Here $\omega(v, u)$ is a nonnegative functional continuous at zero and vanishing only there. We shall then call (22.44) an *asymptotic approximation* (of order or preorder ω) to equation (22.43).

Many exact and approximate methods for scalar equations and systems of type (22.43) require no more than an asymptotic approxima-

tion, of sufficiently high order, to each equation. Asymptotic approximations are therefore often sufficient for investigations of the branching
equation. In this subsection we describe a few constructions for asymptotic approximations of branching equations. In various special cases,
these constructions have been applied (often in other terminology) by
many authors. The general form is due to Zabreiko and Krasnosel'skii
[5, 6].

First consider methods for constructing asymptotic approximations

$$\tilde{\phi}(v, u) = 0 \tag{22.45}$$

to the branching equation (22.23)

$$\phi(v, u) = 0 \tag{22.46}$$

for equation (22.18)

$$Bh = \Omega(v, h), \tag{22.47}$$

based on the matrix representation (22.16) of the operator B.

Equation (22.47) may be written as a system

$$0 = Q_0\Omega(v, u + v), \tag{22.48}$$

$$B_{22}v = Q^0\Omega(v, u + v), \tag{22.49}$$

where $u \in E_0$, $v \in E^0$. To construct the branching equation (22.46), we
need a solution

$$v = R(v, u) \tag{22.50}$$

of equation (22.49); the left-hand side $\phi(v, u)$ of the branching equation
is then defined by

$$\phi(v, u) = Q_0\Omega\left[v, u + R(v, u)\right]. \tag{22.51}$$

To construct asymptotic approximations to the branching equation
(22.46), it suffices to find asymptotic approximations

$$v = \tilde{R}(v, u) \tag{22.52}$$

to the operator (22.50); the left-hand side of the asymptotic approximation (22.45) to the branching equation may be defined by

$$\tilde{\phi}(v, u) = Q_0\Omega\left[v, u + \tilde{R}(v, u)\right]. \tag{22.53}$$

Using a solution $u = \tilde{u}(v)$ of equation (22.45), whose left-hand side is defined by (22.53), we can construct asymptotic approximations $h = \tilde{h}(v)$ to the solutions $h = h(v)$ of equation (22.47):

$$\tilde{h}(v) = \tilde{u}(v) + R\left[v, \tilde{u}(v)\right], \qquad (22.54)$$

where $R(v, u)$ is the operator (22.50), or, alternatively,

$$\tilde{h}(v) = \tilde{u}(v) + R_0\left[v, \tilde{u}(v)\right], \qquad (22.55)$$

where $R_0(v, u)$ is some asymptotic approximation of the operator (22.50). A convenient approximation $R_0(v, u)$ is the operator (22.52); the asymptotic approximation $\tilde{h}(v)$ to the solution $h(v)$ of equation (22.47) is then defined by

$$\tilde{h}(v) = \tilde{u}(v) + \tilde{R}\left[v, \tilde{u}(v)\right]. \qquad (22.56)$$

To estimate the order (preorder) of the asymptotic approximations we need only the order (preorder) of the asymptotic approximation $u = \tilde{u}(v)$ to the solution of equation (22.46) and the smoothness properties of the operator (22.50). To estimate the order (preorder) of the asymptotic approximation (22.56), we also need the order of the asymptotic approximation (22.52) to the operator (22.50).

In the sequel, we shall impose restrictions of the following type on the operator $\Omega(v, h)$:

$$\left\|Q_0\Omega(v, u + v_1) - Q_0\Omega(v, u + v_2)\right\| \leqq q_0(v, u, t)\|v_1 - v_2\| \qquad (22.57)$$

$$(v \in S(0, a); u \in E_0, \|u\| \leqq b; v_1, v_2 \in E^0, \|v_1\|, \|v_2\| \leqq t),$$

$$\left\|Q^0\Omega(v, u + v_1) - Q^0\Omega(v, u + v_2)\right\| \leqq q^0(v, u, t)\|v_1 - v_2\| \qquad (22.58)$$

$$(v \in S(0, a); u \in E_0, \|u\| \leqq b; v_1, v_2 \in E^0; \|v_1\|, \|v_2\| \leqq t).$$

Here $q_0(v, u, t)$ and $q^0(v, u, t)$ $(q_0(0, 0, 0) = q^0(0, 0, 0) = 0)$ are functions continuous at zero, nondecreasing in t, such that

$$q_0\left[v, u, c\omega(v, u)\right] \leqq k_0(c)\,\delta(v, u) \qquad (0 < c < \infty), \qquad (22.59)$$

$$q^0\left[v, u, c\omega(v, u)\right] \leqq k^0(c)\,\omega_0(v, u) \qquad (0 < c < \infty), \qquad (22.60)$$

where the functions $\omega(v, u)$, $\omega_0(v, u)$, $\delta(v, u)$ are nonnegative, continuous at zero, and vanish only at zero.

Lemma 22.1. *Assume that the operator* $Q^0\Omega(v, u + v)$ *satisfies condition* (22.58) *and the inequality*

$$\|Q^0\Omega(v, u)\| \leq c\omega(v, u).$$ (22.61)

Then the operator (22.50) *satisfies the inequality*

$$\|R(v, u)\| \leq c^*\omega(v, u),$$ (22.62)

where c^* *is a constant.*

Proof. By (22.58) and (22.61), the identity

$$B_{22}R(v, u) = \{Q^0\Omega[v, u + R(v, u)] - Q^0\Omega(v, u)\} + Q^0\Omega(v, u)$$

implies that

$$\|R(v, u)\| \leq \|B_{22}^{-1}\|\|q^0[v, u, \|R(v, u)\|]\|\|R(v, u)\| + \|B_{22}^{-1}\|c\omega(v, u).$$ (22.63)

The operator $R(v, u)$ obviously satisfies the condition

$$\lim_{v, u \to 0} \|R(v, u)\| = 0,$$

whence

$$\lim_{v, u \to 0} q^0[v, u, \|R(v, u)\|] = 0.$$

Inequality (22.62) follows from (22.63). ∎

Exercise 22.9. Assume that the operator $Q^0\Omega(v, u + v)$ satisfies conditions (22.58) and (22.61), and also the inequality

$$\|Q^0\Omega(v, u_1 + v) - Q^0\Omega(v, u_2 + v)\| \leq p(v, \rho, \|v\|) \|u_1 - u_2\|$$ (22.64)

$$(v \in S(0, a);\ u_1, u_2 \in E_0;\ \|u_1\|, \|u_2\| \leq \rho;\ v \in E^0,\ \|v\| \leq b),$$

where $p(v, \rho, t)$ is a nonnegative function continuous at zero such that

$$p[v, \rho, c\omega(v, u)] \leq k(c)\pi(v, \rho)\qquad (\|u\| \leq \rho, 0 < c < \infty).$$ (22.65)

Prove that the operator (22.50) satisfies the inequality

$$\|R(v, u_1) - R(v, u_2)\| \leq d\pi(v, \rho)\|u_1 - u_2\|\qquad (\|u_1\|, \|u_2\| \leq \rho),$$ (22.66)

where d is a constant.

By equality (22.53), the construction of asymptotic approximations (22.45) to the branching equation (22.46) amounts to construction of asymptotic approximations (22.52) to the operator (22.50). The order of the approximation (22.53) to the operator (22.51) is then easily estimated in terms of the order of the approximation (22.52) to the operator (22.50).

Lemma 22.2. Assume that the operators $Q_0\Omega(v, u + v)$ and $Q^0\Omega(v, u + v)$ satisfy conditions (22.57)–(22.59), (22.61). Assume that the operator (22.52) satisfies the condition

$$\|\tilde{R}(v, u)\| \leq \tilde{c}\omega(v, u) \tag{22.67}$$

and is an asymptotic approximation of preorder $\tilde{\omega}$ to the operator (22.50):

$$\|\tilde{R}(v, u) - R(v, u)\| = O\left[\tilde{\omega}(v, u)\right]. \tag{22.68}$$

Then the operator (22.53) is an asymptotic approximation of preorder $\delta\tilde{\omega}$ to the operator (22.51):

$$\|\tilde{\phi}(v, u) - \phi(v, u)\| = O\left[\delta(v, u)\,\tilde{\omega}(v, u)\right]. \tag{22.69}$$

The proof follows from the following relations, which are consequences of (22.51), (22.53), (22.57), (22.61), (22.62), (22.67) and (22.68):

$$\|\tilde{\phi}(v, u) - \phi(v, u)\| = \|Q_0\Omega\left[v, u + \tilde{R}(v, u)\right] - Q_0\Omega\left[v, u + R(v, u)\right]\| \leq$$

$$\leq q_0\left[v, u, \|\tilde{R}(v, u)\| + \|R(v, u)\|\right]\|\tilde{R}(v, u) - R(v, u)\| \leq$$

$$\leq q_0\left[v, u, (\tilde{c} + c^*)\,\omega(v, u)\right]\|\tilde{R}(v, u) - R(v, u)\| = O\left[\delta(v, u)\,\tilde{\omega}(v, u)\right].$$

A natural approach to the construction of asymptotic approximations (22.52) to the operator (22.50) is to employ the iterative methods set forth in subsection 20.6. Let us consider in detail asymptotic approximations based on the successive approximations (20.41).

Theorem 22.2. Assume that the operator $\Omega(v, h)$ satisfies conditions (22.57) through (22.61), and that the function $R_0(v, u)$ satisfies the inequality

$$\|R_0(v, u)\| \leq c_0\omega(v, u) \tag{22.70}$$

(for example, it may vanish).

Then the operator $R_n(v, u)$ defined by the recurrence relation

$$R_k(v, u) = B_{22}^{-1}Q^0\Omega\left[v, u + R_{k-1}(v, u)\right] \qquad (k = 1, 2, \ldots), \tag{22.71}$$

is an asymptotic approximation of preorder $\omega\omega_0^n$ to the operator (22.50):

$$\|R_n(v, u) - R(v, u)\| \leq d_n\omega(v, u)\left[\omega_0(v, u)\right]^n, \tag{22.72}$$

while the operator

$$\phi_{n+1}(v, u) = Q_0 \Omega \left[v, u + R_n(v, u) \right] \tag{22.73}$$

is an asymptotic approximation of preorder $\delta \omega \omega_0^n$ to the operator (22.51):

$$\| \phi_{n+1}(v, u) - \phi(v, u) \| \leq \tilde{d}_n \delta(v, u) \, \omega(v, u) \left[\omega_0(v, u) \right]^n . \tag{22.74}$$

Proof. Note first that the assertion of Theorem 20.7 remains valid if inequality (20.62) is replaced by the weaker inequality

$$\| f(\lambda, x_1) - f(\lambda, x_2) - f'_x(\lambda_0, x_0)(x_1 - x_2) \| \leq \phi(\lambda - \lambda_0, t) \| x_1 - x_2 \|$$

$$(\lambda \in S(\lambda_0, a); \| x_1 - x_0 \|, \| x_2 - x_0 \| \leq t) \tag{22.75}$$

(see also Exercise 20.19). We now apply this more general assertion to equation (22.49), combining v and u into a single vector parameter λ. Inequality (20.58) will then coincide with (22.75) if we set

$$\phi(\lambda, t) = q^0(v, u, t),$$

and inequality (20.63) will coincide with (22.60). Condition (20.64) holds by virtue of (22.61). Finally, let the initial approximation $x_0(\lambda)$ be the operator $R_0(v, u)$; this operator satisfies condition (20.59), by (22.70). It now follows that inequality (22.72) holds. Moreover, by Lemma 20.1, the operator $R_n(v, u)$ satisfies the condition

$$\| R_n(v, u) \| \leq c_n \omega(v, u) . \tag{22.76}$$

Now set

$$\tilde{R}(v, u) = R_n(v, u), \quad \omega(v, u) = \omega(v, u) \left[\omega_0(v, u) \right]^n$$

and apply Lemma 22.2. Conditions (22.67)–(22.68) of this lemma hold by virtue of (22.76) and (22.72). Therefore,

$$\| \phi_{n+1}(v, u) - \phi(v, u) \| \leq c_1 \delta(v, u) \, \tilde{\omega}(v, u) = \tilde{d}_n \delta(v, u) \, \omega(v, u) \left[\omega_0(v, u) \right]^n .$$

This proves (22.74). ∎

Theorem 22.2 implies the following method for constructing successive asymptotic approximations of increasing orders to solutions of equation (22.47). For each n, define an asymptotic approximation $h_n(v)$ by

$$h_n(v) = u_n(v) + R_n \left[v, u_n(v) \right], \tag{22.77}$$

where $u_n(v)$ is defined by the asymptotic approximation

$$Q_0\Omega\left[v, u + R_n(v, u)\right] = 0 \tag{22.78}$$

of the branching equation. The operator $R_n(v, u)$ is defined by the recurrence relation

$$R_k(v, u) = B_{22}^{-1}Q^0\Omega\left[v, u + R_{k-1}(v, u)\right] \qquad (k = 1, 2, \ldots ; R_0(v, u) = 0). \tag{22.79}$$

This type of successive approximation was applied by Duffing in his study of periodic solutions of differential equations.

We now consider asymptotic approximations

$$\tilde{\psi}(v, f) = 0 \qquad (f \in F_0) \tag{22.80}$$

to the branching equation (22.28)

$$\psi(v, f) = 0 \qquad (f \in F_0) \tag{22.81}$$

for equation (22.47), based on the operator C_0 defined by (22.30) (assuming, of course, that the operator B has zero index). To construct the branching equation we need a solution

$$h = T(v, f) \tag{22.82}$$

of the equation

$$(B + C)h = \Omega(v, h) + f ; \tag{22.83}$$

then the left-hand side $\psi(v, f)$ of the branching equation (22.81) is defined by

$$\psi(v, f) = CT(v, f) - f , \tag{22.84}$$

where C is the operator (22.31). For an asymptotic approximation (22.80) to the branching equation, we need an asymptotic approximation

$$h = \tilde{T}(v, f) \tag{22.85}$$

to the operator (22.82); then the left-hand side $\tilde{\psi}(v, f)$ of equation (22.80) is defined by

$$\tilde{\psi}(v, f) = C\tilde{T}(v, f) - f . \tag{22.86}$$

The solutions $f = \tilde{f}(v)$ of equation (22.80), whose left-hand side is defined by (22.86), may be used to construct asymptotic approximations to the solutions $h(v)$ of equation (22.47), e.g., by

$$\tilde{h}(v) = \tilde{T}\left[v, \tilde{f}(v)\right], \tag{22.87}$$

where $\tilde{T}(v, f)$ is the operator (22.85). To estimate the order of the asymptotic approximation (22.87) to the exact solution, we need the order of the solutions of equation (22.80) as asymptotic approximations to the corresponding solutions of equation (22.81), the order of the asymptotic approximation (22.85) to the operator (22.82), and the smoothness properties of the operator (22.82).

Exercise 22.10. Let the operator $\Omega(v, h)$ satisfy a Lipschitz condition

$$\|\Omega(v, h_1) - \Omega(v, h_2)\| \leqq q(v, t)\|h_1 - h_2\| \qquad (\|h_1\|, \|h_2\| \leqq t), \tag{22.88}$$

where $q(v, t)$ ($q(0, 0) = 0$) is continuous at zero and nondecreasing in t. Show that the operator (22.82) satisfies a Lipschitz condition

$$\|T(v, f_1) - T(v, f_2)\| \leqq q_1\|f_1 - f_2\|, \tag{22.89}$$

where q_1 is a constant.

A convenient technique for constructing asymptotic approximations (22.85) to the operator (22.82) utilizes the iterative methods of subsection 20.6. We state one theorem.

Theorem 22.3. *Let the operator* $\Omega(v, h)$ *satisfy condition* (22.88), *with*

$$q\left[v, c\omega(v, f) + \|C_0^{-1}f\|\right] \leqq k(c)\,\omega_0(v, f) \qquad (0 < c < \infty), \tag{22.90}$$

where $\omega(v, f)$ *and* $\omega_0(v, f)$ *are nonnegative functions, continuous at zero and vanishing only at zero, such that*

$$\|\Omega(v, C_0^{-1}f)\| \leqq \hat{c}\omega(v, f). \tag{22.91}$$

Assume that the operator $T_0(v, f)$ *satisfies the inequality*

$$\|T_0(v, f) - C_0^{-1}f\| \leqq c_0\omega(v, f). \tag{22.92}$$

Then the operator $T_n(v, f)$ *defined by the recurrence relation*

$$T_k(v, f) = (B + C)^{-1}\Omega\left[v, T_{k-1}(v, f)\right] + C_0^{-1}f \qquad (k = 1, 2, \ldots), \tag{22.93}$$

is an asymptotic approximation of preorder $\omega\omega_0^n$ *to the operator* (22.82),

$$\|T_n(v, f) - T(v, f)\| \leqq c_n\omega(v, f)\left[\omega_0(v, f)\right]^n, \tag{22.94}$$

while the operator

$$\psi_n(v,f) = CT_n(v,f) - f \tag{22.95}$$

is an asymptotic approximation of the same preorder $\omega\omega_0^n$ to the operator (22.86),

$$\|\psi_n(v,f) - \psi(v,f)\| \leq d_n\omega(v,f)[\omega_0(v,f)]^n. \tag{22.96}$$

We conclude this subsection with two remarks.

In constructing asymptotic approximations to branching equations, we have tried to find asymptotic approximations to the operators (22.50) and (22.82), using the ordinary successive approximations (20.41) for equations (22.49) and (22.83), respectively. It is clear that the same purpose would have been served by the iterative processes (20.44) and (20.45). However, the latter are less convenient, since they require inversion of different operators at each step. This seems to imply that the iterative process (20.41) is preferable (despite the fact that (20.44) and (20.45) yield asymptotic approximations of sharply increasing orders).

We have confined our discussion to asymptotic approximations to branching equations of two special types. Analogous constructions yield asymptotic approximations for other types of branching equation (see subsection 22.4).

22.6. *Equations in formal power series.* In subsection 20.7 we studied equation (20.75):

$$g(v, h) = 0, \tag{22.97}$$

where $g(v, h)$ is a formal power series (20.71):

$$g(v, h) = \sum_{k+m=1}^{\infty} V_{k,m}(v, h). \tag{22.98}$$

It was assumed there that the operator $V_{0,1} = g'_h(0, 0)$ has a continuous inverse Γ. Under this assumption, equation (22.97) has a unique formal solution

$$h(v) = h_1 v + h_2 v + \ldots + h_k v + \ldots \tag{22.99}$$

(where h_i are homogeneous forms).

If $V_{0,1}$ is not continuously invertible, equation (22.97) may not have solutions of type (22.97). A simple example is the scalar equation $v - h^2 = 0$. We therefore wish to extend the concept of a solution of equation (22.97), imitating the procedure of §21 for scalar equations in formal power series.

Let $V_{0,1}$ be decomposable, with matrix representation (based on projections P_0, P^0, Q_0, Q^0)

$$V_{0,1} = \begin{pmatrix} 0 & 0 \\ 0 & B_{22} \end{pmatrix}.$$

Then equation (22.97) can be replaced by the equivalent system

$$B_{22}v + Q^0 V_{1,0}v + Q^0 \sum_{k+m=2}^{\infty} V_{k,m}(v, u + v) = 0, \qquad (22.100)$$

$$Q_0 V_{1,0}v + Q_0 \sum_{k+m=2}^{\infty} V_{k,m}(v, u + v) = 0, \qquad (22.101)$$

where $u = P_0 h \in E_0 = P_0 E$, $v = P^0 h \in E^0 = P^0 E$.

Regarding the pair $\mu = \{v, u\}$ ($v \in \Lambda$, $u \in E_0$) as a parameter, we can apply Theorem 20.10 to equation (22.100). It follows that equation (22.100) has a unique solution in the class of formal power series in μ, or, equivalently, a solution

$$v = \sum_{i+j=1}^{\infty} v_{i,j}(v, u) \qquad (22.102)$$

in the class of formal power series in the variables v and u. The homogeneous (i, j)-forms $v_{i,j}$ may then be explicitly defined by recurrence relations, say, by the method of undetermined coefficients (subsection 20.4).

Substituting the series (22.102) in (22.101), we get the branching equation

$$\phi(v, u) = 0 \qquad (22.103)$$

for u. The left-hand side of this equation is a formal power series

$$\phi(v, u) = \phi_0(v) + \sum_{i+j=2}^{\infty} \phi_{i,j}(v, u), \qquad (22.104)$$

any number of whose coefficients (the forms $\phi_{i,j}$) may actually be found.

Equations of type (22.103) were considered in § 21 for finite-dimensional E_0. As a rule, they have no solutions in the class of formal power series in v. The same therefore holds for equation (22.97) if $V_{0,1}$ is not invertible.

For simplicity's sake, the following discussion will treat the case of a scalar parameter v.

Let us look for solutions of equation (22.97) in the form

$$h = h_1 v^{\varepsilon_1} + h_2 v^{\varepsilon_2} + \ldots + h_k v^{\varepsilon_k} + \ldots, \qquad (22.105)$$

where $0 < \varepsilon_1 < \varepsilon_2 < \ldots < \varepsilon_k < \ldots$ and h_i are elements of the space E. Then the projections $u = P_0 h$ and $v = P^0 h$ must also be formal power series of this type, with coefficients in the subspaces E_0 and E^0, respectively. A simple calculation shows that the coefficients of the series $v = P^0 h$ are uniquely determined by the coefficients of the series $u = P_0 h$, when the latter series is substituted in the right-hand side of (22.102). It will therefore suffice to find the series $u = P_0 h$, which is determined by the branching equation (22.103).

Thus, as in the case of equation (22.1), all the properties of equations of type (22.97) in formal power series are fully determined by the branching equation (22.103); solutions $h(v)$ of equation (22.97) are derived from those of the branching equation $u(v)$ by the simple formula

$$h(v) = \sum_{i+j=1}^{\infty} v_{i,j} [v, u(v)] + u(v). \qquad (22.106)$$

Assume that the index of the operator $V_{0,1}$ is zero. Let E_0 be one-dimensional, so that equation (22.103) is a scalar equation in formal power series. If the left-hand side of equation (22.103) does not vanish identically, then (as shown in subsection 21.3), equation (22.103) has finitely many solutions, each of which is a formal power series of some radix p. Consequently, equation (22.97) also has finitely many solutions of type (22.105). Moreover, each solution has the form

$$h = h_1^0 v^{1/p} + h_2^0 v^{2/p} + \ldots + h_k^0 v^{k/p} + \ldots, \qquad (22.107)$$

where the coefficients clearly cannot be expressed in general form, but any finite number thereof may easily be determined (apart from computational difficulties). To this end, our first requirement is a sufficient number of coefficients $\phi_{i,j}$ of the series (22.104) (for example, one might find the first few terms in the power series expansion of the left-hand side of an asymptotic approximation (of sufficiently high order) to the branching equation; see subsection 22.5). Newton polygons (see subsection 21.3) are then used to find the first terms in the expansion of the solution $u(v)$ in fractional powers of v. Finally, formula (22.106) defines the first terms in the expansion of $h(v)$ in fractional powers of v.

Retain the assumption that E_0 is one-dimensional; if the left-hand side of equation (22.103) is identically zero, formula (22.106) defines a solution of equation (22.97) for any

$$u(v) = u_1 v^{\varepsilon_1} + u_2 v^{\varepsilon_2} + \ldots + u_k v^{\varepsilon_k} + \ldots \tag{22.108}$$

If E_0 is many-dimensional, the actual construction of solutions $h(v)$ is possible only insofar as the system (22.103) of scalar equations can be analyzed in terms of finitely many known homogeneous forms $\phi_{i,j}$. In concrete cases, the method of elimination is to be recommended. If we are "lucky," and the system (22.103) turns out to be structurally stable (see subsection 21.5), all solutions will be formal series in fractional powers of v; the procedure described in §21 yields any desired finite number of terms in the expansion of the solutions $u(v)$ in fractional powers of v, and the latter determine solutions $h(v)$ of equation (22.97).

In conclusion, we remark that the arguments of this subsection are also applicable to branching equations other than (22.103). In principle, all the derivations of branching equations presented in subsection 22.4 apply equally well to equations in formal power series.

The first constructions of branching equations for equations in formal power series are apparently due to Melamed.

22.7. *Equations in analytic operators.* We shall now study equation (22.18) for an analytic operator $\Omega(v, h)$. By Theorem 20.11, the operator $R(v, u)$ of formula (22.22) is analytic. Thus the left-hand side $\phi(v, u)$ of the branching equation (22.23) is also an analytic operator.

If $\Omega(v, h)$ is an analytic operator it has a power series representation. Equation (22.18) may therefore be regarded also as an equation in formal power series (the fact that the series converges is irrelevant). To find formal solutions of equation (22.18), one can construct the branching equation (22.103). It is readily seen that the left-hand side of (22.103) is the power series expansion of the left-hand side of (22.23), and this is why we have retained the same notation $\phi(v, u)$ for both equations.

We have already described the procedure for finding formal solutions of an equation in formal power series. This procedure clearly embraces the construction of the expansions of all analytic solutions of equation (22.18) for an analytic operator Ω. Two questions arise naturally in this connection. Is every true solution (function-solution) of equation (22.18) (for an analytic operator Ω) expandible in a fractional power series in v, and is it, therefore, a solution of equation (22.18) when the latter is regarded as an equation in formal power series? Is every formal solution of equation (22.18) in formal series a convergent series whose sum is a true solution of the equation?

Unfortunately, in the general case both questions must be answered in the negative. A simple, though rather artificial example is the case in which the left-hand side of the branching equation for equation (22.18) vanishes identically. In every such case, formula (22.106) defines a formal solution for any (convergent!) series $u(v)$ with coefficients in E_0; this formula defines a true solution $h(v)$ of equation (22.18) with an analytic operator Ω, for any function $u(v)$ (which may even be discontinuous, let alone not expandible in power series!).

Now this by no means implies that one should not construct solutions of the branching equation (in particular, solutions (22.106) of equation (22.18)) as power series. On the contrary, *the approximate method based on the determination of several terms in the expansions of solutions of the branching equation is still the fundamental and most frequently used method.* One must only remember that this method may not achieve the desired goal. Experience has shown that "pathological" cases are rare. One therefore tries to find series solutions of the branching equation, confining the search to a few coefficients, and then attempts to prove that the resulting series converge in some neighborhood of zero; these series then define solutions of the branching equation. Finally, one tries to prove that the branching equation has no other solutions. This procedure is often feasible, in its entirety or at least in part.

Theorem 22.4. Assume that $V_{0,1}$ has index 0 and E_0 is one-dimensional, Assume that the left-hand side of the branching equation is not identically zero and that $\Omega(v, h)$ is an analytic operator.

Then:

1) *Every solution $h(v)$ of equation (22.18) that vanishes at $v = 0$ (if such solutions exist) has some radix p, is expandible in a series*

$$h(v) = h_1 v^{1/p} + h_2 v^{2/p} + \ldots + h_k v^{k/p} + \ldots \qquad (22.109)$$

in a neighborhood of zero, and is therefore a formal solution of equation (22.18) when the latter is considered as an equation in formal series.

2) Every formal solution (22.105) of equation (22.18) (the latter being considered as an equation in formal series) converges in a neighborhood of zero and is a true solution of equation (22.18).

3) Every solution $u(v)$ of the branching equation has radix p and possesses a convergent series expansion in a neighborhood of zero:

$$u(v) = u_1 v^{1/p} + u_2 v^{2/p} + \ldots + u_k v^{k/p} + \ldots \qquad (22.110)$$

4) Every formal solution $u(v)$ of the branching equation, considered as an equation in formal series, has the form (22.110) and converges in a neighborhood of zero.

Theorem 22.5. Assume that $V_{0,1}$ has zero index and E_0 is one-dimensional. Let the operator $\Omega(v, h)$ be analytic and the degree of the branching equation s.

Then the number of solutions of equation (22.18) (counting their multiplicities) is s. If p is the radix of a solution (22.109) of equation (22.18) or a solution (22.110) of the branching equation, then

$$p \leqq s. \qquad (22.111)$$

If there is a solution of radix p, then there are at least p such solutions.

Theorem 22.5 is an obvious corollary of Theorem 21.6.

Parts 3 and 4 of Theorem 22.4 follows from parts 1 and 2. The proof of the latter two assertions is easy. If the function-vector $h(v)$ is a solution of equation (22.18), then the function $u(v) = P_0 h(v)$ is a solution of the branching equation, the latter having analytic left-hand side $\phi(v, u)$ of some degree s in the variable u. It follows from Theorem 21.6 that $u(v)$ has a series expansion in fractional powers of v. The solution $h(v)$ may be recovered from the series $u(v)$ by means of formula (22.109). Thus $h(v)$ has the form (22.11)—this proves part 1. Now let the formal power series (22.109) be a solution of equation (22.18), considered as an equation in formal power series. Then the projection

$$u(v) = P_0 h_1 v^{1/p} + P_0 h_2 v^{2/p} + \ldots + P_0 h_k v^{k/p} + \ldots \qquad (22.112)$$

is a solution of the branching equation. It follows from Theorem 21.8 that the series (22.112) converges in a neighborhood of zero. Equality

(22.106) now implies that the series (22.107) converges in a neighborhood of zero—this proves part 2.

We now describe a procedure for application of Theorems 22.4 and 22.5.

Given equation (22.18) with analytic right-hand side and one-dimensional degeneracy (E_0 one-dimensional). Using, say, the method of subsection 22.5, one can then construct asymptotic approximations of any finite order to the left-hand side

$$\phi(v, u) = a_{1,0}v + \sum_{i+j=2}^{\infty} a_{i,j}v^i u^j \qquad (22.113)$$

of the branching equation. This means that any finite number of coefficients $a_{i,j}$ may be determined.

Case A. Assume that one of the coefficients $a_{0,j}$ is nonzero. Denote the minimum j such that $a_{0,j} \neq 0$ by s; this number s is the degree of the series (22.113). To find the required number of terms in the expansion of the solutions $u(v)$ of the branching equation (22.103) in fractional powers of v, we use Newton polygons. Recall that this requires only finitely many coefficients of the series (22.113). The first terms of the series for $u(v)$ are then used to determine the first terms in the expansion of $h(v)$.

The solution is complete if the Newton polygons yield s different partial sums (of fixed length) of the series expansions of the solutions of the branching equation.

Case B. Suppose that we have computed all coefficients $a_{i,j}$, $i + j \leq \kappa$, and that all the $a_{0,j}$ ($j = 0, 1, \ldots,$) vanish but some of the other coeffients do not. Let i_0 denote the minimum index such that one of the coefficients $a_{i_0,j}$ ($j = 0, 1, \ldots, \kappa$) is nonzero. If $a_{i_0,0} = 0$, the branching equation has solutions, but we can say nothing of their number, since the order s of the series (22.113) remains unknown. We do not know (and cannot determine, on the basis of finitely many coefficients $a_{i,j}$!) whether the right-hand side is divisible by v^{i_0} or v^{i_1} where $0 < i_1 < i_0$.

However, if $j_0 \leq \kappa$, where j_0 is the minimum index such that $a_{i_0,j_0} \neq 0$, then $s \geq j_0$. Thus the number of solutions (counting multiplicities) is at least j_0. The first terms in the expansions of these j_0 solutions in fractional powers of v may be found using Newton polygons.

Case C. Under the assumptions of Case B, assume that $a_{i_0, 0} \neq 0$. It is then not known whether the branching equation has small solutions, since the order of the series (22.113) may be zero.

Case D. All determined coefficients $a_{i, j}$ $(i + j \leq \kappa)$ vanish. Nothing can then be said of the solutions of equation (22.18).

Exercise 22.11. Consider the equation

$$x(t) = (1 + v) \int_a^b K(t, s) [x(s) + \alpha x^k(s)] \, ds + vf(t), \qquad (22.114)$$

where $K(t, s)$ is a continuous symmetric kernel:

$$K(t, s) = \sum_{i=1}^\infty \frac{e_i(t) \, e_i(s)}{\lambda_i},$$

$\lambda_1 = 1, \lambda_1 < \lambda_2 < \ldots < \lambda_n < \ldots, \lambda_n \to \infty, f(t)$ a given function. Investigate the branching equation (which depends on the sign of α and on k). Construct the first two terms of the expansion of the solution of equation (22.114) using Newton polygons.

We now proceed to many-dimensional degeneracies.

Theorem 22.6. *Assume that* $V_{0,1}$ *has index 0 and* E_0 *is of dimension* $m \geq 2$. *Let* $\Omega(v, h)$ *be an analytic operator.*

Then:

1) *If equation* (22.18) *has finitely many solutions, each of them has the form* (22.109).

2) *If equation* (22.18) *has infinitely many solutions, some of these cannot be expanded in series in integral or fractional powers of* v.

3) *If equation* (22.18), *regarded as an equation in formal power series, has only finitely many (formal) solutions, each such solution is convergent in a neighborhood of zero to a solution of equation* (22.18).

4) *If equation* (22.18), *regarded as an equation in formal power series, has infinitely many (formal) solutions, some of these converge only for* $v = 0$.

The proof of this theorem is analogous to that of parts 1 and 2 of Theorem 22.4: each of the assertions to be proved is reduced to an analogous assertion for finite-dimensional equations, and theorems of §21 are then applied.

In actual analysis of equation (22.18), one must bear in mind (as in the case of one-dimensional degeneracy) that in fact only finitely many of the first homogeneous operators in (22.104) are known. The branching equation is therefore tractable only when the straightforward elimination scheme described in § 21 (or other, similar techniques; see, e.g., § 23) is applicable. These schemes also determine the possible radixes of the required solutions.

The following proposition is important in view of Theorem 22.6.

Theorem 22.7. *Let E_0 be finite-dimensional. Assume that Ω is analytic and that zero is an isolated solution of the equation*

$$Bh = \Omega(0, h).\tag{22.115}$$

Then equation (22.18) *has finitely many solutions.*

To prove this it suffices to note that zero is an isolated solution of equation (22.115) if and only if it is an isolated solution of the equation

$$\phi(0, h) = 0,\tag{22.116}$$

where $\phi(v, h)$ is the left-hand side of the branching equation (22.103). It then remains to refer to the corresponding theorem from § 21. ∎

The above proof also indicates a method for verifying that zero is an isolated solution of equation (22.115); it suffices to show that zero is an isolated solution of equation (22.116). Since (22.116) is a finite system of scalar equations, this may be shown, e.g., by elimination.

One can also verify directly that the trivial solution of equation (22.115) is isolated. Let

$$\Omega(0, h) = \Omega_k h + \Omega_{k+1} h + \ldots,\tag{22.117}$$

where Ω_i are homogeneous forms. The arguments employed in proving Theorem 14.5 easily imply the following assertion: the trivial solution of equation (22.115) is isolated if

$$Q_0 \Omega_k u \neq 0 \qquad (u \in E_0, u \neq 0).\tag{22.118}$$

More general criteria for isolated solutions have been indicated by Zabreiko and Krasnosel'skii [2].

The first general theorems on the convergence of formal solutions are due to Pokornyi [1], for the case of integral equations and one-dimensional degeneracy. Much of the substance of this subsection is due to Melamed, some to Vainberg and Trenogin. One of the first authors to apply Newton polygons to study the branching of solutions of operator equations was Stapan.

§ 23. Simple solutions and the method of undetermined coefficients

23.1. *Simple solutions.* Let E, F and Λ be Banach spaces. In this section we shall continue our study of the equation

$$f(\lambda, x) = 0, \tag{23.1}$$

where $f(\lambda, x)$ is an operator defined on the product of the balls $S(\lambda_0, a)$ and $S(x_0, b)$ in Λ and E, with range in F, such that

$$f(\lambda_0, x_0) = 0. \tag{23.2}$$

As before, our interest is in solutions $x = x(\lambda)$ of equation (23.1) which are close to x_0 when λ is close to λ_0, such that $x(\lambda_0) = x_0$.

Assume that the operator $f(\lambda, x)$ has a derivative $f_x'(\lambda, x)$, continuous at the point $\{\lambda_0, x_0\}$. A solution $x = x(\lambda)$ is said to be *simple** if, for all λ close to but different from λ_0, the operator

$$D_*(\lambda) = f_x'[\lambda, x_*(\lambda)] \tag{23.3}$$

has a continuous inverse

$$\Gamma_*(\lambda) = \{f_x'[\lambda, x_*(\lambda)]\}^{-1}. \tag{23.4}$$

If the operator

$$\Gamma = [f_x'(\lambda_0, x_0)]^{-1}$$

is defined, then (see §20) the unique implicit function $x = x_*(\lambda)$ (defined by equation (23.1)) is a simple solution of equation (23.1). In degenerate cases, equation (23.1) may have one or more simple solutions, no simple solutions, and so on. For example, the equation

$$\lambda^3 - \lambda^2 x - \lambda x^2 + x^3 = 0$$

has one simple solution $(x = -\lambda)$ and another which is not simple $(x = \lambda)$.

---

* Simplicity of solutions has in fact played a decisive role in most work on the theory of branching of solutions (beginning with Nekrasov and Nazarov). However, only in the work of Gel'man [4, 5] is the concept actually introduced and studied for certain classes of equations with analytic operators. Gel'man's important papers are the source of many of the constructions in this section. The section will present theorems from the paper of Zabreiko and Krasnosel'skii [7].

The behavior of the function (23.4) in the neighborhood of λ_0 is an important characteristic of a simple solution $x = x_*(\lambda)$. In the non-degenerate case, this function is defined and continuous at λ_0. In degenerate cases, the norm of the function (23.4) increases without bound as λ approaches λ_0. A fairly full description of the behavior of the function (23.4) near λ_0 is given by estimates of the type

$$\|\Gamma_*(\lambda)\| \leqq \sigma(\lambda - \lambda_0), \tag{23.5}$$

where $\sigma(v)$ is positive for $v \neq 0$. A convenient choice for $\sigma(v)$ is a function of the form

$$\sigma(v) = M\|v\|^{-s}, \tag{23.6}$$

where M and s are positive constants. More precise are sets of estimates

$$\|\Gamma_*(\lambda) Q_k\| \leqq \sigma_k(\lambda - \lambda_0) \qquad (k = 1, \dots, l), \tag{23.7}$$

where $Q_1, Q_2, \dots, Q_l$ are projections such that $Q_1 + Q_2 + \dots + Q_l = I$, or even sets of estimates

$$\|\Gamma_*(\lambda) Q_k(\lambda)\| \leqq \sigma_k(\lambda - \lambda_0) \qquad (k = 1, \dots, l). \tag{23.8}$$

A solution $x = x_*(\lambda)$ of equation (23.1) is said to be *isolated* if there exists a function $\omega(v)$ ($\omega(0) = 0$), positive for nonzero v and continuous at zero, such that equation (23.1) has a unique solution $x_*(\lambda)$ in the ball $S[x_*(\lambda), \omega(\lambda - \lambda_0)]$, where λ is sufficiently close to λ_0. Simple examples show that isolated solutions need not be simple. On the other hand, every simple solution of equation (23.1) is isolated.

In fact, let $x = x_*(\lambda)$ be a simple solution of equation (23.1), and assume that inequality (23.5) is true. Since the derivative $f_x'(\lambda, x)$ of the left-hand side of (23.1) is continuous at $\{\lambda_0, x_0\}$, we have

$$\|f_x'(\lambda, x) - f_x'(\lambda_0, x_0)\| \leqq \chi(\lambda - \lambda_0, x - x_0), \tag{23.9}$$

where $\chi(v, h)$ ($\chi(0, 0) = 0$) is some function continuous at zero. Let $\omega_1(v)$ be a function such that

$$\sigma(v) \chi[v, \omega_1(v)] \leqq q < 1 \qquad (\|v\| \leqq a). \tag{23.10}$$

It follows from the obvious inequality

$$\|f(\lambda, x) - f[\lambda, x_*(\lambda)] - f_x'[\lambda, x_*(\lambda)][x - x_*(\lambda)]\| \leqq$$

$$\leqq \|\{f_x'[\lambda, (1 - \theta) x + \theta x_*(\lambda)] - f_x'[\lambda, x_*(\lambda)]\}[x - x_*(\lambda)]\|,$$

where $0 < \theta < 1$, that, if $x \in S[x_*(\lambda), \omega(\lambda - \lambda_0)]$, then

$$\| f(\lambda, x) \| \geq \| f'_x [\lambda, x_*(\lambda)] [x - x_*(\lambda)] \| -$$

$$- \| \{ f'_x [\lambda, (1 - \theta) x + \theta x_*(\lambda)] - f'_x [\lambda, x_*(\lambda)] \} [x - x_*(\lambda)] \| \geq$$

$$\geq \left\{ \frac{1}{\sigma(\lambda - \lambda_0)} - \chi [\lambda - \lambda_0, \omega_1 (\lambda - \lambda_0)] \right\} \| x - x_*(\lambda) \| \geq$$

$$\geq \frac{1 - q}{\sigma(\lambda - \lambda_0)} \| x - x_*(\lambda) \| .$$

This estimate implies that if $x \in S[x_*(\lambda), \omega_1(\lambda - \lambda_0)]$ and $x \neq x_*(\lambda)$ the vectors $f(\lambda, x)$ are different from zero. This means that the solution $x_*(\lambda)$ is indeed isolated.

Exercise 23.1. Let $x_*(\lambda)$ be a simple solution of equation (23.1). Show that there exists a function $\omega(v)$, positive for nonzero v, such that the operator

$$T_*(\lambda, x) = x - \Gamma_*(\lambda) f(\lambda, x)$$

satisfies a Lipschitz condition with constant $q(r)$ ($q(r) \to 0$ as $r \to 0$) in the ball $S[x_*(\lambda), \omega(\lambda - \lambda_0)]$ for $\| \lambda - \lambda_0 \| \leq r$.

We now present a general existence theorem for simple solutions. We shall assume given a function $x_0(\lambda)$ ($x_0(\lambda_0) = x_0$), continuous at λ_0, such that for all λ close to but different from λ_0 the linear operator

$$D_0(\lambda) = f'_x [\lambda, x_0(\lambda)] \tag{23.11}$$

has a continuous inverse

$$\Gamma_0(\lambda) = \{ f'_x [\lambda, x_0(\lambda)] \}^{-1} . \tag{23.12}$$

Theorem 23.1. *Assume that the function $x_0(\lambda)$ and the operator $f(\lambda, x)$ satisfy the condition*

$$\| \Gamma_0(\lambda) f[\lambda, x_0(\lambda)] \| \leq \rho(\lambda - \lambda_0), \tag{23.13}$$

where $\rho(v)$ ($\rho(0) = 0$) is continuous at zero, and

$$\| \Gamma_0(\lambda) \{ f'_x [\lambda, x_0(\lambda) + h] - f'_x [\lambda, x_0(\lambda)] \} \| \leq \phi(\lambda - \lambda_0, \| h \|), \tag{23.14}$$

where $\phi(v, t)$ $(\phi(v, 0) = 0)$ is an increasing function of t. Let $\omega(v)$ $(\omega(0) = 0)$ be a function, continuous at zero, such that

$$\overline{\lim_{v \to 0}} \frac{\rho(v)}{\omega(v)} < \infty, \tag{23.15}$$

$$\lim_{v \to 0} \phi[v, c\omega(v)] = 0 \qquad (0 < c < \infty). \tag{23.16}$$

Then there exists $a_0 \in (0, a]$ such that when $\lambda \in S(\lambda_0, a_0)$ equation (23.1) has a solution $x_*(\lambda)$ satisfying the condition

$$\|x_*(\lambda) - x_0(\lambda)\| = O[\omega(\lambda - \lambda_0)]. \tag{23.17}$$

The function $x_*(\lambda)$ is a simple solution, and

$$\lim_{\lambda \to \lambda_0} \frac{\|\Gamma_0(\lambda)\|}{\|\Gamma_*(\lambda)\|} = 1. \tag{23.18}$$

Proof. Let q be some number on $(0, 1)$, c and a_0 positive numbers such that

$$\rho(v) \leqq (1 - q) c\omega(v), \qquad \phi[v, c\omega(v)] \leqq q \qquad (\|v\| \leqq a_0).$$

Consider the operator

$$T_0(\lambda, x) = x - \Gamma_0(\lambda) f(\lambda, x). \tag{23.19}$$

It is easy to see that this operator is differentiable with respect to x, and

$$T'_{0x}(\lambda, x) = I - \Gamma_0(\lambda) f'_x(\lambda, x).$$

It follows from (23.14) that

$$\|T'_{0x}(\lambda, x)\| \leqq \phi[\lambda - \lambda_0, \|x - x_0(\lambda)\|]. \tag{23.20}$$

If $\lambda \in S(\lambda_0, a_0)$, the operator (23.19) satisfies a Lipschitz condition with constant q in the ball $S[x_0(\lambda), c\omega(\lambda - \lambda_0)]$. Indeed, for $x_1, x_2 \in S[x_0(\lambda), c\omega(\lambda - \lambda_0)]$,

$$\|T_0(\lambda, x_1) - T_0(\lambda, x_2)\| \leqq \|T'_{0x}[\lambda, (1 - \theta) x_1 + \theta x_2] (x_1 - x_2)\|,$$

where $0 < \theta < 1$, and, by (23.20), this implies that

$$\|T_0(\lambda, x_1) - T_0(\lambda, x_2)\| \leqq$$

$$\leqq \phi\left[\lambda - \lambda_0 \left\|(1 - \theta)x_1 + \theta x_2 - x_0(\lambda)\right\|\right]\|x_1 - x_2\| \leqq$$

$$\leqq \phi\left[\lambda - \lambda_0, c\omega(\lambda - \lambda_0)\right]\|x_1 - x_2\| \leqq q\|x_1 - x_2\|.$$

Further, inequality (23.13) implies the estimate

$$\|T_0\left[\lambda, x_0(\lambda)\right] - x_0(\lambda)\| = \|\Gamma_0(\lambda) f\left[\lambda, x_0(\lambda)\right]\| \leqq \sigma(\lambda - \lambda_0).$$

Thus, if $x \in S\left[x_0(\lambda), c\omega(\lambda - \lambda_0)\right]$, then

$$\|T_0(\lambda, x) - x_0(\lambda)\| \leqq$$

$$\leqq \|T_0(\lambda, x) - T_0\left[\lambda, x_0(\lambda)\right]\| + \|T_0\left[\lambda, x_0(\lambda)\right] - x_0(\lambda)\| \leqq$$

$$\leqq q\|x - x_0(\lambda)\| + (1 - q)c\omega(\lambda - \lambda_0) \leqq c\omega(\lambda - \lambda_0).$$

It follows that the operator (23.19) satisfies the assumptions of the contracting mapping principle in the ball $S\left[x_0(\lambda), c\omega(\lambda - \lambda_0)\right]$, for $\lambda \in S(\lambda_0, a_0)$. Consequently, it has a unique fixed point $x_*(\lambda)$ in every ball $S\left[x_0(\lambda), c\omega(\lambda - \lambda_0)\right]$. The function $x_*(\lambda)$ is clearly a solution of equation (23.1).

It follows from (23.20) that

$$\lim_{\lambda \to \lambda_0} \sup_{x \in S\left[x_0(\lambda), c\omega(\lambda - \lambda_0)\right]} \|T'_{0x}(\lambda, x)\| = 0. \qquad (23.21)$$

Since

$$f'_x(\lambda, x) = \Gamma_0^{-1}(\lambda)\left[I - T'_{0x}(\lambda, x)\right],$$

if follows from (23.21) that the operators $f'_x(\lambda, x)$ $(x \in S\left[x_0(\lambda), c\omega(\lambda - \lambda_0)\right])$ have continuous inverses, and

$$\|\left[f'_x(\lambda, x)\right]^{-1}\| = (1 + \varepsilon)\|\Gamma_0(\lambda)\| \quad (x \in S\left[x_0(\lambda), c\omega(\lambda - \lambda_0)\right]), \qquad (23.22)$$

where $\varepsilon \to 0$ as $\lambda \to \lambda_0$. This implies that $x_*(\lambda)$ is a simple solution satisfying (23.18). ∎

Exercise 23.2. Show that the assertion of Theorem 23.1 (with the exception of (23.18)) remains valid if (23.15) is replaced by the weaker relation

$$\overline{\lim_{v \to 0}} \phi\left[v, c\omega(v)\right] < \frac{c - 1}{c}.$$

Exercise 23.3. Under the assumptions of Theorem 23.1, replace (23.15) by the stronger condition

$$\lim_{v \to 0} \frac{\rho(v)}{\omega(v)} = 0.$$

Show that the simple solutions $x_*(\lambda)$ satisfies not only (23.17) but the stronger condition

$$\|x_*(\lambda) - x_0(\lambda)\| = o\left[\omega(\lambda - \lambda_0)\right].$$

The assumptions of Theorem 23.1 are rather inconvenient. A less general, though more convenient theorem is

Theorem 23.2. *Assume that*

$$\|f[\lambda, x_0(\lambda)]\| \leqq \rho_0(\lambda - \lambda_0),$$ (23.23)

where $\rho_0(v)$ $(\rho_0(0) = 0)$ *is continuous at zero,*

$$\|\{f'_x[\lambda, x_0(\lambda)]\}^{-1}\| \leqq \sigma(\lambda - \lambda_0)$$ (23.24)

and

$$\|f'_x[\lambda, x_0(\lambda) + h] - f'_x[\lambda, x_0(\lambda)]\| \leqq \phi_0(\lambda - \lambda_0, \|h\|),$$ (23.25)

where $\phi(v, t)$ $(\phi(v, 0) = 0)$ *is continuous at zero and an increasing function of* t. *Let* $\omega(v)$ $(\omega(0) = 0)$ *be a function, continuous at zero, such that*

$$\overline{\lim_{v \to 0}} \frac{\sigma(v)\rho_0(v)}{\omega(v)} < \infty$$ (23.26)

and

$$\lim_{v \to 0} \sigma(v)\phi_0[v, c\omega(v)] = 0 \qquad (0 < c < \infty).$$ (23.27)

Then the assertion of Theorem 23.1 is valid.

Proof. It follows from (23.23) and (23.24) that

$$\|\Gamma_0(\lambda) f[x_0(\lambda)]\| \leqq \sigma(\lambda - \lambda_0)\rho_0(\lambda - \lambda_0).$$

Therefore (23.13) is satisfied if we take

$$\rho(v) = \sigma(v)\rho_0(v).$$

Further, it follows from (23.23) and (23.25) that

$$\|\Gamma_0(\lambda)\{f'_x[\lambda, x_0(\lambda) + h] - f'_x[\lambda, x_0(\lambda)]\}\| \leqq \sigma(\lambda - \lambda_0)\phi_0(\lambda - \lambda_0, \|h\|).$$

Thus (23.14) is also valid, with

$$\phi(v, t) = \sigma(v) \, \phi_0(v, t).$$

It remains to note that (23.26) implies (23.15), and (23.27) implies (23.16).

Thus, the assumptions of Theorem 23.1 are satisfied, and this proves our theorem. ∎

In most cases, the right-hand sides of the inequalities (23.23) to (23.25) have the form

$$\rho(v) = N\|v\|^r, \quad \sigma(v) = M\|v\|^{-s}, \quad \phi(v, t) = c\|v\|^n t.$$

In this case, the function $\omega(v)$ may be defined as

$$\omega(v) = \|v\|^k.$$

Condition (23.26) is then equivalent to the inequality $r - s \geqq k$, condition (23.27) to the inequality $n + k - s > 0$. These inequalities are consistent if

$$r + n > 2s, \qquad r > s \, ; \tag{23.28}$$

and k may then be defined as $k = r - s$.

Despite the extremely broad requirements of Theorem 23.2, there is a certain sense in which it can be inverted: if $x_*(\lambda)$ is a simple solution of equation (23.1), then any asymptotic approximation $x_0(\lambda)$ to $x_*(\lambda)$, of sufficiently high order, satisfies the assumptions of Theorem 23.2. This is made precise in the following theorem.

Theorem 23.3. Assume that the left-hand side $f(\lambda, x)$ of equation (23.1) satisfies the inequality

$$\|f'_x(\lambda, x_1) - f'_x(\lambda, x_2)\| \leqq \psi(\|x_1 - x_2\|) \, (\lambda \in S(\lambda_0, a); \, x_1, x_2 \in S(x_0, b)),$$

$$\tag{23.29}$$

where $\psi(t) \, (\psi(0) = 0)$ is continuous at zero. Let $x_*(\lambda)$ be a simple solution of equation (23.1).

Then there exist functions $\omega(v)$ and $\omega_0(v)$, continuous at zero and positive for nonzero v, such that any asymptotic approximation $x_0(\lambda)$ of preorder ω_0 satisfies conditions (23.23) to (23.25) of Theorem 23.2, and the right-hand sides of inequalities (23.23) to (23.25) satisfy conditions (23.26) and (23.27).

Proof. Since $x_*(\lambda)$ is a simple solution, there exists a function $\sigma_*(v)$

such that

$$\left\| \left\{ f_x' \left[\lambda, x_*(\lambda) \right] \right\}^{-1} \right\| \leqq \sigma_*(\lambda - \lambda_0).$$

Choose a function $\omega_0(v)$ such that

$$\lim_{v \to 0} \sigma_*(v) \omega_0(v) = 0$$

and such that, for any $c > 0$,

$$\lim_{v \to 0} \sigma_*(v) \psi \left[c \sigma_*(v) \omega_0(v) \right] = 0. \tag{23.30}$$

Let $x_0(\lambda)$ be an asymptotic approximation to the simple solution $x_*(\lambda)$, of preorder ω_0. Then

$$\left\| f \left[\lambda, x_0(\lambda) \right] \right\| = \left\| f \left[\lambda, x_0(\lambda) \right] - f \left[\lambda, x_*(\lambda) \right] \right\| =$$

$$= O \left[\left\| x_0(\lambda) - x_*(\lambda) \right\| \right] = O \left[\omega_0(\lambda - \lambda_0) \right],$$

and so condition (23.23) is satisfied by setting

$$\rho_0(v) = c_1 \omega_0(v),$$

where c_1 is a constant. Now, for any $h \in E$, we have the inequality

$$\left\| f_x' \left[\lambda, x_0(\lambda) \right] h \right\| \geqq \left\| f_x' \left[\lambda, x_*(\lambda) \right] h \right\| - \left\| \left\{ f_x' \left[\lambda, x_*(\lambda) \right] - f_x' \left[\lambda, x_0(\lambda) \right] \right\} h \right\| \geqq$$

$$\geqq \frac{\|h\|}{\sigma_*(\lambda - \lambda_0)} - \psi \left[\left\| x_*(\lambda) - x_0(\lambda) \right\| \right] \|h\| \geqq$$

$$\geqq \frac{1 - \sigma_*(\lambda - \lambda_0) \psi \left[c_1 \omega_0(\lambda - \lambda_0) \right]}{\sigma_*(\lambda - \lambda_0)} \|h\|,$$

which, by (23.30), implies that

$$\left\| \left\{ f_x' \left[\lambda, x_0(\lambda) \right] \right\}^{-1} \right\| = O \left[\sigma_*(\lambda - \lambda_0) \right].$$

Thus condition (23.24) is also satisfied, if we set

$$\sigma(v) = c_2 \sigma_*(v),$$

where c_2 is a constant. Finally, inequality (23.29) implies (23.25) with

$$\phi_0(v, t) = \psi(t).$$

Now set

$$\omega(v) = \sigma_*(v)\,\omega_0(v)\,.$$

Then (23.26) is clearly satisfied, and (23.27) follows from (23.30). ∎

Exercise 23.4. Let $Q_1, \ldots, Q_l$ be projections such that $Q_1 + \ldots + Q_l = I$ (where I is the identity operator). Assume that

$$\|Q_i f[\lambda, x_0(\lambda)]\| \leqq \rho_i(\lambda - \lambda_0) \qquad\qquad (i = 1, \ldots, l)\,,$$

$$\|\{f_x'[\lambda, x_0(\lambda)]\}^{-1} Q_i\| \leqq \sigma_i(\lambda - \lambda_0) \qquad\qquad (i = 1, \ldots, l)\,,$$

$$\|Q_i \{f_x'[\lambda, x_0(\lambda) + h] - f_x'[\lambda, x_0(\lambda)]\}\| \leqq \phi_i(\lambda - \lambda_0, \|h\|) \qquad (i = 1, \ldots, l)\,.$$

and, for some function $\omega(v)$ $(\omega(0) = 0)$ continuous at zero,

$$\overline{\lim_{v \to 0}} \; \frac{\sigma_i(v)\rho_i(v)}{\omega(v)} < \infty \qquad\qquad (i = 1, \ldots, l)\,,$$

$$\lim_{v \to 0} \sigma(v)\,\phi_i[v, c\omega(v)] = 0 \qquad (i = 1, \ldots, l)\,.$$

Show that the assertion of Theorem 23.1 is valid.

Exercise 23.5. Define all simple solutions $x(t; \lambda)$ $(x(t; 0) = 0)$ of the equation

$$x(t) = (1 + \lambda) \int_0^{2\pi} G(t, s)\,[x(s) + \beta x^3(s)]\,ds \; ;$$

$G(t; s)$ is the Green's function of the problem

$$\ddot{x} + 2x = 0\,, \qquad x(0) = x(2\pi) = 0\,.$$

Exercise 23.6. Replace (23.15) by the stronger assumption $\rho(v) = o\,[\omega(v)]$, retaining all other assumptions of Theorem 23.1 unchanged. Show that the equation

$$\Phi(\lambda, u) = 0 \qquad (\Phi(\lambda_0, 0) = 0)\,,$$

where

$$\Phi(\lambda, u) = \frac{1}{\omega(\lambda - \lambda_0)} \{f_x'[\lambda, x_0(\lambda)]\}^{-1} f[\lambda, x_0(\lambda) + \omega(\lambda - \lambda_0)\,u]\,,$$

satisfies the assumptions of the classical implicit function theorem (Theorem 20.1 or Theorem 1.7). Show that the unique solution $u = u_*(\lambda)$ of this equation such that $u(\lambda_0) = 0$ defines a simple solution $x_*(\lambda)$ of equation (23.1), by the formula

$$x_*(\lambda) = x_0(\lambda) + \omega(\lambda - \lambda_0)\,u_*(\lambda)\,.$$

23.2. *Asymptotic approximations to simple solutions.* Theorem 23.3 implies a general method for constructing simple solutions.

If $x_0(\lambda)$ is an asymptotic approximation of order k to a simple solution $x_*(\lambda)$, then the function $f[\lambda, x_0(\lambda)]$ is obviously an asymptotic approximation to zero, of the same order k. The converse is of course false. However, one can try to find asymptotic approximations $x_0(\lambda)$ to simple solutions $x_*(\lambda)$ (which may not exist!) among those functions $x_0(\lambda)$ such that

$$f[\lambda, x_0(\lambda)] = o\left(\|\lambda - \lambda_0\|^k\right). \tag{23.31}$$

This may be done, e.g., by the method of undetermined coefficients, which will be studied in subsection 23.5.

Suppose that we have been "lucky," having found a function $x_0(\lambda)$ satisfying the assumptions of Theorem 23.1. This theorem then implies that $x_0(\lambda)$ is an asymptotic approximation to the unknown simple solution $x_*(\lambda)$.

Now the proof of Theorem 23.1 was based on the fact that this un-known simple solution $x_*(\lambda)$ is a fixed point of the operator (23.17), which satisfies the assumptions of the contracting mapping principle in some ball (which is variable, depending on λ) containing the point $x_0(\lambda)$. The contracting mapping principle implies that the simple solution $x_*(\lambda)$ is the limit of successive approximations

$$x_n(\lambda) = x_{n-1}(\lambda) - \Gamma_0(\lambda) f[\lambda, x_{n-1}(\lambda)] \qquad (n = 1, 2, \ldots). \tag{23.32}$$

Exercise 23.7. Estimate the rate of convergence of (23.32) to the simple solution.

Formulas (23.32) may be regarded (see subsections 20.6 and 22.5) as a source of asymptotic approximations of increasing orders to the simple solution $x_*(\lambda)$.

Theorem 23.4. Assume that conditions (23.13) to (23.16) are satisfied, and moreover

$$\phi[v, c\omega(v)] \leqq k(c)\, \omega_0(v) \qquad (\|v\| \leqq a, 0 < c < \infty). \tag{23.33}$$

Then the function $x_n(\lambda)$ defined by the recurrence relation (23.32) is an asymptotic approximation to the simple solution $x_*(\lambda)$ at λ_0, of pre-order $\omega_0^n\omega$:

$$\|x_n(\lambda) - x_*(\lambda)\| = O\left\{[\omega_0(\lambda - \lambda_0)]^n\, \omega(\lambda - \lambda_0)\right\}. \tag{23.34}$$

Proof. In the proof of Theorem 23.1, we showed that, if $\lambda \in S(\lambda_0, a)$, the operator (23.19) satisfies the assumptions of the contracting mapping principle in every ball $S\left[x_0(\lambda), c\omega(\lambda - \lambda_0)\right]$. The successive approximations (23.32) are therefore defined, and moreover

$$\mu_n(\lambda) = \left\|x_n(\lambda) - x_0(\lambda)\right\| \leqq c\omega(\lambda - \lambda_0) \qquad (n = 1, 2, \ldots). \qquad (23.35)$$

Similarly, the simple solution $x_*(\lambda)$ to which (23.32) converges satisfies the inequality

$$\mu_*(\lambda) = \left\|x_*(\lambda) - x_0(\lambda)\right\| \leqq c\omega(\lambda - \lambda_0). \qquad (23.36)$$

We now prove (23.34) by induction. Assume that (23.34) is true for $x_0(\lambda), \ldots, x_{n-1}(\lambda)$. Then the equality

$$x_n(\lambda) - x_*(\lambda) = T_0\left[\lambda, x_{n-1}(\lambda)\right] - T_0\left[\lambda, x_*(\lambda)\right]$$

and the estimate (23.30) imply that

$$\left\|x_n(\lambda) - x_*(\lambda)\right\| \leqq$$

$$\leqq \phi\left[\lambda - \lambda_0, \left\|(1 - \theta)x_n(\lambda) + \theta x_*(\lambda) - x_0(\lambda)\right\|\right] \left\|x_{n-1}(\lambda) - x_*(\lambda)\right\|.$$

By (23.35) and (23.36), this inequality implies that

$$\left\|x_n(\lambda) - x_*(\lambda)\right\| \leqq \phi\left[\lambda - \lambda_0, c\omega(\lambda - \lambda_0)\right]\left\|x_{n-1}(\lambda) - x_*(\lambda)\right\|. \qquad (23.37)$$

By assumption,

$$\left\|x_{n-1}(\lambda) - x_*(\lambda)\right\| = O\left\{\left[\omega_0(\lambda - \lambda_0)\right]^{n-1} \omega(\lambda - \lambda_0)\right\};$$

and so (23.37) and (23.33) imply (23.34). ∎

Formulas (23.32) describe the modified Newton–Kantorovich method for approximation of simple solutions $x_*(\lambda)$ of equation (23.1). Given an initial approximation $x_0(\lambda)$, we can also apply the ordinary Newton–Kantorovich method, which leads to the successive approximations

$$x_n(\lambda) = x_{n-1}(\lambda) - \left\{f_x'\left[\lambda, x_{n-1}(\lambda)\right]\right\}^{-1} f\left[\lambda, x_{n-1}(\lambda)\right]. \qquad (23.38)$$

We now state, without proof, an assertion concerning the order of the asymptotic approximation (23.38) to the simple solution $x_*(\lambda)$.

Theorem 23.5. Assume that condition (23.15) is satisfied, and also

$$\left\|\left\{f_x'\left[\lambda, x_0(\lambda)\right]\right\}^{-1}\right\| \leqq \sigma(\lambda - \lambda_0), \qquad (23.39)$$

$$\left\|f_x'(\lambda, x_1) - f_x'(\lambda, x_2)\right\| \leqq c\left\|x_1 - x_2\right\|^\delta, \qquad (23.40)$$

where $0 < \delta < 1$. *Let* $\omega(\lambda)$ $(\omega(0) = 0)$ *be a function, continuous at zero, which satisfies condition* (23.15) *and the condition*

$$\lim_{v \to 0} \sigma(v) \left[\omega(v) \right]^\delta = 0. \tag{23.41}$$

Then the function $x_n(\lambda)$ *defined by the recurrence relation* (23.38) *is an asymptotic approximation to the simple solution* $x_*(\lambda)$ *at* λ_0, *of preorder* $\sigma^{[(1+\delta)^n - 1]/\delta} \cdot \omega^{(1+\delta)^n}$.

$$\left\| x_n(\lambda) - x_*(\lambda) \right\| = O \left\{ \left[\sigma(\lambda - \lambda_0) \right]^{[(1+\delta)^n - 1]/\delta} \left[\omega(\lambda - \lambda_0) \right]^{(1+\delta)^n} \right\}. \tag{23.42}$$

In particular, if

$$\rho(v) = N \|v\|^r, \qquad \sigma(v) = M \|v\|^{-s},$$

where $r\delta > s$, a possible choice for the function $\omega(v)$ is $\omega(v) = \|v\|^r$. It follows from Theorem 23.5 that the function $x_n(\lambda)$ is an asymptotic approximation to the simple solution $x_*(\lambda)$, of preorder $(1 + \delta)^n r - [(1 + \delta)^n - 1] s\delta^{-1}$:

$$\left\| x_n(\lambda) - x_*(\lambda) \right\| = O \left[\|\lambda - \lambda_0\|^{(1+\delta)^n r - [(1+\delta)^n - 1]s/\delta} \right].$$

23.3. Simple solutions and branching equations. In looking for simple solutions, it is sometimes convenient to replace equation (23.1) by an equivalent equation. In particular, one can construct the Lyapunov branching equation (22.23) or the Schmidt branching equation (22.28) for equation (23.1), and try to find their simple solutions. This procedure is justified by the following theorem, which we state without proof.

Theorem 23.6. *A solution* $x = x_*(\lambda)$ *of equation* (23.1) *is simple if and only if the corresponding solution of the* (*Lyapunov or Schmidt*) *branching equation is simple.*

We have already mentioned that a similar concept of simple solution was introduced in the interesting papers of Gel'man, for a special class of equations (in other terminology). Gel'man also proved the analog of Theorem 23.6 for his definition of simple solutions. The proof of Theorem 23.6 is cumbersome, though the basic idea is quite simple: computation of the derivatives of the relevant operators at the simple solutions by the standard rules of differential calculus.

Branching equations are especially convenient when they are scalar equations or systems of n equations in n unknowns. In the former case the estimates (23.15) reduce to an estimate of a single scalar function, and in the latter to estimation of a determinant.

Exercise 23.8. Consider equation (23.1) in the form (22.18), with B normally solvable and of finite nonzero Noether index. Show that equation (22.18) has no small simple solutions.

In conclusion, note that to verify the assumptions of Theorem 23.2 (in proving inequalities (23.14) and (23.15)) we need only asymptotic approximations of the operators $f(\lambda, x)$ and $f'_x(\lambda, x)$, of sufficiently high orders, not the operators themselves. This simple observation is particularly important when simplicity of the solutions is to be checked via branching equations; we have repeatedly emphasized that in practice one must be satisfied with asymptotic approximations of the left-hand sides of the branching equations.

23.4. *Quasi-solutions of operator equations.* Continuing our study of equation (23.1), let us call a function $x(\lambda)$ a *quasi-solution* of equation (23.1) of *order* (*preorder*) ω if

$$f[\lambda, x(\lambda)] = o\left[\omega(\lambda - \lambda_0)\right] \quad (f[\lambda, x(\lambda)] = O\left[\omega(\lambda - \lambda_0)\right]) \qquad (23.43)$$

where $\omega(v)$ is a function, continuous at zero, such that $\omega(0) = 0$. If $\omega(v) = \|v\|^k$, where $0 < k < \infty$, $x(\lambda)$ is said to be a k-th order (k-th preorder) quasi-solution. Every solution $x_*(\lambda)$ of equation (23.1) is obviously also a quasi-solution, of arbitrary order or preorder. An asymptotic approximation $x(\lambda)$ to a solution $x_*(\lambda)$, of order ω, is a quasi-solution of at least the same order. The converse is clearly false. Nevertheless, it is clearly natural to look for asymptotic approximations to the true solutions of an equation among the quasi-solutions of various orders. A basic criterion for a given quasi-solution $x_0(\lambda)$ to be an asymptotic approximation to a true solution is Theorem 23.1.

The following statement is important in applications: *In order to construct quasi-solutions of equation (23.1), we do not need the operator $f(\lambda, x)$ itself ; it is sufficient to have an asymptotic approximation of sufficient high order.*

Various methods may be employed to determine quasi-solutions of equation (23.1). A general technique for finite systems of analytic equations, is elimination (see § 21). Recall that if

$$f_\mu(v, h_1, \ldots, h_n) = 0 \qquad (\mu = 1, \ldots, m) \qquad (23.44)$$

is a structurally stable system, its solutions can be found by first solving a system of scalar equations

$$f_\mu^{(n-1)}(v, h_1) = 0 \qquad (\mu = 1, \ldots, m_{n-1}) \qquad (23.45)$$

i.e., the $(n-1)$-th elimination system; one then substitutes each solution in the $(n-2)$-th elimination system and finds the second components of the solutions, and so on, until all solutions of the system (23.44) are determined. We have already indicated that the main difficulty is to solve the system (23.45): the solutions cannot be found on the basis of finitely many coefficients of the series expansions of the left-hand sides. Nevertheless, one can apply Newton polygons to this system to find quasi-solutions of arbitrary high order; the same method is then re-used to find quasi-solutions of the preceding elimination system, and so on, until the quasi-solutions of (23.44) have been determined.

This method may also be applied to a branching equation to find quasi-solutions of equation (23.1). One must remember that construction of quasi-solutions based on elimination involves prohibitive computations. There are other methods: undetermined coefficients, Newton–Graves diagrams (see Graves [1]). We shall consider only the former, confining ourselves to equations with a scalar parameter λ.

23.5. *The method of undetermined coefficients.* Throughout this subsection we shall assume that the left-hand side of equation (23.1) has derivatives of order up to N at the point $\{\lambda_0, x_0\}$. This implies that the operator $f(\lambda, x)$ may be expressed as a sum

$$f(\lambda, x) = \sum_{i+j=1}^{N} [V_{i,j}(x - x_0)](\lambda - \lambda_0)^j + \omega(\lambda, x), \qquad (23.46)$$

where $V_{i,j}$ are (i, j)-forms and $\omega(\lambda, x)$ is an operator consisting of terms of order of smallness higher than N:

$$\omega(\lambda, x) = o(|\lambda - \lambda_0|^N + \|x - x_0\|^N).$$

We wish to determine quasi-solutions of order N of equation (23.1). We first consider the question of analytic quasi-solutions of order N.

If the function

$$x(\lambda) = \sum_{l=1}^{\infty} c_l(\lambda - \lambda_0)^l$$

is analytic at $\{\lambda_0, x_0\}$ and is a quasi-solution of equation (23.1) of order N, it is obvious that the function

$$x_N(\lambda) = \sum_{l=1}^{N} c_l(\lambda - \lambda_0)^l$$

is also a quasi-solution, of the same order. We may thus try to find analytic quasi-solutions $x(\lambda)$ of equation (23.1) in the form

$$x_0(\lambda) = \sum_{l=1}^{N} c_l \mu^l, \tag{23.47}$$

where $\mu = \lambda - \lambda_0$.

To determine the coefficients $c_1, \ldots, c_N$, substitute (23.47) in the operator (23.46); the result is the equality

$$f\left(\lambda, \sum_{l=1}^{N} c_l \mu^l\right) = \sum_{l=1}^{N} W_i(c_1, \ldots, c_l)\,\mu^l + o(\mu^l), \tag{23.48}$$

where

$$W_l(c_1, \ldots, c_l) = \sum_{k_0 + k_1 + 2k_2 + \ldots + lk_l = l} \tilde{V}_{k_0, k_1 + \ldots + k_l} c_1^{k_1} \ldots c_l^{k_l}; \tag{23.49}$$

obviously,

$$W_l(c_1, \ldots, c_l) = V_{0,1} c_l + \hat{W}_l(c_1, \ldots, c_{l-1}) \tag{23.50}$$

where

$$\hat{W}_l(c_1, \ldots, c_{l-1}) = \sum_{k_0 + k_1 + 2k_2 + \ldots + (l-1)k_{l-1} = l} \tilde{V}_{k_0, k_1 + \ldots + k_{l-1}} c_1^{k_1} \ldots c_{l-1}^{k_{l-1}}.$$

$$\tag{23.51}$$

It follows that (23.47) is a quasi-solution of order N if and only if

$$W_l(c_1, \ldots, c_l) = 0 \qquad (l = 1, \ldots, N). \tag{23.52}$$

The system (23.52) is called the *Nekrasov–Nazarov system of order* N. We thus have a system of N equations for the coefficients $c_1, \ldots, c_N$ of the quasi-solution (23.47). Formally speaking, the system (23.52) is recurrent in form: the first equation contains only one unknown c_1, the second contains the unknowns c_1 and c_2, etc. It can therefore be shown that the unknowns $c_1, c_2, \ldots$ can be determined successively, one after the other. However, this is true only if the operator $V_{0,1}$ is continuously invertible (see § 20).

Exercise 23.9. Let $f(\lambda, x)$ be a scalar analytic function such that $f(0, 0) = 0$. Let

$$f'_x(0,0) = \ldots = f^{(s-1)}_{x^{s-1}}(0,0) = 0, \qquad f^{(s)}_{x^s}(0,0) \neq 0.$$

Show that the equation $f(\lambda, x) = 0$ has analytic quasi-solutions of order 1 if and only if $f'(0,0) = 0$. Show that it has analytic quasi-solutions of order 2 only if one of the following three conditions is satisfied:

a) $s = 2$, $f'_\lambda(0,0) = 0$;
b) $s > 2$, $f'_\lambda(0,0) = 0$, $f''_{\lambda x}(0,0) \neq 0$;
c) $s > 2$, $f'_\lambda(0,0) = f''_{\lambda x}(0,0) = f''_{\lambda^2}(0,0) = 0$.

Find necessary and sufficient conditions for the existence of analytic quasi-solutions of order k, where k is a fixed integer.

We shall consider the system (23.52) in detail for the case in which the operator $B = V_{0,1}$ is decomposable (see subsection 22.2) and its Noether index vanishes. It is immediate that in this case the system may have uncountably many solutions, and so equation (23.1) may have uncountably many quasi-solutions of order N. Assume known a regular matrix representation of the operator B of type (22.16):

$$B = \begin{pmatrix} 0 & 0 \\ 0 & B_{22} \end{pmatrix},$$

and let P_0, P^0, Q_0, Q^0 be the corresponding projections (the range of P_0 is the null subspace of B, $P^0 = I - P_0$, the range of Q^0 is the range of B, $Q_0 = I - Q^0$). Consider the first of equations (23.52):

$$V_{0,1}c_1 + V_{1,0} = 0. \tag{23.53}$$

This equation is solvable only if

$$Q_0 V_{1,1} = 0.$$

If this condition holds, equation (23.53) has infinitely many solutions

$$c_1 = -B_{22}^{-1}Q^0 V_{1,0} + a_1, \tag{23.54}$$

where a_1 is any element of the null subspace E_0 of $V_{0,1}$. Thus, in this case the first equation of (23.52) does not determine the first coefficient uniquely. Now consider the second equation of (23.52):

$$V_{0,1}c_2 + V_{2,0} + 2V_{1,1}c_1 + \tilde{V}_{0,2}c_1^2 = 0. \tag{23.55}$$

A necessary condition for this equation to be solvable for c_2 is that c_1 satisfies the additional condition

$$Q_0V_{2,0} + 2Q_0V_{1,1}c_1 + Q_0V_{0,2}c_1^2 = 0$$

or, by (23.55), the condition

$$\{Q_0V_{2,0} - 2Q_0V_{1,1}B_{2,2}^{-1}Q^0V_{1,0} + Q_0\tilde{V}_{0,2}[B_{22}^{-1}Q^0V_{1,0}]^2\} +$$

$$+ \{2Q_0V_{1,1}a_1 - 2Q_0\tilde{V}_{0,2}[B_{22}^{-1}Q^0V_{1,0}, a_1]\} + Q_0\tilde{V}_{0,2}a_1^2 = 0. \tag{23.56}$$

The element a_1 is generally determined by this equation. Note that equation (23.56) may also be unsolvable; if so, equation (23.1) obviously has no analytic quasi-solutions of order 2. If equation (23.56) has several solutions, equation (23.1) will also have several analytic quasi-solutions; substituting each of these in (23.55), we get the element c_2:

$$c_2 = -B_{22}^{-1}(V_{2,0} + 2V_{1,1}c_1 + \tilde{V}_{0,2}c_1^2) + a_2, \tag{23.57}$$

where a_2 is any element of E_0.

To determine a_2 (and a_1, if it is not uniquely determined by equation (23.56)), the expression (23.57) for c_2 must be substituted in the third equation of the system (23.52); the solvability conditions for the latter equation give a_2, then c_3 (more precisely, its projection P^0c_3) is found, and so on.

This scheme for equations (23.52) is due to Nekrasov and Nazarov, who used it for integral equations. It has been studied in detail by many authors (of whom we mention only Akhmedov [1] and Trenogin [1]).

We now consider the Nekrasov–Nazarov scheme in greater detail. To this end, we introduce the notation

$$a_l = P_0c_l, \quad b_l = P^0c_l \quad (l = 1, \dots, N).$$

Each equation

$$W_l(c_1, \dots, c_l) = 0 \tag{23.58}$$

of the system (23.52) may then be expressed as a system of two equations

$$\left.\begin{array}{l} Q_0 \hat{W}_l(a_1 + b_1, \ldots, a_{l-1} + b_{l-1}) = 0, \\ B_{22} b_l + Q^0 \hat{W}_l(a_1 + b_1, \ldots, a_{l-1} + b_{l-1}) = 0. \end{array}\right\} \qquad (23.59)$$

The second of these equations may be rewritten as

$$b_l = - B_{22}^{-1} Q^0 \hat{W}_l(a_1 + b_1, \ldots, a_{l-1} + b_{l-1}). \qquad (23.60)$$

The Nekrasov–Nazarov procedure is essentially as follows: first, regarding the elements $a_1, a_2, \ldots$ as unknowns, one uses the equalities (23.60) to determine the elements $b_1, b_2, \ldots$:

$$b_1 = \mathfrak{N}_1, \quad b_2 = \mathfrak{N}_2(a_1), \ldots, b_l = \mathfrak{N}_l(a_1, \ldots, a_{l-1}), \ldots, \qquad (23.61)$$

and then determines the elements $a_1, a_2, \ldots$ from the sequence of equations

$$Q_0 \hat{W}_l [a_1 + \mathfrak{N}_1, \ldots, a_{l-1} + \mathfrak{N}_{l-1}(a_1, \ldots, a_{l-2})] = 0 \quad (l = 1, \ldots, N). \qquad (23.62)$$

We shall call (23.62) the *Nekrasov–Nazarov system of branching equations*.

In replacing equation (23.1) by the Nekrasov–Nazarov branching equations we have considerably simplified the investigation of analytic quasi-solutions. First, this system is finite-dimensional (for one-dimensional degeneracy it is simply a scalar system); second, it is easily seen that, for each $l = 1, 2, \ldots$, equations (23.62) form a system of algebraic equations; third, the left-hand sides of the equations are given explicitly.

It is interesting to determine the relations between the branching equations of subsection 22.3 and the Nekrasov–Nazarov branching equations. In this connection, the following assertion, due to Pokornyi [2], is true for integral equations.

Theorem 23.7. Let $\phi(v, u)$ be the left-hand side of the Lyapunov branching equation (22.23), based on a matrix representation (22.16) for the operator $B = V_{0,1}$. Define operators $\Delta_1, \Delta_2(a_1), \ldots, \Delta_l(a_1, \ldots, a_{l-1}), \ldots$ by the equalities

$$\phi\left(v, \sum_{l=1}^{N-1} a_l v^l \right) = \sum_{l=1}^{N} \Delta_l(a_1, \ldots, a_{l-1}) v^l + \Delta_*(v, a_1, \ldots, a_{N-1}) v^N,$$

$$\qquad (23.63)$$

where

$$\Delta_*(0, a_1, \ldots, a_{N-1}) = 0.$$

Then

$$\Delta_l(a_1, \ldots, a_{l-1}) \equiv$$

$$\equiv Q_0 \hat{W}_l \left[a_1 + \mathfrak{N}_1, \ldots, a_{l-1} + \mathfrak{N}_{l-1}(a_1, \ldots, a_{l-2}) \right]. \qquad (23.64)$$

We leave the proof to the reader.

Theorem 23.7 implies that the Nekrasov–Nazarov branching equations are precisely the Nekrasov–Nazarov equations for the Lyapunov branching equation.

The system (23.52) may be replaced by a finite-dimensional system in other ways. Let C be the operator (22.31). Then each of equations (23.52) is equivalent to a system

$$(B + C)c_l + \overset{\circ}{W}_l(c_1, \ldots, c_{l-1}) = f_l, \qquad Cc_l = f_l, \qquad (23.65)$$

where f_l is a new unknown, in the space F. It follows from (23.65) that

$$c_l = (B + C)^{-1} \overset{\circ}{W}_l(c_1, \ldots, c_{l-1}) + C_0^{-1} f_l. \qquad (23.66)$$

Thus the elements $c_1, c_2, \ldots$ may be expressed in terms of $f_1, f_2, \ldots$:

$$c_l = \mathfrak{I}_l(f_1, \ldots, f_l) \qquad (l = 1, \ldots, N), \qquad (23.67)$$

where

$$\mathfrak{I}_l(f_1, \ldots, f_l) = \overset{\circ}{\mathfrak{I}}_l(f_1, \ldots, f_{l-1}) + C_0^{-1} f_l. \qquad (23.68)$$

To determine $f_1, f_2, \ldots$, we must use equations (23.65), with the functions (23.67) substituted for the elements c_l. The result is a system

$$C \overset{\circ}{\mathfrak{I}}_l(f_1, \ldots, f_{l-1}) = 0 \qquad (l = 1, 2, \ldots, N). \qquad (23.69)$$

The following analogue of Theorem 23.7 is valid.

Theorem 23.8. *Let* $\psi(v, f)$ *be the left-hand side of the Schmidt branching equations based on the operator* (22.30). *Define operators* $\nabla_1, \nabla_2(f_1), \ldots,$ $\nabla_N(f_1, \ldots, f_{N-1})$ *by the equalities*

$$\psi\left(v, \sum_{l=1}^{N-1} f_l v^l \right) = \sum_{l=1}^{N} \nabla_l(f_1, \ldots, f_{N-1}) v^l + \nabla_*(v, f_1, \ldots, f_{N-1}) v^N,$$

$$(23.70)$$

where

$$\nabla_*(0, f_1, \ldots, f_{N-1}) = 0.$$

Then

$$\nabla_l(f_1,\ldots,f_{l-1}) = C\ddot{3}_l\,(f_1,\ldots,f_{l-1}) \qquad (l = 1, 2,\ldots, N)\,. \tag{23.71}$$

Theorem 23.8 implies that the branching equations (23.69) are precisely the Nekrasov–Nazarov equations for the Schmidt branching equation.

Exercise 23.10. Show that the solutions $a_1^0,\ldots,a_{N-1}^0$ of the system (23.62) and the solutions $f_1^0,\ldots,f_{N-1}^0$ of the system (23.69) satisfy the relation

$$f_1^0 = C_0 a_1^0,\ldots,f_{N-1}^0 = C_0 a_{N-1}^0\,.$$

Having replaced the Nekrasov–Nazarov equations (23.52) by branching equations, we must now study the finite systems (23.62) or (23.69). This is comparatively easy if the null subspace of the operator $B = V_{0,1}$ is one-dimensional, for then each of equations (23.62) (or (23.69)) is scalar. We present an assertion concerning finite-dimensional Nekrasov–Nazarov equations, which usually facilitates investigation of the systems (23.62) and (23.69).

Consider the finite system

$$g(v, h) = 0\,, \tag{23.72}$$

where the operator $g(v, h)$ is differentiable at the point $\{0, 0\}$ up to order N:

$$g(v, h) = \sum_{i+j=1}^{N} Z_{i,j}(h)\, v^i + \omega(v, h)\,, \tag{23.73}$$

where $Z_{i,j}(h)$ is a j-form and $\omega(v, h)$ an operator consisting of terms of order of smallness higher than N:

$$\omega(v, h) = o(|\,v\,|^N + \|h\|^N)\,.$$

Assume that $Z_{0,1} = 0$. The Nekrasov–Nazarov equations for the system (23.72) are then

$$\Gamma_l(d_1,\ldots,d_{l-1}) = 0 \qquad (l = 1, 2,\ldots, N)\,; \tag{23.74}$$

the operators $\Gamma_l(d_1,\ldots,d_{l-1})$ are determined from the expansion

$$g\left(v,\ \sum_{l=1}^{N-1} d_l v^l\right) = \sum_{l=1}^{N} \Gamma_l(d_1,\ldots,d_{l-1})\, v^l + \Gamma_*(v, d_1,\ldots,d_{N-1})\, v^N\,,$$

$$\tag{23.75}$$

where $\Gamma_*(0, d_1,\ldots,d_{N-1}) = 0$.

The system (23.74) is said to be *normal* if there exist numbers N_0 and k_0 ($N_0 \leqq N$, $k < N$) such that each operator Γ_l ($l \geqq N_0$) is independent of the variables $d_{l-k+1}, \ldots, d_{l-1}$ and depends on the variable d_{l-k}. The number k is known as the *shift* of the system (23.74). If the system (23.74) is normal, its solution $d_1^0, \ldots, d_{N-k}^0$ is said to be *nondegenerate* if the derivative of the operator $\Gamma_N(d_1, \ldots, d_{N-k})$ with respect to d_{N-k} at the point $\{d_1^0, \ldots, d_{N-k}^0\}$ is nondegenerate (i.e., the determinant of the matrix of partial derivatives of the operators $\Gamma_N(d_1, \ldots, d_{N-k})$ with respect to the variables d_{N-k} at the point $\{d_1^0, \ldots, d_{N-k}^0\}$ does not vanish).

Let the system (23.74) be normal, with a nondegenerate solution $d_1^0, \ldots, d_{N-k}^0$. Consider the equation

$$\Gamma_N(d_1^0, \ldots, d_{N-k-1}^0, z) + \Gamma_*(v, d_1^0, \ldots, d_{N-k-1}^0, z, 0, \ldots, 0) = 0. \quad (23.76)$$

Let $z = z(v)$ be a solution of this equation such that $z(0) = d_{N-k}^0$. It then follows from (23.75) that

$$h(v) = d_1^0 v + \ldots + d_{N-k-1}^0 v^{N-k-1} + z(v) v^{N-k} \quad (23.77)$$

is a solution of equation (23.72).

Assume that the operator $\Gamma_*(v, d_1, \ldots, d_{N-k-1}, d_{N-k}, 0, \ldots, 0)$ is smooth with respect to the variable d_{N-k}, in the sense that

$$\|\Gamma_*(v, d_1^0, \ldots, d_{N-k-1}^0, z_1, 0, \ldots, 0) -$$
$$- \Gamma_*(v, d_1^0, \ldots, d_{N-k-1}^0, z_2, 0, \ldots, 0)\| \leqq q(v)\|z_1 - z_2\|, \quad (23.78)$$

where $q(v)$ ($q(0) = 0$) is a function which is continuous at zero (we leave it to the reader to prove that this condition follows from the inequality

$$\|\omega(v, h_1) - \omega(v, h_2)\| \leqq q(v, t)\|h_1 - h_2\| \quad (\|h_1\|, \|h_2\| \leqq t),$$

where $q(v, t) = o(|v|^k + t^k)$. We can then apply the classical implicit function theorem (Theorem 20.1) to equation (23.76); it follows that this equation defines a unique solution $z = z(v)$ such that $z(0) = d_{N-k}^0$. Therefore, the nondegenerate solution $d_1^0, \ldots, d_{N-k}^0$ of the normal system (23.74) defines an asymptotic approximation

$$h_0(v) = f_1^0 v + \ldots + d_{N-k}^0 v^{N-k} \quad (23.79)$$

of order $N - k$ to the solution

$$h_*(v) = h_0(v) + [z(v) - d_{N-k}^0] v^{N-k} \quad (23.80)$$

of equation (23.72). The solution (23.80) turns out to be simple. Indeed, if follows from (23.75) that

$$g_h'\left[v, h_*(v)\right] = v^k H + o(v^k),$$

where H is the derivative of the operator $\Gamma_N(d_1, \ldots, d_{N-k})$ with respect to d_{N-k} at the point $\{d_1^0, \ldots, d_{N-k}^0\}$. Since H is invertible, the operator $g_h'\left[v, h_*(v)\right]$ is also invertible for small v.

We thus arrive at the following theorem.

Theorem 23.9. Assume that the Nekrasov–Nazarov equations of order N for equation (23.72) form a normal system with shift k. Let $d_1^0, \ldots,$ d_{N-k}^0 be a nondegenerate solution of the Nekrasov–Nazarov system satisfying condition (23.78).

Then equation (23.72) has a unique simple solution $h_*(v)$ such that

$$\left\| h_*(v) - h_0(v) \right\| = o(v^{N-k}), \tag{23.81}$$

where $h_0(v) = d_1^0 v + \ldots + d_{N-k}^0 v^{N-k}$.

Exercise 23.11. Under the assumptions of Theorem 23.9, let the operator $g(v, h)$ (the left-hand side of equation (23.72)) be differentiable at zero up to order $N_1 > N$. Show that the operators Γ_l $(l > N)$ in the N_1-th order Nekrasov–Nazarov system satisfy the identity

$$\Gamma_l(d_1^0, \ldots, d_{N-k}^0, d_{N-k+1}, \ldots, d_{l-k}) = H d_{l-k} + \gamma_l(d_{N-k+1}, \ldots, d_{l-k-1}).$$

We have been considering analytic quasi-solutions. Based on the results of §21, one naturally expects equations (23.1) to have quasi-solutions which are analytic functions of the parameter $\mu = (\lambda - \lambda_0)^{1/p}$, where p is an integer. It is easy to reduce the problem of finding such quasi-solutions to the determination of analytic quasi-solutions. To this end, it suffices to set

$$\lambda = \lambda_0 + \mu^p \tag{23.82}$$

in the original equation: this yields a new equation

$$f_{(p)}(\mu, x) = 0, \tag{23.83}$$

where

$$f_{(p)}(\mu, x) = f(\lambda_0 + \mu^p, x).$$

The drawback of this method is that the number p is not known in advance.

23.6. *Equations with analytic operators.* We now consider equation (23.1) with the left-hand side $f(\lambda, x)$ an analytic operator at the point $\{\lambda_0, x_0\}$. Let

$$f(\lambda, x) = \sum_{m+n=1}^{\infty} [V_{m,n}(x - x_0)](\lambda - \lambda_0)^n. \qquad (23.84)$$

The method of undetermined coefficients, as described in subsection 23.5, may then be applied to determine analytic quasi-solutions, of arbitrary orders, of equation (23.1) with the operator (23.84). Moreover, the same method can be used to determine all the coefficients of the formal solutions

$$x(\lambda) = x_0 + \sum_{l=1}^{\infty} c_l(\lambda - \lambda_0)^l \qquad (23.85)$$

of this equation. To determine finitely many coefficients of the series (23.85), one must use the system (23.52). Of course, the number of equations in this system depends on how many coefficients of the series are needed. To determine *all* the coefficients, we need a countably infinite system of equations. This system is derived by substituting the series (23.85) in the operator (23.84), and then re-expanding the resulting series in integral powers of $\lambda - \lambda_0$:

$$f(\lambda, x) = \sum_{l=1}^{\infty} W_l(c_1, \ldots, c_l)(\lambda - \lambda_0)^l, \qquad (23.86)$$

where

$$W_l(c_1, \ldots, c_l) = \sum_{k_0 + k_1 + 2k_2 + \ldots + lk_l = l} \tilde{V}_{k_0, k_1 + \ldots + k_l}(c_1^{k_1}, \ldots, c_l^{k_l}) \qquad (23.87)$$

$$(l = 1, 2, \ldots).$$

Finally, the functions (23.87) are equated to zero. Thus, the coefficients $c_1, c_2, \ldots$ of the solutions (23.85) of equation (23.1) with the operator (23.84) must satisfy the equations

$$W_l(c_1, \ldots, c_l) = 0 \qquad (l = 1, 2, \ldots), \qquad (23.88)$$

which it is natural to term the *Nekrasov–Nazarov system*.

We have already studied the system (23.88) (see subsection 20.8) when the operator $V_{0,1}$ is continuously invertible. Here (and in all subsequent sections) we shall consider the degenerate case. One approach to the Nekrasov–Nazarov system in this case is via the branching equations of subsection 23.5:

$$\Delta_l(a_1, \ldots, a_{l-1}) = 0 \qquad (l = 1, 2, \ldots) \tag{23.89}$$

or

$$\nabla_l(f_1, \ldots, f_{l-1}) = 0 \qquad (l = 1, 2, \ldots), \tag{23.90}$$

which, by Theorems 23.7 and 23.8, are the Nekrasov–Nazarov systems for the Lyapunov and Schmidt branching equations, respectively.

Suppose that we have determined a solution of the system (23.88). It determines a formal power series (23.85) which is a formal solution of equation (23.1). It is natural to ask whether this series converges to a solution $x(\lambda)$ of equation (23.1) for small $\lambda - \lambda_0$. As far as we know, this problem has not yet received a complete treatment.

The situation is quite simple for one-dimensional degeneracy. If the infinite system (23.88) has finitely many solutions, the series (23.85) determined thereby converge to solutions of equation (23.1) for small $\lambda - \lambda_0$. But if (23.88) has infinitely many solutions, there are some for which the series (23.85) converges only at $\lambda = \lambda_0$. Recently, Zabreiko and Kats have verified that these statements remain valid for two-dimensional degeneracy; their results for higher dimensions are less complete.

Two fundamental methods should be recommended for the many-dimensional case. The first is based on Theorems 23.1 and 23.2 concerning simple solutions. Suppose that a partial sum

$$x_0(\lambda) = x_0 + \sum_{l=1}^{N_0} c_l(\lambda - \lambda_0)^l \tag{23.91}$$

of the series (23.85) can be shown to be an asymptotic approximation, of suitable order, to some simple solution $x_*(\lambda)$; one can then state that the series (23.85) converges to this simple solution for small $\lambda - \lambda_0$. The second method constructs majorant number series for (23.85). A full description of these majorant series lies beyond the scope of this book. They have been constructed by various authors, of whom the first were Lyapunov and Schmidt. Also worthy of mention is a paper by Akhmedov (who was apparently the first to consider problems of majorant series for operator equations). The most refined and precise results known to us are those of Gel'man [1-3]. We emphasize that the method of majorant series not only enables one to prove convergence of formal solutions for small $\lambda - \lambda_0$; it also yields estimates of the convergence domains. In other words, it yields estimates of the domains in which one of the solutions of equation (23.1) is defined.*

We saw in §21 and §22 that in the degenerate case, equation (23.1) with analytic left-hand side generally has solutions of the type

$$x(\lambda) = x_0 + \sum_{l=1}^{\infty} c_l(\lambda - \lambda_0)^{l/p} \qquad (23.92)$$

(where λ is a scalar parameter). Determination of solutions of type (23.92) is easily reduced to determination of solutions with integral power series expansions, by setting

$$\lambda - \lambda_0 = \mu^p. \qquad (23.93)$$

Thus, all the above arguments concerning solutions of type (23.85) found by the method of undetermined coefficients are also valid for solutions of type (23.92).

The radixes p of the solutions (23.92) are most frequently found by trial and error: one first tries to find solutions expandible in integral powers of $\lambda - \lambda_0$, then solutions expandible in powers of $(\lambda - \lambda_0)^{1/2}$, and so on. We have already discussed the problem of estimating the possible radixes p. Suffice it to remark that for complex spaces a good estimate for the radix p is the index of a singular point x_0 of the vector field $f(\lambda_0, x)$, provided this singular point is isolated; estimates of this type lie outside the scope of our book.

§ 24. The problem of bifurcation points

24.1. *Statement of the problem.* In this section we shall study a special problem concerning small solutions, which plays a major role in many applications.

---

* Many of the other methods described in this chapter for studying small solutions are also easily adapted to estimating the domains of definition of these solutions. The problem of obtaining best possible (nonrefinable) estimates for the domains of definition of small solutions has received almost no attention.

Let $A(\lambda, x)$ be an operator defined in a real Banach space E (where λ is a real parameter) such that

$$A(\lambda, \theta) \equiv \theta, \qquad (24.1)$$

where θ is the zero of E. Then the equation

$$x = A(\lambda, x) \qquad (24.2)$$

has the trivial solution θ for all λ. For simplicity, let us assume that λ may assume any real value (the reasoning is also valid when λ is confined to a finite interval). In the sequel we shall need values of the operator $A(\lambda, x)$ only in a certain ball $\|x\| < \rho$.

A number λ_0 will be called a *bifurcation value of the parameter* λ, or a *bifurcation point** of equation (24.2) if, for every $\varepsilon > 0$, there exists at least one λ in the interval $A(\lambda, x)$ for which equation (24.2) has at least one nontrivial solution in the ball $\|x\| < \varepsilon$.

Equations of type (24.2) are obtained in various problems concerning types of loss of stability in elastic systems; in these problems, the bifurcation points determine the so-called *critical loads*. They are also involved in many problems of wave theory: the bifurcation points determine the properties a flow must have in order to give rise to waves. There are many more examples.

Throughout this section we shall assume that

$$A(\lambda, x) = B(\lambda) x + T(\lambda, x), \qquad (24.3)$$

where $B(\lambda)$ is a continuous linear operator; $B(\lambda)$ is continuous in λ in the operator topology, and $T(\lambda, x)$ consists of terms whose order of smallness in x exceeds unity, i.e.,

$$\|T(\lambda, x)\| \leqq \omega(\|x\|; \lambda_1, \lambda_2) \qquad (\lambda_1 \leqq \lambda \leqq \lambda_2) \qquad (24.4)$$

where

$$\lim_{\|x\| \to 0} \frac{\omega(\|x\|; \lambda_1, \lambda_2)}{\|x\|} = 0 \qquad (24.5)$$

for fixed λ_1, λ_2.

---

* The term "bifurcation point" has been used by various authors in different senses. The present definition is due to Krasnosel'skii. Many theorems on bifurcation points and various related problems are proved in Krasnosel'skii's monographs [8, 11].

Theorem 24.1. *If* 1 *is not in the spectrum of the linear operator* $B(\lambda_0)$, *then* λ_0 *is not a bifurcation point of equation* (24.2).

The proof is trivial. If the assertion is false, there exist a sequence of numbers λ_n and nonzero elements $x_n \in E$ such that $\lambda_n \to \lambda_0$, $\|x_n\| \to 0$ and

$$x_n = B(\lambda_n) x_n + T(\lambda_n, x_n) \qquad (n = 1, 2, \dots).$$

By this equality,

$$x_n = [I - B(\lambda_0)]^{-1} [B(\lambda_n) - B(\lambda_0)] x_n + [I - B(\lambda_0)]^{-1} T(\lambda_n, x_n),$$

whence

$$1 = \left\| \frac{x_n}{\|x_n\|} \right\| \leq$$

$$\leq \|[I - B(\lambda_0)]^{-1}\| \left\{ \|B(\lambda_n) - B(\lambda_0)\| + \frac{\omega(\|x_n\|; \lambda_0 - 1, \lambda_0 + 1)}{\|x_n\|} \right\}.$$

This is a contradiction, since the right-hand side of the inequality is arbitrarily small for large n. ∎

According to Theorem 24.1, only values of λ_0 such that 1 is in the spectrum of $B(\lambda_0)$ can be bifurcation points. We shall not discuss the problem of how to find these values of λ_0; this is a special and, in general, very difficult problem of the theory of linear operators. If $B(\lambda)$ has the form

$$B(\lambda) = \lambda B_0 \tag{24.6}$$

where B_0 is a compact linear operator, then 1 is a point of the spectrum (eigenvalue) of $B(\lambda_0)$ if λ_0^{-1} is an eigenvalue of B_0.

Exercise 24.1. Construct a compact operator $A(\lambda, x) = \lambda A x$ such that $A\theta = \theta$, $B_0 = A'(\theta)$ has a nonzero real eigenvalue μ_0, but $\lambda_0 = \mu_0^{-1}$ is not a bifurcation point of equation (24.2).

If we have somehow found a value of λ_0 such that 1 is in the spectrum of $B(\lambda_0)$, additional analysis is needed to determine whether λ_0 is a bifurcation point. In many cases, this may be done by the methods described in §22 and §23. However, if additional information is available concerning the existence of a solution of equation (24.2), known for all λ, this considerably simplifies the problem, and often leads to a simple answer in cases where the previous methods become unwieldy and inapplicable in practice.

24.2. *Equations with compact operators.* We recall that an operator $A(\lambda, x)$ is compact if it is jointly continuous in all variables and maps any bounded set $\lambda \in [\lambda_1, \lambda_2]$, $\|x\| \leq \rho$ onto a compact set. Under our assumptions, θ is a zero of the compact vector field

$$\Phi_\lambda x = x - A(\lambda, x). \tag{24.7}$$

If this point is an isolated zero of the field (24.7) for some fixed λ, its index $\gamma(\lambda; \theta)$ is defined (see § 14).

Lemma 24.1. *If λ_0 is not a bifurcation point, the index $\gamma(\lambda; \theta)$ is constant for all λ in some neighborhood of λ_0.*

Proof. Since λ_0 is not a bifurcation point, there exists ε such that the ball $\|x\| \leq \varepsilon$ contains no nontrivial solutions of equation (24.2) for $\lambda \in [\lambda_0 - \varepsilon, \lambda_0 + \varepsilon]$. It follows that the fields (24.7) are all homotopic to each other on the sphere $\|x\| = \varepsilon$ for $\lambda_0 - \varepsilon \leq \lambda \leq \lambda_0 + \varepsilon$. By Lemma 14.3 (p. 173) the rotation of the field (24.7) on this sphere is constant for $\lambda_0 - \varepsilon \leq \lambda \leq \lambda_0 + \varepsilon$. It remains to note that the rotation on the boundary of the ball coincides with the index of the zero point, since the field does not vanish at other points of the ball. ∎

Lemma 24.1 implies

Theorem 24.2. *Assume that every neighborhood of the number λ_0 contains two points λ_1 and λ_2 such that*

$$\gamma(\lambda_1, \theta) \neq \gamma(\lambda_2, \theta). \tag{24.8}$$

Then λ_0 is a bifurcation point of equation (24.2).

The following two corollaries of Theorem 24.2 are convenient in applications.

Let 1 be an eigenvalue of the operator $B(\lambda_0)$, of multiplicity m. It then follows from the general theory of perturbations of linear operators (see, e.g., Riesz and Nagy [1], Kato [1]) that when λ is close to λ_0 the operator $B(\lambda)$ has eigenvalues $\mu_1, \ldots, \mu_s$ near 1, and the sum of their multiplicities is m. Let $\pi(\lambda)$ denote the sum of multiplicities of the real eigenvalues of $B(\lambda)$ which are close to and greater than 1. Assume that if λ is close to (but different from) λ_0 the number 1 is not an eigenvalue of $B(\lambda)$. Then the number $(-1)^{\pi(\lambda)}$ is constant for λ close to and greater than λ_0; denote it by $\kappa^+(\lambda_0)$. Similarly, define the number $\kappa^-(\lambda_0)$ as the common value of $(-1)^{\pi(\lambda)}$ for λ close to but smaller than λ_0. The numbers $\kappa^+(\lambda_0)$ and $\kappa^-(\lambda_0)$ may be found by methods of perturbation theory (determine the eigenvalues $\mu_1, \ldots, \mu_s$ and find their multi-plicities). In the important special case when the operator $B(\lambda)$ has the form (24.6) (e.g., if $A(\lambda, x) \equiv \lambda A x$), we have the obvious formula

$$\kappa^+(\lambda_0) \kappa^-(\lambda_0) = (-1)^m. \tag{24.9}$$

Theorem 24.3. Let 1 be an eigenvalue of the operator $B(\lambda_0)$. Define the numbers $\kappa^+(\lambda_0)$, $\kappa^-(\lambda_0)$ as above, and let

$$\kappa^+(\lambda_0) \cdot \kappa^-(\lambda_0) = -1 \tag{24.10}$$

(e.g., $B(\lambda)$ may have the form (24.6) and the eigenvalue 1 of $B(\lambda_0)$ has odd multiplicity).

Then λ_0 is a bifurcation point of equation (24.2).

Proof. If 1 is not an eigenvalue of $B(\lambda)$, then the index $\gamma(\lambda, \theta)$ of the zero θ is defined (see § 14) by

$$\gamma(\lambda, \theta) = (-1)^{\beta(\lambda)}, \tag{24.11}$$

where $\beta(\lambda)$ is the sum of multiplicities of the real eigenvalues of $B(\lambda)$ greater than 1.

Let λ_0 denote the sum of multiplicities of the eigenvalues of $B(\lambda_0)$ greater than 1. Then, obviously,

$$\beta(\lambda) = \beta_0 + \pi(\lambda) \,(\text{mod } 2).$$

It follows from this equality and from (24.11) that

$$\gamma(\lambda_1, \theta) \gamma(\lambda_2, \theta) = \kappa^+(\lambda_0) \cdot \kappa^-(\lambda_0),$$

if λ_1 and λ_2 are close to λ_0, $\lambda_1 < \lambda_0$ and $\lambda_2 > \lambda_0$. Thus (24.10) implies (24.8), and Theorem 24.2 completes the proof. ∎

It should be emphasized that application of Theorem 24.3 is especially convenient, in that it requires analysis of the linearized equation alone.

Suppose that we have been unable to prove (24.10), or that $\kappa^+(\lambda_0) = \kappa^-(\lambda_0)$, but we know that there exist λ arbitrarily close to λ_0 such that 1 is not an eigenvalue of $B(\lambda)$. It then follows from (24.11) that any neighbor-hood of λ_0 contains λ such that $\| \gamma(\lambda) \| = 1$. Thus condition (24.8) will hold, provided $| \gamma(\lambda_0) | \neq 1$. Simple theorems on calculation of the index $\gamma(\lambda_0)$ were given in § 14; an algorithm reducing the general com-putation of the index $\gamma(\lambda_0)$ to analysis of finite-dimensional fields has been worked out by Zabreiko and Krasnosel'skii [2].

Exercise 24.2. Let the index $\gamma(\lambda, \theta)$ be equal to γ^+ for all λ close to and greater than λ_0 (equal to γ^- for all λ close to and smaller than λ_0). Show that equation (24.2) has nontrivial small solutions for all λ close to and greater (smaller) than λ_0, provided $\gamma(\lambda_0) \neq \gamma^+$ ($\gamma(\lambda_0) \neq \gamma^-$).

24.3. *Use of the branching equation.* The theorems of subsection 24.2 may be generalized to the case when the operator $A(\lambda, x)$ of (24.3) is not compact, but the continuous operator $T(\lambda, x)$ not only satisfies the estimate (24.4) but also has the natural smoothness property:

$$\|T(\lambda, x) - T(\lambda, y)\| \leqq q(\rho)\|x - y\| \quad (\lambda_1 \leqq \lambda \leqq \lambda_2; \|x\|, \|y\| \leqq \rho),$$

where $q(\rho) = o(1)$, i.e.,

$$\lim_{\rho \to 0} q(\rho) = 0.$$

Let 1 be an isolated eigenvalue of the operator $B(\lambda_0)$, of finite multiplicity m. Assume that the numbers $\kappa^+(\lambda_0)$ and $\kappa^-(\lambda_0)$ are defined as above. Then the statement of Theorem 24.3 remains valid. This may be proved by going over to the branching equation (and subsequently unsing Theorem 24.3, since the branching equation involves operators in finite-dimensional spaces).

For more general theorems on bifurcation points for equations with sufficiently smooth operators (not assumed to be compact), see the paper of Zabreiko, Krasnosel'skii and Pokrovskii [1].

24.4. *Amplitude curves.* In setting up an operator equation describing some physical phenomenon, real process, etc., one must always ignore certain aspects of the phenomenon, disregard relations which are inessential in the opinion of the investigator, neglect "small" forces, and so on. In this connection, it is always important to know how the solutions of operator equations vary under small perturbations of the equations.

Exercise 24.3. Let A be a compact operator which maps the ball $\|x\| \leqq \rho$ into itself and has there a unique fixed point x^* such that $\|x^*\| < \rho$. Show that for every $\varepsilon > 0$ there exists $\delta > 0$ such that every compact operator B with

$$\|Bx - Ax\| < \delta \quad (\|x\| \leqq \rho),$$

has at least one fixed point in the ball $\|x\| \leqq \rho$, and moreover all fixed points x^{**} of B satisfy the inequality $\|x^* - x^{**}\| < \varepsilon$.

Let us see what happens to bifurcation points when the equation undergoes small perturbations, confining ourselves to the simple equation

$$x = \lambda Ax + \varepsilon f, \tag{24.12}$$

where $A\theta = \theta$, f is a fixed element, ε is a small scalar parameter. Assume that

$$Ax = Bx + C_k x + Tx, \tag{24.13}$$

where B is a continuous linear operator, C_k a homogeneous k-form, Tx comprises terms of order of smallness higher than k.

Let λ_0^{-1} be an isolated simple eigenvalue of B, e_0 a corresponding normalized eigenvector:

$$\lambda_0 B e_0 = e_0, \qquad \|e_0\| = 1. \tag{24.14}$$

Let l denote an eigenvector of the dual operator B^* corresponding to the same eigenvalue λ_0^{-1}: $\lambda_0 B^* l = l$, i.e., $\lambda_0 l(Bx) \equiv l(x)$ for all $x \in E$; let l be normalized by the condition $l(e_0) = 1$. The linear operator

$$P_0 x = l(x) e_0 \qquad (x \in E) \tag{24.15}$$

is a projection whose range is the one-dimensional subspace E_0 of all elements $t e_0$ ($-\infty < t < \infty$). The operator $P^0 = I - P_0$ is a projection with range E^0 invariant under B, and the restriction of $I - B$ to E^0 has an inverse Γ^0.

As in previous sections, we use the notation $\lambda = \lambda_0 + v$, $u = P_0 x$, $v = P^0 x$. Then equation (24.12) becomes a system

$$0 = vu + (\lambda_0 + v) P_0 [C_k(u + v) + T(u + v)] + \varepsilon P_0 f, \tag{24.16}$$

$$v = (\lambda_0 + v) Bv + (\lambda_0 + v) P^0 [C_k(u + v) + T(u + v)] + \varepsilon P^0 f. \tag{24.17}$$

We shall assume that the element $P_0 f$ is not zero—this is the general case. Without loss of generality, we can then assume that

$$P_0 f = e_0, \tag{24.18}$$

since otherwise we replace the parameter ε by a new scalar parameter $\varepsilon_1 = \varepsilon l(f)$. It follows from (24.16) and (24.18) that

$$\varepsilon = -vl(u) - (\lambda_0 + v) l [C_k(u + v) + T(u + v)].$$

Thus equation (24.12) is equivalent to the system consisting of equation (24.16) plus the equation

$$v = (\lambda_0 + v)\, Bv - vl(u)\, P^0 f + (\lambda_0 + v)\, P^0 \left[C_k(u + v) + T(u + v)\right] -$$

$$- (\lambda_0 + v)\, l\left[C_k(u + v) + T(u + v)\right] P^0 f. \qquad (24.19)$$

Throughout this subsection, we shall assume that the operators C_k and T satisfy a natural smoothness condition:

$$\left\|C_k x + Tx - C_k y\right\| \leqq q_0 r^{k-1} \|x - y\| \qquad (\|x\|, \|y\| \leqq r \leqq r_0). \qquad (24.20)$$

Equation (24.19) may be rewritten as

$$v = \Gamma^0(v)\, Q(v; u, v), \qquad (24.21)$$

where $\Gamma^0(v)$ is the operator inverse to the restriction of $I - (\lambda_0 + v)\, B$ to E^0, and

$$Q(v; u, v) = -vl(u)\, P^0 f + (\lambda_0 + v)\, P^0 \left[C_k(u + v) + T(u + v)\right] -$$

$$- (\lambda_0 + v)\, l\left[C_k(u + v) + T(u + v)\right] P^0 f. \qquad (24.22)$$

It follows from (24.20) that

$$\left\|Q(v; u, v_1) - Q(v; u, v_2)\right\| \leqq$$

$$\leqq \left\|\Gamma^0(v)\right\| \cdot |\lambda_0 + v| \left(\left\|P^0\right\| + \|l\| \cdot \left\|P^0 f\right\|\right) 2^{k-1} q_0 r^{k-1} \|v_1 - v_2\| \qquad (24.23)$$

$$(\|u\| \leqq r; \|v_1\|, \|v_2\| \leqq r; r \leqq r_0).$$

Now, for small v,

$$\left\|\Gamma^0(v)\right\| = \left\|(I - \lambda_0 B - vB)^{-1}\right\| = \left\|\Gamma^0(I - vB\Gamma^0)^{-1}\right\| \leqq$$

$$\leqq \left\|\Gamma^0\right\| \cdot \left\|I + vB\Gamma^0 + v^2 (B\Gamma^0)^2 + \dots\right\| \leqq \frac{\left\|\Gamma^0\right\|}{1 - |v| \cdot \left\|B\Gamma^0\right\|},$$

and so it follows from (24.23) that the operator $\Gamma^0(v)\, Q(v; u, v)$ is a contraction in any small ball $\|v\| \leqq r$ (for small u and v)—it satisfies a Lipschitz condition with arbitrarily small constant. It follows from

(24.20) that

$$\|\Gamma^0(v)\,Q(v;u,\theta)\| \leqq \|\Gamma^0(v)\| \cdot |v| \cdot \|P^0f\| \cdot \|l\| \cdot \|u\| \; +$$

$$+ \; |\lambda_0 + v| \, (\|P^0\| + \|l\| \cdot \|P^0f\|) \, q_0 \|u\|^k,$$

i.e., for small v and u,

$$\|\Gamma^0(v)\,Q(v;u,\theta)\| \leqq a\,|v| \cdot \|u\| + b\|u\|^k, \tag{24.24}$$

where a and b are constants. Thus, for small v and u the operator $\Gamma^0(v)$ $Q(v;u,v)$ has an invariant ball $\|v\| \leqq r$ for some small fixed radius r. It follows from the contracting mapping principle that equation (24.21) (or, equivalently, equation (24.19)) is solvable in v; the solution

$$v = R(v, u) \tag{24.25}$$

is single-valued for small v and u. Inequality (24.24) implies the following estimate for small v and u:

$$\|R(v,u)\| \leqq a_0\,|v| \cdot \|u\| + b_0\|u\|^k. \tag{24.26}$$

Since equation (24.19) is equivalent to (24.25), we see that equation (24.12) is equivalent to the system consisting of equation (24.16) together with equation (24.25).

Substituting (24.25) in the right-hand side of (14.16), we get the equation

$$0 = vu + (\lambda_0 + v)\,P_0\,\{C_k[u + R(v,u)] + T[u + R(v,u)]\} + \varepsilon e_0, \tag{24.27}$$

from which we must determine the projection $u = P_0 x$ of the solution x of equation (24.12), provided, of course, that a solution exists. We can then use formula (24.25) to find the other projection $v = P^0 x$ of the solution.

In effect, this is merely a repetition of the derivation of the branching equation for the special equation (24.12) in two parameters λ and ε.

It follows from the estimate (24.26) that, for small v, the projection $u = P_0 x$ of a nontrivial solution x comprises the principal part of the solution. Consequently, a general description of all small solutions of equation (24.12) may be derived from an analysis of the single equation (24.27).

Setting $u = \xi e_0$ (or, equivalently, $\xi = l(u)$), we can express equation

(24.27) as a scalar equation:

$$0 = v\xi + (\lambda_0 + v)l\{C_k[\xi e_0 + R(v, \xi e_0)] + T[\xi e_0 + R(v, \xi e_0)]\} + \varepsilon.$$
(24.28)

The principal role in the subsequent arguments is allotted to the number

$$\kappa_0 = l(C_k e_0).$$
(24.29)

In the general case this number is not zero, and we shall assume this. Rewrite equation (24.28) as

$$0 = v\xi + \lambda_0\kappa_0\xi^k + \varepsilon + \phi(v, \xi),$$
(24.30)

where

$$\phi(v, \xi) = \kappa_0 v\xi^k +$$

$$+ (\lambda_0 + v)l\{C_k[\xi e_0 + R(v, \xi e_0)] - C_k(\xi e_0) + T[\xi e_0 + R(v, \xi e_0)]\}.$$
(24.31)

Since C_k is a homogeneous k-form,

$$\|C_k x - C_k y\| \leqq q_1 (\|x\| + \|y\|)^{k-1} \|x - y\|.$$

Therefore,

$$\|C_k[\xi e_0 + R(v, \xi e_0)] - C_k(\xi e_0)\| \leqq q_1 [|\xi| + \|R(v, \xi e_0)\|]^{k-1}\|x - y\|$$

and, by (24.26),

$$\|C_k[\xi e_0 + R(v, \xi e_0)] - C_k(\xi e_0)\| \leqq$$

$$\leqq q_1 [(1 + a_0 |v|)|\xi| + b_0 |\xi|^k]^{k-1} (a_0\|v\| \cdot |\xi| + b_0 |\xi|^k);$$

it follows that, for small v and ξ,

$$\|C_k[\xi e_0 + R(v, \xi e_0)] - C_k(\xi e_0)\| \leqq q_2(|v| \cdot |\xi|^k + |\xi|^{2k-1}). \quad (24.32)$$

The operator T consists solely of terms whose order of smallness is higher than k. Assume that, for small x,

$$\|Tx\| \leqq q_2\|x\|^{k+1}.$$

It then follows from (24.26) that

$$\| T[\xi e_0 + R(v, \xi e_0)] \| \leqq q_2 [| \xi | + a_0 | v | \cdot | \xi | + b_0 | \xi |^k]^{k+1}$$

and, moreover,

$$\| T[\xi e_0 + R(v, \xi e_0)] \| \leqq q_3(\xi) | \xi |^k, \tag{24.33}$$

where $q_3(\xi) \to 0$ as $\xi \to 0$.

It follows from (24.32) and (24.33) that for small v and ξ the function (24.31) satisfies the estimate

$$| \phi(v, \xi) | \leqq | v | \cdot | \kappa_0 | \cdot | \xi |^k + q_2(| v | \cdot | \xi |^k + | \xi |^{2k-1}) + q_3(\xi) \cdot | \xi |^k.$$

Thus equation (24.30) may be simplified as follows:

$$0 = v\xi + \lambda_0 \kappa_0 \xi^k [1 + \psi(v, \xi)] + \varepsilon, \tag{24.34}$$

where $\psi(v, \varepsilon)$ satisfies the condition

$$\lim_{v, \xi \to 0} \psi(v, \xi) = 0. \tag{24.35}$$

When $\varepsilon = 0$, equation (24.34) obviously has the trivial solution $\xi = 0$; for every small ξ, the equation also has a nontrivial solution $v \approx - \lambda_0 \kappa_0 \xi^{k-1}$. Thus λ_0 is a bifurcation point.

When considered as functions of v for various fixed values of ε, the solutions $\xi = \xi(v; \varepsilon)$ of equation (24.34) are known as *amplitude functions*; they describe *amplitude curves* [in the (ξ, v)-plane], since in the theory of nonlinear oscillations their construction is of basic importance. Amplitude functions are clearly multiple-valued.

Figure 24.1 illustrates the amplitude curves $\xi = \xi(v; \varepsilon)$. For a similar case ($A$ compact, without assuming the smoothness of the operator T), all these curves have been described by Krasnosel'skii [6].

There is a general procedure for constructing amplitude curves; the main idea is to look for v as a function of the variable ξ.

Equation (24.34) implies the equality

$$0 = v + \frac{\varepsilon}{\xi} \lambda_0 \kappa_0 \xi^{k-1} [1 + \psi(v, \xi)]. \tag{24.36}$$

By (24.35),

$$- \delta < \psi(v, \xi) < \delta,$$

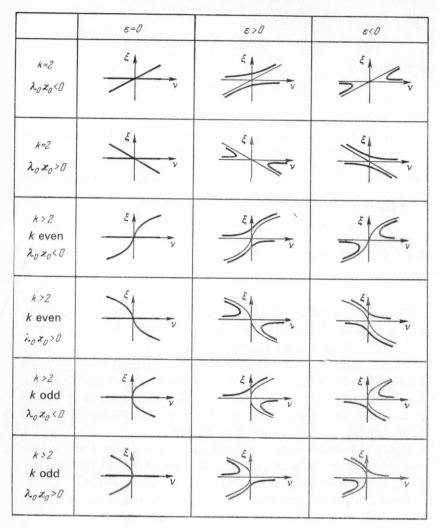

Fig. 24.1

for any fixed and arbitrarily small δ in some fixed neighborhood of zero. Construct the two curves

$$v_1(\xi) = -\frac{\varepsilon}{\xi} - \lambda_0 \kappa_0 (1 - \delta)\, \xi^{k-1},$$

$$v_2(\xi) = -\frac{\varepsilon}{\xi} - \lambda_0\kappa_0(1 + \delta)\,\xi^{k-1};$$

these curves "almost coincide" for small δ. Let

$$\left.\begin{array}{l} v_+(\xi) = \max\{v_1(\xi), v_2(\xi)\}, \\ v_-(\xi) = \min\{v_1(\xi), v_2(\xi)\}. \end{array}\right\} \tag{24.37}$$

Again, the curves described by these functions "almost coincide" — they are both close to the curve

$$v_0(\xi) = -\frac{\varepsilon}{\xi} - \lambda_0\kappa_0\xi^{k-1}. \tag{24.38}$$

Obviously, when $v \geqq v_+(\xi)$ the right-hand side of equation (24.36) is positive, while it is negative for $v \leqq v_-(\xi)$. There is thus a solution $v(\xi)$ of equation (24.36) for every small ξ.

Exercise 24.4. Show that the function $v(\xi)$ (for small ξ) is single valued in a fixed small neighborhood of zero.

The curve $v(\xi)$ lies between the curves (24.37), and is therefore close to the curve (24.38). The curves are illustrated in Figure 24.1. There is obviously no need to remember all these curves; all one needs is the final rule: If the number (24.29) is not zero, the amplitude curves $\xi = \xi(v)$ may be constructed (up to terms of higher orders of smallness) by plotting the inverse functions $v = v_0(\xi)$ according to (24.38).

In conclusion, note that equation (24.38) is easily derived from equation (24.12): Determine small (for small v and ε) solutions of equation (24.12), of type $x = \xi e_0$, from the equation

$$\xi = (\lambda_0 + v)\,l[A(\xi e_0)] + \varepsilon l(f)$$

or, equivalently, from the equation

$$0 = v\xi + \lambda_0 l\,[C_k(\xi e_0)] + vl\,[C_k(\xi e_0)] + (\lambda_0 + v)l\,[T(\xi e_0)] + \varepsilon l(f),$$

dropping in this equation all terms of higher orders of smallness. The resulting equation

$$0 = v\xi + \lambda_0 l(C_k e_0)\,\xi^k + \varepsilon l(f)$$

is precisely (24.38).

This subsection follows the paper of Zabreiko, Krasnosel'skii and Pokrovskii [1].

Bibliography

The bibliography has been rearranged in English alphabetical order. Items marked by an asterisk are in Russian. Abbreviations are those used in *Mathematical Reviews* (allowing for the different system of transliteration), except for the following:

DAN TadzhSSR—Doklady Akademii Nauk Tadzhikskoi SSR
DAN SSSR—Doklady Akademii Nauk SSSR
DAN UkrSSR—Doklady Akademii Nauk Ukrainskoi SSR
Problemy Mat. Anal. Slozh. Sis.—Problemy Matematicheskogo Analiza Slozhnykh Sistem (Voronezh)
SMZh—Sibirskii Matematicheskii Zhurnal
Trudy Sem. Funk. Anal.—Trudy Seminara po Funktsional'nomu Analizu (Voronezh)
UMN—Uspekhi Matematicheskikh Nauk.

*1. A. A. ABRAMOV   On a method for acceleration of iterative processes. *DAN SSSR* **74**, No. 6 (1950).

*1. P. G. AIZENGENDLER   Some questions in the theory of branching of solutions of nonlinear equations. *UMN* **21**, No. 1 (1966).

*1. P. G. AIZENGENDLER and M. M. VAINBERG   Theory of branching of solutions of nonlinear equations in the multidimensional case. *DAN SSSR* **163**, No. 3 (1965).

*1. N. I. AKHIEZER and I. M. GLAZMAN   *Theory of linear operators in Hilbert space.* "Nauka," Moscow, 1966.

*1. K. T. AKHMEDOV   Analytical Nekrasov-Nazarov method in nonlinear analysis. *UMN* **12**, No. 4 (1957).

*1. P. S. ALEKSANDROV   *Combinatorial topology.* Gostekhizdat, Moscow, 1947.

1. A. ALEXIEWICZ and W. ORLICZ   Analytical operations in real Banach spaces. *Studia Math.* **14** (1953).

1. M. ALTMAN   On the approximate solution of non-linear functional equations. *Bull. Acad. Polon. Sci.* **5**, No. 5 (1957).

2. —   Concerning approximate solutions of non-linear functional equations. *Bull. Acad. Polon. Sci.* **5**, No. 5 (1957).

3. —   On the approximate solutions of operator equations in Hilbert space. *Bull. Acad. Polon. Sci.* **5**, No. 6 (1957).

4. —   Concerning the approximate solutions of operator equations in Hilbert space. *Bull. Acad. Polon. Sci.* **5**, No. 7 (1957).

5. — On the generalisation of Newton's method. *Bull. Acad. Polon. Sci.* **5**, No. 8 (1957).

*1. M. YA. ANTONOVSKII, V. G. BOLTYANSKII and T. A. SARYMSAKOV *Topological semifields.* Tashkent, 1960.

*2. —, — and — Metric spaces over semifields. *Nauchnye Trudy Tashkent. Gos. Univ.,* No. 191 (1961).

*3. —, — and — Finite-dimensional modules over semifields. *Nauchnye Trudy Tashkent. Gos. Univ.,* No. 208 (1962).

*4. —, — and — *Topological Boolean algebras.* Izdatel'stvo AN UzSSR, Tashkent, 1963.

*1. I. A. BAKHTIN On a certain class of equations with positive operators. *DAN SSSR* **117**, No. 1 (1957).

*2. — On a test for the normality of a cone. *Trudy Sem. Funk. Anal.,* No. 6, Voronezh, 1958.

*3. — On a class of positive operators. *Trudy Sem. Funk. Anal.,* No. 6, Voronezh, 1958.

*4. — On nonlinear equations with concave and uniformly concave operators. *DAN SSSR* **126**, No. 1 (1959).

*5. — On a class of nonlinear integral equations. *Trudy Sem. Funk. Anal.,* Nos. 3–4, Rostov University and Voronezh University, 1960.

*6. — On equations with uniformly concave operators. In: *Functional analysis and its applications,* Izdatel'stvo AN AzerSSR, Baku, 1961.

*7. — *A study of equations with positive operators.* Dissertation, Leningrad State University, 1966.

*1. I. A. BAKHTIN and M. A. KRASNOSEL'SKII On the problem of the buckling of a rod of variable rigidity. *DAN SSSR* **105**, No. 4 (1955).

*2. — and — On the theory of equations with concave operators. *DAN SSSR* **123**, No. 1 (1956).

*3. — and — The method of successive approximations in the theory of equations with concave operators. *SMZh* **2**, No. 3 (1961).

1. S. BANACH Über homogene Polynome in L². *Studia Math.* **7** (1938).

2. — *Théorie des opérations linéaires.* Monografje Matematyczne, Warsaw, 1932.

1. R. G. BARTLE Singular points of functional equations. *Trans. Amer. Math. Soc.* **75**, No. 2 (1953).

*1. G. A. BEZSMERTNYKH On the simultaneous determination of two eigenvalues of a selfadjoint operator. *DAN SSSR* **128**, No. 6 (1959).

*2. — On the approximate methods for the solution of operator equations in Hilbert space. *DAN SSSR* **131**, No. 2 (1960).

*3. — On the behavior of the corrections in some methods for the approximate computation of the boundaries of the spectrum of a selfadjoint operator. *DAN SSSR* **136**, No. 6 (1961).

*1. I. S. BEREZIN and N. P. ZHIDKOV *Computing methods.* 2 Vols. Fizmatgiz, Moscow, 1959–1960.

*1. M. SH. BIRMAN Some estimates for the method of steepest descent. *UMN* **5**, No. 3 (1950).

*2. — On the computation of eigenvalues by the method of steepest descent. *Uch. Zap. Leningr. Gornogo Instituta* **27**, No. 1 (1952).

*1. S. V. BOBROV *The enchanted bicorn.* Detgiz, Moscow, 1949.

1. S. BOCHNER and W. T. MARTIN *Several complex variables.* Princeton University Press, 1948.

*1. O. K. BOGORYAN On the convergence of residuals in the Bubnov–Galerkin and Ritz methods. *DAN SSSR* **141**, No. 2 (1961).

1. F. BONSALL Linear operators in complete positive cones. *Proc. London Math. Soc.* **3**, No. 8 (1958).

1. N. BOURBAKI *Eléments de mathématique.* Livre IV: *Fonctions d'une variable réelle.* Hermann, Paris, 1949–1951.

1. F. E. BROWDER Nonexpansive nonlinear operators in Banach space. *Proc. Nat. Acad. Sci. U.S.A.* **54** (1965).

1. F. E. BROWDER and W. V. PETRYSHYN The solution by iteration of nonlinear functional equations in Banach spaces. *Bull. Amer. Math. Soc.* **72**, No. 3 (1966).

2. — The solution by iteration of linear functional equations in Banach spaces. *Bull. Amer. Math. Soc.* **72**, No. 3 (1966).

*1. N. V. BUGAEV Various applications of the principle of maximum and minimum exponents in the theory of algebraic functions. *Mat. Sb.* **14** (1888).

1. H. CARTAN *Elementary theory of analytic functions of one or several complex variables.* Addison–Wesley, Reading, Mass., 1963.

*1. N. G. CHEBOTAREV The Newton polygon and its role in the modern development of mathematics. In collection of articles on the tercentenary of Newton's birth. Izdatel'-stvo AN SSSR, 1943.

*2. — *Theory of algebraic functions.* Gostekhizdat, 1948.

1. L. COLLATZ *Funktionalanalysis und numerische Mathematik.* Springer-Verlag, Berlin, 1964.

1. J. CRONIN Branch points of solutions of equations in Banach space. *Trans. Amer. Math. Soc.: I,* **69** (1950); *II,* **76** (1954).

2. — *Fixed points and topological degree in nonlinear analysis.* Amer. Math. Soc., Providence, R. I., 1964.

*1. I. K. DAUGAVET Application of the general theory of approximate methods to studying the convergence of the Galerkin method for some boundary problems of mathematical physics. *DAN SSSR* **98**, No. 6 (1954).

*2. — On the rate of convergence of the Galerkin method for ordinary differential equations. *Izv. Vuzov, Matematika,* No. 5 (1958).

*3. — On the method of moments for an ordinary differential equation. *SMZh* **6**, No. 1 (1965).

*1. D. F. DAVIDENKO On application of the method of variation of parameters to inversion of matrices. *DAN SSSR* **131**, No. 3 (1960).

*2. — On application of the method of variation of parameters to computation of determinants. *DAN SSSR* **131**, No. 4 (1960).

*3. — On a method for numerical solution of systems of nonlinear equations. *DAN SSSR* **88**, No. 4 (1953).

*4. — On the approximate solution of systems of nonlinear equations. *Ukr. Mat. Zh.* **5**, No. 2 (1953).

*5. — On application of the method of variation of parameters to the theory of nonlinear functional equations. *Ukr. Mat. Zh.* **7**, No. 1 (1955).

*6. — On application of the method of variation of parameters to construct iterative formulas increasing the precision in numerical solution of nonlinear integral equations. *DAN SSSR* **162**, No. 3 (1965).

*7. — Application of the method of variation of parameters to construct iteration formulas for increasing precision in matrix inversion. *DAN SSSR* **162**, No. 4 (1965).

*1. L. N. DOVBYSH Stability of the Ritz method for problems of the spectral theory of operators. *Trudy Mat. Inst. AN SSSR* **84** (1965).

*2. — *Stability of the Ritz method for problems of the spectral theory of operators.* Dissertation, Leningrad, 1965.

1. N. DUNFORD and J. T. SCHWARTZ. *Linear operators,* Part I: *General theory.* Interscience Publishers, New York, 1958.

*1. A. V. DZHISHKARIANI Error estimates in the Ritz method for eigenvalues and eigenfunctions of a differential equation. *Soobshch. AN GruzSSR* **25**, No. 1 (1960).

*2. — Error estimates in the Ritz method for a nonhomogeneous differential equation. *Soobshch. AN GruzSSR* **25**, No. 3 (1960).

*3. — On the rate of convergence of the Bubnov–Galerkin method. *Zh. Vychisl. Mat. i Mat. Fiz.* **4**, No. 2 (1964).

1. M. EDELSTEIN A remark on a theorem of M. A. Krasnosel'skii. *Amer. Math. Monthly* **73**, No. 5 (1966).

*1. N. P. ERUGIN *Implicit functions.* Izdatel'stvo Leningradskogo Gosudarstvennogo Univ., 1956.

*1. A. R. ESAYAN On a bound for the spectrum of a certain class of linear operators. *Trudy Sem. Funk. Anal.,* No. 7, Voronezh, 1963.

*2. — *Application of the theory of cones to estimation of the spectrum of nonpositive operators.* Dissertation, Voronezh, 1964.

*3. — On a bound for the spectrum of the sum of positive semicommutative operators. *SMZh* **7**, No. 2 (1966).

*1. A. R. ESAYAN and T. SABIROV On the incompatibility of certain semiorder relations. *DAN TadzhSSR* **6**, No. 4 (1963).

*1. A. R. ESAYAN and V. YA. STETSENKO Bounds for the spectrum of integral operators and infinite matrices. *DAN SSSR* **157**, No. 2 (1964).

*2. —and— On the convergence of successive approximations for equations of the second kind. *DAN TadzhSSR* **7**, No. 2 (1964).

*3. —and— Lower bounds for the spectral radius of linear operators. *UMN* **20**, No. 5 (1965).

*1. D. K. FADDEEV and V. N. FADDEEVA *Numerical methods of linear algebra.* Fizmatgiz, 1963.

*1. V. N. FADDEEVA *Numerical methods of linear algebra.* Gostekhizdat, 1950.

1. I. FENYÖ Über die Lösung der im Banachschen Raume definierten nichtlinearen Gleichungen. *Acta Math. Acad. Sci. Hung.,* Nos 1–2 (1954).

468   BIBLIOGRAPHY

*1. V. M. FRIDMAN   The method of successive approximations for the Fredholm integral of the first kind. *UMN* **11**, No. 1 (1956).

*2. — On an approximate method for defining the frequencies of oscillations. In: *Oscillations in turbodynamos,* Izdatel'stvo AN SSSR, 1956.

*3. — New methods for solution of a linear operator equation. *DAN SSSR* **128**, No. 3 (1959).

*4. — An iterative process with minimal errors for a nonlinear operator equation. *DAN SSSR* **139**, No. 5 (1961).

*5. — Method of minimal iterations with minimal errors for a system of linear algebraic equations with a symmetric matrix. *Zh. Vychisl. Mat. i Mat. Fiz.* **2**, No. 2 (1962).

*6. — On the convergence of methods of steepest descent. *UMN* **17**, No. 3 (1962).

*1. B. A. FUKS   *Theory of analytic functions of several complex variables.* Gostekhizdat, 1948.

*1. B. A. FUKS and B. V. SHABAT   *Functions of a complex variable and some of their applications.* Fizmatgiz, 1959.

*1. F. R. GANTMAKHER   *Theory of matrices.* Nauka, 1966.

*1. F. R. GANTMAKHER and M. G. KREIN   *Oscillation matrices and kernels and small oscillations of mechanical systems.* Gostekhizdat, 1950.

*1. M. K. GAVURIN   Application of polynomials of best approximation for improving the convergence of iterative processes. *UMN* **5**, No. 3 (1950).

*2. — Analytical methods for investigation of nonlinear functional transforms. *Uch. Zap. Leningr. Gos. Univ., Ser. Mat.* **19** (1950).

*3. — Nonlinear functional equations and continuous analogs of iterative methods. *Izv. Vuzov, Matematika,* No. 5 (1958).

*1. A. E. GEL'MAN   Small-parameter method for operator equations. *DAN SSSR* **123**, No. 5 (1958).

*2. — Theorems on implicit abstract functions and stability questions for operator equations. *DAN SSSR* **127**, No. 5 (1959).

*3. — Theorems on an implicit abstract function. *DAN SSSR* **132**, No. 3 (1960).

*4. — On analytic solutions of essentially nonlinear equations. *DAN SSSR* **144**, No. 1 (1962).

*5. — On simple solutions of operator equations in the branching case. *DAN SSSR* **152**, No. 5 (1963).

*6. — On analytic solutions of a class of operator equations. *DAN SSSR* **153**, No. 6 (1963).

*1. I. Ts. GOKHBERG and M. G. KREIN   Fundamental propositions on defect numbers, generalized eigenvalues and indices of linear operators. *UMN* **12**, No. 2 (1957).

*1. V. L. GONCHAROV   *Theory of interpolation and approximation of functions.* Gostekhizdat, 1954.

1. E. GOURSAT   *Cours d'analyse mathématique.* Gauthier–Villars, Paris, 1942.

*1. D. A. GRAVE   *Elements of higher algebra.* Kiev, 1911.

*2. — *Treatise on algebraic analysis.* Part II. Kiev, 1939.

1. L. M. GRAVES   Remarks on singular points of functional equations. *Trans. Amer. Math. Soc.* **79**, No. 1 (1955).

*1. N. N. GUDOVICH On an abstract scheme of the difference method. *Zh. Vychisl. Mat. i Mat. Fiz.* **6**, No. 5 (1966).

*2. — On the convergence of dual methods. *Trudy Sem. Funk. Anal.,* No. 10, Voronezh, 1968.

1. A. HAMMERSTEIN Nichtlineare Integralgleichungen nebst Anwendungen. *Acta Math.* **54**, (1930).

1. E. HEINZ Beiträge zur Störungen Theorie der Spektralzerlegung. *Math. Ann.* **123** (1951).

1. T. H. HILDEBRANDT and L. M. GRAVES Implicit functions and their differentials in general analysis. *Trans. Amer. Math. Soc.* **29** (1927).

1. E. HILLE and R. S. PHILLIPS *Functional analysis and semigroups* (revised edition). American Mathematical Society, New York, 1957.

1. W. V. D. HODGE and D. PEDOE *Methods of algebraic geometry.* Cambridge University Press, 1947 (Vol. I), 1952 (Vol. II), 1954 (Vol. III).

1. R. IGLISCH Zur Theorie der reellen Verzweigungen von Lösungen nichtlinearer Integralgleichungen. *J. Reine Angew. Math.* **164**, No. 3 (1931).

2. — Zur Theorie der Schwingungen. *Monatsh. Math. Phys.* **37** (1930), **39** (1932), **42** (1935).

*1. I. A. IL'F and E. P. PETROV *Collected works.* Vol. 2. Goslitizdat, 1961.

*1. B. IMOMNAZAROV An *a priori* estimate for the round-off error in the method of successive approximations. *DAN TadzhSSR* **8**, No. 11 (1965).

*1. B. IMOMNAZAROV and V. YA. STETSENKO On the solvability of equations of the second kind and an *a priori* estimate for the round-off error in the method of successive approximations. *Conference on the application of functional analysis, Baku,* 1965.

*1. R. I. KACHUROVSKII Nonlinear equations with monotone and other operators. *DAN SSSR* **173**, No. 3 (1967).

*2. — Nonlinear monotone operators in Banach spaces. *UMN* **23**, No. 2 (1968).

*1. L. V. KANTOROVICH On the method of steepest descent. *DAN SSSR* **56**, No. 3 (1947).

*2. — Functional analysis and applied mathematics. *UMN* **3**, No. 6 (1948).

*3. — On Newton's method. *Trudy Mat. Inst. Steklov* **28** (1949).

*4. — The majorant principle and Newton's method. *DAN SSSR* **76**, No. 1 (1951).

*5. — Further applications of the majorant method. *DAN SSSR* **80**, No. 6 (1951).

*6. — Approximate solution of functional equations. *UMN* **11**, No. 6 (1956).

*7. — Further applications of Newton's method. *Vestnik Leningrad. Gos. Univ.* **7**, *Ser. Mat. Mekh. Astronom.,* No. 2 (1957).

*1. L. V. KANTOROVICH and G. P. AKILOV *Functional analysis in normed spaces.* Fizmatgiz, 1959.

*1. L. V. KANTOROVICH and V. I. KRYLOV *Approximate methods of higher analysis.* Fizmatgiz, 1962.

*1. L. V. KANTOROVICH, B. Z. VULIKH and A. G. PINSKER *Functional analysis in semiordered spaces.* Gostekhizdat. 1950.

1. S. KARLIN Positive operators. *J. Math. Mech.* **8** (1955).

470 BIBLIOGRAPHY

*1. E. B. KARPILOVSKAYA   On the convergence of the interpolation method for ordinary differential equations. *UMN* **8**, No. 3 (1953).

*2. — On the convergence of the method of collocation. *DAN SSSR* **151**, No. 4 (1963).

*3. — On the convergence of the collocation method for certain boundary-value problems of mathematical physics. *SMZh* **4**, No. 3 (1963).

1. T. KATO   On the perturbation theory of closed linear operators. *J. Math. Soc. Japan* **4** (1952).

*1. M. V. KELDYSH   On the method of B. G. Galerkin for the solution of boundary-value problems. *Izv. AN SSSR, Ser. Mat.* **6** (1942).

1. W. A. KIRK   A fixed point theorem for mappings which do not increase distance. *Amer. Math. Monthly* **72** (1965).

*1. O. KIS   On the convergence of the iterative method for differential and integral equations. *Magyar Tud. Akad. Mat. Kutató Int. Közl.* **3**, Nos. 1–2 (1958).

*2. — On the convergence of the coincidence method. *Acta Math. Acad. Sci. Hung.* **17**, Nos. 3–4 (1966).

*1. L. V. KIVISTIK   On the method of steepest descent for the solution of nonlinear equations. *Izv. AN EstSSR, Ser. Fiz.-Mat. i Tekhn. Nauk* **9**, No. 2 (1960).

*2. — On certain iterative methods for the solution of operator equations in Hilbert spaces. *Izv. AN EstSSR, Ser. Fiz.-Mat. i Tekhn. Nauk* **9**, No. 3 (1960).

*3. — On a modification of the iteration method with minimal residuals for the solution of nonlinear operator equations. *DAN SSSR* **136**, No. 1 (1961).

*1. A. I. KOSHELEV   Newton's method and generalized solutions of elliptic nonlinear equations. *DAN SSSR* **91**, No. 6 (1953).

*1. V. N. KOSTARCHUK   On a method for the solution of systems of linear equations and determination of the eigenvectors of matrices. *DAN SSSR* **98**, No. 4 (1954).

*2. — Method of normal chords for the solution of linear operator equations. *Trudy Sem. Funk. Anal.*, Nos. 3–4, Rostov Univ., Voronezh Univ., 1960.

*1. V. N. KOSTARCHUK and B. P. PUGACHEV   Exact estimate for the decrease in the error at a single step of the method of steepest descent. *Trudy Sem. Funk. Anal.*, No. 2, Voronezh, 1956.

*1. M. A. KRASNOSEL'SKII   On defect numbers of closed operators. *DAN SSSR* **56** (1947).

*2. — Convergence of the Galerkin method for nonlinear equations. *DAN SSSR* **73**, No. 6 (1950).

*3. — On the problem of bifurcation points. *DAN SSSR* **79**, No. 3 (1951).

*4. — Some problems of nonlinear analysis. *UMN* **9**, No. 3 (1954).

*5. — Two remarks on the method of successive approximations. *UMN* **10**, No. 1 (1955).

*6. — On the Nekrasov equation from the theory of waves on the surface of a heavy liquid. *DAN SSSR* **109**, No. 3 (1956).

*7. — On some methods for approximate computation of eigenvalues and eigenvectors of a positive definite matrix. *UMN* **11**, No. 3 (1956).

*8. — *Topological methods in the theory of nonlinear integral equations.* Gostekhizdat, 1956.

*9. — On application of the methods of functional analysis to problems of nonlinear oscillations. *Proc. Third All-Union Mathematical Conference,* Vol. 3 (1958).

*10. — On solution of equations with selfadjoint operators by the method of successive approximations. *UMN* **15**, No. 3 (1961).

*11. — *Positive solutions of operator equations.* Fizmatgiz, 1962.

*12. — *The shift operator along trajectories of differential equations.* Nauka, 1966.

*13. — On *a priori* error estimates. *DAN SSSR* **178**, No. 5 (1968).

*1. M. A. KRASNOSEL'SKII and V. A. CHECHIK On a theorem of L. V. Kantorovich. *Trudy Sem. Funk. Anal.,* Nos. 3–4, Rostov Univ., Voronezh Univ., 1960.

*1. M. A. KRASNOSEL'SKII and S. G. KREIN An iterative process with minimal residuals. *Mat. Sb.* **31** (73) (1952).

*2. —and— Note on the distribution of errors in solution of a system of linear equations by an iterative process. *UMN* **7**, No. 4 (1952).

*1. M. A. KRASNOSEL'SKII and L. A. LADYZHENSKII Structure of the spectrum of positive inhomogeneous operators. *Trudy Moskov. Mat. Obshch.* **3** (1954).

*1. M. A. KRASNOSEL'SKII, A. I. PEROV, A. I. POVOLOTSKII and P. P. ZABREIKO *Vector fields on the plane.* Fizmatgiz, 1963.

*1. M. A. KRASNOSEL'SKII and YA. B. RUTITSKII *Convex functions and Orlicz spaces.* Fizmatgiz, 1958.

*2. —and— On some applications of solution method for nonlinear operator equations, based on linearization. *DAN SSSR* **141**, No. 4 (1961).

*3. —and— On some approximate solution methods for nonlinear operator equations. *Proceedings of the Fourth All-Union Mathematical Conference,* Vol. 2 (1964).

*4. —and— On some approximate solution methods for nonlinear operator equations. *Trudy Sem. Funk. Anal.,* No. 11 *(Problemy Mat. Anal. Slozh. Sis.),* Voronezh, 1968.

*1. M. A. KRASNOSEL'SKII and P. E. SOBOLEVSKII On a nonnegative eigenfunction of the first boundary-value problem for an elliptic equation. *UMN* **16**, No. 1 (1961).

*1. M. A. KRASNOSEL'SKII and V. YA. STETSENKO A note on Seidel's method. *Zh. Vychisl. Mat. i Mat. Fiz.* **9**, No. 1 (1969).

*2. —and— On the theory of equations with concave operators. *SMZh* **10**, No. 3 (1969).

*1. M. A. KRASNOSEL'SKII, P. P. ZABREIKO, E. I. PUSTYL'NIK and P. E. SOBOLEVSKII *Integral operators in spaces of summable functions.* Nauka, 1966.

*1. N. N. KRASOVSKII *Some problems in the theory of stability of motion.* Fizmatgiz, 1959.

*1. M. G. KREIN and M. A. KRASNOSEL'SKII Fundamental theorems on the extension of Hermitian operators and some of their applications to the theory of orthogonal polynomials and the moment problem. *UMN* **2**, No. 3 (1947).

*2. —and— Stability of the index of an unbounded operator. *Mat. Sb.* **30**, No. 1 (1952).

*1. M. G. KREIN, M. A. KRASNOSEL'SKII and D. P. MIL'MAN On defect numbers of linear operators in Banach space and certain geometrical questions. *Sb. Trudov Inst. Mat. AN UkrSSR,* No. 11 (1948).

*1. M. G. KREIN and M. A. RUTMAN Linear operators with an invariant cone in Banach space. *UMN* **3**, No. 1 (1948).

*1. S. G. KREIN   *Linear differential equations in Banach space.* Nauka, 1968.
*1. S. G. KREIN and O. I. PROZOROVSKAYA   An analog of Seidel's method for operator equations. *Trudy Sem. Funk. Anal.,* No. 5, Voronezh, 1957.
*1. N. M. KRYLOV   *Selected works.* Vols. I–III. Kiev, 1961.
 1. G. KUREPA   Tableaux ramifiés d'ensembles, Espaces pseudodistanciés. *C.R. Acad. Sci. Paris* **198** (1934).
*1. N. S. KURPEL'   *Projective-iterative methods for solution of operator equations.* Naukova Dumka, Kiev, 1968.
*1. I. A. KUSAKIN   On the convergence of certain iterative methods. *Uch. Zap. Azerb. Univ., Ser. Fiz-Mat.,* No. 6 (1965).
 2. —   On the convergence of some methods of approximate solution of operator equations. *DAN UkrSSR,* No. 7 (1965) (in Ukrainian).
 1. K. LAMSON   A general implicit function theorem with an application to problems of relative minima. *Amer. J. Math.* **42** (1920).
 1. C. LANCZOS   An iteration method for the solution of the eigenvalue problem of linear differential and integral operators. *J. Res. Nat. Bur. Standards* **45**, No. 4 (1950).
 1. S. LEFSCHETZ   *Differential equations: geometric theory.* Interscience Publishers, New York, 1957.
 1. N. J. LEHMANN   Eine Fehlerschätzung zum Ritzschen Verfahren für inhomogene Randwertaufgaben. *Numer. Math.* **2**, No. 2 (1960).
 1. J. LERAY and J. SCHAUDER (YU. SHAUDER)   Topologie et équations fonctionelles. *Ann. Sci. Ecole Norm. Sup.* **51**, No. 3 (1934).
*1. A. YU. LEVIN   *On the many-point boundary-value problem.* Candidate Dissertation, Leningrad, 1962.
*1. A. YU. LEVIN and E. A. LIFSHITS   On the general contraction principle of M. A. Krasnosel'skii. *Problemy Mat. Anal. Slozh. Sis.,* No. 1, Voronezh Univ., 1967.
*1. A. YU. LEVIN and V. V. STRYGIN   On the rate of convergence of the Newton–Kantorovich method. *UMN* **17**, No. 3 (1962).
 1. L. LICHTENSTEIN   *Vorlesungen uber einige Klassen nichtlinearer Integralgleichungen und Integro-Differentialgleichungen nebst Anwendungen.* Berlin, 1931.
*1. S. M. LOZINSKII   Inverse functions, implicit functions and the solution of equations. *Vestnik Leningr. Gos. Univ.,* No. 7 (1957).
*1. A. YU. LUCHKA   *Theory and application of the method of averaging of functional corrections.* Izdatel'stvo AN UkrSSR, Kiev, 1963.
*1. A. I. LUR'E and A. I. CHEKMAREV   Forced oscillations in a nonlinear system with characteristic consisting of two straight-line segments. *Prikl. Mat. Mekh.* **1**, No. 3 (1938).
*1. A. M. LYAPUNOV   *Collected works.* Vol. IV. Izdatel'stvo AN SSSR, 1959.
*1. L. A. LYUSTERNIK   On the convergence of an iterative process for solution of a system of algebraic equations, with random initial values and accumulation of errors. *Vychisl. Mat. i Vychisl. Tekhn.* **1** (1953).
*2. —   Remarks on the numerical solution of boundary-value problems for the Laplace equation and the computation of eigenvalues by the method of nets. *Trudy Mat. Inst. AN SSSR* **20** (1947).

*1. L. A. LYUSTERNIK and V. I. SOBOLEV   *Elements of functional analysis.* Nauka, 1965.
*1. N. V. MARCHENKO   Existence of solutions for a certain class of nonlinear integral equations. *DAN SSSR* **137**, No. 3 (1961).
*1. A. I. MARKUSHEVICH   *Theory of analytic functions.* Gostekhizdat, 1950.
 1. S. MAZUR and W. ORLICZ   Grundlegende Eigenschaften der polynomischen Operationen, I, II. *Studia Math.* **5** (1934).
*1. V. B. MELAMED   On computation of the index of the fixed point of a compact vector field. *DAN SSSR* **126**, No. 3 (1959).
*2.   —   On computation of the rotation of a compact vector field in the critical case. *SMZh* **2**, No. 3 (1961).
*3.   —   On analytic solutions of certain nonlinear integral equations. *DAN SSSR* **140**, No. 4 (1961).
*4.   —   On the problem of branching of solutions of a nonlinear analytic equation. *DAN SSSR* **145**, No. 3 (1962).
*5.   —   On formal solutions of certain nonlinear equations. *SMZh* **6**, No. 1 (1965).
*1. V. B. MELAMED and A. I. PEROV   Generalization of M. A. Krasnosel'skii's theorem on the compactness of the Fréchet derivative of a compact operator. *SMZh* **4**, No. 3 (1963).
 1. A. D. MICHAL and A. H. CLIFFORD   Fonctions analytiques implicites dans les espaces vectoriels abstraits. *C.R. Acad. Sci. Paris*, **197** (1933).
*1. S. G. MIKHLIN   Some sufficient conditions for convergence of Galerkin's method. *Uch. Zap. Leningr. Gos. Univ., Ser. Mat. Nauk* **135**, No. 21 (1950).
*2.   —   *Direct methods in mathematical physics.* Gostekhizdat, 1950.
*3.   —   *The problem of the minimum of a quadratic functional.* Gostekhizdat, 1952.
*4.   —   On Ritz's method. *DAN SSSR* **106**, No. 3 (1956).
*5.   —   *Variational methods in mathematical physics.* Gostekhizdat, 1957.
*6.   —   On the stability of Ritz's method. *DAN SSSR* **135**, No. 1 (1960).
*7.   —   Some conditions for the stability of Ritz's method. *Vestnik Leningr. Gos. Univ.*, No. 13 (1961).
*8.   —   *Numerical implementation of variational methods.* Nauka, 1966.
*1. E. MUKHAMADIEV   On the theory of periodic compact vector fields. *UMN* **22**, No. 2 (1967).
*1. E. MUKHAMADIEV and YU. V. POKORNYI   On monotone operators with several fixed points. *DAN TadzhSSR* **10**, No. 4 (1967).
*1. E. MUKHAMADIEV and V. YA. STETSENKO   The fixed-point principle in a generalized metric space. *Izv. AN TadzhSSR* **9**, No. 4 (1969).
*1. S. N. MUKHTAROV   On equations with indecomposable operators. *DAN TadzhSSR* **9**, No. 10 (1966).
*1. I. P. MYSOVSKIKH   On the question of the convergence of Newton's method. *Trudy Mat. Inst. Steklov* **28** (1949).
*2.   —   On the convergence and applications of Kantorovich's method for solving nonlinear functional equations. *Vestnik Leningr. Gos. Univ.*, No. 11 (1953).
*1. I. P. NATANSON   On the theory of approximate solution of equations. *Uch. Zap. Leningr. Gos. Ped. Inst. Gertsena* **64** (1948).

*2. — *Constructive theory of functions.* Gostekhizdat, 1951.

*1. N. N. NAZAROV Study of the solution of the equation $\phi(x) = \lambda \int_0^1 \Gamma(x, y, \phi(y))\,dy$
in the neighborhood of a branch point. *Trudy Uzb. Filiala AN SSSR* 4, No. 2 (1941).

*2. — Nonlinear integral Hammerstein equations. *Trudy Sredneaziatskii Gos. Univ.*,
5, No. 33, Tashkent (1941).

*3. — Branch points of periodic solutions of nonlinear equations of oscillation theory.
Trudy Uzb. Filiala AN SSSR 4, No. 2 (1941).

*4. — Branch points of the solutions of nonlinear integral equations. *Trudy Inst.
Mat. AN UzSSR,* No. 4 (1948).

*1. A. I. NEKRASOV On waves of stationary mode. *Izv. Ivanovsk. Politekhn. Inst.* 6 (1922)

*2. — *Exact theory of stationary waves on the surface of a heavy liquid.* Izdatel'stvo
AN SSSR, 1951.

1. F. NIIRO On indecomposable operators in $l_p (1 < p < \infty)$ and a problem of
H. H. Schaefer. *Sci. Papers Coll. Gen. Educ. Univ. Tokyo* 14, No. 2 (1964).

2. — On indecomposable operators in $L_p (1 < p < \infty)$ and a problem of H. H.
Schaefer. *Sci. Papers. Coll. Gen. Educ. Univ. Tokyo* 16, No. 1 (1966).

1. Z. OPIAL Weak convergence of the sequence of successive approximations for
nonexpansive mappings. *Bull. Amer. Math. Soc.* 73, No. 4 (1967).

2. — *Nonexpansive and monotone mappings in Banach spaces.* Center for Dynamical
systems, Brown Univ., 1967.

*1. D. YU. PANOV On application of B. G. Galerkin's method for the solution of
certain problems of elasticity theory. *Prikl. Mat. Mekh.* 3, No. 2 (1939).

*1. A. I. PEROV On the Cauchy problem for a system of ordinary differential equations.
In: *Approximate methods for the solution of differential equations,* Kiev, 1964.

*1. A. I. PETROV and A. V. KIBENKO On a general method for investigation of boundary
value problems. *Izv. AN SSSR* 30, No. 2 (1966).

*1. I. PETERSEN On the convergence of approximate interpolation methods for ordinary
differential equations. *Izv. AN EstSSR* 1 (1961).

*1. V. V. POKORNYI On the convergence of formal solutions of nonlinear integral
equations. *DAN SSSR* 120, No. 4 (1958).

*2. — On two analytic methods in the theory of small solutions of nonlinear integral
equations. *DAN SSSR* 133, No. 5 (1960).

*3. — On analytic solutions of some nonlinear equations. *Trudy Sem. Funk. Anal.*,
No. 2, Voronezh, 1956.

*4. — On the construction of the branching equation. *Trudy Sem. Funk. Anal.,* No. 5,
Voronezh, 1957.

*1. V. V. POKORNYI and P. P. RYBIN On stabilizing the process of looking for formal
implicit functions. *UMN* 15, No. 4 (1960).

1. YU. V. POKORNYI Some compression and tensile moduli of a cone. *DAN UkrSSR,*
No. 6 (1963) (in Ukrainian).

*2. — On *B*-positive and *B*-monotone operators. *Problemy Mat. Anal. Slozh. Sis.*,
No. 1, Voronezh Univ., 1967.

*3. — *On some conditions for existence of solutions of nonlinear operator equations in a space with a cone*. Dissertation, Voronezh, 1957.

*4. — On certain estimates for the Green's function of a many-point boundary-value problem. *Mat. Zametki* **3**, No. 5 (1968).

1. N. I. POL'SKII On the convergence of B. G. Galerkin's method. *Dokl. AN UkrSSR*, No. 6 (1949) (in Ukrainian).

*2. — On the convergence of certain approximate methods in analysis. *Ukr. Mat. Zh.* **7**, No. 1 (1955).

*3. — Galerkin's method and its justification. *Proceedings of the Third All-Union Mathematical Conference*, Vol. 1 (1956).

*4. — On a general scheme for application of approximate methods. *DAN SSSR* **111**, No. 6 (1956).

*5. — On necessary and sufficient conditions for the convergence of the generalized Galerkin method. *Nauchn. Zap. Zhitomir. Ped. Inst.* (1957).

*6. — Projection methods in applied mathematics. *DAN SSSR* **143**, No. 4 (1962).

*7. — Projection methods for the solution of linear problems. *UMN* **18**, No. 2 (1963).

*1. B. T. POLYAK Gradient methods for solution of equations and inequalities. *Zh. Vychisl. Mat. i Mat. Fiz.* **4**, No. 6 (1964).

*1. L. S. PONTRYAGIN *Foundations of combinatorial topology*. Gostekhizdat, 1947.

*1. B. P. PUGACHEV On two methods for approximation of eigenvalues and eigenvectors. *DAN SSSR* **110**, No. 3 (1956).

*2. — On a method for simultaneous computation of both end-points of the spectrum. *Trudy Sem. Funk. Anal.*, No. 5, Voronezh, 1957.

*3. — On the question of the rate of convergence of the normal-chord method. *Trudy Sem. Funk. Anal.*, Nos. 3–4, Rostov Univ., Voronezh Univ., 1960.

*4. — Investigation of a method for approximation of eigenvalues and eigenvectors. *Trudy Sem. Funk. Anal.*, Nos. 3–4, Rostov Univ., Voronezh Univ., 1960.

*5. — On accelerating the rate of convergence of iterative processes of the second degree. *Zh. Vychisl. Mat. i Mat. Fiz.* **2**, No. 4 (1962).

*6. — Remarks on the justification of certain iterative processes. *Zh. Vychisl. Mat. i Mat. Fiz.* **2**, No. 5 (1962).

*7. — On the convergence of methods which are locally close to the Newton–Kantorovich method. *Trudy Sem. Funk. Anal.*, No. 7, Voronezh, 1963.

1. V. PUISEUX Rech. sur les fonctions algebraïques. *J. Math. Pures Appl.* **15** (1950).

1. F. RIESZ and B. SZ.-NAGY *Leçons d'analyse fonctionnelle* (Second edition). Akademiai Kiadó, Budapest, 1953.

*1. P. P. RYBIN On the convergence of series obtained in solving nonlinear integral equations. *DAN SSSR* **115**, No. 3 (1957).

*2. — Singular solutions of a perturbed linear integral equation. *Vestnik Leningr. Gos. Univ.*, **19** (1957).

*1. T. SABIROV On the question of perturbation of the eigenvalues and eigenvectors of linear operators. *DAN TadzhSSR* **9**, No. 3 (1966).

*1. K. SAMADOV and V. YA. STETSENKO Bounds for the solutions of linear operator equations. *DAN TadzhSSR* **11**, No. 3 (1968).

1. H. H. SCHAEFER Spektraleigenschaften positiver linearer Operatoren. *Math. Z.* **82**, No. 4 (1963).

1. E. SCHMIDT Zur Theorie der linearen und nichtlinearen Integralgleichungen. 3. Teil: Über die Auflosung der nichtlinearen Integralgleichungen und Verzweigung ihrer Lösungen. *Math. Ann.* **65** (1908).

*1. N. V. SMIRNOV *Introduction to the theory of nonlinear integral equations.* ONTI, 1936.

*1. V. I. SMIRNOV *Course of higher mathematics.* Vol. III. Gostekhizdat, 1957.

*2. — *Course of higher mathematics.* Vol. V. Gostekhizdat, 1959.

*1. S. L. SOBOLEV *Some applications of functional analysis in mathematical physics.* Leningrad University, 1950.

*1. YU. D. SOKOLOV *The method of averaging functional corrections.* Naukova Dumka, Kiev, 1967.

*1. A. E. STAPAN Branching of the solutions of nonlinear integral equations. *Uch. Zap. Rizhsk. Ped. Inst.* **4** (1957).

1. M. STEIN Sufficient conditions for the convergence of Newton's methods in complex Banach space. *Proc. Amer. Math. Soc.* **3** (1952).

*1. V. YA. STETSENKO On bounds for the spectra of certain classes of linear operators. *DAN SSSR* **157**, No. 5 (1964).

*2. — On a method for estimating the spectrum of a linear operator. *UMN* **19**, No. 2 (1964).

*3. — On an iterative method for determining the spectral radius of positive linear operators. *Mat. Sb.* **67**, No. 2 (1965).

*4. — Indecomposability criteria for linear operators. *UMN* **21**, No. 5 (1966).

*5. — On a spectral property of an indecomposable operator. *UMN* **22**, No. 3 (1967).

*6. — On spectral properties of indecomposable operators. *DAN SSSR* **178**, No. 3 (1968).

*7. — On a method for accelerating the convergence of iterative processes. *DAN SSSR* **178**, No. 5 (1968).

*1. V. YA. STETSENKO and A. R. ESAYAN On the question of reducing the norm of an operator to its spectral radius. *DAN TadzhSSR* **6**, No. 9 (1963).

*1. V. YA. STETSENKO and B. IMOMNAZAROV On the existence of eigenvectors of nonlinear noncompact continuous operators. *SMZh* **8**, No. 1 (1967).

1. M. H. STONE *Linear transformations in Hilbert space and their applications to analysis.* New York, 1932.

1. G. SZEGÖ *Orthogonal polynomials.* Am. Math. Soc., New York, 1939.

1. B. SZ.-NAGY Perturbations des transformations autoadjointes dans l'espace de Hilbert. *Comm. Math. Helv.* **19** (1946–1947).

*1. E. E. TAMME On implicit operators. *DAN SSSR* **120**, No. 2 (1958).

1. A. E. TAYLOR Analytic functions in general analysis. *Ann. Scuola Norm. Sup. Pisa* **2**, No. 6 (1936).

2. — Analysis in complex Banach spaces. *Bull. Amer. Math. Soc.* **49** (1943).

1. A. C. THOMPSON On certain contraction mappings in a partially ordered vector space. *Proc. Amer. Math. Soc.* **14**, No. 3 (1963).

*1. V. A. TRENOGIN   Branching of solutions of nonlinear equations in Banach space. *UMN* **13**, No. 4 (1958).

*2. — On branching of solutions of nonlinear equations in the analytic case. *Trudy Moskov. Fiz.-Tekh. Inst.*, No. 3 (1959).

*3. — The branching equation and the Newton polygon. *DAN SSSR* **131**, No. 5 (1960).

*4. — Perturbation of a linear equation by small nonlinear terms. *DAN SSSR* **131**, No. 2 (1961). ....

*5. — Perturbation of eigenvalues and eigenelements of linear operators. *DAN SSSR* **167**, No. 3 (1966).

*1. E. A. TROITSKAYA   On eigenvalues and eigenvectors of compact operators. *Izv. Vuzov, Matematika*, No. 3 (1961).

*1. P. S. URYSON   *Works on topology and other fields of mathematics.* Vol. I. Gostekhizdat, 1951.

*1. M. M. VAINBERG   On the convergence of a process of steepest descent for nonlinear equations. *SMZh* **2**, No. 2 (1961).

*1. M. M. VAINBERG and V. A. TRENOGIN   The methods of Lyapunov and Schmidt in the theory of nonlinear equations, and their further development. *UMN* **17**, No. 2 (1962).

*2. —and— On the theory of implicit functions. *UMN* **17**, No. 2 (1962).

*3. —and— On the theory of branching of solutions of nonlinear equations. *UMN* **18**, No. 5 (1963).

*1. G. M. VAINIKKO   Error estimates in the Galerkin method for a linear differential equation. *Uch. Zap. Tartusk. Univ.* **129** (1962).

*2. — Some error estimates in the Bubnov–Galerkin method, I. Asymptotic estimates. *Uch. Zap. Tartusk. Univ.* **150** (1964).

*3. — Some error estimates in the Bubnov–Galerkin method, II. Estimates of the *n*-th approximation. *Uch. Zap. Tartusk. Univ.* **150** (1964).

*4. — Asymptotic error estimates for projection methods in the eigenvalue problem. *Zh. Vychisl. Mat. i Mat. Fiz.* **4**, No. 3 (1964).

*5. — Error estimates in the Bubnov–Galerkin method in the eigenvalue problem. *Zh. Vychisl. Mat. i Mat. Fiz.* **5**, No. 4 (1965).

*6. — On the convergence and stability of the collocation method. *Differentsial'nye Uravneniya* **1**, No. 2 (1965).

*7. — Necessary and sufficient conditions for stability of the Galerkin–Petrov method. *Uch. Zap. Tartusk. Univ.* **177** (1965).

*8. — On the question of the convergence of the Galerkin method. *Uch. Zap. Tartusk. Univ.* **177** (1965).

*9. — On the convergence of the collocation method for nonlinear differential equations. *Zh. Vychisl. Mat. i Mat. Fiz.* **6**, No. 1 (1966).

*10. — On the rate of convergence of certain approximate methods of the Galerkin type in the eigenvalue problem. *Izv. Vuzov, Matematika*, No. 2 (1966).

*11. — Perturbed Galerkin method and general theory of approximate methods for nonlinear equations. *Zh. Vychisl. Mat. i Mat. Fiz.* **7**, No. 4 (1967).

*12. — On the stability of the Galerkin–Petrov method for nonlinear equations. *Problemy Mat. Anal. Slozh. Sis.*, No. 1, Voronezh, 1967.

*13. — On the rate of convergence of approximate methods in the eigenvalue problem. *Zh. Vychisl. Mat. i Mat. Fiz.* **7**, No. 5 (1967).

*14. — On the rate of convergence of the moment method for ordinary differential equations. *SMZh* **9**, No. 1 (1968).

*15. — On similar operators. *DAN SSSR* **179**, No. 5 (1968).

*16. — On the connection between the methods of mechanical quadratures and finite differences. *Zh. Vychisl. Mat. i Mat. Fiz.* **9**, No. 2 (1969).

*17. — Compact approximation of linear compact operators by operators in factor spaces. *Uch. Zap. Tartusk. Univ.* **220** (1968).

*18. — The principle of compact approximation in the theory of approximate methods. *Zh. Vychisl. Mat. i Mat. Fiz.* **9**, No. 4 (1969).

*19. — *On the approximation of linear and nonlinear operators and approximate solution of operator equations.* Doctoral dissertation, Voronezh, 1969.

*1. G. M. VAINIKKO and A. M. DEMENT'EVA On the rate of convergence of the method of mechanical quadratures in the eigenvalue problem. *Zh. Vychisl. Mat. i Mat. Fiz.* **8**, No. 5 (1968).

*1. G. M. VAINIKKO and YU. B. UMANSKII Regular operators. *Funktsional'nyi Analiz i ego Prilozheniya* **2**, No. 2 (1968).

1. B. L. VAN DER WAERDEN *Einführung in die algebraische Geometrie.* Berlin, 1938.

2. — *Modern algebra,* Vols. I & II. Frederick Ungar Publ. Co., New York, 1949.

*1. B. A. VERTGEIM On conditions for the applicability of Newton's method. *DAN SSSR* **110**, No. 5 (1956).

*2. — On some conditions for convergence of Newton's method and on application of the method to the solution of equation systems. *Nauchn. Trudy Permsk. Gornogo Inst.*, No. 1 (1956).

*3. — On certain methods of approximate solution of nonlinear functional equations in Banach spaces. *UMN* **12**, No. 1 (1967).

*1. M. I. VISHIK and L. A. LYUSTERNIK Solution of some perturbation problems for matrices and selfadjoint and nonselfadjoint differential operators. *UMN* **15**, No. 3 (1960).

*1. YU. V. VOROB'EV Operator-valued orthogonal polynomials and approximate methods for determining the spectra of linear bounded operators. *UMN* **9**, No. 1 (1954).

*2. — Application of operator-valued orthogonal polynomials to the solution of inhomogeneous linear equations. *UMN* **10**, No. 1 (1955).

*3. — *The method of moments in applied mathematics.* Fizmatgiz, 1958.

*1. B. Z. VULIKH *Introduction to the theory of semiordered spaces.* Fizmatgiz, 1961.

1. R. J. WALKER *Algebraic curves.* Princeton Univ. Press, 1950.

1. A. WEIL *Foundations of algebraic geometry.* New York, 1948.

1. H. WIARDA *Integralgleichungen.* Teubner, Leipzig, 1930.

*1. G. N. YASKOVA and M. N. YAKOVLEV Some conditions for stability of the Petrov–Galerkin method. *Trudy Mat. Inst. Steklov* **66** (1962).

1. A. C. ZAANEN *Linear analysis*, New York—Amsterdam, 1953.

*1. P. P. ZABREIKO On computation of the Poincaré index. *DAN SSSR* **145**, No. 5 (1962).

*2. — Nonlinear integral operators. *Trudy Sem. Funk. Anal.*, No. 8, Voronezh, 1966.

*1. P. P. ZABREIKO and M. A. KRASNOSEL'SKII On computation of the index of an isolated fixed point of a compact vector field. *DAN SSSR* **141**, No. 2 (1961).

*2. —and— Computation of the index of a fixed point of a vector field. *SMZh* **5**, No. 3 (1964).

*3. —and— On the theory of implicit functions in Banach spaces. *UMN* **21**, No. 3 (1966).

*4. —and— On branching equations. *Trudy Sem. Funk. Anal.*, No. 11 *(Problemy Mat. Anal. Slozh. Sis.)*, Voronezh, 1968.

*5. —and— Asymptotic approximations in the theory of implicit functions. *Trudy Sem. Funk. Anal.*, No. 11 *(Problemy Mat. Anal. Slozh. Sis.)*, Voronezh, 1968.

*6. —and— On asymptotic approximations in the theory of implicit functions. *Abstracts of Conference on the Applications of Functional Analysis, Baku*, 1965.

*7. —and— Simple solutions of operator equations. *Problemy Mat. Anal. Slozh. Sis.*, No. 2, Voronezh, 1968.

*1. P. P. ZABREIKO, M. A. KRASNOSEL'SKII and A. V. POKROVSKII On the problem of bifurcation points. *Problemy Mat. Anal. Slozh. Sis.*, No. 2, Voronezh, 1968.

*1. P. P. ZABREIKO, M. A. KRASNOSEL'SKII and V. YA. STETSENKO On bounds for the spectral radius of linear positive operators. *Mat. Zametki* **1**, No. 4 (1967).

1. O. ZARISKI and P. SAMUEL *Commutative algebra*. Vols. I & II. Van Nostrand Co., Inc., New York, 1958 & 1960.

1. A. I. ZINCHENKO On certain methods for approximate solution of equations with nondifferentiable operators. *Dokl. AN UkrSSR*, No. 2 (1963) (in Ukrainian).

*2. — On the stability of a method of approximate solution of equations with nondifferentiable operators. *Trudy Sem. Funk. Anal.*, No. 7, Voronezh, 1963.

*1. V. I. ZUBOV On the question of the existence and approximate representation of implicit functions. *Vestnik Leningr. Gos. Univ., Ser. Mat. Mekh. Astronom.* **19**, No. 4 (1956).

Subject index